Ecological Studies

Analysis and Synthesis

Edited by

J. Jacobs, München · O. L. Lange, Würzburg

J. S. Olson, Oak Ridge · W. Wieser, Innsbruck

Volume 12

Alfred M. Stockard Lakeside Laboratory, The University of Michigan Biological Station on Douglas Lake, Pellston, Michigan

(Photograph courtesy of Gary R. Williams)

Perspectives of Biophysical Ecology

Edited by

David M. Gates
and
Rudolf B. Schmerl

Springer-Verlag New York · Heidelberg · Berlin 1975

David M. Gates
Director
Biological Station
The University of Michigan
Ann Arbor, Michigan 48104

Rudolf B. Schmerl
Assistant Dean for Research
School of Education
The University of Michigan
Ann Arbor, Michigan 48104

Library of Congress Cataloging in Publication Data

Main entry under title:
Perspectives of biophysical ecology.
(Ecological studies, v. 12)
Papers presented at a symposium held at The University of Michigan, Biological Station, Aug. 20–24, 1973.
Includes bibliographies.
1. Ecology—Simulation methods—Congresses.
2. Biological physics—Congresses. I. Gates, David Murray, 1921– ed. II. Schmerl, Rudolf B., ed.
III. Michigan. University. Biological Station.
IV. Series.
QH541.15.S5P47 574.5 74-17493

Printed in the United States of America.

ISBN 0-387-06743-4 Springer-Verlag New York · Heidelberg · Berlin
ISBN 3-540-06743-4 Springer-Verlag Berlin · Heidelberg · New York

Preface

A symposium on biophysical ecology was held at The University of Michigan Biological Station on Douglas Lake August 20–24, 1973. Biophysical ecology is an approach to ecology which uses fundamental principles of physics and chemistry along with mathematics as a tool to understand the interactions between organisms and their environment. It is fundamentally a mechanistic approach to ecology, and as such, it is amenable to theoretical modeling.

A theoretical model applied to an organism and its interactions with its environment should include all the significant environmental factors, organism properties, and the mechanisms that connect these things together in an appropriate organism response. The purpose of a theoretical model is to use it to explain observed facts and to make predictions beyond the realm of observation which can be verified or denied by further observation. If the predictions are confirmed, the model must be reasonably complete except for second or third-order refinements. If the predictions are denied by further observation, one must go back to the basic ideas that entered the model and decide what has been overlooked or even what has been included that perhaps should not have been. Theoretical modeling must always have recourse to experiment in the laboratory and observation in the field.

For plants, a theoretical model might be formulated to explain the manner and magnitude by which various environmental factors affect leaf temperature. A theoretical model might explain how wind, radiation, and air temperature affect the rate of water loss from a leaf or precisely how photosynthesis depends upon carbon dioxide concentration for a given light intensity, air temperature, wind speed, and relative humidity. A theoretical model might be formulated to explain how various environmental factors such as wind, air temperature, radiation intensity, and humidity affect the temperature of an animal. The intent in biophysical ecology is to obtain not simply a qualitative description of cause and effect, but a precise quantitative relationship between the final event and the causative factors.

As a general rule, it has been extremely difficult for the person trained primarily in biology to engage in theoretical analysis or modeling based on mechanisms from physics and chemistry. The biologist is usually more thoroughly trained in chemistry than in physics and, unfortunately, often lacks adequate advanced training in mathematics. The physicist or chemist often has little interest in a biological problem, and, when there has been interest, it has been at the molecular or cellular level. To fulfill the needs of the science of ecology, one must work with whole organisms and with the full scientific spectrum of significant factors, features, and mechanisms. This requires knowing biology well, but also knowing quite a lot concerning physics, chemistry, and mathematics. To achieve this, we have tried to find biologists with a facility for physics and mathematics and to lead them into the field of biophysical ecology. This has not been easy to do. Very few people trained in biology are able to originate and advance new methods of theoretical analysis. We have also tried to interest people trained primarily in physics and mathematics in the problems of ecology, and this also has not been easy. It is often difficult for such people to understand the biological problem, particularly the more subtle aspects of a problem, and to see the directions they should go in ecological science to work on significant and not on trivial matters. The dichotomy that continues to exist among people trained in the biological or the physical sciences is truly amazing. Yet to address certain questions in modern ecology, one must, as much as possible, be the complete scientist.

The purpose of the Biophysical Ecology Symposium held at Douglas Lake was to review our achievements to date in this field and to determine in which directions we should go to advance the state of the science as rapidly as possible. To do this, we brought together most of the Ph.D. students and post-doctorates who had trained with me during the last decade. To this group we added two others: a group of physiological ecologists who had a strong interest in the methodology of biophysical ecology, and a group of graduate students who have a serious interest in the subject. During the symposium the plant and animal papers were alternated in sequence. Nearly everyone in attendance participated in the entire set of presentations. During the last day the symposium participants formed four workshop groups, two around problems of plant ecology and two around animal ecology. The four groups met in working sessions and were asked to formulate recommendations for future work in biophysical ecology. These recommendations are summarized at the end of the book.

Not all the papers presented at the symposium were strictly biophysical ecology. Many were standard observational, empirical plant or animal ecology or physiological ecology. Some of the papers had nothing to do with modeling or analysis but formed significant contributions in terms of data. All the papers contributed ideas of importance to those working in biophysical ecology.

The symposium was supported by funds generously made available by The University of Michigan's Institute of Science and Technology, and the publication of this volume has been facilitated by funds from the Class of 1962–Institute of Science and Technology Publishing Fund. The Institute's Director and Assistant Director, James T. Wilson and Jay Katz, deserve special thanks. The staff of the Biological Station was responsible for transportation, housing, and commissary arrangements for all guests at the Station, and did a superb job. I am deeply grateful to all who participated in this symposium for their contributions.

David M. Gates

Contributors

ALBERTE, RANDALL S., Ph.D., Post-Doctoral Fellow, Department of Biology, University of California, Los Angeles, California 90024.

ALDERFER, RONALD G., Ph.D., Assistant Professor of Biology, University of Chicago, Chicago, Illinois 60610.

BAKER, DONALD N., Ph.D., Soil Scientist, Cotton Production Unit, Agricultural Research Station, U.S. Department of Agriculture, Mississippi State University, State College, Mississippi 39762.

BAKKEN, George S., Ph.D., Post-Doctoral Fellow, Department of Botany, The University of Michigan, Ann Arbor, Michigan 48104.

BAKKER, ROBERT T., B.S., Junior Fellow, Society of Fellows, Department of Vertebrate Biology, Museum of Comparative Zoology, Harvard University, Cambridge, Massachusetts 02163.

BARTHOLOMEW, GEORGE A., Ph.D., Professor of Zoology, University of California, Los Angeles, California 90024.

BECKMAN, WILLIAM A., Ph.D., Professor of Mechanical Engineering, University of Wisconsin, Madison, Wisconsin 53706.

BIRKEBAK, RICHARD C., Ph.D., Professor of Mechanical Engineering, University of Kentucky, Lexington, Kentucky 40506.

BJÖRKMAN, OLLE, Ph.D., Biologist, Department of Plant Biology, Carnegie Institution of Washington, Stanford, California 94305.

BUSCHBOM, UWE, Ph.D., Akad. Rat, Botanisches Institut II der Universität Würzburg, 87 Würzburg, Nittlerer Dallenbergweg 64, West Germany.

CALDER, WILLIAM A., III, Ph.D., Associate Professor of Zoology, University of Arizona, Tucson, Arizona 85721.

DAVIS, L. BERKLEY, JR., Ph.D., Development Engineer, Gas Turbine Division, General Electric Company, Schenectady, New York 12305.

DAWSON, WILLIAM R., Ph.D., D.Sc., Professor of Zoology, The University of Michigan, Ann Arbor, Michigan 48104.

DUNN, E. LLOYD, Ph.D., Assistant Professor of Botany, University of Georgia, Athens, Georgia 30602.

EPTING, ROBERT J., M.A., Research Assistant, Department of Biology, University of California, Los Angeles, California 90024.

EVENARI, MICHAEL, Ph.D., Professor of Botany, Hebrew University, Jerusalem, Israel.

GATES, DAVID M., Professor of Botany and Director, Biological Station, The University of Michigan, Ann Arbor, Michigan 48104.

HALL, ANTHONY E., Ph.D., Assistant Professor of Plant Physiology, University of California, Riverside, California 92502.

HESKETH, JOHN D., Ph.D., Plant Physiologist, Cotton Production Unit, Agricultural Research Service, U.S. Department of Agriculture, Mississippi State University, State College, Mississippi 39762.

HILL, STEPHEN C., M.S., Research Associate, Division of Forestry, West Virginia State University, Morgantown, West Virginia 26506.

HUTSON, WILLIAM G., M.S., Forester, Georgia-Pacific Corporation, Rainelle, West Virginia 25962.

JACOBSEN, NADINE K., Ph.D., Research Specialist, Biothermal Laboratory, Department of Natural Resources, State College of Agriculture and Life Sciences, Cornell University, Ithaca, New York 14850.

JOHNSON, HYRUM B., Ph.D., Assistant Professor of Biology, University of California, Riverside, California 92502.

JOHNSON, SAMUEL E., II, Ph.D., Assistant Professor of Zoology, Clark University, Worcester, Massachusetts 01610.

KAPPEN, LUDGER, Ph.D., Akad. Oberrat. Botanisches Institut II der Universität Würzburg, 87 Würzburg, Nittlerer Dallenbergweg 64, West Germany.

KAUFMANN, MERRILL R., Ph.D., Associate Professor of Plant Physiology, University of California, Riverside, California 92502.

KLUGER, MATTHEW J., Ph.D., Assistant Professor of Physiology, The University of Michigan, Ann Arbor, Michigan 48104.

LANGE, OTTO L., Ph.D., Professor der Botanik, Botanisches Institut II der Universität Würzburg, 87 Würzburg, Nittlerer Dallenbergweg 64, West Germany.

LEE, RICHARD, Ph.D., Professor of Forest Hydrology, West Virginia University, Morgantown, West Virginia 26506.

LOMMEN, PAUL W., Ph.D., Research Associate, Ecology Center, Utah State University, Logan, Utah 84322.

MITCHELL, JOHN W., Ph.D., Professor of Mechanical Engineering, University of Wisconsin, Madison, Wisconsin 53706.

MOEN, AARON N., Ph.D., Associate Professor of Wildlife Ecology, Cornell University, Ithaca, New York 14850.

MORHARDT, J. EMIL, Ph.D., Assistant Professor of Biology, Washington University, St. Louis, Missouri 63130.

MORHARDT, SYLVIA STAEHLE, Ph.D., Ecologist, Harland Bartholomew and Associates, 165 N. Meramec, Clayton, Missouri 63105.

O'LEARY, JAMES W., Ph.D., Professor of Biological Sciences, University of Arizona, Tucson, Arizona 85721.

PORTER, WARREN P., Ph.D., Professor of Zoology, University of Wisconsin, Madison, Wisconsin 53706.

SCHMERL, RUDOLF B., Ph.D., Assistant Dean for Research, School of Education, The University of Michigan, Ann Arbor, Michigan 48104.

SCHULZE, ERNST D., Ph.D., Dozent der Botanik, Botanisches Institut II der Universität Würzburg, 87 Würzburg, Nittlerer Dallenbergweg 64, West Germany.

SKULDT, DALE J., M.S.M.E., Engineer, Combustion Engineering Inc., Prospect Hill Rd., Windsor, Connecticut 06095.

SMITH, SANDRA K., B.A., Graduate Student, Biology Program, The University of Michigan, Ann Arbor, Michigan 48104.

SOUTHWICK, EDWARD E., Ph.D., Lecturer, Department of Biology, Georgetown, University, Washington, D.C. 20007.

SPOTILA, JAMES R., Ph.D., Assistant Professor of Biology, State University College at Buffalo, Buffalo, New York 14222.

STRAIN, BOYD R., Ph.D., Associate Professor of Botany, Duke University, Durham North Carolina 27706.

TAYLOR, S. ELWYNN, Ph.D., Biometeorologist, Environmental Study Service Center, National Oceanic and Atmospheric Administration, National Weather Service, U.S. Department of Commerce, Auburn University, Auburn, Alabama 36830.

TENHUNEN, JOHN D., Ph.D. Candidate, Department of Botany, The University of Michigan, Ann Arbor, Michigan 48104.

TRACY, C. RICHARD, Ph.D., Assistant Scientist, Institute for Environmental Studies, University of Wisconsin, Madison, Wisconsin 53706.

YOCUM, CONRAD S., Ph.D., Professor of Botany, The University of Michigan, Ann Arbor, Michigan 48104.

Contents

1

Introduction: Biophysical Ecology

David M. Gates

Prefatory Remarks

Biophysical ecology, a subdiscipline of ecology, uses an analytical approach involving the laws of physics and chemistry to understand the mechanisms by which plants and animals interact with their environment. Ecology is the study of organisms and their interactions with their environment and with one another. Taking the first part of this definition, one recognizes that we are dealing with the interaction of biology with the physical world. If ecology is to be done well, the biological aspect of the science must be understood well and the physical aspect must be understood equally well.

Biological and physical sciences each have had their descriptive phases, but the physical sciences have been more amenable to incorporating mathematical analysis. Mathematics abbreviates the lengthy thought processes involved in logic and extends these thought processes to extrapolation and prediction. It is for this reason that mathematics is applied as an analytical tool in the solution of biological problems.

Not only do all organisms live in a physical world; in every respect, they utilize basic physical mechanisms for their viability and reproducibility. As remarkable as the biological world seems to us, I do not believe that its workings are more than an incredible number of physical mechanisms interacting in a large number of subtle combinations. The complexities involved are enormous and our ability to understand these is limited. Nevertheless, certain mechanisms, forces, and processes may dominate the performance and behavior of organisms. Our task in the study of biology is to understand these and to recognize those of primary importance first, then those of secondary or tertiary importance. This viewpoint does not deny that every possible kind of cell-to-cell, organ-to-organ, or organism-to-organism interaction may exist. A community of organisms has many remarkable properties, some of which may not be characteristic of any other kind of assemblage in the universe.

A reductionist approach to biology, or specifically to ecology, by no means excludes a holistic viewpoint. I am convinced that a great deal of biological understanding will be achieved through analysis based on mechanisms; at the same time, other approaches are worthwhile and necessary.

My initial approach to the subject of biophysical ecology was strongly microclimatological, as shown by the references given in Gates (1959). Because of my training in physics and my work in atmospherics from 1947 through 1959, I was not familiar with the literature of biological science, except superficially. The 1959 article, given as an address at the Semicentennial Celebration of The University of Michigan Biological Station, was largely concerned with the radiation environment of plant and animal habitats. My interest in this subject and my feeling that it was of great importance to ecology was evident in a much earlier publication by Gates and Tantraporn (1952), concerned with the infrared reflectivity of vegetation. During the intervening years, I often thought about the problems challenging ecologists, e.g., problems of adaptation, productivity, succession, competition, and distribution among organisms. Concern with these problems had been firmly imprinted on my mind by my many years as a youth at the Biological Station and close association with my ecologist father, Frank C. Gates, and his colleagues. My interest in biology was intensive at that time, but I wished to work in a branch of science more analytical than biology then appeared to be. Physics was compatible with that desire and, as is clear in retrospect, was an excellent route to ecology. I had the good fortune to learn a good deal about biology from close association with the many great biologists at the Biological Station.

I spoke of the dichotomy between physical and biological sciences in my Semicentennial Address in 1959 and suggested some means to close this gap in the training of students. Some changes have occurred in university curricula in this respect during the last 15 years, but the changes are not nearly sufficient. Once again I wish to emphasize that to do the science of ecology well, one must do the biological and the physical sciences related to it equally well. Usually the training of ecologists stresses biology very heavily and neglects the physical sciences. This is not adequate preparation for a subject inherently extremely difficult. Relatively few people have come to grips with the most difficult and challenging ecological problems. For the most part, they have been satisfied with the qualitative aspects of ecology, or with the quantitative aspects in terms of numbers and rates. The new, extremely worthwhile work concerning systems ecology provides insights to the interrelations among many components of ecosystems. Yet within such interrelations of trophic levels, the flow of energy, the flow and cycling of minerals, and the gains and losses of biomass are the fundamental mechanisms that control, regulate, and influence them. These fundamental mechanisms are physiological and physical; they involve organisms and their environment. Once the coupling mechanisms between an organism and its environment are thoroughly understood, an extremely critical domain of physiological ecology will still remain to be worked out. This domain is in the biochemistry of metabolism and growth, resistance to heat and cold, fertility, germination, and a whole complex of important biological events. Many of these events are mediated through enzymes, and the incredible number of complex, closely related, biochemical reactions staggers the imagination.

Yet eventually the ecologist must confront these problems and bring as much biochemistry as necessary into the analysis of organisms and their interaction with their environment.

The process of understanding the interactions of plants and animals with their environment and their response to various factors involves every aspect of the organism and the environment. No scientific problem could be more difficult or more challenging, yet modern science is fully capable of addressing it. The analysis of this problem involves taxonomy, systematics, physiology, biochemistry, biophysics, physics, meteorology, climatology, mathematics, engineering, and other disciplines. Clearly a single investigator cannot learn all these things well, but given the proper initial training, one can do much to address the problems of autecology rigorously and competently, and this must be done if ecology is to advance rapidly as a modern science.

Organism–Environment Interaction

An organism interacts with or is coupled to its environment through the exchange of energy and materials, but also through sensory perceptions of sight, sound, touch, taste, and emotion. Only a few of these processes are addressed in this volume. The problem we face when attempting to understand the interactions of organisms with their environment is to choose where to begin. This choice may be made on the basis of the most dominant or first-order factors, or it may be made through some other rationale. Since any event that involves life requires the expenditure of energy, I decided some years ago that I would approach the problem first through the flow of energy. Once I understood energy exchange, I would concentrate on understanding mass exchange, and finally become involved with the internal physiology and chemistry of the organisms. However, energy and mass exchange are closely related, and one cannot deal with the one without immediate involvement with the other.

I also made another choice very early. Given a choice of working with either plants or animals, I decided to work with plants first. Not only are plants the primary producers, but in many respects they are easier to work with than animals. They do not bite, their metabolic rates are relatively low, they do not move about very much, and for the most part they have a more agreeable odor.

Energy exchange between an organism and its environment may involve radiation, convection, conduction, evaporation, and chemical reactions. Each of these is fairly complex and must be dealt with in detail. The flow of energy between an organism and the environment is time-dependent and often involves rather rapid rates of change. Amid frequent and irregular variations of energy flow are periods of steady state when an organism is neither gaining nor losing a net amount of energy. With full recognition of the ubiquity and importance of time-dependent energy flow, I made an early decision to solve the steady-state problem first. To approximate time-dependent or transient events, one can consider a series of incremental changes in steady-state energy flow. But beyond that approximation, one must come to grips with the complete time-dependent analysis to realize as accurately as possible how the real world of plants and animals works in response to environmental factors.

The temperature of a plant or an animal is a manifestation of its energy state, which in turn is determined by the rate of flow of energy between the organism and its environment. Many physical and physiological events that occur in organisms, such as metabolism, water loss, mobility of ions, fluidity of fats, permeability of membranes, and gas exchange, depend upon temperature. Hence the temperature of a plant or an animal, and precisely how the environment influenced it, were of primary interest in our initial research.

Without going into great detail here, I shall summarize the advances made by our research group during the past 15 years. A complete bibliography concerning this work is included. References in this paper to publications of our research group are in this list. A separate bibliography is attached for references to papers published outside our group.

Plants

Since the temperature of a plant is a measure of its energy status, which is the result of energy exchange between the plant and its environment, my first major effort was to identify the mechanisms involved in the exchange of energy. Once this was accomplished, a research program evolved which led to the explanation of many ecological events involving plants.

The energy budget of a plant leaf in steady-state condition is written in the form

$$Q_a = \varepsilon\sigma T_l^4 + H(T_l - T_a) + LE + P \tag{1}$$

where Q_a = amount of radiation absorbed
P = energy consumed by photosynthesis
ε = emissivity of the leaf
σ = Stefan–Boltzmann coefficient for blackbody radiation
T_l = leaf-surface temperature
H = convection coefficient
T_a = air temperature
E = amount of water consumed by transpiration
and L = heat of evaporation (about 580 cal g^{-1} at 30°C) and converts the rate of water loss in grams to energy units (actually, L is a function of the leaf temperature)

All terms in Eq. (1) are expressed in cal cm^{-2} min^{-1} or ergs cm^{-2} s^{-1}. The photosynthetic term is negligible for most heat-budget calculations, and only in rare cases, such as with some of the arums, does the respiration rate have a significant effect on plant temperature. In Eq. (1) all surfaces of the leaf are considered at the same temperature, and heat exchange by conduction is considered negligible.

These are clearly approximations that may be more or less true, depending upon the particular leaf. For most thin, broad leaves, these approximations are very good. For a thick leaf or, for example, the blade of a prickly pear cactus, the upper and lower surface temperatures are usually very different, and heat is exchanged between them. Equation (1) is not adequate when considering heat flow in stems, branches, trunks, etc.; other terms, mainly conduction, must be added.

The rate of evaporation of water from a leaf is determined by the amount of energy available and by the presence of a water-vapor gradient between the leaf and the air. The rate of water loss by transpiration is given by

$$E = \frac{{}_sd_l(T_l) - (\mathrm{rh})_sd_a(T_a)}{r_l + r_a} \tag{2}$$

where ${}_sd_l(T_l)$ = saturation water-vapor density of the air in the mesophyll of the leaf as a function of the leaf temperature

${}_sd_a(T_a)$ = saturation water-vapor density of the air beyond the leaf and its boundary layer and is a function of the air temperature

rh = relative humidity of the air

r_l = diffusion resistance for water vapor in air within the stomatal cavity of the leaf and neighboring passages

r_a = diffusion resistance for water vapor in air in the boundary layer adhering to the leaf surface

In retrospect, the energy-budget relationship given by Eq. (1) and the water-vapor exchange concept as expressed by Eq. (2) are very obvious formulations, but at the time, first in Gates (1959, 1961) and then in Gates (1963), the ideas were new to me. My first papers dealt primarily with radiation in the environment and its importance as an environmental factor. In fact, quite early I published a paper concerning the infrared reflectivity of vegetation (Gates and Tantraporn, 1952) which became the primary reference in this area. Raschke (1955, 1956, 1958, 1960) had also developed in considerable detail the concepts of energy and gas exchange for a leaf. When I first worked out my ideas, I was not aware of Raschke's papers. Only when I presented a paper on this subject to the Denver meeting of the American Association for the Advancement of Science in 1961 were these called to my attention, by James Bonner. I had been working for 15 years in physics and had not been reading the biological journals. My first ideas concerning energy exchange were influenced considerably by a paper of Budyko's (1956) and by discussions with him during a visit I made to Russia in 1958.

To apply Eq. (1) effectively to the determination of leaf temperatures and transpiration rates, one must solve it simultaneously with Eq. (2), and the various coefficients and parameters in each equation must be known accurately for the particular leaf. For example, the following parameters are required: leaf absorptivity to sunlight, absorptivity and emissivity to thermal radiation, actual surface area of the leaf, effective areas for the absorption of solar and thermal radiation and for the emission of thermal radiation, leaf width and length (which enters the convection coefficient), the diffusion resistance to water vapor within the leaf (i.e., the stomatal and substomatal resistance), and the diffusion resistance of the boundary layer of air adhering to the leaf surface. There are ways to determine each of the quantities mentioned above, and these are discussed in the various references given here. It should not be surprising that the properties of a leaf must be well known if one is to determine the leaf temperature and transpiration rate with accuracy. The photosynthetic rate of a leaf does not significantly affect the leaf temperature and, although it is important within itself for metabolic energy, it can be dropped from the energy-budget determination. In addition to knowing the properties of the leaf, one must also have a good determination of the environmental variables in the vicinity of the leaf. This requires knowledge concerning radiation intensities, wavelength distribution, geometrical arrangement (point source, extended source, etc.), air speed, air temperature, and humidity. Each of these quantities must be properly determined to evaluate the response of a leaf to them.

During the years immediately following 1960, a great deal of effort was spent on a determination of the convection coefficient for leaves of broad-leaved plants and of conifers. One night while driving to South Dakota to launch a skyhook balloon for some stratospheric measurements, I saw a magnificent aurora borealis on the northern horizon. For some strange reason, this evanescent phenomenon created in my imagination an image of warm air dancing about on a leaf surface. As a student of physics I knew well the technique of schlieren photography used for photographing shock waves from high-speed bullets and air flow over airplane wings, and decided to try some on the boundary layers of plant leaves. With me on that drive was C. M. Benedict, photographer of the Boulder Laboratories of the National Bureau of Standards. I told him I would buy the optical equipment if he would help shoot the pictures. We did this and the result was the magnificent series of still shots and movies of convection phenomena around plant leaves that were published (Gates and Benedict, 1963) and shown often in lectures given throughout the world.

At this time we used the schlieren pictures together with some very delicate measurements of temperature in the ambient air and in the convection plumes to determine the convection coefficients for leaves. The values obtained were consistent with the values known for objects of similar shape, such as flat plates, cylinders, or arrays of cylinders, depending upon whether the leaf is broad and flat or needle-like or a whole branch of needles. This method for obtaining the convection coefficient was not particularly accurate. To make more accurate measurement of convection coefficient for leaves, we built a wind tunnel and used silver casts of leaves (Tibbals *et al.*, 1964).

A leaf is illuminated by sunlight, which is nearly a point source of radiation, whereas skylight, reflected light, and thermal radiation from ground and atmosphere come from extended sources. This means that a leaf, depending upon its orientation, presents various effective areas to these sources of radiation. We built a radiation chamber into which the silver casts of leaves were placed and determined the leaf areas affected by absorbed and emitted radiation.

Following many years of effort to develop methods for the determination of leaf parameters and the various coefficients that enter the energy-budget analysis, I decided to publish tables and graphs that facilitate the analysis and interpretation.

I did this in collaboration with Laverne Papian, who was responsible for the computer programming of Eqs. (1) and (2). This was published in book form (Gates and Papian, 1971) as an *Atlas of Energy Budgets of Plant Leaves*. Not only can one read from the charts and tables the leaf temperatures and water-loss rates for a leaf of any specified characteristics, but one can vary the parameters as desired and obtain a new set of temperatures or transpiration rates for a leaf.

By use of these methods we were able to explain many kinds of observations not well accounted for previously. Observations of leaf temperatures by Lange (1959) and by Gates, Hiesey, Milner, and Nobs (1964) showed that under identical environmental conditions the leaves of some plants fully exposed to sunlight were at temperatures above air temperature, others below air temperature. These observations are readily explained by means of calculations, using Eqs. (1) and (2). One example is shown in Fig. 1.1. Not only is it important to explain such observations, but it is absolutely necessary to be able to make predictions of the temperature and water-loss status of leaves in environmental situations much beyond the particular conditions used for the necessarily limited observations. Such predictions

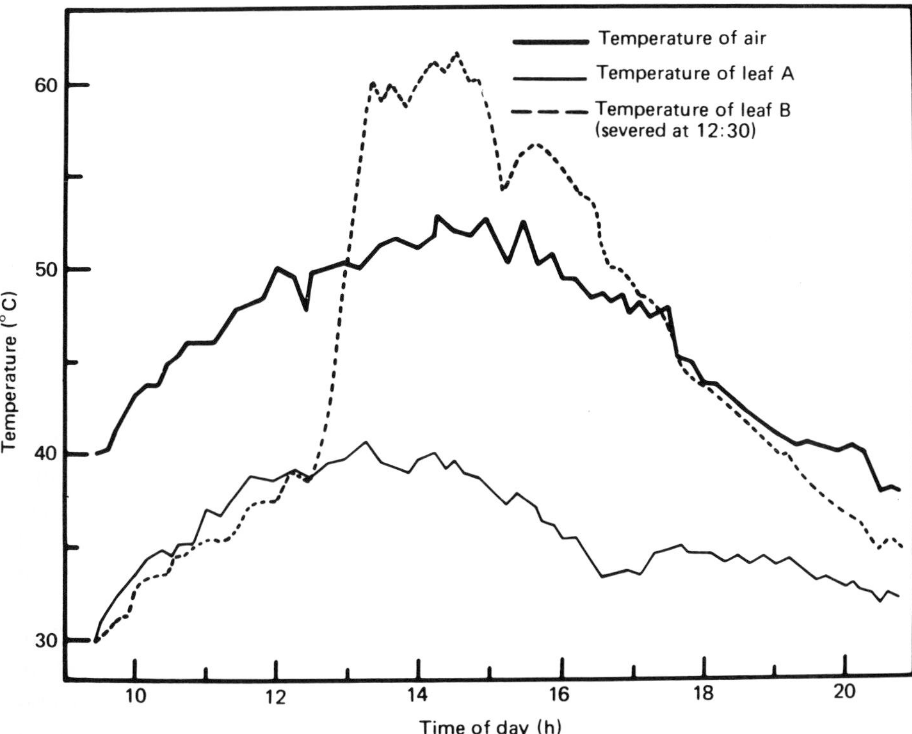

Fig. 1.1. The temperatures of leaves *A* and *B* are below the air temperature. At 12:30 leaf *B* was severed and the flow of water through the stem was interrupted. Without the influence of transpiration leaf B became warmer than the air. (Measurements with *Citrullus colocynthis* under desert conditions in Mauretania, Africa; after Lange, 1959.)

can be made with great quantitative accuracy using the analysis developed around Eqs. (1) and (2).

The question is often asked whether transpiration rates of leaves increased or decreased with increase of wind speed. This question cannot readily be answered by experiment because of the large number of combinations possible for the many variables involved. However, by means of analysis it is possible to address the question for any set of assumed or actual environmental conditions and plant properties. The results are given in Gates (1968) and are illustrated in Fig. 1.2. If the radiation load on a leaf is great, an increase of wind speed will reduce the leaf temperature, but if the radiation load is relatively weak, an increase of wind speed will warm the leaf since the air temperature is warmer than the leaf temperature. At an intermediate condition of radiation load on a leaf, the leaf temperature is not notably influenced by the wind speed since the air temperature equals the leaf temperature. It is not only necessary to understand this phenomenon qualitatively, it is also important to know exactly the conditions that produce a given result.

Another example of the use of analysis to explain an observation and also to predict all combinations of plant properties and environmental conditions related to a phenomenon is that of the observations reported by Gates, Alderfer, and Taylor (1968) concerning leaf temperatures of desert plants. We had measured the temperatures of leaves of many desert plants growing in the southwestern United States. We realized that we were never getting leaf temperatures more than about 2°C above the air temperature for all the plants with small leaves. This was in striking contrast to many earlier observations concerning leaf temperatures, such as those reported by Gates (1963), for which the temperatures of oak leaves in full sunshine became as much as 20°C above air temperature, and differences of 10°C were very frequent. Recourse to analysis very quickly explained the observation and allowed us to extrapolate the limits of the phenomenon, which was the result of strong convective coupling of the leaves with the air. The dependence of leaf temperature on leaf size for a given set of environmental conditions is illustrated in Fig. 1.3.

Another example that shows the critical importance of analysis to the understanding of a complicated and somewhat subtle phenomenon is the matter of water-vapor diffusion through the stomates of a leaf. Ting and Loomis (1963) had used perforated membranes for a series of water-diffusion experiments from the underlying water to the free air outside the membranes, and concluded that the conductance of a stomate pathway was proportional to the diameter of the pore. The theoretical analysis by Lee and Gates (1964) showed that although Ting and Loomis would expect to get, and in fact did get, a linear relationship between water-vapor-diffusion rate and pore diameter, this conclusion could not be extrapolated to stomates. The reason is quite simple. The experiments were done with membranes whose ratio of pore diameter to length is large. Theory shows that, for this condition, one will get a linear relationship, but for stomates the ratio of pore diameter to length is generally very small and the linear relationship simply does not hold. All previous experiments with septa in which the pore diameter/thickness ratio was large confirm the linear law, but the theory shows clearly the inapplicability of the law to stomates and leaf transpiration.

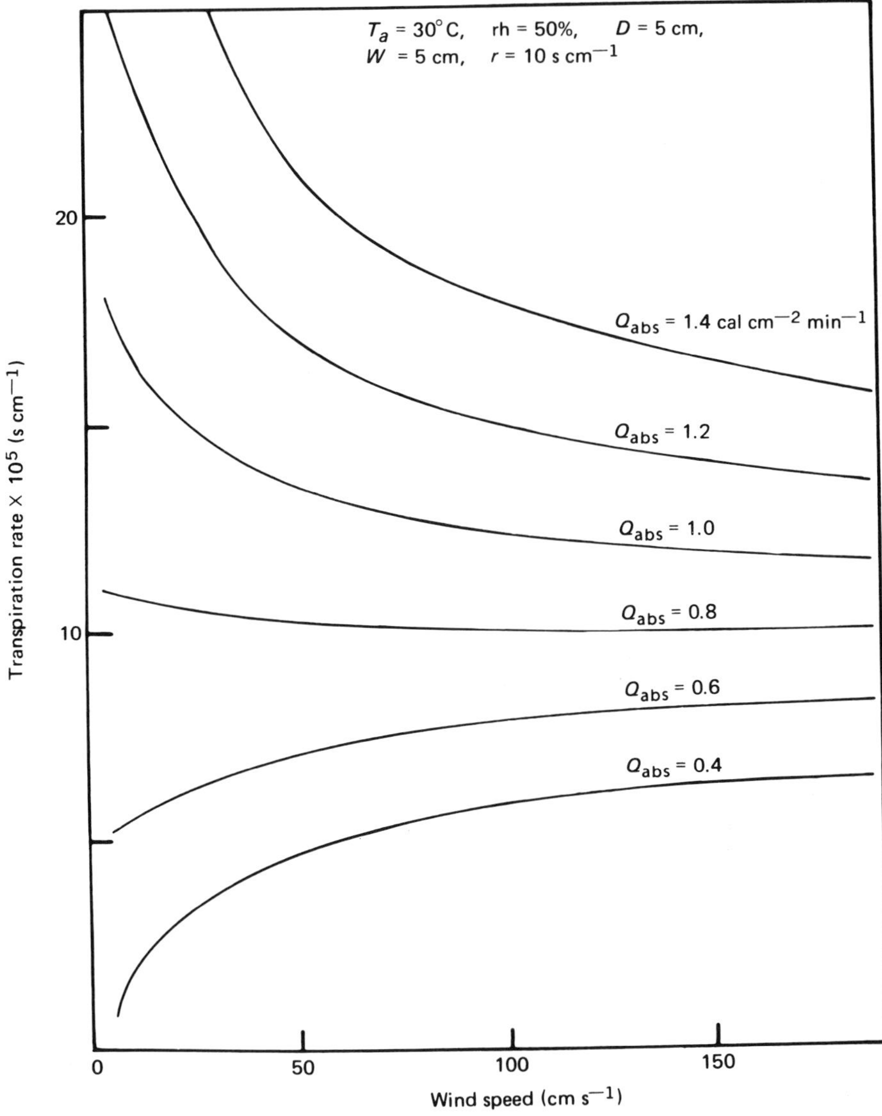

Fig. 1.2. Calculated transpiration rate of a leaf as a function of wind speed and the amount of radiation absorbed. The air temperature was 30°C and the relative humidity was 50 percent. The leaf measured 5 by 5 cm and had an internal resistance to water loss of 10 s cm^{-1}.

One of the most complicated phenomena in ecology is the interaction of radiation with organisms. Radiation is ever-present and ubiquitous in the terrestrial environment. The intensity and nature of radiation is important not only from a photochemical standpoint but also in terms of the thermodynamics of organisms. The energy level at which an organism may function is largely determined by the

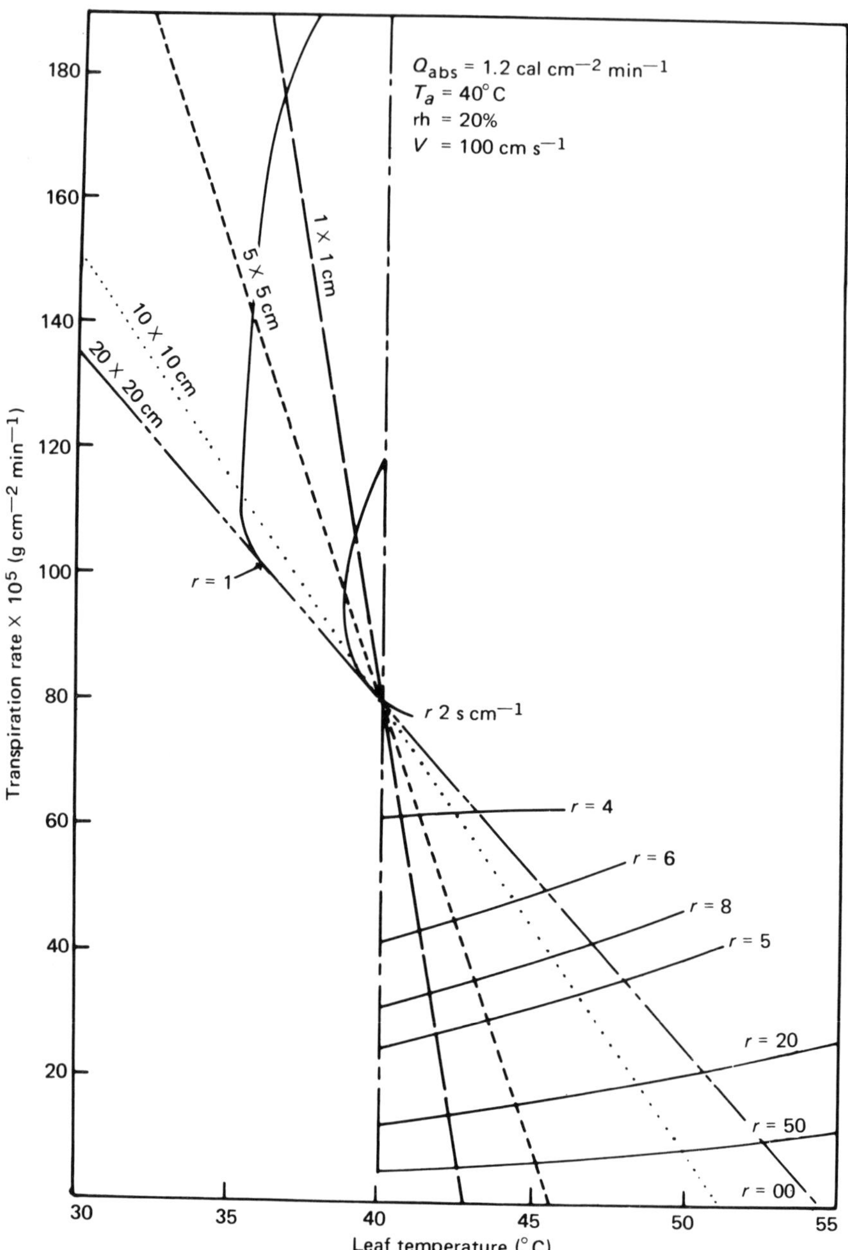

Fig. 1.3. Transpiration rate and leaf temperature as a function of leaf size and internal diffusion resistance for very warm, dry, and intense radiation conditions. There is little air movement.

radiation field of its environment. For this reason we have given careful consideration to all aspects of the radiation problem. Low-energy thermal radiation is important to an organism because of its influence on temperature. Intermediate energy radiation of the visible spectrum is useful for photochemistry, and high-energy ultraviolet, x-ray, or gamma-ray radiation is destructive to life, although some near-ultraviolet radiation is necessary to life.

An organism is coupled to the radiation field by the absorptivity of its surface. A high value of absorptivity means a tight coupling; a low value indicates a loose coupling. The degree of coupling determines the extent to which the organism's temperature and photochemical reactions are influenced by the intensity of the incident radiation. The absorptivity of any surface various enormously with the wavelength or frequency of the radiation. A characteristic set of spectral absorptivity curves as a function of wavelength is shown in Fig. 1.4. A plant leaf, for example, absorbs most of the ultraviolet and blue radiation incident upon it, somewhat less of the green, a large fraction of red light, and a relatively small amount of the near-infrared radiation because of high reflectivity and transmissivity at these wavelengths. Beyond 3.0 μm the absorptivity of a leaf becomes very strong again. A leaf absorbs well those wavelengths that it requires for photochemical events such as photosynthesis.

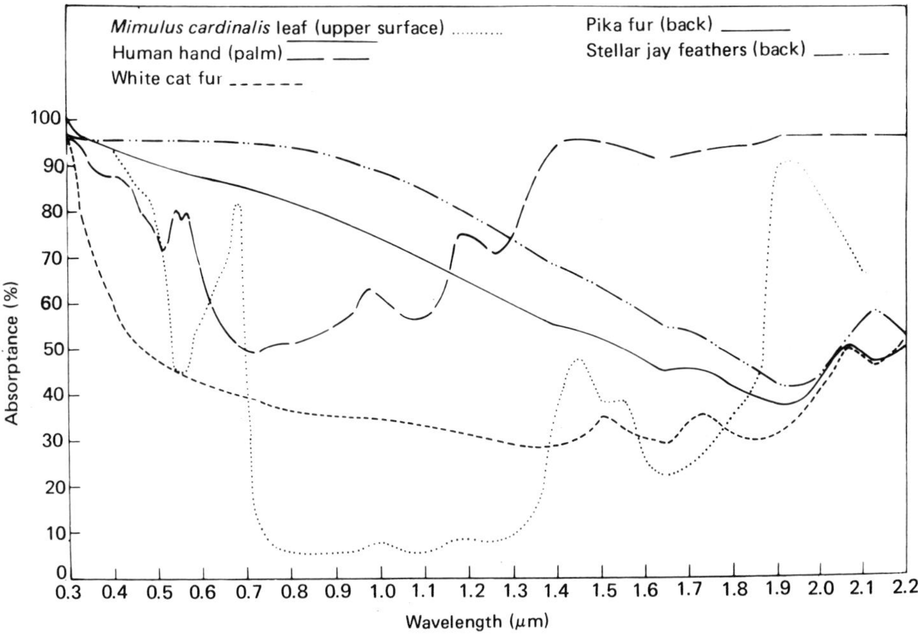

Fig. 1.4. Spectral absorptivity of various surfaces as a function of wavelength. *Mimulus cardinalis* is the leaf of a small plant known as the "monkey flower." The other surfaces are those of animals.

The action spectra of plant processes in general show a strict dichotomy between absorption in the blue and red, as shown in Fig. 1.5. Approximately half of the energy from the sun which reaches the earth's surface is in the infrared at wavelengths greater than 700 nm. Broad leaves reflect about 45 percent more or less of the near infrared, transmit approximately the same percentage of incident radiation, and absorb about 10 percent more or less. The abrupt change between a very high absorptivity in the visible and very low absorptivity in the near infrared occurs at 700 nm and is a dramatic transition. It more or less coincides with a transfer from energy transitions that are primarily electronic to those which are primarily vibrational for the interaction of molecules and radiation. Gates, Keegan, Schleter, and Weidner (1965) reported in considerable detail the spectral reflectivities of vegetation in the visible and near infrared. Beyond a wavelength of about 3000 nm the absorptivity of plant leaves becomes very strong; the values range from 0.90 to 0.98 (Gates and Tantraporn, 1952). This phenomenon of leaves becoming very "black" in the far infrared is extremely important. A high absorptivity at these wavelengths gives leaves a very high emissivity to infrared thermal radiation and permits them to remain cool by effectively emitting radiation. Hence, in summary, a leaf absorbs very well those wavelengths of solar radiation from the ultraviolet to about 700 nm which are needed for photochemistry, reflects a great deal of the incident sunlight at wavelengths from 700 nm to about 3000 nm to

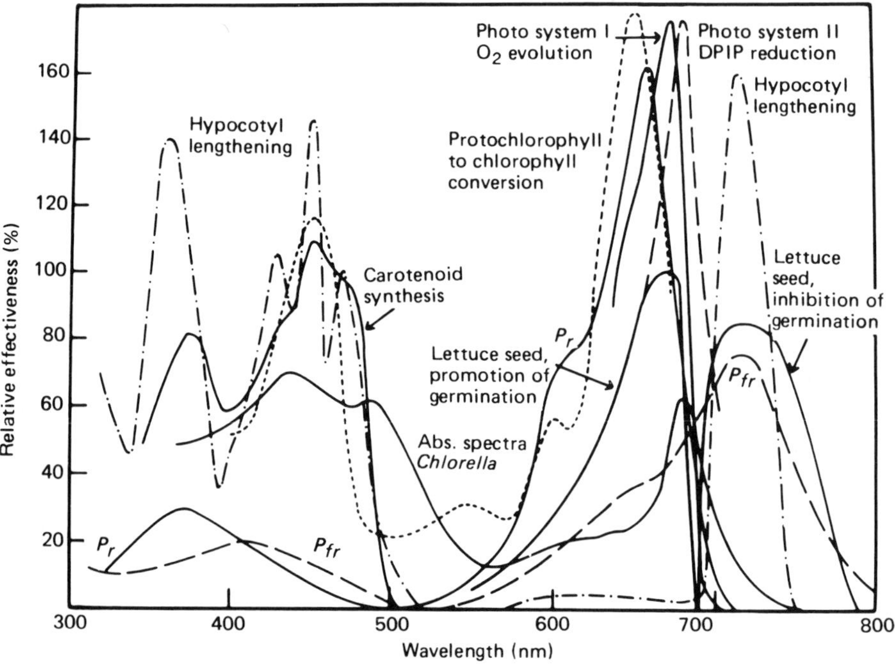

Fig. 1.5. Action spectra for various plant processes as a function of wavelength.

reduce the heat input, and then absorbs very well the far infrared beyond 3000 nm where relatively little sunlight reaches the earth's surface, but it is these wavelengths where a leaf emits energy effectively. This is a fine example of the fact that a detailed understanding of the physics of interaction between radiation and matter and a simultaneous understanding of the biology of the organism is critical to an understanding of the ecology of plants and radiation. Nothing is more important to a plant than its interaction with radiation; however, other phenomena, such as fluid flow and gas exchange, are equally vital.

A plant leaf requires the flow of materials as well as energy to function properly within its environment. In a sense, the flow of matter has already entered into the consideration of energy exchange for a leaf since the evaporation and diffusion of water vapor were important components. However, in addition to the flow of water molecules, there is a requirement for carbon dioxide and oxygen gas exchange and of minerals in solution. To advance our understanding of how a whole leaf functions in its environment, we decided to focus on the metabolic events, specifically the photosynthetic and respiration processes. We assumed that the leaf was properly supplied with nutrients through the petiole and concentrated our analysis on the energy exchange, gas exchange, and chemical kinetics of the metabolic events. This work is described in detail in the paper by Lommen, Schwintzer, Yocum, and Gates (1971) and is summarized in Chapter 2 of this volume. The 1971 *Planta* article highlights many of the objectives we have had in our research of the past decade concerning the physiological ecology of plants.

Advances in the understanding of a complex subject are made a step at a time. Although we realized that these complexities existed, we did not include all of them in our initial analysis of leaf autecology. The complex biochemistry of photosynthesis and respiration is lumped into a few simple functional relations which depend upon light intensity, carbon dioxide or oxygen concentration, and temperature. Stomatal action as a function of several variables was not adequately included, although the mechanism for doing so was there. However, the main point is that various kinds of refinements of the analysis can readily be made as rapidly as they are developed.

An analytical framework is necessary when working with many variables simultaneously to understand the vast number of interrelationships among them. I hesitate to use the word "model" because many people conceive of a model as something very synthetic and far removed from the basic mechanisms that govern the living system. Our task is to describe the mechanisms by which a plant, or a plant leaf, interacts with its environment, and from these to produce an analysis that will enable us to make predictions about the future course of the plant leaf–environment system. Out of this analysis will come an understanding of how a leaf functions within the environment.

The analytical model that described photosynthesis by a whole leaf in terms of gas diffusion and enzyme kinetics led to the following expression:

$$P = \frac{[C_A + K + S_1(P_M - W) - WS_2] - \{[C_A + K + S_1(P_M - W) - WS_2]^2 - 4S_1[(C_A - WS_2)(P_M - W) - WK]\}^{1/2}}{2S_1} \tag{3}$$

where P = net photosynthesis
C_A = carbon dioxide concentration in the air
S_1, S_2 = equivalent resistances of various kinds
P_M = gross photosynthesis at saturating carbon dioxide concentration and light for a given temperature
W = respiration rate
K = constant equal to the chloroplast concentration of carbon dioxide at which $P = P_M/2$

The maximum photosynthetic capacity, P_M, is a function of light and temperature. The stomatal resistance, which enters S_1 and S_2, is a function of light and temperature. The respiration rate, W, is a function of light, temperature, and oxygen concentration, depending in detail on whether it is respiration from the peroxisomes or the mitochondria.

Equations (1)–(3) give a rather complete description of the influence of the environment on the metabolic performance of a leaf. If one considers the net photosynthetic rate and transpiration rate of the leaf to be important dependent variables, one can understand their variation as a function of the following independent variables: total radiation intensity, light intensity, air temperature, wind speed, relative humidity, carbon dioxide concentration of the air, and the following leaf parameters: absorptivity to radiation; convection coefficient, depending upon leaf size, shape, and orientation; diffusion resistances to water vapor and carbon dioxide, including boundary layer of surface air, stomatal, substomatal, and mesophyll; P_M; W; and various K's. Clearly, a great deal of information concerning a leaf must be known to proceed with an understanding of the productivity of a whole leaf and of an entire plant. This should not be surprising. To answer many ecological questions, we must have detailed information concerning each particular plant.

When dealing with a phenomenon that involves many variables and parameters, it is difficult to deal with all the variations simultaneously, but for the same reason it is absolutely necessary to work with them within a coherent analytical framework. Once the proper connection of the variables is made, one can plot the variation in any selected two-dimensional slice through the multidimensional space represented by all the variables and parameters. This we have done, and Figs. 1.6 and 1.7 represent two examples. The detailed numbers used for these calculations are given in the article by Lommen, Schwintzer, Yocum, and Gates (1971). Here we have the photosynthetic rate and the transpiration rate of a leaf, which are the dependent variables. In Fig. 1.6 the light intensity and relative humidity are the independent variables and in Fig. 1.7 the air temperature and the relative humidity are the independent variables. There is no way to obtain an evaluation of any or all interrelationships among the variables other than by analysis using a theoretical model based upon mechanisms. Any other set of environmental conditions and leaf properties can be selected and the photosynthetic rate and transpiration rate determined as a function of various independent variables. All values calculated may not agree exactly with the performance by a given plant leaf because of vagaries that involve stomate action, water supply, nutrient status, etc., but as a

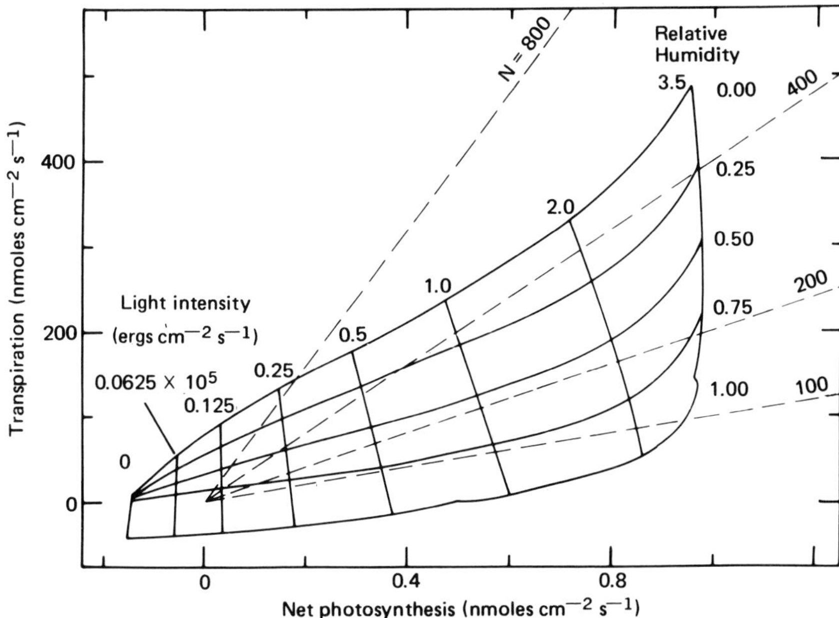

Fig. 1.6. Calculated transpiration rate and photosynthetic rate of a leaf as a function of light intensity and relative humidity. N is the ratio of number of water molecules transpired to carbon atoms assimilated. The leaf measured 5 by 5 cm. Air temperature was 30°C and wind speed was 10 cm s^{-1}.

general rule the relationship should be a reasonably accurate description of what a plant leaf of this kind will do.

Animals

Climate is coupled to an animal through the flow of energy. I reasoned that, by understanding the complete energy budget of an animal, we could predict the microclimate within which an animal would be constrained to live in order to survive. An animal derives energy from external physical sources and from internal physiological sources. An animal has a preferred range of body temperatures; in some animals this range is very small, for others it is very broad. In addition, an animal has an extended temperature range that it will tolerate beyond its preferred range, and lethal limits to body temperature, which it would rather not approach too closely. By means of an analysis of the exchange of energy between an animal and its environment, one can make an approximate estimate of the values of the variables that make up the limiting microclimate necessary for survival.

From the standpoint of animal physiology, information concerning metabolic rate and evaporative-water-loss rate is very interesting, but such information by itself is not particularly useful except in the context of the total energy budget of

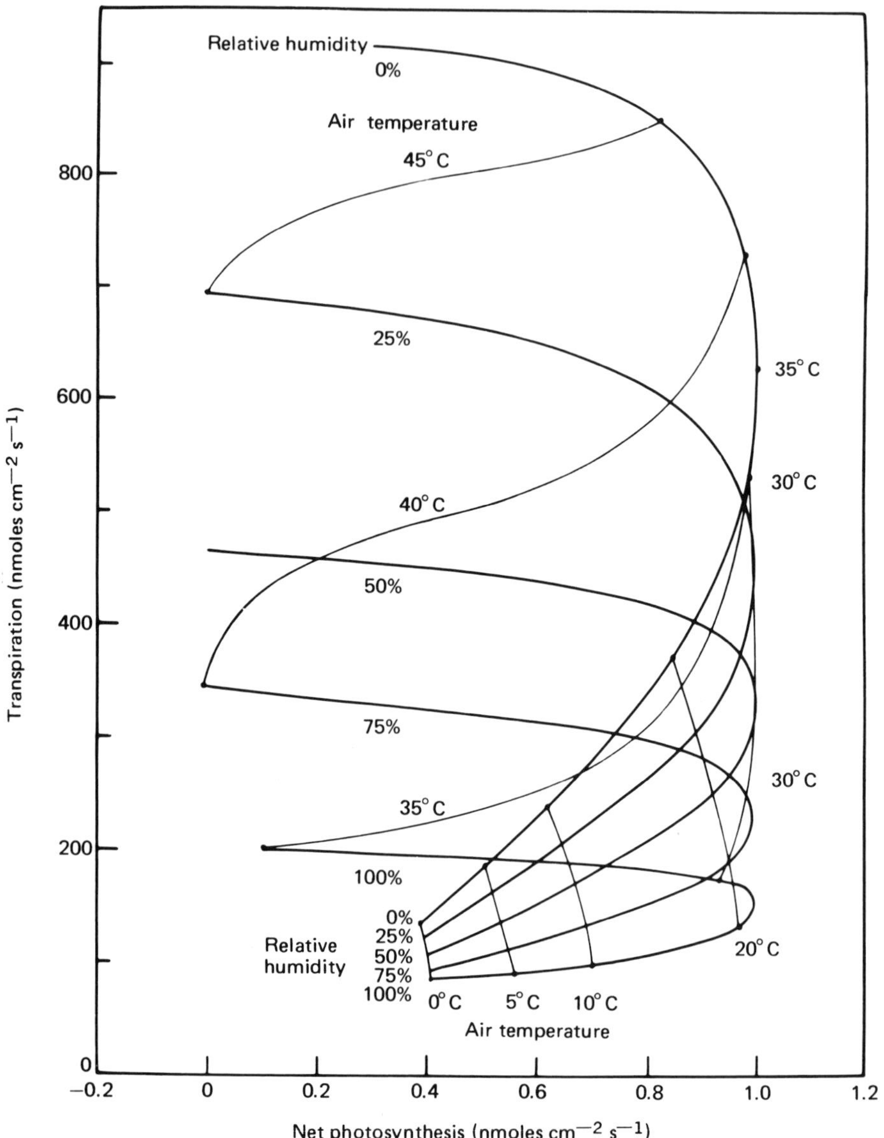

Fig. 1.7. Calculated transpiration rate and photosynthetic rate of a leaf as a function of air temperature and relative humidity. Light intensity in the wavelength interval 400–700 nm was 4×10^5 ergs cm^{-2} s^{-1}, and total incident sunlight was 8.9×10^5 ergs cm^{-2} s^{-1}. Wind speed was 10 cm s^{-1}.

an animal. Knowledge concerning one energy source or sink for an animal is not particularly useful except in its appropriate context of all energy sources or sinks for an animal. These may include, in addition to metabolic rate and evaporative-water-loss rate, the exchange of energy by conduction, convection, and radiation.

It was clear to those of us working with the idea of energy exchange for animals that there were several complexities and complications with which we could not deal in our first approximation to a solution of the problem. Animals are often in transient states rather than in steady-state situations, yet for the initial analysis, it made sense to approach the problem as if it were time-independent. Animals have many parts, such as head, body, limbs, tail, and ears, of various shapes, sizes, colors, orientations, and surface roughnesses, each of which influences the exchange of energy between an animal and its environment. These complexities naturally make analysis of the precise rates of energy transfer for an animal more difficult. As a first approximation we took animals to be represented by simple cylinders with insulation quality and surface characteristics representative of the animal. This is referred to as the "lumped-parameter model." Something approaching the true animal shape—a sphere for a head; long, slender cylinders for the appendages; a larger cylinder for the body—is referred to as the "distributed-parameter model." It is straightforward to do the distributed parameter model, but it requires additional information or assumptions concerning the partitioning of heat within the animal. The steady-state lumped-parameter model gave a good approximation to the initial problem and adequate insight to the rates of energy exchange for an animal.

In 1964 and 1965, Peter Bartlett and I did research concerning the energy budget of the fence lizard, *Scleropolous occidentalis* (Bartlett and Gates, 1967). In this work we made metal castings of the lizards and derived convection coefficients for them by means of experiments in a wind tunnel, and compared the predicted location of a lizard on a tree trunk as a function of the time of day with the locations observed in the field. The agreement was very gratifying.

Warren Porter came from the University of California at Los Angeles, where he had done his Ph.D. work under Kenneth Norris, to work with me for two years beginning in September 1967. That marked the beginning of our intensive work on the energy-budget problem for animals. Porter and I wanted to do as thorough an analysis as possible of a variety of animals. We were constrained to work with those animals for which there were good physiological data. These animals included the desert iguana (*Dipsoarus dorsalis*), the masked shrew (*Sorex cinereus*), the zebra finch, the Kentucky cardinal (*Richmondena cardinalis*), the sheep (*Ovis aries*), the pig (*Sus scrofa*), and the jack rabbit (*Lepus californicus*).

While doing this research, we derived the concept of "climate space." It occurred to us that climate as represented by four variables, radiation, air temperature, wind speed, and humidity, was indeed a four-dimensional volume or space. We decided as a first approximation not to include the influence of humidity on the energy-exchange rate for an animal, but to leave humidity as a refinement of the analysis to be made later. This allowed us to work with a three-dimensional space that could be projected on a plane whose axes would be two of the three variables and change in the plane itself would be the third variable. We elected to use the air temperature as the ordinate, the flux of radiation as the abscissa, and the wind speed as the third variable. A "climate space" diagram is derived in a very simple manner.

A "blackbody" is simply a cavity, box, room, or cave where the thermal radiation of the interior is in equilibrium with the wall and air temperature. A blackbody is very precisely described and defined by the laws of radiation physics. The result is that blackbody radiation is represented as a very precise line in the climate-space diagram (Fig. 1.8). If we are in a room whose wall and air temperatures are 20°C, we know that the thermal radiation flux is 0.6 cal cm^{-1} min^{-1}. If, instead of being within an enclosure, we are outdoors on a clear night, we receive blackbody thermal radiation from the ground surface and graybody thermal radiation from the semitransparent sky. If the sky were totally transparent, the observer at the surface of the earth outdoors at night would receive no radiation from the atmosphere and in turn would radiate to the cosmic cold of space at a few degrees absolute. With a completely transparent sky, the observer would receive on the average from below and from above an amount of flux equal to half of the blackbody radiation characteristics of the earth's surface. However, as stated above, the earth's atmosphere is semitransparent and does emit some thermal radiation toward the ground. The result is readily determined from atmospheric radiation observations and theory. An observer at the surface of the earth outdoors on a clear night receives, on the average, according to the air temperature near the surface, the amount of radiation given by the left-hand line in Fig. 1.8. This is a lower limit to the radiation environment of our planet, with its semitransparent atmosphere.

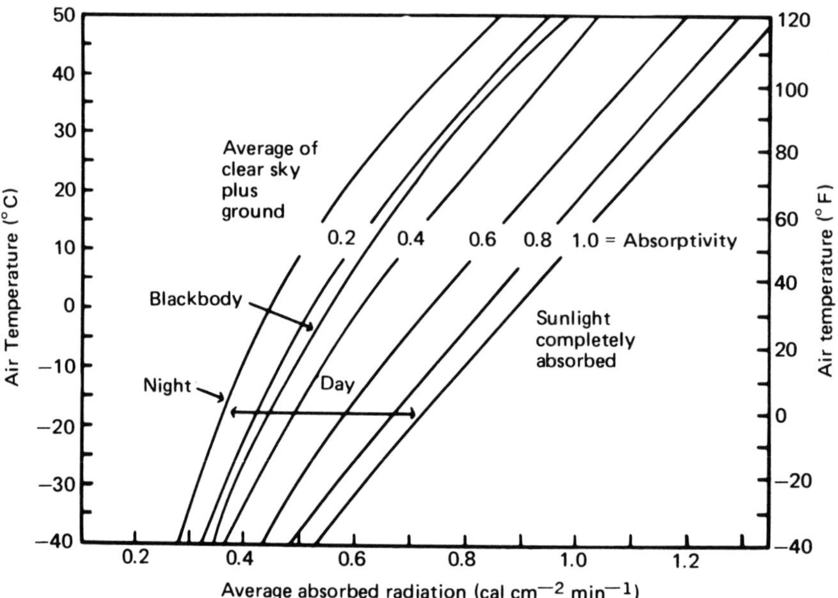

Fig. 1.8. Relationship between the total amount of radiation flux incident on an object as a function of the air temperature. At night the object receives thermal radiation from ground and atmosphere. In the daytime the object receives direct, reflected, and scattered sunlight, in addition to the thermal radiation from ground and atmosphere. If the object has an absorptivity to sunlight, the amount of radiation flux absorbed is given according to the particular line.

The maximum amount of radiation flux incident upon an animal occurs at midday or early afternoon on a clear, sunny day. It is more difficult to estimate with accuracy the amount of incident flux during the daytime with solar and sky radiation in addition to the thermal radiation fluxes from ground and atmosphere. On the general assumption that warmer air temperatures usually occur at noon and during the summer, it is possible to estimate the total incident radiation flux on an animal in the open. This gives the right-hand line shown in Fig. 1.8 and once again is purely a limit of the physical world. We spend all our lives in conditions described by the upper and lower bounds shown in Fig. 1.8 of the two-dimensional climate space, air temperature and radiation. If the animal surface absorbs some fraction of solar radiation, the total amount of radiation absorbed by the animal is given by the lines marked 0.8, 0.6, etc. These lines are based on the assumption that the animal absorptivity to thermal radiation is 1.0. If an animal has an absorptivity to sunlight of 0.6, the maximum amount of radiation that can enter its surface for any given air temperature is represented by the line marked 0.6 in Fig. 1.8.

The wind speed, a dimension or variable of the climate, can be added to the climate-space diagram only when we consider its interaction with the animal. To do this we must write the equation for the energy budget of the animal. The wind interacts with the animal through the convective heat exchange, which takes the form

$$C = k \frac{V^{1/2}}{D^{1/2}} (T_r - T_a) \tag{4}$$

where k = constant
V = wind speed (cm s^{-1})
D = characteristic dimension, or diameter of the cylinder used to approximate the animal (cm)
$T_r - T_a$ = is the difference between the radiant surface temperature of the animal and the air temperature

The energy budget for an animal is written as follows, if we ignore thermal conduction to a substrate:

$$Q_{\text{abs}} + M - E = \varepsilon\sigma\left(T_b - \frac{M - E}{K}\right)^4 + k \frac{V^{1/2}}{D^{1/2}} T_b - T_a - \frac{M - E}{K} \tag{5}$$

where Q_{abs} = total amount of radiation absorbed by the animal
M = metabolic heat production
E = evaporative heat loss
ε = emissivity of the animal surface
σ = Stefan–Boltzmann constant
T_b = body or core temperature of the animal
K = thermal conductivity of fur, feathers, or fat

An animal has a metabolic and a water-loss rate that vary with the ambient temperature. For a given air speed V, there is a line in the climate diagram (Q versus T_a plot) which is the locus of those points that will balance the energy budget for the animal. At each high air temperature and high radiation value, the appropriate $M - E$ value must enter the equation and a balance is found for a specific set of Q and T_a. At the high-energy-input limit, this is given by the upper line in the climate-space diagram and at the low-energy-input limit, by the lower line in the climate space. Hence the upper and lower lines are primarily physiological limits to the climate space, whereas the left- and right-hand lines are strictly physical limits.

Figure 1.9 is the climate-space diagram for the desert iguana. Note that the lines representing the upper and lower limits in the climate-space diagram become more horizontal with increasing wind speed. Also note that, for blackbody conditions, the lower air-temperature limit for this poikilotherm is about 3°C and the upper limit is about 46°C.

Figure 1.10 is the climate-space diagram for a homeotherm, the pig. First note that the left-hand and right-hand lines representing the physical limit are essentially the same as for the desert iguana, but the physiological lines representing the upper and lower temperature limits are much steeper. The large diameter of the pig produces a decoupling of the convection term from the air temperature for a given air speed. This shows that the desert iguana can withstand substantially higher air temperatures and higher radiation intensity than the pig. But as a homeotherm, the pig can survive very much colder conditions than the desert iguana. These climate-space diagrams appear to correspond closely to the actual limits of these animals within their microclimates as experienced in the real world.

While Porter and I were completing this initial work, Craig Heller was finishing his Ph.D. dissertation at Yale University on the altitudinal zonation of the chipmunk in the Sierra Nevada mountains. He had the physiological measurements necessary for a thorough energy-budget analysis of the chipmunk in its native

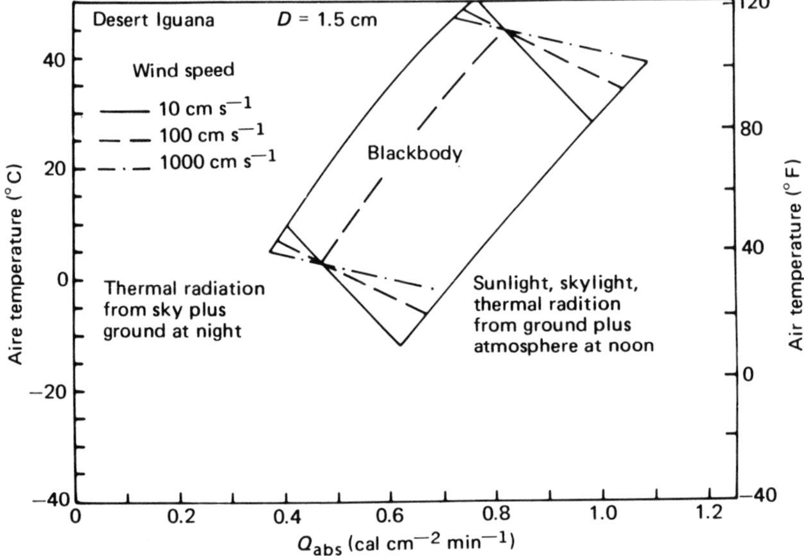

Fig. 1.9. Climate space for a desert iguana. Absorptivity to sunlight is 0.8.

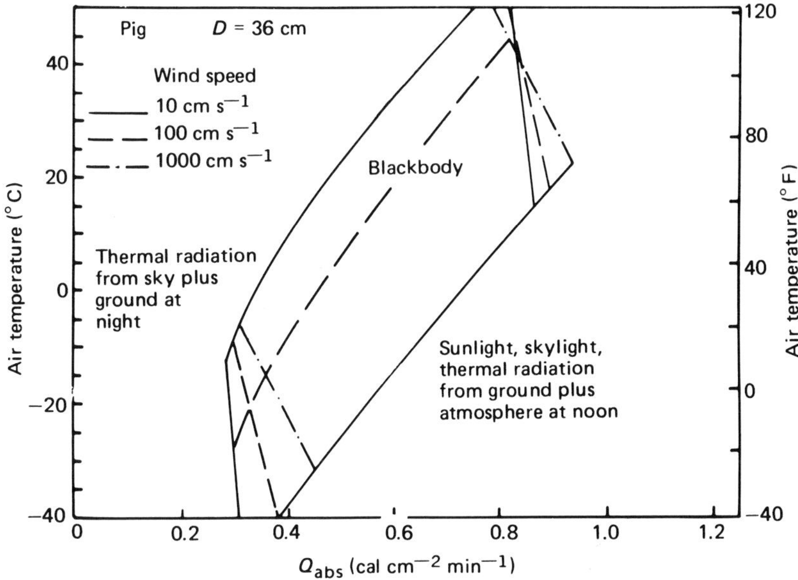

Fig. 1.10. Climate space for a pig. Absorptivity to sunlight is 0.8.

habitat. He visited us in St. Louis and we worked out the energy-budget diagrams for four species of chipmunk, as reported in Heller and Gates (1971). These are shown in Fig. 1.11. The great similarity of the climate spaces of *Eutamius alpinus*, *E. speciosus*, *E. amoenus*, and *E. minimus* lent credence to the conclusion arrived at earlier by Heller that the fundamental niches of these species overlap greatly. The lower tolerance of *E. speciosus* to high levels of Q_{abs} would be a serious limiting factor if it were to enter the more open habitats of *E. alpinus*, *E. amoenus*, or *E. minimus*. However, *E. speciosus* is excluded from the neighboring habitats by the aggressive behavior of *E. alpinus* and *E. amoenus*.

Energy-budget analysis and the derivation of the climate-space diagram applied to the adaptation and behavior of animals in their natural habitats are very nicely illustrated by the work of Heller. It is worthwhile quoting in some detail from Heller and Gates (1971):

> The maximum T_a in the climate space of *E. minimus* is consistent with field observations. The sage brush desert habitat of *E. minimus* clearly corresponds to conditions at the upper limits of the climate spaces of the species. This habitat has little shade, high levels of incident radiation, maximum summer temperatures between 35° and 40°C, and soil surface temperatures up to 60°C. *E. minimus* can minimize its Q_{abs} by being active in the morning and by seeking small patches of shade under sage-brush, but at noon on a clear day the maximum T_a that *E. minimus* could tolerate even at the lowest possible Q_{abs}, would be 36°C if it maintained T_b at 40°C. Unpublished field data record that the T_a of the sagebrush habitat in the summer generally reaches 33°C to 36°C by 11 A.M. As predicted by the climate space,

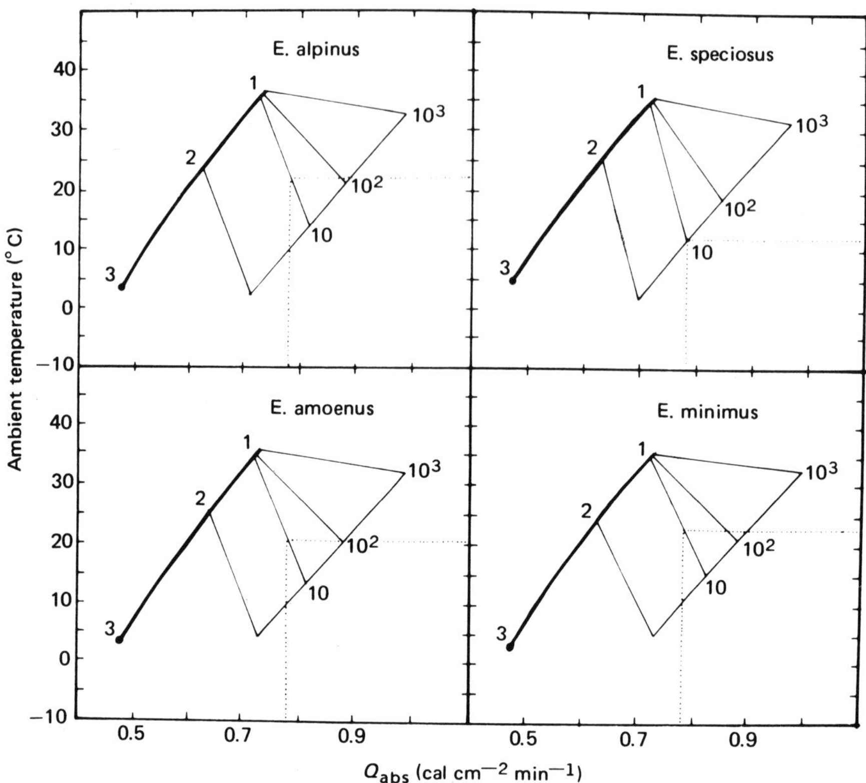

Fig. 1.11. Climate spaces for four species of chipmunks: *Eutamius alpinus, E. speciosus, E. amoenus,* and *E. minimus.* The chipmunk is diurnal. Therefore, the left-hand limit to the climate space is the blackbody line characteristic of burrow temperatures and radiation.

the above-ground activity of *E. minimus* in this habitat drops off sharply at 11 A.M.

E. minimus goes underground as T_a reaches a critical value, but *E. amoenus* remains active all day in spite of daily maximum air temperatures around 35°C. The climate space would predict that *E. amoenus* could endure only such a high T_a if it minimizes Q_{abs} and/or maximizes convective cooling. *E. amoenus* does keep its Q_{abs} low while active by limiting its activity to the large patches of shade under the piñon pines. Around midday *E. amoenus* is frequently observed to sit on the well-shaded branches of the piñon pine, thereby maximizing convection and minimizing Q_{abs} so that the maximum daily temperatures can be endured without going underground.

The maximum daily temperatures in *E. alpinus* habitat generally do not exceed 20°C, but most of this habitat is exposed to full sun. *E. alpinus* forages among and between rocks where the air is quite still in spite of breezes of 100 to 500 cm sec^{-1} above the rocks. The climate space indicates that in full sun and still air, 20°C is too high an ambient temperature for *E. alpinus*. Around midday *E. alpinus* was observed to forage between the rocks for short periods and then run to the top of a rock and sit in the full sun. The climate space reveals that even at maximum Q_{abs}, moving from still air (10 cm sec^{-1}) to a gentle breeze (100 cm sec^{-1}) is sufficient to allow *E. alpinus* to endure a T_a slightly over 20°C.

More recently, Sylvia Morhardt has applied the steady-state energy-budget analysis and climate space to an interpretation of field observations concerning the behavior of the Belding ground squirrel in its native habitat in the Sierra Nevada mountains. This work (Morhardt and Gates, 1974) adds further support to the validity and applicability of the energy-budget analysis to field observations.

The steady-state energy budget and climate space of the American alligator, *Alligator mississippiensis*, were determined by Spotila, Soule, and Gates (1972). In the case of the alligator, the conductance of heat to and from the ground surface by direct contact becomes significant. When the alligator is at its preferred body temperature, it is restricted to a small portion of its climate space. Large alligators can operate outside their climate space because they can store large amounts of heat, which they dump when they return to cooler conditions. Water is the most important part of an alligator's habitat because it acts as a substitute for blackbody conditions and ensures maximum utilization of an alligator's potential climate space.

It is apparent that the steady-state, time-independent analysis of animal energetics is not sufficient for a detailed accounting of animal behavior. It is appropriate to quote from Heller and Gates (1971) concerning the importance of time-dependent transient states of animals:

> Transient states are extremely important in adapting animals to microclimates which sometimes exceed the limits defined by their climate spaces. The climate space describes average steady-state limits, but the animal may make transient excursions beyond these limits balanced by retreats to within the limits. For example, *E. minimus* is able through the use of transients to colonize habitats which are beyond the fundamental niche of the aggressively dominant *E. amoenus*. *E. minimus* has a greater tolerance to hyperthermia and maintains a lower evaporative water loss than *E. amoenus*. *E. minimus* probably adapts to the dry, hot sagebrush desert by undergoing short periods of hyperthermia while active and then returning to the burrow to dissipate the heat load. This behavior is envisioned in the climate space as excursions from the blackbody line at the temperature of the burrow to the right and/or up and then retreats back to the blackbody line. The steady-state boundaries of the climate space are defined by the average body temperatures and the wind speed. The shorter the time spent outside of the boundary, the farther the animal could go beyond its average steady state.

All animals spend considerable periods of time in steady-state situations, but transient states are a frequent and significant part of their behavior. Because we recognize the importance of time-dependent states, we have endeavored to improve our analysis in this respect. Analytical solutions to the time-dependent energy-budget equation are complicated.

Our first approach to the solution of the time-dependent heat-transfer problem was a numerical iterative procedure using a computer and applied to large reptiles as reported by Spotila, Lommen, Bakken, and Gates (1973). The animal was approximated by a large cylinder with a central core of tissue kept at uniform temperature by blood flow and surrounded by six layers of fat. The calculations showed that a large reptile (diameter about 100 cm) would have a relatively constant high body temperature when exposed to warm, diurnally fluctuating environmental conditions, even with a low metabolic rate, as long as the average values of the physical parameters result in a body temperature within tolerable limits. Changes in fat thickness are of minor importance in determining the constancy of body temperature in a giant reptile. Gigantism helps to provide an equable internal temperature in a stable warm climate with strong diurnal variations but with relatively little seasonal change.

A time constant was determined for large reptiles by allowing the model to warm up until no further change in core temperature occurred. The core temperature as a function of time was described by

$$T = T_f + (T_0 - T_f)e^{-t/\tau} \tag{6}$$

where T = core temperature at any time t
T_0 = initial core temperature
T_f = final core temperature
τ = time constant [numerically equal to the length of time for $(1 - e^{-1})$ or 63.7 percent of the temperature change to occur in a standard heating or cooling-curve experiment]

The time constant of an animal is usually considered to be the product of its heat capacity and insulation. Assuming a reptile to be a cylinder of diameter D with fat thickness equal to $0.05D$, we were able to obtain the following expression for the time constant:

$$\tau = 0.085D^{1.38} \tag{7}$$

where τ is in hours if D is given in centimeters. This equation must be used with great care since the time constant is sensitive to a number of variables. It is sensitive to the wind speed, to the shape and size of the appendages, to the properties of the fat layers, and to the net metabolic heat production $(M - E)$, which was taken to be negative and monotonically increasing with core temperature. The slope of the $M - E$ relationship with temperature is more important in its influence on τ than the magnitude of $M - E$ because of the feedback loop involved. $M - E$ becomes more negative with increased T_b and hence tends to decrease T_b. The net result is that the equilibrium temperature is reached more quickly than if there were no feedback and hence a smaller time constant.

Some time in 1971 it became clear that we would have to extend the mathematical solution to the time-dependent energy-budget equation to get out a greater amount of information. George Bakken began to address this problem and has made some extremely significant advances, both theoretical and experimental. These are given in Chapter 16. Although we are still working with lumped-parameter models, the time-dependent analysis is sufficiently detailed to give much greater insight into the strategies animals use for survival in a world of thermal extremes. Not only must the theory be advanced as far as possible, but clever methods of measurement and observation in the field must be devised. When this work is carried to its logical conclusion, refinements will then be necessary that will require attention to a distributed parameter model.

Since the whole issue of understanding animal behavior, adaptation, competition, and physiological ecology involves many variables, it is essential to have a theoretical model that brings the variables together in consistent manner. The benefits to be derived from a theoretical model based on physical mechanisms that incorporate physiological processes are enormous indeed. Theory alone will not give us all the answers, just as observation alone will not do so, but the proper balance among theory, observation, and experiment will ultimately give us the answers to our questions concerning animal ecology.

Publications on Biophysical Ecology by David M. Gates and his Colleagues

Alderfer, R. G., Gates, D. M.: 1971. Energy exchange in plant canopies. Ecology *52*, 855–861.

Bartlett, P. N., Gates, D. M.: 1967. The energy budget of a lizard on a tree trunk. Ecology *48*, 315–322.

Derby, R., Gates, D. M.: 1966. The temperature of tree trunks, calculated and observed. Am. J. Bot. *53*, 580–587.

Gates, D. M.: 1959. The physical environment as affected by radiation. Semicentennial Celebration Proc. Univ. Michigan Biol. Sta., pp. 31–52. Ann Arbor: Univ. Michigan.

———: 1962. Energy exchange in the biosphere. New York: Harper & Row.

———: 1963. The energy environment in which we live. Am. Scientist *51*, 327–348.

———: 1963. Leaf temperature and energy exchange. Arch. Meteorol., Geophys., and Bioklimatol., Ser. B *12*, 321–336.

———: 1963. The measurement of water vapor boundary layers in biological systems with a radio refractometer. *In* Humidity and moisture; measurement and control in science and industry, vol. 2, pp. 33–38. New York: Van Nostrand Reinhold.

———, Benedict, C. M.: 1963. Convection phenomena from plants in still air. Am. J. Bot. *50*, 563–573.

———, Vetter, M. J., Thompson, M. C.: 1963. Measurement of moisture boundary layers and leaf transpiration with a microwave refractometer. Nature *197*, 1070–1072.

———: 1964. Leaf temperature and transpiration. Agronomy J. *56*, 273–277.

———, Lee, R.: 1964. Diffusion resistance in leaves as related to their stomatal anatomy and micro-structure. Am. J. Bot. *51*, 963–975.

———, Hiesey, W. M., Milner, H. W., Nobs, M. A.: 1964. Temperature of mimulus leaves in natural environments and in a controlled chamber. Carnegie Inst. Wash. Yearbook *63*, 418–430.

———: 1965. Energy, plants, and ecology. Ecology *46*, 1–13.

———: 1965. Heat, radiant and sensible, Chap. 1, Radiant energy, its receipt and disposal. Meteorol. Monogr. *6*, 1–26.

———: 1965. Heat transfer in plants. Scientific American *213*, 76–84.
———, Keegan, H. J., Schleter, J. C., Weidner, V. R.: 1965. Spectral properties of plants. Appl. Optics *4*, 11–20.
———, Tibbals, E. C., Kreith, F.: 1965. Radiation and convection for ponderosa pine. Am. J. Bot. *52*, 66–71.
———: 1966. Spectral distribution of solar radiation at the earth's surface. The spectral quality of sunlight, skylight, and global radiation varies with atmospheric conditions. Science *151*, 523–529.
———: 1966. Transpiration and energy exchange. Quart. Rev. Biol. *41*, 353–364.
———, Janke, R.: 1966. The energy environment of the alpine tundra. Oecologia Plantarum *1*, 39–61.
———: 1967. Remote sensing for the biologist. BioScience *17*, 303–307.
———: 1967. Thermal balance of the biosphere. Proc. Fourth Internat. Biometeorol. Congr. Biometeorol. *3*, 29–39 (Suppl. Vol. 11 Internat. J. Biometeorol.).
———, Hanks, R. J.: 1967. Plant factors affecting evapo-transpiration. *In* Irrigation of agricultural lands. Am. Soc. Agronomy Monogr. *11*, 506–521.
———: 1968. Energy exchange and ecology. BioScience *18*, 90–95.
———: 1968. Energy exchange between organism and environment. Proc. 28th Ann. Biol. Colloquium, Oregon State Univ. Biometeorol. *4*, 1–22.
———: 1968. Energy exchange between organisms and environment. Austral. J. Sci. *31*, 67–74.
———: 1968. Relationship between plants and atmosphere. Am. Naturalist *23*, 133–140.
———: 1968. Sensing biological environments with a portable radiation thermometer. Appl. Optics *7*, 1803–1809.
———: 1968. Toward understanding ecosystems. Adv. Ecol. Res. *5*, 1–35.
———: 1968. Transpiration and leaf temperature. Ann. Rev. Plant Physiol. *19*, 211–238.
———, Alderfer, R., Taylor, S. E.: 1968. Leaf temperatures of desert plants. Science *159*, 994–995.
———: 1969. Climate and stability. *In* Diversity and stability in ecological systems, Brookhaven Symposia in Biology 22, pp. 115–127. Upton, N.Y.: Brookhaven Natl. Lab.
———: 1969. Infrared measurement of plant and animal surface temperature and their interpretation. *In* Remote sensing in ecology (ed. P. L. Johnson), pp. 95–107. Athens: Univ. Georgia Press.
———: 1969. Transpiration rates and temperatures of leaves in cool humid environment. *In* The ecology of an elfin forest in Puerto Rico 4. J. Arnold Arboretum, Harvard Univ. *50*, 93–98.
———: 1970. Animal climates (where animals must live). Environmental Research *3*, 132–144.
———: 1970. Physical and physiological properties of plants. *In* Remote sensing: with special reference to agriculture and forestry, pp. 224–252. Washington, D.C.: J. Natl. Acad. Sci.
———: 1971. Biophysical ecology. Missouri Bot. Garden Bull. *59*, 4–16.
———: 1971. The flow of energy in the biosphere. Scientific American *225*, 89–100.
———: 1971. Microclimatology. *In* Topics in the study of life, Bio source book, pp. 386–396. New York: Harper & Row.
———: 1972. Man and his environment: climate. New York: Harper & Row.
———, Johnson, H. B., Yocum, C. S., Lommen, P. W.: 1972. Geophysical factors affecting plant productivity. *In* Theoretical foundations of the photosynthetic productivity, pp. 406–419. Moscow: Nauka.

Heller, C., Gates, D. M.: 1971. Altitudinal zonation of chipmunks (*Eutamias*): energy budgets. Ecology *52*, 424–453.

Hoffman, G. R., Gates, D. M.: 1970. An energy budget approach to the study of water loss in cryptogams. Bull. Torrey Bot. Club *97*, 361–366.

Hoffman, G. R., Gates, D. M.: 1971. Transpiration water loss and energy budgets of selected plant species. Oecologia Plantarum *6*, 115–131.

Idso, S., Baker, D. G., Gates, D. M.: 1966. The energy environment of plants. Adv. Agronomy *18*, 171–218.

Lommen, P. W., Schwintzer, C. R., Yocum, C. S., Gates, D. M.: 1971. A model describing photosynthesis in the terms of gas diffusion and enzyme kinetics. Planta *98*, 195–220.

MacBryde, B., Jefferies, R. L., Alderfer, R., Gates, D. M.: 1971. Water and energy relations of plant leaves during period of heat stress. Oecologia Plantarum *6*, 151–162.

Miller, P., Gates, D. M.: 1967. Transpiration resistance of plants. Am. Midland Naturalist *77*, 77–85.

Parkhurst, D. F., Gates, D. M.: 1966. Transpiration resistance and energy budget of *Populus sargentii* leaves. Nature *210*, 172–174.

———, Duncan, P. R., Kreith, F., Gates, D. M.: 1968. Wind tunnel modelling of convection of heat between air and broad leaves of plants. Agric. Meteorol. *5*, 33–47.

———, Duncan, P. R., Kreith, F., Gates, D. M.: 1968. Convection heat transfer from broad leaves of plants. J. Heat Transfer *90*, 71–76.

Porter, W. P., Gates, D. M.: 1967. Thermodynamic equilibria of animals with environment. Ecol. Monogr. *39*, 245–270.

Spotila, J. R., Soule, O., Gates, D. M.: 1972. The biophysical ecology of the alligator: heat energy budgets and climate spaces. Ecology *53*, 1094–1102.

———, Lommen, P. W., Bakken, G. S., Gates, D. M.: 1973. A mathematical model for body temperatures of large reptiles: implications for dinosaur ecology. Am. Naturalist *107*, 391–404.

Taylor, E. S., Gates, D. M.: 1970. Some field methods for obtaining meaningful leaf diffusion resistances and transpiration rates. Oecologia Plantarum *5*, 103–111.

Tibbals, E. C., Carr, E. K., Kreith, F., Gates, D.M.: 1964. Radiation and convection in conifers. Am. J. Bot. *51*, 529–538.

Other References

Budyko, M. T.: 1956. The heat balance of the earth's surface, Gidrometeorologicheskoe izdatel'stro (Leningrad). Available as a translation, PB 131692. Washington, D.C.: U.S. Dept. Comm., Off. Tech. Serv.

Lange, O. L.: 1959. Untersuchungen über Wärmehaushalt und Hitzeresistenz mauretanischer Wüsten- und Savannen pflanzen. Flora *147*, 595–651.

Raschke, K.: 1955. A method for the micro-meteorological measurement of the energy budget of a leaf *in situ* (as exemplified with *Alcora indica* during a 12-hour cycle). India: Univ. Poona. Ph.D. diss.

———: 1956. Über die Physikalischen Beziehungen zwischen Wärmeübergängszahl, Strahlungsaustausch, Temperatur und Transpiration eines Blattes. Planta *48*, 200–238.

———: 1958. Über den Einfluss der Diffusionswiderstände auf die Transpiration und die Temperatur eines Blattes. Flora *146*, 546–578.

———: 1960. Heat transfer between the plant and the environment. Ann. Rev. Plant Physiol. *11*, 111–126.

Ting, I. P., Loomis, W. E.: 1963. Diffusion through stomates. Am. J. Bot. *50*, 866–872.

PART I

Analytical Models of Plants

Introduction

David M. Gates

To understand the multitude of processes that influence primary productivity by plants, it is necessary to construct analytical models that link these processes together in the same manner as does the plant itself. These processes are dependent upon fundamental mechanisms that are an integral part of the physical world. The mechanisms involved are well founded in physical theory and include energy flow, gas exchange, and chemical kinetics. The morphology, anatomy, and physiology of the plant influence the magnitude of each physical event as it interacts with the particular plant. This influence is mediated through various plant parameters, such as leaf absorptivity to radiation, leaf size, shape, orientation, diffusion resistance to gas exchange, and various coefficients that characterize the rates of groups of chemical reactions that involve photosynthesis and respiration.

To try to understand a complex collection of events which influence primary productivity of entire plants by purely empirical methods is not only laborious but inadequate. On the other hand, if one pursues a theoretical approach without recourse to observation and experiment in the real world, one will never even approximate an understanding of the living system, with all its complexities and subtleties. What is urgently needed in ecology is a progressive evolution of theory, experiment, and observation in close coordination. One must develop a concept of the phenomena under study, in this case primary productivity by higher plants, and formulate a model or set of relationships about the mechanisms involved. Then one must test the hypotheses that enter the model. This usually requires careful experiments concerning various components of the total system. Many years of work are often necessary to check out each subsystem of the overall model. We must remind ourselves that we are dealing here with ecology. This, by definition, implies that we must understand plant–environment interaction. The environment is a subsystem of the total plant–environment system, as is the plant itself. Each and every aspect of the environment and of the plant must be formulated and understood.

There is a propitious time for the advancement of a particular science and there is a premature time. There have been sufficient advances in our understanding of energy exchange for plants, of gas flow between a plant and its environment, and of the biochemistry of the metabolic events that the time is now at hand to formulate appropriate sets of relationships, a model, among these processes to understand how it is that a whole plant functions. As with any branch of science, when there is sufficient growth and maturity, a sufficient backlog of data and observed facts, there will be a sudden convergence of ideas, and researchers in various parts of the world will simultaneously develop similar concepts.

This has been the situation with the biophysical ecology of plants. P. Chartier in France has approached the subject of primary productivity by a single leaf in a very analytical manner based on physical mechanisms. Paul Waggoner of the United States has similarly worked out an analytical approach based on mechanistic modeling of a whole leaf. About the same time, O. Björkman and A. E. Hall were working out a theoretical model primarily directed at the gas exchange and chemical kinetics within a plant leaf. Simultaneously, P. Lommen, C. Yocum, C. Schwintzer, and I were working out a detailed theoretical model of whole leaf productivity based on energy and gas exchange and chemical kinetics.

Beyond this microscale of modeling of primary productivity, one gets into the mesoscale, which involves the exchange of energy, moisture, and momentum between plant canopies and the atmosphere. The agricultural meteorologists have contributed very strongly to the advancement of this subject. On the one hand, it yields much useful information; on the other hand, it leaves some very fundamental questions about energy and gas exchange and metabolic events at the microscale level. It is very likely that the ultimate answers to difficult ecological questions concerning plant productivity, competition, adaptation, and succession will be resolved only at the biochemical–metabolic level when combined with the full micro and mesoscale analysis.

It is appropriate to begin with a discussion of the theoretical models, which will give us a basis for understanding experimental and observational data. The model developed by Lommen, Schwintzer, Yocum, and me is strongly biophysical, whereas the model originated by Hall and Björkman is somewhat more biochemical in character. Conrad Yocum combines the two approaches with his work on the gas–liquid phase in the leaf mesophyll. Once a reasonably good theoretical model is formulated and tested, it is important to exploit it as fully as possible for the making of predictions concerning the response of plant leaves to environmental factors. Elwynn Taylor has done this in an attempt to understand the influence of leaf size and diffusion resistance on net photosynthesis, water use, and photosynthetic efficiency. His results are based on an earlier version of the model, and therefore can be considered to indicate significant broad relationships but not precise quantitative predictions. Nevertheless, the ecologist must be willing to use a theoretical model to explore the full domain of plant–environment interactions and, at the risk of making some mistakes, see what it will suggest in the way of evolutionary strategies.

2

Photosynthetic Model

PAUL W. LOMMEN, SANDRA K. SMITH, CONRAD S. YOCUM, AND DAVID M. GATES

Introduction

Since our original model was described (Lommen *et al.*, 1971), we have designed and built a system capable of simultaneous measurement of net photosynthesis and transpiration for a single attached leaf. The data obtained using this system have generally confirmed the earlier description. When oxygen concentration is $\gtrsim 5$ percent, our original model has been revised in line with our data and the findings of others (Björkman, 1971). In addition, evidence has appeared for two unexpected effects. First, the temperature optimum of a leaf appears to increase with an increase in ambient carbon dioxide concentration. Second, indirect evidence suggests that mesophyll resistance increases with increasing oxygen concentration. Stomatal resistance and the factors affecting it will not be discussed here.

We will summarize briefly the main idea of our model. Most of the main effects originate from one or both of two fundamental relationships:

Fick's law for gas diffusion:

$$F = \frac{C_2 - C_1}{R} \tag{1}$$

Michaelis–Menten relationship describing enzyme–substrate reactions:

$$P_g = \frac{P_m}{1 + K/C_c} \tag{2}$$

where F = flux of gas diffusing from region 1 to region 2 (nmoles cm^{-2} s^{-1})
C_1, C_2 = gas concentrations in regions 1 and 2, respectively (nmoles cm^{-3})
R = resistance to diffusion (s cm^{-1})
P_g = gross photosynthesis rate (nmoles cm^{-2} s^{-1})
K = Michaelis–Menten constant (nmoles cm^{-3})
P_M = maximum rate of photosynthesis (nmoles cm^{-2} s^{-1})
C_c = concentration of CO_2 at the chloroplast (nmoles cm^{-3})

Two Michaelis–Menten relationships are combined in our basic equations for photosynthesis and respiration:

$$P_g = \frac{\dfrac{P_{MLT}[G(T)]}{1 + K_L/L}}{1 + \dfrac{K(1 + C_{0A}/K_{C_{0A}})}{C_c}} \tag{3}$$

Here the light- and temperature-dependent numerator is equivalent to P_m of Eq. (2). $G(T)$ is a temperature factor whose value is between 0 and 1 with an optimum typically at 30°C. K in Eq. (2) has been replaced by $K(1 + C_{0A}/K_{C_{0A}})$ (where C_{0A} is the oxygen concentration), which is the standard mathematical approach for describing competitive inhibition between carbon dioxide and oxygen. P_{MLT}, K_L, K, $K_{C_{0A}}$ are all parameters that must be determined experimentally. The equation for respiration, W, assumed to be entirely photorespiration, is the same as for photosynthesis but with a different family of parameters:

$$W = \frac{\dfrac{W_{LT0}[G(T)]}{1 + K_{WL}/L}}{1 + \dfrac{K_{W0}(1 + C_W/K_{WC})}{C_{0A}}} \tag{4}$$

There is good evidence that carbon dioxide and oxygen are competitive inhibitors of each other in both photosynthesis and respiration, as will be indicated.

The resistance network used is shown in Fig. 2.1. If we apply Fick's law across each resistance, using the Michaelis–Menten equation at the chloroplast for the uptake of CO_2, we get an analytical solution for P, net photosynthesis. The algebra is fully described in Lommen (1971). The equation is

$$P = \frac{[C_A + K + S_1(P_M - W) - WS_2] - \{[C_A + K + S_1(P_M - W) - WS_2]^2 - 4S_1[(C_A - WS_2)(P_M - W) - WK]\}^{1/2}}{2S_1} \tag{5}$$

where

$$S_1 \equiv R_1 + \frac{R_2(R_3 + R_4)}{R_2 + R_3 + R_4}$$

$$S_2 \equiv \frac{R_2 R_4}{R_2 + R_3 + R_4} \tag{5a}$$

This is the first of two key equations we use. The second is a leaf-energy-budget equation from Gates (1968) and elsewhere in this volume.

Figure 2.2 is a diagram with boxes and arrows which indicate the various effects we have considered in our model. When it comes to calculating what this means or predicts, the boxes and arrows translate into variables and equations. Enough factors are considered so that it is impossible to write one equation for, e.g., the internal CO_2 concentration (C_{INT}). For example, to calculate the value of net photosynthesis:

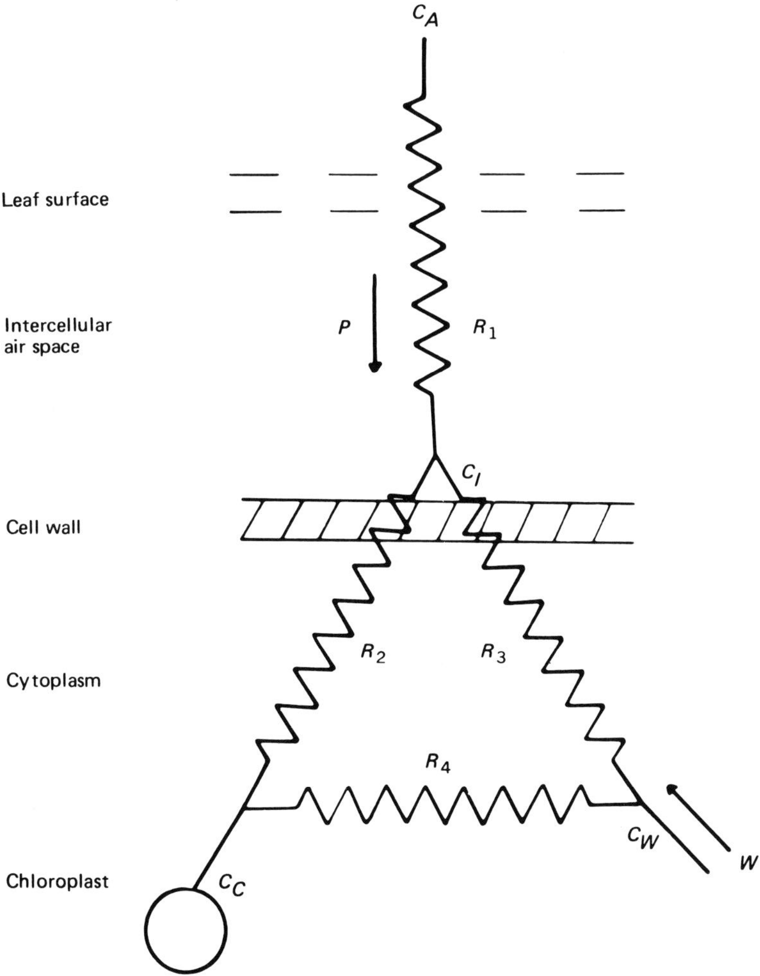

Fig. 2.1. Resistance network, sources of CO_2, and site of CO_2 uptake.

1. Choose initial conditions: radiation, air temperature, relative humidity, wind speed, leaf spectral characteristics, leaf shape and size, oxygen concentration, carbon dioxide concentration in external air.
2. Choose minimum stomatal and mesophyll resistances.
3. Choose biochemical parameters: temperature dependencies, maximum rates, rate constants.
4. Make reasonable guesses for leaf temperature and C_{INT} (=air temperature and =0, respectively, are reasonable).

Following are the steps to follow in the calculation of leaf temperature and C_{INT}:

1. Calculate (iteratively) a new value of leaf temperature so that the energy-budget equation is satisfied.

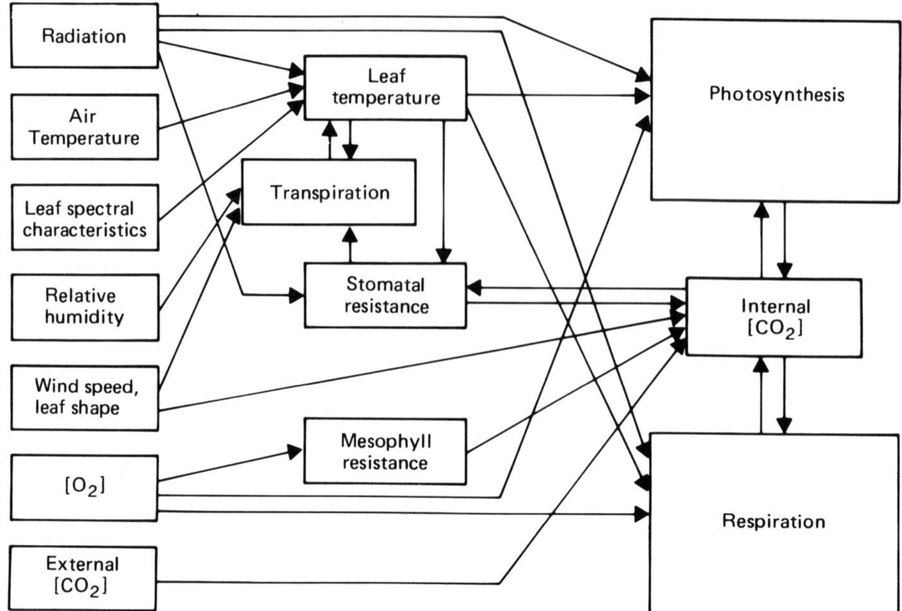

Fig. 2.2. Various effects considered in the model. Note the several loops within the diagram.

2. Calculate W using the current leaf temperature and C_{INT} using Eq. (4).
3. Calculate net P and a new value for C_{INT} using Eq. (5) for net P and Fick's law, Eq. (1), for C_{INT}.
4. Go back to step 2 and keep going from 2 to 3 until the new and old values for C_{INT} are the same.
5. Go back to step 1 using the improved value for C_{INT}.
6. Stop when step 1 no longer produces a changed value for leaf temperature. The values of all the variables have now been determined.

Measuring System

The system was originally designed and built by E. L. Dunn in St. Louis, Missouri, in 1971. Since then the apparatus has been dismantled, moved, and revised three times. An "open" system design was used. Gas was taken from standard pressurized steel cylinders and allowed to flow through flow meters; temperature- and moisture-conditioning processes; a small chamber ($\sim 500\ \text{cm}^3$) that contained a single attached leaf, water vapor, and CO_2 analyzers; and then the gas was vented to room air.

Leaf and incoming dew-point temperatures were controlled to $\pm 0.1°\text{C}$ by two Honeywell Model 111 dual "electro-pulse" controlling recorders. The moisture and CO_2 content of the gas stream were sampled upstream and downstream from the chamber using an EG&G Model 880 dew-point hygrometer, a Beckman 215 infrared gas analyzer (IRGA) (operated in absolute mode), and a Beckman 315B IRGA with 13.5-in. cells (operated in differential mode). Light intensity was provided by 2 quartz-iodide lamps, nominally 500 W each, operated at 100 ± 2 V. Filtering was done crudely by a 5-cm layer of water and finely by a 10-cm^2 Bausch & Lomb 90-22 filter, which effectively transmitted only the photosynthetically active portion of the spectrum. Intensity was changed by various combinations of neutral white plastic filters. We used intensities from 0 to ~ 110 mW cm^{-2} ($\sim$twice full sun).

Desired oxygen concentrations were obtained by appropriate mixtures in the steel cylinders. Carbon dioxide concentration was varied by mixtures in the steel cylinders and scrubbing part of the gas stream with ascarite. The continuous temperature readings of room air, leaf, chamber air, and hot and cold water baths were recorded on a West Marksman 24 point recorder, which also recorded millivolt readings from a Silicon light sensor and the absolute IRGA. An Esterline-Angus dual pen recorder Model L1102S recorded dew point and differential IRGA outputs. Six Brooks and Matheson flow meters were used, with capacities from 137 to 2200 cm^3 min^{-1}. Once the system and plant reached operating level, each data point typically took 15–20 min to obtain. About 20 data points were obtained in a typical run of 10 h.

Each of the calibration curves used was fitted with a polynominal (of degree 2, 3, 4, or 5) and included in our main data-analysis computer program. This was a total of 29 curves (multipoint recorder temperature, 215 IRGA, Silicon light sensor, water-vapor density as a function of temperature, 6 for differential IRGA, and 19 for flow meters). Since the differential IRGA output depended on both PPM difference and absolute concentration, the calibration resulted in a family of curves rather than a single curve. On a graph of (Δppm) versus (ppm in reference side of differential IRGA), we had a curve for each of the following values of Δ (chart units): 5, 10, 20, 30, 50, 70, where Δ(charts units) = 100 corresponds to a full-scale deflection. Thus for each Δppm value, the computer needed millivolt outputs of both the differential and absolute IRGAs and then would interpolate between adjacent calibration curves. For the flow meters the computer would read codes to determine which meter and which float (three meters had two floats each) corresponded to the data; calculate a nominal rate with the polynomial fit to the manufacturer's calibration curve; and correct the rate for temperature, pressure, and oxygen concentration (procedure adapted from undated Matheson publication, "Tube Flowmeter Correction Factors—Calculation Procedure"). About half an hour was needed to type in a day's data on the computer terminal and have the computer calculate and type back for each point transpiration, stomatal resistance, net photosynthesis, and carbon dioxide concentration in external air and in intercellular air spaces.

The standard deviations of our net photosynthesis values are typically 5–10 percent. Along any single curve, however, the relative positions of the points are usually known to within 2 or 3 percent, since many of the errors are the same for every point on the curve.

Results

Figure 2.3 shows *Phaselois vulgaris* (kidney bean) data of net photosynthesis plotted versus C_I, the CO_2 concentration in the intercellular air spaces. Plotting versus C_I removes any possible stomatal effects from the curves. Light intensity was saturating (160 kergs cm^{-2} s^{-1}), leaf temperature was optimal (34.5°C), and oxygen concentration (constant along each curve) was 2, 24, and 55 percent. The solid curves are fits to the data using the following procedure. At very low oxygen concentration, Eq. (5) simplifies greatly and only three parameters are necessary to fit the curve of P versus C_I; P_M, mesophyl resistance (which in our notation is R_2 in parallel with $R_3 + R_4$); and K. We used this value of P_M to calculate new values of resistance and K for each of the next two curves, along with a value of W_{LTO}. The results are shown in Table 2.1.

The data are fitted very nicely by curves having the same P_m. This can happen only if (1) CO_2 and O_2 are competitive inhibitors, and (2) respiration is turned off by the high CO_2 concentration.

Figure 2.4 shows two curves. The data do not quite merge, but the points are

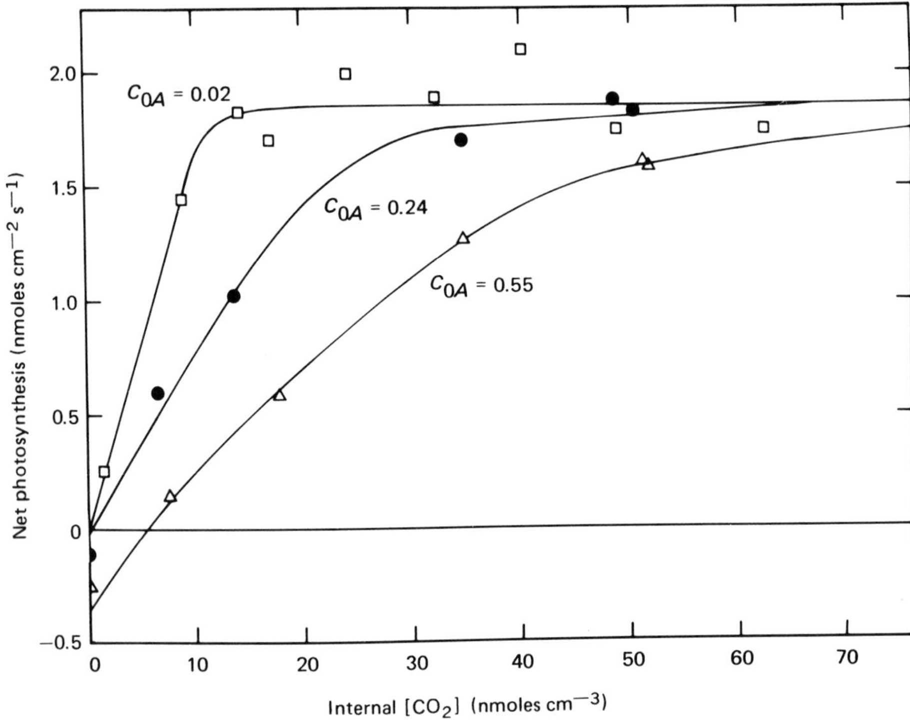

Fig. 2.3. Net photosynthesis versus internal carbon dioxide concentration for kidney bean 1. Data taken at saturating light and optimum temperature.

Table 2.1 Parameters for Kidney Bean 1 for Various Oxygen Concentrations

Parameter	Oxygen Concentration (%)		
	2	24	55
P_M (nmoles cm^{-2} s^{-1})	1.86 ± 0.07	1.86	1.86
K (nmoles cm^{-3})	0.043 ± 0.45	0.70 ± 0.5	1.5 ± 0.3
R (s cm^{-1})	6.0 ± 1.8	12 ± 1.5	23 ± 1
W_{LTO} (nmoles cm^{-2} s^{-1})	—	0	4.5 ± 1

fairly dense as far as they go. The 55 percent curve is fitted very nicely by the same P_M as for the 2 percent curve (suggesting that they ultimately merge), a K value 20 to 60 times larger (suggesting competitive inhibition), and resistance 2.5 to 5.5 times larger (suggesting an O_2 effect on the mesophyll resistance). This kidney bean was grown at a lower light level than the previously mentioned plant and has a smaller P_M and a larger R. Values are summarized in Table 2.2.

The magnitudes of the effects on K and R are the same for the two sets of data; oxygen concentration changing from 0 to 50 percent increases K by a factor of

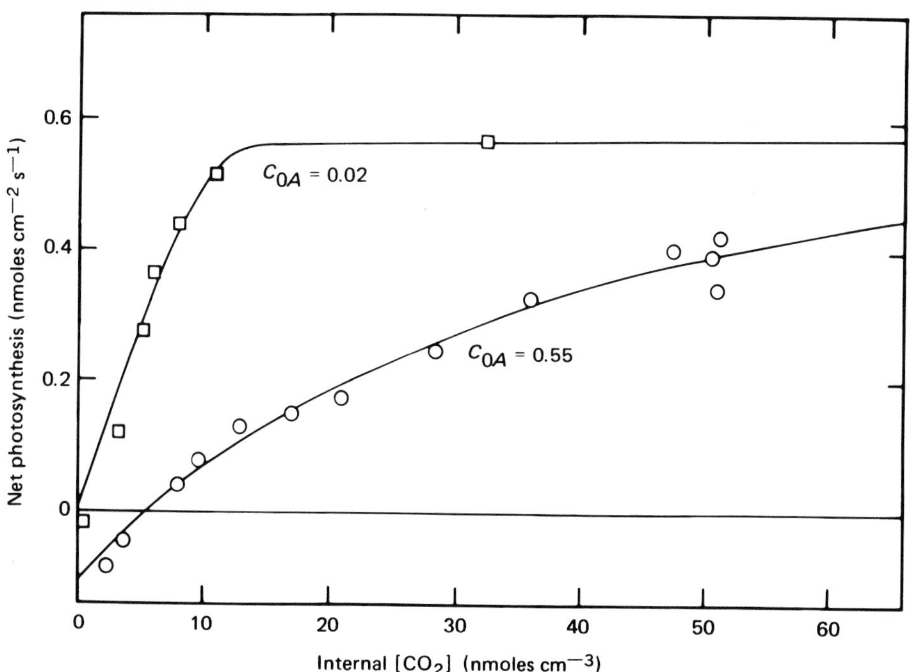

Fig. 2.4. Net photosynthesis versus internal carbon dioxide concentration for kidney bean 2. Data taken at saturating light and optimum temperature.

Table 2.2 Parameters for Kidney Bean 2 for Various Oxygen Concentrations

Parameter	Oxygen Concentration (%)	
	2	55
P_M (nmoles cm^{-2} s^{-1})	0.57 ± 0.015	0.57
K (nmoles cm^{-3})	0.20 ± 0.23	7.8 ± 4
R (s cm^{-1})	17 ± 2	67 ± 25
W_{LTO} (nmoles cm^{-2} s^{-1})	0	1.45 ± 0.67

about 40 and increases R by a factor of about 4. The resistance effect is probably due to the effect of oxygen on carbonic anhydrase in the mesophyll cells (see Chapter 3). This is not an effect of what is often called "biochemical resistance." Mesophyl resistance, as used by, e.g., Lloyd Dunn (Chapter 10), includes a bio-

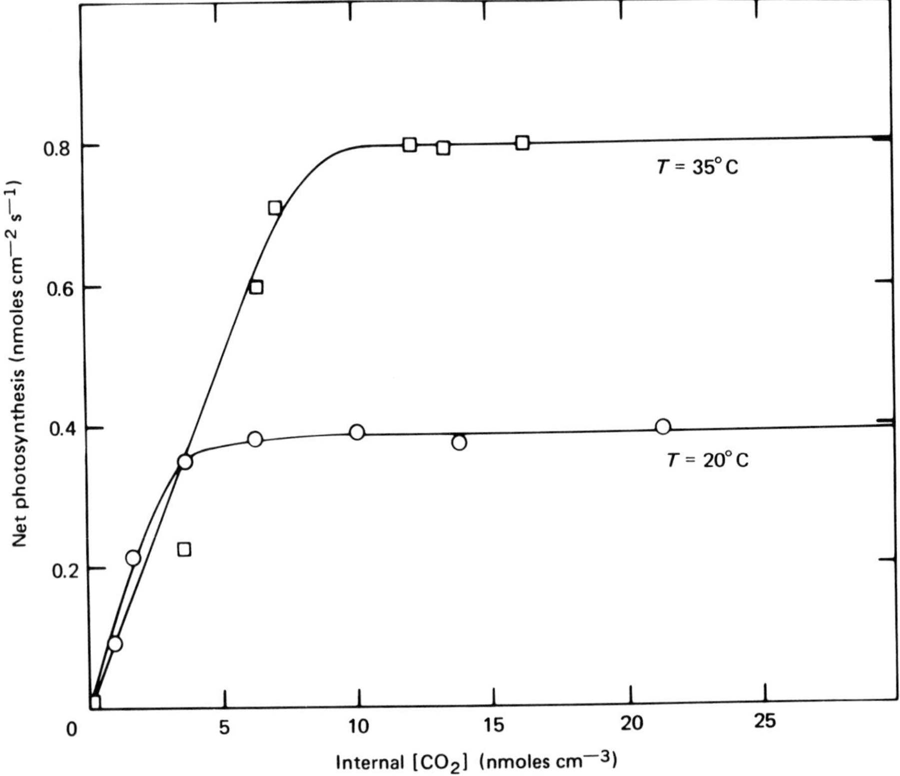

Fig. 2.5. Net photosynthesis versus internal carbon dioxide concentration for 20°C and 35°C. Data taken at saturating light conditions.

chemical portion. Our use of the Michaelis–Menten formulation means that we do not need the concept of biochemical resistance.

Figures 2.3 and 2.4 show that the model is able to describe the data satisfactorily and that the curve-fitting procedure is useful in determining values of parameters. P_M values are usually determined to a standard deviation of ~5 percent; mesophyll resistance values, ~20 percent; K values, from 20 to 50 percent. If the elbow is sharp in the P versus C_I curve, however, only an upper limit can be obtained for K.

Figure 2.5 shows P versus C_I for 20°C and 35°C for kidney bean. Oxygen concentration was 2 percent. Light intensity was saturating (7.4 kergs $cm^{-2}\ s^{-1}$). Table 2.3 lists the values of the parameters for each curve. The apparently faster-rising 20°C curve is probably not a real effect. The difference in P_M is about what you would expect from a curve of P versus T. R values are close enough to each other so that a temperature dependence of R cannot be suggested. One of the

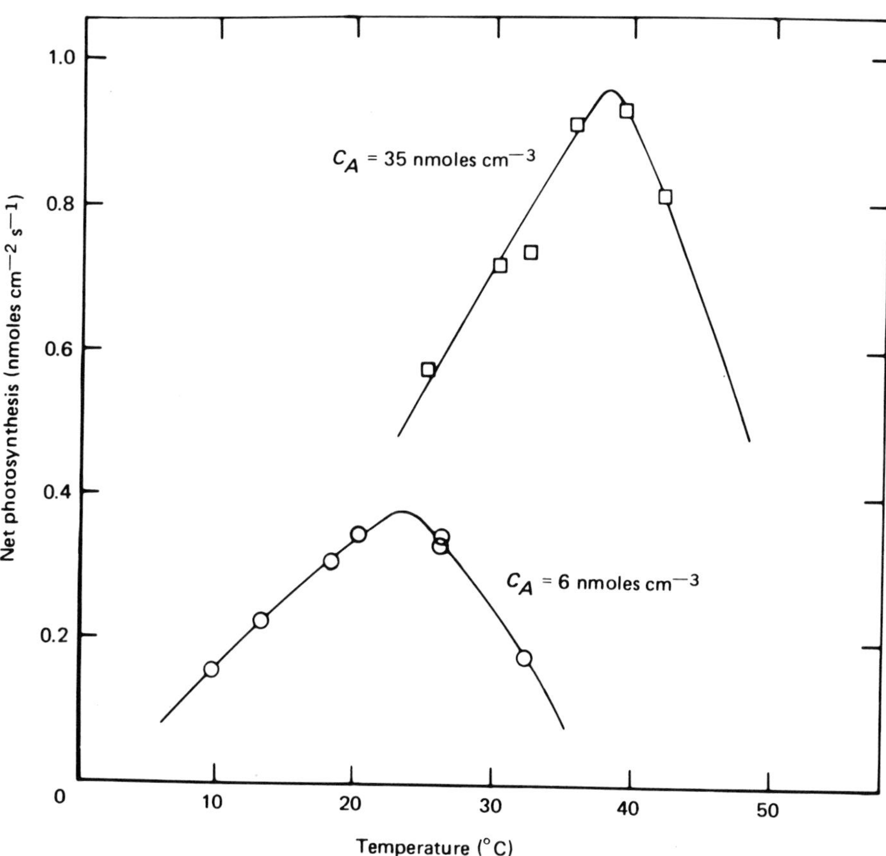

Fig. 2.6. Net photosynthesis versus leaf temperature at high and low carbon dioxide concentration.

Table 2.3 Parameters for Kidney Bean at Different Temperatures

Parameter	Leaf Temperature (°C)	
	20	35
P_M (nmoles cm^{-2} s^{-1})	0.39 ± 0.01	0.80 ± 0.02
K (nmoles cm^{-3})	0.1 ± 0.1	$< 10^{-5}$
R (s cm^{-1})	7.7 ± 1.2	10.5 ± 0.5

things that we wanted to look for was a temperature dependence on K. The curves' elbows are much too sharp here, and so for this plant no temperature information can be inferred. The sharp elbows and consequent small values of K indicate also that the resistances control the photosynthesis rate rather than the biochemistry (at least for this plant).

Figure 2.6 shows an unexpected effect of kidney bean with low oxygen and nearly saturating light. [I have recently learned from Lange and Schultze (personal communication), however, that this effect was noted several years ago.] It is not clear what causes this effect. Possibly a different enzyme or set of enzymes is controlling each curve. (From water-vapor data it is clear that the peak in the lower is not a stomatal effect.)

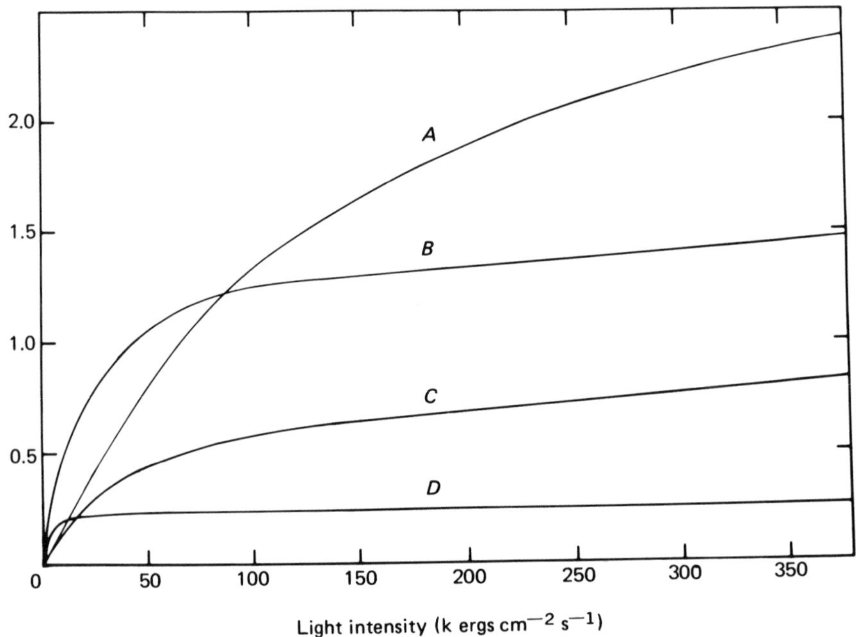

Fig. 2.7. Net photosynthesis versus light intensity for low oxygen concentration and optimum temperature for four species.

Table 2.4 Parameters from P versus L Curves

Species	P_M (nmoles cm^{-2} s^{-1})	K_L (kergs cm^{-2} s^{-1})
A. Zea mays	2.8 ± 0.2	115 ± 40
B. Amaranthus retroflexus	1.52 ± 0.05	22 ± 4
C. Phaseolis vulgaris	0.84 ± 0.1	44 ± 20
D. Acer saccharum	0.25 ± 0.005	2.6 ± 0.5

Figure 2.7 shows P versus light intensity for low oxygen and optimum temperature for several species. A Michaelis–Menten curve is plotted for each species. Table 2.4 lists appropriate parameters.

Summary

The data we have so far indicate the model's usefulness for describing quantitatively the relations between the variables involved, providing a framework for new data, and suggesting further experiments.

Acknowledgments

This work was supported by a grant from the National Science Foundation to Paul W. Lommen and by grants from the Atomic Energy Commission and the Ford Foundation to David M. Gates.

References

Björkman, O.: 1971. Interaction between the effects of oxygen and CO_2 concentration on quantum yield and light-saturated rate of photosynthesis in leaves of *Atriplex patula* ssp. *spicata*. *In* Carnegie Inst. Wash. Yearbook *70*, 520–526.

Gates, D. M.: 1968. Transpiration and leaf temperature. Ann. Rev. Plant Physiol. *19*, 211–238.

Lommen, P. W., Schwintzer, C. R., Yocum, C. S., Gates, D. M.: 1971. A model describing photosynthesis in terms of gas diffusion and enzyme kinetics. Planta *98*, 195–220.

3

Mesophyll Resistances

CONRAD S. YOCUM AND PAUL W. LOMMEN

Introduction

Recent attempts to model the rate of photosynthesis in green plant leaves have included a diffusion-resistance term, the sum of the boundary-layer (or atmospheric) resistance and the "leaf" resistance. The leaf resistance, in turn, represents the sum of the stomatal resistance and the "mesophyll" resistance, the latter consisting of the gaseous diffusion path through the intercellular air spaces and the liquid diffusion path from the cell walls of the mesophyll cells to the sites of photosynthesis within the chloroplasts (Hall, 1971; Lommen *et al.*, 1971). Modeling of the leaf-resistance components has received relatively little attention, perhaps for two reasons. First, the resistance to diffusion of water vapor out of the leaf can now be measured readily. Then, on the assumptions (1) that the liquid paths for both H_2O and CO_2 add negligible resistance, and (2) that the gaseous paths for H_2O and CO_2 are identical, the leaf resistance for diffusion of CO_2 has been estimated from the ratio of their diffusion coefficients. Additional liquid resistance has been either neglected or estimated from the internal surface of a leaf and the mean distance from the intercellular air spaces to the center of the chloroplasts (Hall, 1971). Here it has been assumed that CO_2 is the only diffusing form of carbon and that there is no consumption in the diffusion path. Furthermore, plane-parallel surfaces are assumed for both the source and the site of uptake of CO_2. The second reason is the complex geometry of a leaf. The chloroplasts are embedded within part of the diffusing medium, their distances from the leaf surface stomata varying by a factor of 10 or more. Furthermore, their concentration is not uniform, the palisade layer containing more per unit volume of the leaf than does the spongy layer. We discuss here an attempt to quantify the transport of CO_2 within a leaf, the kinetics of chloroplast enzymes, and the interaction of transport and kinetics, which may, in part, regulate the rate of photosynthesis.

We consider a leaf to be an infinite slab supplied uniformly with CO_2 only from the lower stomata-containing surface and to contain chloroplasts uniformly distributed close to the surfaces of the cells, which are, in turn, uniformly distributed between the leaf surfaces. The rate of CO_2 uptake and the concentration gradients within such a system were estimated from solutions to the diffusion-enzyme kinetics equation for the case where the enzymes are uniformly distributed in a diffusing medium without carriers for facilitated transport. Such solutions were first provided by Murray (1968) and Keller (1968) for spherical coordinates where the substrate is furnished to the sphere surface containing uniformly distributed enzymes that consume the substrate. Murray (1968) showed that the

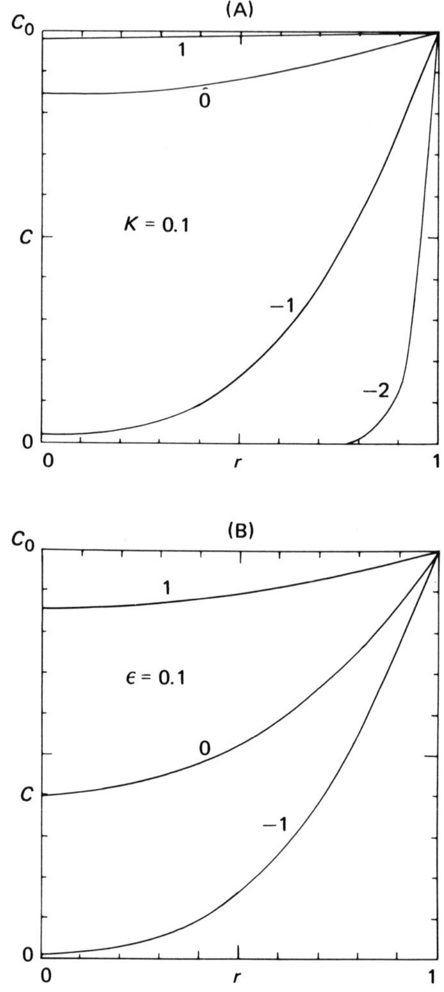

Fig. 3.1. Substrate concentration, C, versus radius, r, of a sphere where the substrate is furnished to the surface of the sphere ($C_0 = 1$ at $r = 1$) and where the substrate removal sites with Michaelis–Menten kinetics are uniformly distributed within the sphere. (a): K constant at 0.1 with variable ε; (b): ε constant at 0.1 with variable K. The curves have been labeled with the exponent (base 10) of the variable.

concentration gradients within and the rate of uptake by the sphere depend upon two dimensionless ratios, K and ε. Here

$$K = K_m C_0^{-1} \qquad \text{or} \qquad K = \frac{K_m}{C_0}$$

where K is the ratio of K_m, the Michaelis constant of the enzyme, and C_0 is the concentration of its substrate (CO_2 in the present case) at the surface of the system through which it is diffusing.

$$\varepsilon = \frac{DC_0}{nql^2}$$

where D = diffusion coefficient of the substrate
n = number of uptake sites
q = maximum uptake rate for each site
l = maximum length of the diffusion path (radius of a sphere, thickness of a slab)

We have applied the solutions for the spherical case to that of a slab with the realization that some error, but assurance that no large error, is introduced. Solutions for the slab and cylinder will be published shortly, together with the details of the procedure and the computer programming for the three geometries (Lommen and Yocum, unpublished). A family of concentration versus radius curves was plotted for several K and ε combinations. Concentration of CO_2 versus l at several values of K and ε have been plotted in Fig. 3.1. The rate of uptake by the system is the slope of the curve [dC/dr at the surface ($r = 1$)]. Note that $dC/dr = 0$ when $r = 0$. K and ε were estimated from known or measurable parameters for 1 cm^2 (one side) of leaf surface under conditions optimal for photosynthesis when the atmosphere contained 1.25×10^{-8} mole CO_2/cm^3 air. Since a leaf contains both gas and liquid transport paths, we consider the diffusion through each separately.

Gas-Phase CO_2

Using a buoyant density technique, we found a 150-μm-thick ivy leaf to contain 40 percent or 6×10^{-3} cm^3 gas phase per cm^2 of leaf. K and ε for such a leaf were estimated as follows:

$$K \cong \frac{1 \times 10^{-9}\ \text{mole cm}^{-3}}{1 \times 10^{-8}\ \text{mole cm}^{-3}} = \frac{K_m}{C_0} = 0.1$$

$$\varepsilon \cong \frac{0.1\ \text{cm}^2\ \text{s}^{-1} \times 1 \times 10^{-8}\ \text{mole cm}^{-3} \times 6 \times 10^{-3}\ \text{cm}^3\ \text{cm}^{-2}}{5 \times 10^{-9}\ \text{mole cm}^{-2}\ \text{s} \times (150 \times 10^{-4}\ \text{cm})^2} = 5$$

When K and ε are 0.1 and 5, respectively, the concentration decrease [Fig. 3.1] was estimated to be 9 percent due to CO_2 uptake along the thickness of the leaf, or $C/C_0 = 0.91$.

Liquid-Phase CO_2

As in the gas phase, both the volume and the diffusion path length of the liquid phase are required. From Fig. 3.1 we see that most, although not all, of the liquid diffusion path is within the chloroplasts themselves. Hence a rough estimate of the volume of the system would be equal to the chloroplast volume (per cm^2 of leaf) and the diffusion path length equal to the chloroplast thickness.

The chloroplast volume was estimated by two independent methods.

1. The number of chloroplasts per cm^2 of leaf (1×10^8) was obtained by grinding a known leaf area in isotonic sucrose and counting a known fraction of the resulting suspension. The volume of a chloroplast was obtained by measuring the diameter and the thickness. The discoid shape of a chloroplast approximates that of a segment of a sphere and the equation for the volume v of such a segment was used.

$$v = \tfrac{1}{6}\pi h(3a^2 + h^2)$$

$$= 15.7\ \mu\text{m}^3/\text{chloroplast} \times 1 \times 10^8\ \mu\text{m cm}^{-2} = 1.6 \times 10^{-3}\ \text{cm}^3\ \text{cm}^{-2}$$

where h = diameter of the chloroplast base and a = chloroplast height. To this volume was added the cytoplasm and cell wall volume:

$$0.5\ \mu\text{m} \times 5\ \mu\text{m diameter} = 1 \times 10^{-3}\ \text{cm}^3\ \text{cm}^{-2}$$

$$1.6 \times 10^{-3}\ \text{cm}^3\ \text{cm}^{-2}\ \text{leaf} + 1.0 \times 10^{-3}\ \text{cm}^3\ \text{cm}^{-2}\ \text{leaf}$$

$$= 2.6 \times 10^{-3}\ \text{cm}^3\ \text{cm}^{-2}\ \text{leaf}$$

2. An estimate of the relative chloroplast volume was obtained from the relative cross-sectional area of the chloroplasts to that of the mesophyll cell (Fig. 3.2) perpendicular to the long axis. The result here was about 25 percent (Fig. 3.1) of the mesophyll cell volume or 2.3×10^{-3} cm^3 cm^{-2} leaf.

The maximum diffusion path length was obtained (1) by direct measurement from an electron micrograph (Fig. 3.1), and (2) by dividing the chloroplast volume by the chloroplast area. Both methods gave from 1.5 to 2.0 μm. A conducting (teledeltos) paper model showed the diffusion from the mesophyll wall to the convex surface to be negligible. From the above data, and assuming the medium from intercellular air spaces to the site of photosynthesis to be homogeneous, we estimated K and ε as follows:

$$K = \frac{1 \times 10^{-9}\ \text{mole cm}^{-3}}{1 \times 10^{-8}\ \text{mole cm}^{-3}} = 0.1$$

$$\varepsilon = \frac{1 \times 10^{-5}\ \text{cm}^2\ \text{s}^{-1} \times 5 \times 10^{-9} \times 212 \times 10^{-3}\ \text{cm}^3\ \text{cm}^{-2}}{5 \times 10^{-9}\ \text{mole cm}^{-2}\ \text{s} \times (2 \times 10^{-4}\ \text{cm})^2} = 0.5$$

From Fig. 3.1(a) we find a relatively large CO_2 concentration decrease, of about one-half of the surface value at the back of the chloroplast; that is, $C/C_0 = 0.5$.

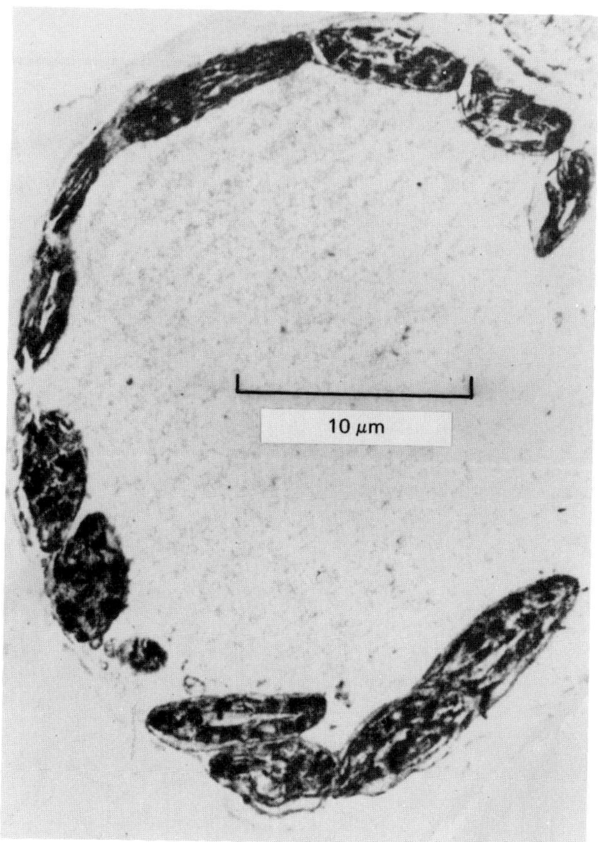

Fig. 3.2. Cross section of a mesophyll cell from a bean leaf. (After Stocking and Ongun, 1962.)

Bicarbonate and Carbonic Anhydrase

All the above estimates were based on the assumption that the main form of diffusing carbon is free CO_2 or H_2CO_3. We submit the following evidence to indicate that this assumption is neither necessary nor valid.

1. Ribulose diphosphate carboxylase turns over most rapidly at pH 8 (Tolbert, 1973). At this pH, the predominant form of *C* is HCO_3^-. H_2O or OH^- hydrates CO_2 at a rate 10^{-2} to 10^{-3} that of photosyntheses (Burr, 1936).

2. Theory and experiments on model membranes show that, under physiological conditions ($CO_2 = 10^{-5}$ *M* or less and pH 7–8), the rate of diffusion by HCO_3^- exceeds that by free CO_2 one to two orders of magnitude because a much larger HCO_3^- than CO_2 concentration gradient is established (Enns, 1967; Ward and Robb, 1967).

3. However, ribulose diphosphate carboxylase uses CO_2 rather than HCO_3^- as substrate (Cooper *et al.*, 1969).

4. But plants contain an active carbonic anhydrase which catalyzes the hydration and dehydration of CO_2 (Tobin, 1970).

5. Carbonic anhydrase is largely a chloroplast enzyme (Poincelot, 1972).

6. *Chlorella* carbonic anhydrase is an inducible enzyme found in low concentrations for cells grown in 5 percent CO_2 (2×10^{-6} mole cm^{-3}) and in high concentrations for cells grown in air (1.2×10^{-8} mole cm^{-3}) (Graham and Reed, 1971).

7. *Spinach* chloroplasts, in vitro, accumulate HCO_3^- in the light and maintain a calculated pH of 7.9 (Werdan and Heldt, 1972).

From the above evidence, we conclude that the transport of carbon within the chloroplast (and possibly the cell wall and cytoplasm) proceeds via bicarbonate diffusion and that carbonic anhydrase catalyzes the formation and dissociation of HCO_3^- within the chloroplasts. We now ask whether the calculated rates of diffusion and catalysis are large enough to account for the rate of photosynthesis.

Carbon Dioxide, Carbonic Acid, Bicarbonate, and Carbonate

At 20°C the solubility coefficient of CO_2 in an aqueous medium approaches 1. Hence, the molar concentration in air and in the liquid in equilibrium with it are approximately the same ($=10^{-8}$ mole cm^{-3}). The dissociation constants for bicarbonate and carbonate as well as their ratios are small (MacInnes and Belcher, 1933).

$$K_1 = \frac{[H^+][HCO_3^-]}{[H_2CO_3]} = 5 \times 10^{-7}\ M$$

$$K_2 = \frac{[H^+][CO_3^{2-}]}{[HCO_3^-]} = 2.5 \times 10^{-11}\ M$$

$$\frac{K_2}{K_1} = 2.2 \times 10^{-4}$$

These show that under physiological conditions of pH 8, both bicarbonate and carbonate are largely undissociated, and that the diffusing species will be largely $KHCO_3$ with some K_2CO_3, where K is the major cation:

$$[H^+] = 10^{-8}\ M \text{ and } [CO_2] = [H_2CO_3] = 10^{-5}\ M$$

$$[HCO_3^-] = \frac{5 \times 10^{-7}\ M \times 10^{-5}\ M}{10^{-8}\ M} = 5 \times 10^{-4}\ M$$
$$= 5 \times 10^{-7} \text{ mole cm}^{-3}$$

and

$$[CO_3^{2-}] = \frac{2.2 \times 10^{-4}(5 \times 10^{-4}\ M)^2}{10^{-5}\ M} = 5 \times 10^{-6}\ M$$
$$= 5 \times 10^{-9} \text{ mole cm}^{-3}$$

Thus, under these conditions, a 50-fold larger gradient can be established for HCO_3^- than for free CO_2. However, the maximum CO_3^{2-} gradient, 5×10^{-6} M, in the opposite direction is small and hence may limit diffusion of the system for CO_2. At pH 9 the equilibrium concentration of carbonate is 5×10^{-4} M and clearly adequate. But ribulose diphosphate carboxylase activity at pH 9 is one-tenth that at pH 8 (Tolbert, 1973). Thus it appears that a delicate balance between $[H^+]$, $[HCO_3^-]$, $[CO_3^{2-}]$ and enzyme activity must be maintained in the chloroplasts, or possibly that the photosynthetic enzymes are compartmented from the diffusing medium by lipid membranes permeable to CO_2 and impermeable to cations, including H^+.

It is also possible that one of the roles for glycolate in photorespiration and/or malate production in photosynthesis may be to prevent the carboxylation sites from becoming too alkaline. Thus the potassium glycolate produced by photorespiration may diffuse out from the chloroplast and after oxidation furnish one K^+ that would be available for inward CO_2 transport via $KHCO_3$. However, organic acid flux alone cannot account for all the C transport required for photosynthesis.

Now we calculate K, ε, and the resulting concentration gradient for HCO_3^- when carbonic anhydrase and/or ribulose diphosphate carboxylase removes carbon from the system.

Carbonic Anhydrase

According to Tobin (1970) 1 percent of the leaf protein consists of carbonic anhydrase, molecular weight 30,000, $K_m = 10^{-2}$ M. Assuming that the chloroplast is 10 percent protein and contains all the leaf proteins, then

$$\tfrac{1}{100} \times \tfrac{1}{30000} \times \tfrac{10}{100} = 3 \times 10^{-8}\ M$$
$$= 3 \times 10^{-11}\ \text{mole cm}^{-3}$$

From the carbonic anhydrase concentration, turnover number of 10^5 s^{-1} (Tobin, 1970), and the chloroplast volume per cm^2 calculated above, we obtain

$$3 \times 10^{-11}\ M \times 10^5\ \text{s}^{-1} \times 2 \times 10^{-3}\ \text{cm}^3\ \text{cm}^{-2}$$
$$= 6 \times 10^{-9}\ \text{mole CO}_2\ \text{cm}^{-2}\ \text{s}^{-1}$$

for maximum rate of CO_2 production by dissociation of bicarbonate (nq). Then

$$\varepsilon = \frac{1 \times 10^{-5}\ \text{cm}^2\ \text{s}^{-1} \times 5 \times 10^{-7}\ \text{mole cm}^{-3} \times 3 \times 10^{-3}\ \text{cm}^3\ \text{cm}^{-2}}{6 \times 10^{-9}\ \text{mole cm}^{-2}\ \text{s}^{-1} \times (2 \times 10^{-4}\ \text{cm})^2}$$
$$= 70$$

$$K = \frac{10^{-2}\ M}{5 \times 10^{-4}\ M} = 20$$

and, from Fig. 3.1, we find a concentration decrease of 0.01 at $r = 0.1$, $C/C_0 \geqq 0.99$.

Ribulose Diphosphate Carboxylase

Here we make an assumption that the carbonic anhydrase dissociates HCO_3^- rapidly enough to neglect the kinetics. The data for C/C_0 in the scction above permit this assumption.

$$\varepsilon = \frac{1 \times 10^{-5}\ \text{cm}^2\ \text{s}^{-1} \times 1 \times 10^{-8}\ \text{mole cm}^{-3} \times 3 \times 10^{-3}\ \text{cm}^3\ \text{cm}^{-2}}{5 \times 10^{-9}\ \text{mole cm}^{-2}\ \text{s}^{-1} \times (2 \times 10^{-4})^2}$$

$$\cong 1.5$$

$$K = \frac{4.5 \times 10^{-4}}{10^{-5}\ M}$$

$$\frac{C}{C_0} = 0.99$$

Thus we calculate a small concentration decrease of inorganic carbon across the chloroplasts, whether carbonic anhydrase or RDP carboxylase removes the carbon during photosynthesis.

Discussion

In our present attempt to estimate the mesophyll resistance, a number of simplifying assumptions have been made and information has necessarily been gathered from a number of sources. Clearly, a more thorough analysis must be made—and of a single species, using more realistic models (e.g., slab instead of a sphere). However, it seems rather unlikely that our calculated values of K, ε, and C differ from the "true" values by more than an order of magnitude.

The following tentative conclusions appear to be justified.

1. In living systems where the sites of substrate removal are distributed within the diffusing medium itself, the term "resistance" is not adequate. Instead, the dimensionless ratio ε, relating the rate of supply to the potential rate of uptake, and K, the dimensionless ratio of the Michaelis constant to the external concentration, must be established (Murray, 1968).

2. When the product of $K \times \varepsilon$ is about 1 or larger, the concentration of the diffusing substrate at the most remote distance from the source approaches that at the surface ($C/C_0 \cong 1$). That is, during metabolism a small decrease in substrate concentration develops. Thus the sites of removal may then be considered to be effectively in "parallel," and the diffusion paths between them to be resistances in parallel. Of course the overall resistance of a number of uptake sites in parallel is less than for each site alone: a larger diffusional flux to a population of sites than to one becomes possible.

3. K and ε for the intercellular air spaces and within the chloroplasts for $HCO_3^- \cong 1$, and the resulting concentration decreases are small. Thus the major

diffusion paths in leaves both to and within the chloroplasts are effectively in "parallel."

4. From the data of Lemon *et al.* (1971), a small concentration decrease develops ($C/C_0 = 0.99$) between the atmosphere and a photosynthesizing canopy. Therefore, $K\varepsilon \cong 1$ and arguments 1 to 3 also apply to a population of leaves in the atmosphere.

5. Thus we see essentially parallel diffusion paths for CO_2 at all levels of organization from the plant–atmosphere down to the enzyme–substrate.

6. The calculations presented here were based on "optimal conditions," e.g., saturating light for photosynthesis and wide-open stomata. If the radiant flux decreases, ε increases, and the supply becomes even more favorable. Temperature will have a small effect on the rate of supply because the diffusion coefficient increases and solubility of CO_2 decreases with increasing temperature. Below the temperature optimum, nq will increase, ε decrease, and vice versa above the temperature optimum. If carbonic anhydrase is as important as we believe, the temperature dependence of photosynthesis may be regulated by the temperature dependence of carbonic anhydrase (Downton and Slatyer, 1972). The carbonate–bicarbonate dissociation constants and their ratios change insignificantly over the physiologically important temperature range. We have little or no information on the effect of temperature on the K_m of carbonic anhydrase or photosynthetic enzymes; hence we can say little about the temperature dependence of K.

Stomatal closure will decrease C_0 one or more orders of magnitude. K will increase and ε decrease, both linearly, and the resulting concentration decreases are altered slightly. The concentration of bicarbonate decreases and carbonate increases. If these are significant, the concentration of H^+ may decrease to inhibitory values. Perhaps one role of photorespiration may be to maintain an internal CO_2 concentration high enough to prevent the chloroplasts from becoming too alkaline when stomatal resistance is large.

Acknowledgments

The initial phases of this work were carried out when the senior author was on leave of absence at the Missouri Botanical Garden and while the junior author was in the Biophysical Ecology Group, Missouri Botanical Garden. The authors thank David M. Gates for his enthusiastic support and encouragement.

References

Burr, G. O.: 1936. Carbonic anhydrase and photosynthesis. Proc. Roy. Soc. London, Ser. B *120*, 42–47.

Cooper, T. G., Filmer, D., Wishnick, M., Lane, M. D.: 1969. The active species of "CO_2" utilized by ribulose diphosphate carboxylase. J. Biol. Chem. *224*, 1081–1083.

Downton, J., Slatyer, R. O.: 1972. Temperature dependence of photosynthesis in cotton. Plant Physiol. *50*, 518–522.

Enns, T.: 1967. Facilitation by carbonic anhydrase of carbon dioxide transport. Science *155*, 44–47.

Graham, D., Reed, M. L.: 1971. Carbonic anhydrase and the regulation of photosynthesis. Nature New Biol. *231*, 81–83.

Hall, A.: 1971. A model of leaf photosynthesis and respiration. Carnegie Inst. Wash. Yearbook *70*, 530–540.

Keller, H. B.: 1968. Numerical methods for two-point boundary-value problems. Lexington, Mass.: Xerox.

Lemon, E., Stewart, D. W., Shawcroft, R. W.: 1971. The sun's work in a corn field. Science *174*, 371–378.

Lommen, P. W., Schwintzer, C. R., Yocum, C. S., Gates, D. M.: 1971. A model describing photosynthesis in terms of gas diffusion and enzyme kinetics. Planta *98*, 195–220.

MacInnes, D. A., Belcher, D.: 1933. The thermodynamic ionization constants of carbonic acid. J. Am. Chem. Soc. *55*, 2630–2646.

Murray, J. D.: 1968. A simple method for obtaining approximate solutions for a large class of diffusion–kinetic enzyme problems. I. General class and illustrative examples. Math. Biosci. *2*, 379–411.

Poincelot, R. P.: 1972. Intracellular distribution of carbonic anhydrase in spinach leaves. Biochem. Biophys. Acta *258*, 637–642.

Stocking, C. R., Ongun, A.: 1962. The intracellular distribution of some metallic elements in leaves. Am. J. Bot. *49*, 284–289.

Tobin, A.: 1970. Carbonic anhydrase from parsley leaves. J. Biol. Chem. *245*, 2656–2666.

Tolbert, N. E.: 1973. Glycolate biosynthesis. *In* Current topics in cellular regulation (eds. B. L. Horecker, E. R. Stadtman), vol. 7, pp. 21–50. New York: Academic Press.

Ward, W. J., Robb, W. L.: 1967. Carbon dioxide–oxygen separation: facilitated transport of carbon dioxide across a liquid film. Science *156*, 1481–1484.

Werdan, K., Heldt, H. W.: 1972. Accumulation of bicarbonate in intact chloroplasts following a pH gradient. Biochem. Biophys. Acta *283*, 430–441.

4

Model of Leaf Photosynthesis and Respiration

ANTHONY E. HALL AND OLLE BJÖRKMAN

Introduction

The flow of chemical energy into ecosystems is due to net photosynthesis. This flow depends on interactions between photosynthetic and respiratory processes and the constraints set by leaf anatomy and morphology. Further complexity arises if one considers ecologically important interactions between net photosynthesis, energy balance, and the soil–plant–atmosphere continuum of water. Progress in our understanding of these components of ecosystems will be highly dependent upon the development of effective mathematical models.

In developing models, the relationships between different models should be considered (Clymer, 1969). Models of systems at the same level of organization may be interfaced if they have common parameters and the necessary functional capabilities. Incorporating gas-flow resistance terms and temperature dependencies into models of net photosynthesis facilitates the interfacing of these models with models of energy balance (Lommen *et al.*, 1971) and plant–water relations. Models dealing with systems at different levels of organization may be interfaced hierarchically if they have the requisite substructure and are adequately theoretical. Models simulating photosynthesis in plant communities (Duncan *et al.*, 1967) may utilize models of leaf photosynthesis if they both evaluate fluxes on a unit projected leaf-area basis (Hall and Loomis, 1972a). This would be harder to accomplish if one of the models considered fluxes on a unit chlorophyll basis. Models of leaf photosynthesis may be updated using information concerning enzyme function if they have submodels that identify major molecular constituents and reactions. This interfacing is possible only when models are sufficiently theoretical rather than highly empirical. Further, theoretical models have the dual capacity of analysis (testing component interaction within the system) and prediction (testing system response to a changing environment).

An initial step toward reducing empiricism in models of net photosynthesis by leaves was taken by Chartier (1966), who combined the theoretical treatment of diffusive transport developed by Gaastra (1959), with equations describing flows of carbon and energy in metabolism. Subsequent research indicated the necessity for incorporating photorespiration and the oxygen effect on net photosynthesis into leaf models of C_3 species. Laisk (1970) and Hall (1970) achieved this by developing models that included a single hypothesis suggested by Whittingham *et al.* (1967) that appeared to account for both photorespiration and the oxygen effect. It was hypothesized that the enzyme ribulose-1,5-diphosphate carboxylase (RudPcase) has both carboxylase and oxygenase functions, depending upon its environment. The carboxylation would produce 2 molecules of 3-phosphoglycerate (PGA), whereas the oxygenation would use molecular oxygen and produce 1 molecule of PGA and 1 molecule of phosphoglycolate. It was also proposed that CO_2 and O_2 would compete for the same substrate–enzyme complex RudP–RudPcase. Subsequent biochemical studies have provided direct evidence supporting this hypothesis (Bowes *et al.*, 1971; Berry, 1971; Bowes and Ogren, 1972; Bowes and Berry, 1972; Andrews *et al.*, 1973; Lorimer *et al.*, 1973). Hall (1971) incorporated the hypothesis for the dual role of RudPcase into an extension of the model developed by Chartier (1966). Concurrently, Björkman (1971) extended his earlier experimental studies of net photosynthetic responses to O_2 and CO_2 with C_3 species. According to the mathematical model of Hall (1971), the results obtained by Björkman are consistent with the dual role of RudPcase. This paper extends these comparisons between leaf photosynthesis data and the mathematical model of Hall (1971). To facilitate these comparisc ıs, the model was extended by incorporating temperature dependencies to the existing capabilities for predicting the responses of net photosynthesis to quantum flux, $[CO_2]$ and $[O_2]$.

Model Development

Submodel for the Transport of CO_2 *from the External Air to the Protoplast*

The relational model used (Fig. 4.1) is the simpler of the two models developed by Lake (1967) based upon earlier work by Gaastra (1959). The fluxes of CO_2 (P_n, P_t, R_1, and R_d), their pathways, and a number of resistors (r'_a, r'_1, r'_w, and r'_k) are described. P_n is the net flux of CO_2, R_1 is CO_2 evolution in the light (photorespiration), R_d is CO_2 evolution due to the Krebs cycle, and P_t represents CO_2 fixation within the cells. C_t is the CO_2 concentration in the external air, C_i is the concentration in the intercellular spaces, and C_w is the concentration at the protoplast. Resistances are not required in the physical respiratory pathways, because there are no branches in the model and it is likely that the decarboxylations do not depend on $[CO_2]$. This particular model applies to isolateral leaves, where the boundary-layer resistance (r'_a) and the leaf resistances (r'_1 includes stomatal and cuticular components) of the upper and lower epidermises are equal. Methods for dealing with anisolateral leaves were described by Hall (1970, App. A). Leaf resistance (r'_1) is dependent upon irradiance; however, earlier studies by Hall (1970) with *Beta vulgaris* indicated that the dependence of stomatal aperture on irradiance

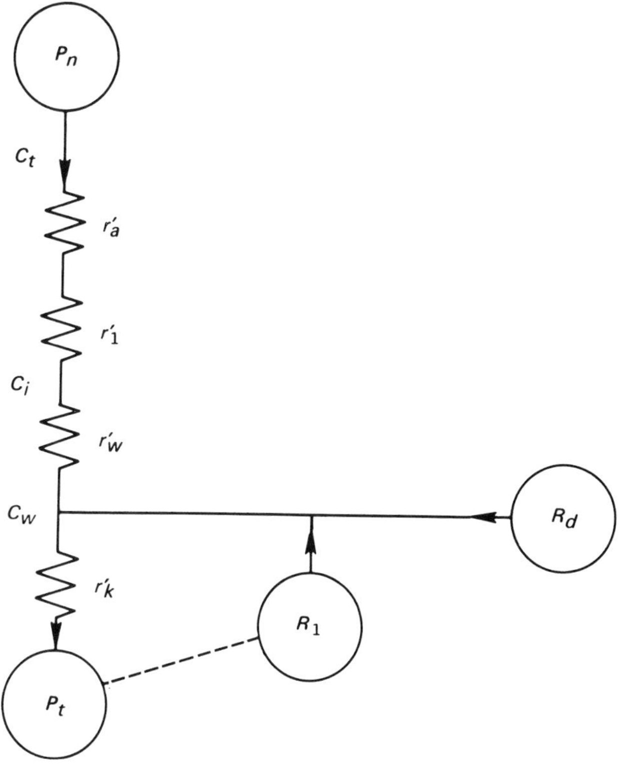

Fig. 4.1. Relational model for the transport of CO_2 from the external air into the leaf, and the relative locations of CO_2 release and fixation within the leaf.

may have little influence on CO_2 exchange. Stomatal apertures begin to close at low irradiances where P_n is low; and decreases in concentration of CO_2 as the gas enters the leaf are relatively small. The studies of Gifford (1971) indicate that this conclusion may not be valid with C_4 species. The dependence of r'_1 on leaf temperature should be approached with caution, as it has recently become apparent that the dependence interacts with the influence on r'_1 of the absolute humidity gradient between leaf and air. As Hall and Kaufmann report in Chapter 11, r'_1 may be insensitive to changes in leaf temperature at low absolute humidity gradients. Contrary to many statements in the literature, in some species r'_1 may not be influenced by moderate changes in $[CO_2]$ (250 ± 150 ppm) at moderate to high irradiances, provided that absolute humidity gradients, remain constant and small (Gaastra, 1959; Parkinson, 1968; Hall, 1970; Boyer, 1971; Hall and Kaufmann, Chap. 11). Consequently, r'_1 is modeled as a constant until an effective submodel for stomatal responses to environment is developed. An additional physical parameter was included to describe the resistance to transport of CO_2 from the evaporating surfaces inside the leaf to the protoplast (r'_w). r'_w may be estimated from leaf anatomical characteristics using the method described by Hall (1971).

In many cases the value of r'_w will be relatively small; a value of 0.1 to 0.2 s cm^{-1} was calculated for *Atriplex patula.*

The mathematical submodel, Eq. (1), for the transport of CO_2 from the external air to the protoplast was derived (Hall, 1971) by assuming, following the suggestion of Jacobs (1967), that there is a rapid equilibration of CO_2 at gas–liquid phase boundaries:

$$P_n = [s(T)C_t - C_w][s(T)(r'_a + r'_1) + r'_w]^{-1} \tag{1}$$

$s(T)$ is the temperature-dependent, dimensionless solubility factor for CO_2 air to water:

$$s = 2.33 - 1.126 \log_{10} T \tag{2}$$

This is an empirical equation, with T as leaf temperature in °C, and is valid above 5°C.

The nature of the biochemical limitation to CO_2 fixation (r'_k) and the relationship between photosynthesis and photorespiration are described by the following submodel. It was assumed that membranes within the protoplast do not represent significant barriers to the movement of CO_2, whether this is achieved by passive or active processes.

Submodel of Photosynthetic and Respiratory Metabolism in C_3 *Species*

The relational model used (Fig. 4.2) consists of three interacting cycles: (1) an abbreviated Calvin cycle, (2) a cycle for photorespiration that involves the oxygenation of RudP to form PGA and eventually CO_2, and (3) a cycle for the reduction of NADP to NADPH that depends upon the photosystems and the quantum flux (E).

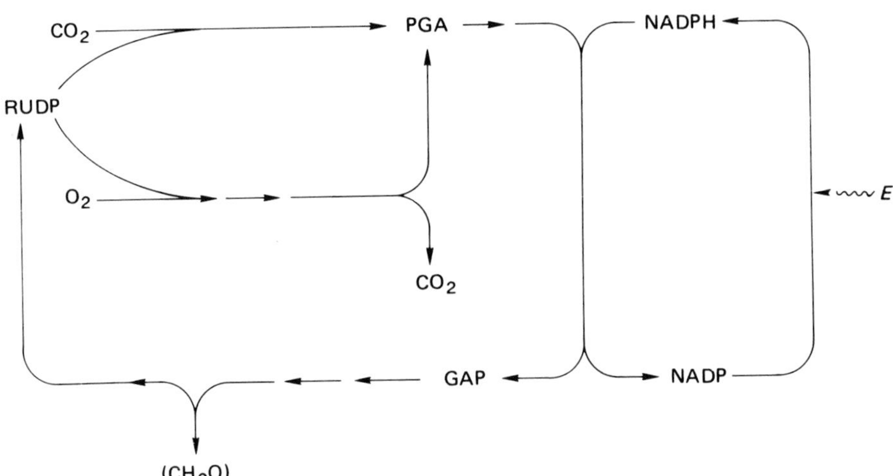

Fig. 4.2. Relational model for photosynthetic and photorespiratory metabolism.

The mathematical model includes a number of assumptions, both explicit and implicit in nature, which are described.

1. The conversion of PGA to GAP is linearly dependent on [NADPH] and [PGA] in conditions where CO_2 fixation is limited by the quantum flux. In steady-state conditions 2 moles of PGA must be reduced per mole of CO_2 fixed. Three moles of PGA must be reduced per mole of CO_2 evolved in photorespiration. This assumes that 2 moles of RudP produces on oxygenation 2 moles of PGA and 2 moles of *P*-glycolate, and that the *P*-glycolate is further metabolized to produce 1 mole of CO_2 and 1 mole of PGA (Bruin *et al.*, 1970). With the convention that CO_2 fixation is positive and CO_2 evolution is negative, this produces the following equation for steady-state conditions:

$$2P_t - 3R_1 = k_2(T)[\mathrm{PGA}][\mathrm{NADPH}] \tag{3}$$

$k_2(T)$ is a temperature-dependent rate constant for the reduction reaction.

2. The model is compared with leaf-gas-exchange data and is designed with a relaxation time in the order of minutes. Thus it was assumed that the total quantity of pyridine nucleotides is constant ($=N_0$):

$$N_0 = [\mathrm{NADPH}] + [\mathrm{NADP}] \tag{4}$$

3. The regeneration of NADPH is linearly dependent upon the incident quantum flux between 400 and 700 nm, and upon [NADP]:

$$2P_t - 3R_1 = k_3[\mathrm{NADP}]E \tag{5}$$

k_3 is a rate constant related to photosystem efficiency in reducing NADP (and leaf absorbancy if the incident flux is used) and is considered to be independent of temperature. This is an improvement compared with the models of Chartier (1966) and Hall (1970), which assumed that the concentration of NADPH was linearly related to E. It is more reasonable to consider that the rate of reduction NADP is related to E. Laisk (1970) adopted a simpler approach by assuming that the rate of formation of RudP is linearly related to E. It may not be necessary to include photophosphorylation and the utilization of ATP in models of leaf photosynthesis, provided that NADPH is always more limiting than ATP, owing to the availability of alternative sources of ATP and its mobility within the cell. ATP production and utilization, however, cannot be ignored in more detailed treatments of Calvin cycle regulation.

4. The total size of the pool of certain Calvin cycle components is constant ($=A_0$) over the relaxation time of the model:

$$A_0 = [\mathrm{RudP}] + [\mathrm{PGA}] \tag{6}$$

This assumption may be valid if the $(CH_2O)_n$ output from the Calvin cycle is regulated so that it does not deplete levels of intermediate compounds.

5. The carboxylation of RudP was modeled by

$$P_t = k_1(T)P_m(T)[\mathrm{RudP}]C_w(k_1(T)A_0C_w + P_m(T) + P_m(T)k^*O_w)^{-1} \tag{7}$$

$k_1(T)$ is a temperature-dependent rate constant. $P_m(T)$ is the CO_2-saturated, and light-saturated rate of CO_2 fixation in vivo (where $[\mathrm{RudP}] = A_0$), which is also temperature-dependent. k^* is a constant that takes into account the complexing of RudPcase with O_2. Further analyses were presented by Hall (1971).

6. The oxygenation of RudP is first order with respect to oxygen concentration (O_w) and [RudP]:

$$R_1 = -0.5k_4(T)[\text{RudP}]O_w \tag{8}$$

k_4 is a temperature-dependent rate constant.

7. It was assumed that CO_2 release by the Krebs cycle (R_d) is temperature-dependent and unaffected by light (Ludwig and Canvin, 1971).

8. Applying the continuity principle to the system at steady state produces Eq. (9):

$$P_n = P_t + R_d + R_1 \tag{9}$$

The mathematical model may be solved by eliminating the unknown variables ([PGA], [NADPH], [NADP], and [RudP]). This was accomplished without further assumptions by simultaneous solution of Eqs. (3)–(9), producing a quadratic equation that is presented in Appendix A. Hall (1971) analyzed this equation by taking appropriate partial differentials and limits. A verbal summary of this analysis is presented here.

1. The quantum efficiency (α) depends upon a rate constant and the pool size N_0:

$$\alpha = 0.5k_3N_0 \tag{10}$$

α may be obtained experimentally and is equivalent to the slope of the response of P_n to E at low values of E, and $[O_2]$.

2. The carboxylation resistance (r_k') depends upon a temperature-dependent rate constant and the pool size A_0:

$$r_k' = [k_1(T)A_0]^{-1} \tag{11}$$

r_k' may be obtained experimentally and is equivalent to the inverse of the slope of the response of P_n to C_w at low values of C_w, saturating light, and low $[O_2]$. Experimental responses of P_n to C_t may be transformed to C_w dependencies, using Eqs. (1) and (2), provided that r_a', r_1', and r_w' are known.

3. A photorespiration parameter was derived (R) that depends upon a temperature-dependent rate constant and the pool size A_0:

$$R = -0.5k_4(T)A_0 \tag{12}$$

R may be obtained from the rate of CO_2 evolution (adjusted for R_d), when C_w is very small and light is saturating, where

$$(P_n - R_d) \rightarrow RO_w \tag{13}$$

Experimentally, P_n responses to C_t are transformed to dependencies on C_w using Eqs. (1) and (2). RO_w is the intercept on the ordinate obtained by extrapolating to obtain $(P_n - R_d)$ at $C_w = 0$. k^* is obtained from the initial slope of this same curve using the following equation:

$$\left(\frac{\partial P_n}{\partial C_w}\right)_{O_w,T} = [r_k'(T)(1 + k^*O_w)]^{-1} \tag{14}$$

4. R_d is a temperature-dependent function obtained from the response of CO_2 evolution in the dark to temperature.

5. $P_m(T)$ is the temperature-dependent, maximum in vivo rate of the carboxylase reaction [Eq. (7)]. The model is fairly insensitive to the value of P_m, provided that a value is chosen that is somewhat larger than the maximum P_n obtained experimentally at low $[O_2]$, saturating light, and saturating $[CO_2]$.

6. $P_x(T)$ is the temperature-dependent, maximum in vivo rate of the reduction reaction [Eq. (3)]:

$$P_x(T) = 0.5k_2(T)A_0N_0 \tag{15}$$

The model is also fairly insensitive to the value of $P_x(T)$, provided that the value is not too small. It was assumed that the reduction reaction has 10 times the capacity of the carboxylase reaction under optimal conditions for both reactions in vivo.

The same basic Arrhenius function was used for all the temperature-dependent parameters (r'_k, R, R_d, P_m, and P_x). It was assumed that the activation energy is 16 kcal mole^{-1} below 27°C and 10 kcal mole^{-1} above 27°C, with a discontinuous change in direction at 27°C. The change in activation energy prevented unreasonable increases in parameter values at high temperatures. The temperature response functions and the values of the other input parameters for *Atriplex patula* are presented in Appendix B.

Computer Operations with the Model

A computerized iteration procedure was used to solve simultaneously Eqs. (1), (2), the quadratic equation from Appendix A, and the temperature-dependent functions from Appendix B. The iteration procedure finds the value of C_w that will satisfy both the transport submodel and the metabolism submodel. This procedure automatically incorporates the phenomenon of CO_2 recycling within the protoplasm into the model. The model may be summarized by the following functional relationship:

$$P_n = f(C_t, O_t, T, E)_{r'_a, r'_1, r'_w, r'_k, k^*, \alpha, P_m, P_x, R, R_d}$$

Model Analyses

The internal functioning of the model was analyzed by differential calculus (Hall, 1971) and compared with experimental observations.

1. Light-saturated P_n should be related to RudPcase activity. Björkman and co-workers have reported positive correlations between light-saturated P_n and RudPcase activity for different C_3 species subjected to different pretreatments (Björkman, 1968a, 1968b; Björkman *et al.*, 1969, 1972; Gauhl, 1969; Medina,

1970, 1971). r_k' will be inversely related to RudPcase activity per unit leaf area, provided that A_0 is constant. Analyses by Hall (1971) showed that this relationship was apparent in the data of Medina (1970).

2. Levels of photorespiration should be related to CO_2-fixing capabilities. More specifically, R will be inversely related to r_k', as they are both dependent upon RudPcase activity. Analyses by Hall (1971) showed that this relationship was apparent in the data of Gauhl (1969), Medina (1970), Slatyer (1970), and Hall and Loomis (1972b). Also, as the CO_2 compensation point is strongly dependent upon the balance between R and r_k' at saturating light, it should be independent of both the RudPcase level and the CO_2 transport systems; therefore, it should be fairly constant for C_3 species at equivalent temperatures and [O_2]. The observations of Moss *et al.* (1969) are consistent with this prediction.

3. Analyses that support the model are presented in Hall (1971).

Model Predictions and Experimental Observations

Photosynthetic Responses to Light and Carbon Dioxide

The photosynthetic responses to light and [CO_2] predicted by the model are reasonable simulations of observed responses (Hall, 1971). However, the model predicted a dependency of the apparent quantum efficiency on [CO_2] at 21 percent [O_2] that was not expected, as many workers have stated that the initial slope of the photosynthetic response to light is independent of [CO_2]. This point was investigated in detail both experimentally and with the model. Both approaches gave the same responses (Fig. 4.3), indicating that the quantum efficiency does increase with increases in [CO_2] at 21 percent [O_2] but is unaffected by [CO_2] at low [O_2]. The influence of [CO_2] on net photosynthesis at low irradiances was investigated with different species and methods. Reductions in [CO_2] inside the leaf caused by drastic variations in stomatal aperture should also influence P_n at low light intensities and 21 percent O_2. Stomatal oscillations were induced in *Phaseolus vulgaris* (pinto bean) at low light intensities (PAR $= 1.7 \times 10^4$ ergs cm^{-2} s^{-1}) and P_n oscillated in phase with transpiration and 90° out of phase with leaf-water potential. This indicates that the variations in P_n were due to variations in [CO_2] inside the leaf and not directly due to variations in leaf-water stress. The relative effects of [CO_2] on P_n at 21 percent [O_2] and 1 percent [O_2] were investigated at low light intensities and 27°C using *Gossypium hirsutum* (Acala-SJ1). Measurements were conducted with different plants to reduce variation due to pretreatment, and resistances to water vapor were determined to correct for stomatal influences (Table 4.1). With moderate increases in [CO_2], large relative increases in P_n were observed at 21 percent [O_2], whereas small relative increases were observed at 1 percent [O_2]. The data were corrected for differences in leaf resistance to water-vapor transfer by calculating the ratio percent change in P_n/percent change in [CO_2] inside the leaf. The relative influence of [CO_2] internal on P_n was also much

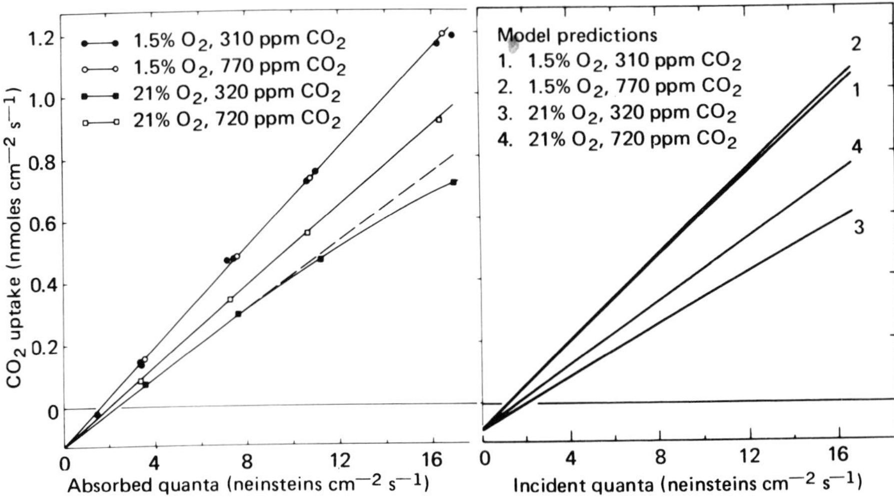

Fig. 4.3. (a) Effects of O_2 and CO_2 concentration on light-limited photosynthetic CO_2 uptake by leaves of *Atriplex patula* ssp. *spicata*. Leaf temperature, 26°C, λ_{max}, 650 nm. (b) Photosynthetic responses to light predicted by the mathematical model for *Atriplex patula* in conditions similar to those of part (a).

Table 4.1 Influence of $[CO_2]$ and $[O_2]$ on Net Photosynthesis by Cotton at Low Irradiances

Plant No.	PAR[a]	$[O_2]$ (%)	C_t[b] (ppm)	P_n[c]	ΔP_n	%	C_i[d] (ppm)	$\%P_n/\%C_t$	$\%P_n/\%C_i$
			255	0.178			242		
1	16,700	21			0.070	39		1.02	1.20
			353	0.248			321		
			257	0.217			239		
2	20,900	21			0.110	51		1.43	1.48
			348	0.327			321		
			268	0.430			242		
3	16,700	1			0.066	15		0.46	0.45
			358	0.496			324		
			254	0.653			212		
4	20,900	1			0.111	17		0.41	0.41
			359	0.764			299		

[a] PAR, irradiance between 400 and 700 nm in ergs cm^{-2} s^{-1}.
[b] C_t, external $[CO_2]$.
[c] P_n, net CO_2 flux in nmoles cm^{-2} s^{-1}.
[d] C_i, internal $[CO_2]$.

larger at 21 percent [O_2] than at 1 percent [O_2]. The absolute changes in P_n brought about by changes in [CO_2] were, however, very similar at 21 percent [O_2] and 1 percent [O_2]. These observations were conducted at higher light intensities (to achieve adequate stomatal opening) than those used by Björkman (1971), and they are also consistent with the model predictions.

Photosynthetic Responses to Temperature

Photosynthetic responses to light at 12 nmoles cm^{-3} [CO_2], and different temperatures were predicted by the model (Fig. 4.4). They exhibit a constant quantum efficiency, with a translation effect due to enhanced CO_2 evolution by the Krebs cycle at higher temperatures. P_n rates were higher and more light was required for saturation at higher temperatures than at lower temperatures.

Photosynthetic responses to temperature at saturating light, 12 nmoles cm^{-3} [CO_2], 1.5 percent [O_2], and 21 percent [O_2] were compared with observed values for *Atriplex patula* (Fig. 4.5). The model simulations are qualitatively reasonable in predicting an increased O_2 effect on P_n at higher temperatures. Quantitative aspects

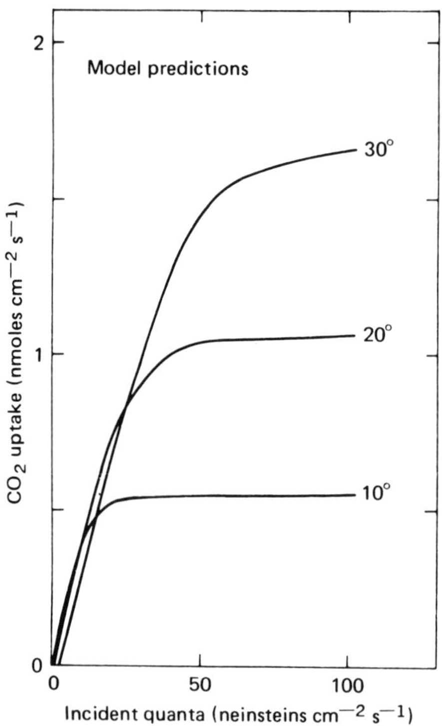

Fig. 4.4. Photosynthetic responses to light at 12 nmoles cm^{-3} [CO_2], and different temperatures in °C as predicted by the mathematical model.

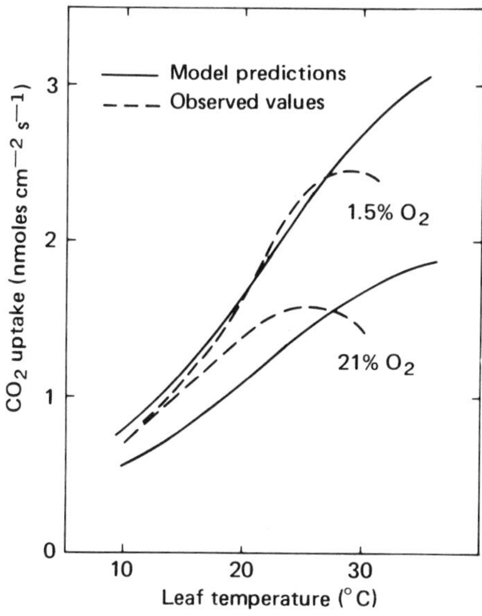

Fig. 4.5. Predicted and observed photosynthetic responses to temperature at saturating irradiances, 12 nmoles cm^{-3} [CO_2], 1.5 percent [O_2], and 21 percent [O_2] with *Atriplex patula.*

Table 4.2 Influence of Temperature on the Inhibitory Effect of O_2 on P_n at Different Light Intensities in Leaves of *Atriplex patula*
Observed Values (Björkman, 1971)

Irradiance (PAR) (ergs cm^{-2} s^{-1} × 10^3)	Rate at 21% O_2, 300 ppm CO_2 / Rate at 1.5% O_2, 300 ppm CO_2		
	25°C	**16.5°C**	**12.5°C**
400	0.70	0.80	0.98
130	0.67	0.69	0.92
50	0.66	0.69	0.77

Predicted Values

Quantum Flux (neinsteins cm^{-2} s^{-1})	Rate at 21% O_2, 12 nmoles cm^{-3} CO_2 / Rate at 1.5% O_2, 12 nmoles cm^{-3} CO_2		
	25°C	**15°C**	**10°C**
200	0.64	0.67	0.70
60	0.63	0.67	0.70
20	0.57	0.65	0.69

of the simulations are not adequate, as the model predicts a substantial O_2 effect on P_n at low temperatures which is not usually observed. Also, the model does not exhibit the pronounced optimum frequently observed.

The influence of temperature and $[O_2]$ on the CO_2 compensation point were predicted by the model and compared with experimental observations (Fig. 4.6). The predicted and observed responses were similar.

The influence of temperature on the inhibitory effect of O_2 on P_n was predicted for different light intensities and compared with observed values (Table 4.2). With both predicted and observed values, the O_2 inhibition of P_n was greatest at 25°C and low light intensities and smallest at lower temperatures and high light intensities. Predicted levels of inhibition were consistently higher than observed values, especially at low temperatures and high light intensities.

The model was also used to predict responses that are difficult to measure. The extent to which enhanced CO_2 evolution contributes to the O_2 effect on P_n was predicted for different light intensities and temperatures (Table 4.3). The model predicts that 41 to 57 percent of the O_2 effect is due to enhanced CO_2 evolution, with the smallest percentage occurring at low light intensities and moderate temperatures. The model predicts the change in CO_2 evolution rates in the protoplast; thus it is free of the CO_2 recycling error that makes experimental verification extremely difficult. Ludwig and Canvin (1971) compared short-term $^{14}CO_2$ uptake

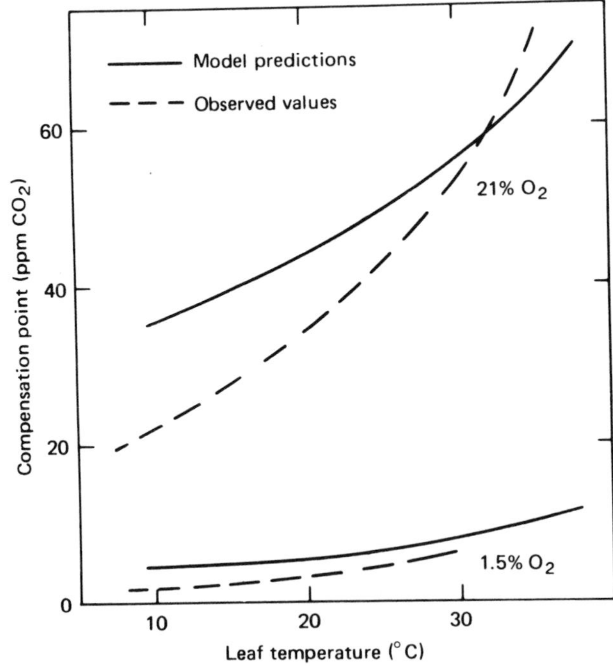

Fig. 4.6. Predicted and observed effects of temperature and $[O_2]$ on the CO_2 compensation point at saturating irradiances with *Atriplex patula*.

Table 4.3 Contribution of Enhanced CO_2 Evolution to the O_2 Effect on P_n as Predicted by the Model[a]

Quantum Flux (neinsteins cm^{-2} s^{-1})	% Contribution of Enhanced CO_2 Evolution to the O_2 Effect on P_n		
	25°C	15°C	10°C
200	57	48	45
60	55	47	46
20	41	43	45

[a] Oxygen concentrations, 21 and 1.5%; CO_2 concentration, 12 nmoles cm^{-3}.

with net CO_2 uptake in sunflower at 21 percent $[O_2]$ and 1 percent $[O_2]$. From these data they estimated the contribution to the oxygen effect on P_n of enhanced CO_2 evolution as being 32 percent at 300 ppm $[CO_2]$. Adjusting this value for the underestimate of P_t by $^{14}CO_2$ uptake due to recycling, one obtains values similar to those in Table 4.3. Their data also indicated that the CO_2 evolution rate in the light was independent of $[CO_2]$ between 50 and 250 ppm at moderate light intensities and temperatures. The model predicts a 2 percent reduction in photorespiration when the $[CO_2]$ is raised from 50 to 300 ppm at 27°C, 21 percent $[O_2]$, and saturating light with *Atriplex patula*. The $^{14}CO_2$ labeling studies of Osmond and Björkman (1972) suggest, however, that the effect of 21 percent $[O_2]$ on P_n is not largely due to the release of photorespiratory CO_2.

Discussion

Photorespiration and the Oxygen Effect on Net Photosynthesis

The mathematical model of photosynthesis and respiration simulates in a reasonable manner observed responses of P_n to light intensity, $[CO_2]$, and $[O_2]$. Interactions between $[CO_2]$ and $[O_2]$ in their effects on P_n are obvious in the experimental data. The model simulates these interactions, photorespiration, and the $[O_2]$ effect on P_n. A central assumption of the model is the hypothesis that the enzyme RudPcase has both carboxylase and oxygenase functions. Consequently, the experimental data and theoretical analyses support this hypothetical dual role of RudPcase. A practical consequence of this hypothesis is that selecting for low levels of either photorespiration or the O_2 effect on P_n within cultivated C_3 species may produce genotypes that also have low photosynthetic rates. The advisability of selecting for C_3 genotypes with low intrinsic rates of photorespiration is also questionable, since a possible essential function for photorespiration has been proposed (Osmond and Björkman, 1972).

Influence of CO_2 and O_2 on the Quantum Efficiency

The model simulates the suppression of quantum efficiency in normal $[O_2]$ when compared with low $[O_2]$. The model assumption that RudPcase has dual functions is responsible for these responses. Consequently, it does not appear necessary to invoke effects of O_2 on the photosystems to explain the O_2 effect on P_n at low light intensities.

The enhancement of P_n by increases in $[CO_2]$ at low light intensities has now been established to occur in several species. The model predicted these effects without assuming effects of $[CO_2]$ on the photosystems. The responses of P_n to $[CO_2]$ at low light intensities provides a physiological explanation for the yield increases obtained in greenhouses by CO_2 enrichment at low light intensities.

Limitations of the Model

Model simulations deviate from observed responses at high temperatures (Fig. 4.5) and high $[CO_2]$ (Hall, 1971). The deviations from reality at high temperatures may be partially eliminated by the inclusion of a submodel relating stomatal response to temperature and the absolute humidity gradient. Deviations from Arrhenius responses at high temperatures and toxic effects of high $[CO_2]$ may also be partially responsible for the deviations in P_n, but they are not understood adequately at this time.

The equations used to model the dual functions of RudPcase [Eqs. (7) and (8)] function adequately, but rigorous derivation was not possible; thus the reaction mechanisms that they describe are only vaguely apparent. For example, what reaction mechanism would result in a first-order competitive influence of CO_2 and O_2 on carboxylation, but only an O_2 effect on oxygenation of RudP? Detailed kinetic treatments for the model must function and be solvable within the constraints imposed by the totality of the model. In evaluating models of this type, it should be recognized that improvements of many of the rough approximations must be solvable within the context of the total model. Further, sensitivity analysis indicates that many changes can be made in a complex interacting system of this type with little influence on P_n predictions, whereas some changes in the model have dramatic influences.

Many of the potential deficiencies in the model may be removed by experimental studies at the organelle and molecular level, provided that they are designed to produce quantitative information that is relevant to the functioning of the integrated system in the leaf.

Appendix A: Mathematical Solution to the Submodel of Metabolism

$$P_n - R_d = [-b - (b^2 - 4ac)^{0.5}]2a^{-1}$$

where
$$
\begin{aligned}
a &= P_x Z(2P_m C_w - 3RO_w Z)\\
b &= -(P_m C_w + RO_w Z)[2P_m C_w(P_x + \alpha E) + Z(2P_x \alpha E - 3RO_w(P_x + \alpha E))]\\
c &= 2P_x \alpha E(P_m C_w + RO_w Z)^2\\
Z &= C_w + P_m r'_k + P_m r'_k k^* O_w\\
r'_k &= (k_1 A_0)^{-1}\\
\alpha &= 0.5k_3 N_0\\
R &= -0.5k_4 A_0\\
P_x &= 0.5k_2 A_0 N_0
\end{aligned}
$$

The following simplification may be used when obtaining partial differentials provided that the limiting conditions are such that $4acb^{-2} \rightarrow 0$.

$$P_n - R_d = -cb^{-1}$$

Appendix B: Model Input Parameters for *Atriplex patula*

Symbol	Description	Value	Unit
r'_a	Boundary-layer resistance	0.5	s cm^{-1}
r'_1	Minimal leaf resistance	1.0	s cm^{-1}
r'_w	Wall resistance	0.15	s cm^{-1}
α	Quantum efficiency	0.073	moles einstein^{-1}
k^*	Parameter for O_2 effect on CO_2 fixation	2.5	Dimensionless when $[O_2]$ is a ratio.

Temperature-Dependent Parameters

The general equation is

$$\text{parameter} = \text{constant} \times \exp(\pm \text{activation energy}/RT)$$

where R = international gas constant
T = absolute temperature in °K

Equations for Leaf Temperatures Less Than **27°C**

$$r'_k = 3.41 \times 10^{-12} \times \exp(8000/T)$$

$$P_x = \frac{3.05 \times 10^{13}}{\exp(8000/T)}$$

$$P_m = \frac{3.05 \times 10^{12}}{\exp(8000/T)}$$

$$R = \frac{-1.14 \times 10^{12}}{\exp(8000/T)}$$

$$R_d = \frac{-3.43 \times 10^{10}}{\exp(8000/T)}$$

Equations for Leaf Temperatures Greater Than **27°C**

$$r'_k = 7.51 \times 10^{-8} \times \exp(5000/T)$$

$$P_x = \frac{1.38 \times 10^{9}}{\exp(5000/T)}$$

$$P_m = \frac{1.38 \times 10^{8}}{\exp(5000/T)}$$

$$R = \frac{-5.19 \times 10^{7}}{\exp(5000/T)}$$

$$R_d = \frac{-1.56 \times 10^{6}}{\exp(5000/T)}$$

References

Andrews, T. J., Lorimer, G. H., Tolbert, N. E.: 1973. Ribulose diphosphate oxygenase. I. Synthesis of phosphoglycolate by fraction-1 protein of leaves. Biochemistry *12*, 11–18.

Berry, J. A.: 1971. The effect of oxygen on CO_2 fixation by carboxydismutase *in vitro* and an examination of a possible reaction of ribulose diphosphate with oxygen. Carnegie Inst. Wash. Yearbook *70*, 526–530.

Björkman, O.: 1968a. Carboxydismutase activity in shade-adapted and sun-adapted species of higher plants. Physiol. Plantarum *21*, 1–10.

———: 1968b. Further studies on differentiation of photosynthetic properties in sun and shade ecotypes of *Solidago virgaurea*. Physiol. Plantarum *21*, 84–99.

———: 1971. Interaction between the effects of oxygen and CO_2 concentration on quantum yield and light-saturated rate of photosynthesis in leaves of *Atriplex patula* ssp. *spicata*. Carnegie Inst. Wash. Yearbook *70*, 520–526.

———, Nobs, M. A., Hiesey, W. M.: 1969. Growth, photosynthetic, and biochemical response of contrasting *Mimulus* clones to light intensity and temperature. Carnegie Inst. Wash. Yearbook *68*, 614–620.

———, Boardman, N. K., Anderson, J. M., Thorne, S. W., Goodchild, D. J., Pyliotis, N. A.: 1972. Effect of light intensity during growth of *Atriplex patula* on the capacity of photosynthetic reactions, chloroplast components and structure. Carnegie Inst. Wash. Yearbook *71*, 115–134.

Bowes, G., Berry, J. A.: 1972. The effect of oxygen on photosynthesis and glycolate excretion in *Chlamydomonas reinhardtii.* Carnegie Inst. Wash. Yearbook *71*, 148–158.

———, Ogren, W. L.: 1972. Oxygen inhibition and other properties of soybean ribulose 1,5-diphosphate carboxylase. J. Bio. Chem. *247*, 2171–2176.

———, Ogren, W. L., Hageman, R. H.: 1971. Phosphoglycolate production catalyzed by ribulose diphosphate carboxylase. Biochem. Biophys. Res. Commun. *45*, 716–722.

Boyer, J. S.: 1971. Nonstomatal inhibition of photosynthesis in sunflower at low leaf water potentials and high light intensities. Plant Physiol. *48*, 532–536.

Bruin, W. J., Nelson, E. B., Tolbert, N. E.: 1970. Glycolate pathway in green algae. Plant Physiol. *46*, 386–391.

Chartier, P.: 1966. Étude théorique de l'assimilation brute de la feuille. Ann. Physiol. Vég. *8*, 167–196.

Clymer, A. B.: 1969. The modeling of hierarchical systems. *In* Proc. Conf. Appl. Continuous System Simulation Languages, pp. 1–16. San Francisco.

Duncan, W. G., Loomis, R. S., Williams, W. A., Hanau, R.: 1967. A model for simulating photosynthesis in plant communities. Hilgardia *38*, 181–205.

Gaastra, P.: 1959. Photosynthesis of crop plants as influenced by light, carbon dioxide, temperature and stomatal diffusion resistance. Meded. Landbouwk. Wageningen *59*, 1–68.

Gauhl, E.: 1969. Leaf factors affecting the rate of light-saturated photosynthesis in ecotypes of *Solanum dulcamara.* Carnegie Inst. Wash. Yearbook *68*, 633–636.

Gifford, R. M.: 1971. The light response of CO_2 exchange: on the source of differences between C_3 and C_4 species. *In* Photosynthesis and photorespiration (eds. M. D. Hatch, C. B. Osmond, R. O. Slatyer), pp. 51–56. New York: Wiley-Interscience.

Hall, A. E.: 1970. Photosynthetic capabilities of healthy and beet yellows virus infected sugar beets (*Beta vulgaris* L.). Davis, Calif.: Univ. California. Ph.D. diss.

———: 1971. A model of leaf photosynthesis and respiration. Carnegie Inst. Wash. Yearbook *70*, 530–540.

———, Loomis, R. S.: 1972a. Photosynthesis and respiration by healthy and beet yellows virus-infected sugar beets (*Beta vulgaris* L.). Crop Sci. *12*, 566–572.

———, Loomis, R. S.: 1972b. An explanation for the difference in photosynthetic capabilities of healthy and beet yellows virus-infected sugar beets (*Beta vulgaris* L.). Plant Physiol. *50*, 576–580.

Jacobs, M. H.: 1967. Diffusion processes. New York: Springer-Verlag.

Laisk, A.: 1970. A model of leaf photosynthesis and photorespiration. *In* Prediction and measurement of photosynthetic productivity, pp. 295–306. Proc. IBP/PP Tech. Meeting, Tȓeboň, Pudoc, Wageningen.

Lake, J. V.: 1967. Respiration of leaves during photosynthesis. I. Estimates from an electrical analogue. Austral. J. Biol. Sci. *20*, 487–493.

Lommen, P. W., Schwintzer, C. R., Yocum, C. S., Gates, D. M.: 1971. A model describing photosynthesis in terms of gas diffusion and enzyme kinetics. Planta *98*, 195–220.

Lorimer, G. H., Andrews, T. S., Tolbert, N. E.: 1973. Ribulose diphosphate oxygenase. II. Further proof of reaction products and mechanisms of action. Biochemistry *12*, 18–23.

Ludwig, L. J., Canvin, D. T.: 1971. The rate of photorespiration during photosynthesis and the relationship of the substrate of light respiration to the products of photosynthesis in sunflower leaves. Plant Physiol. *48*, 712–719.

Medina, E.: 1970. Relationships between nitrogen level, photosynthetic capacity, and carboxydismutase activity in *Atriplex patula* leaves. Carnegie Inst. Wash. Yearbook *69*, 655–662.

———: 1971. Effect of nitrogen supply and light intensity during growth on the photosynthetic capacity and carboxydismutase activity of leaves of *Atriplex patula* ssp. *hastata*. Carnegie Inst. Wash. Yearbook *70*, 551–559.

Moss, D. W., Krenzer, E. G., Brun, W. A.: 1969. CO_2 compensation points in related plant species. Science *164*, 187–188.

Osmond, D. B., Björkman, O.: 1972. Simultaneous measurements of oxygen effects on net photosynthesis and glycolate metabolism in C_3 and C_4 species of *Atriplex*. Carnegie Inst. Wash. Yearbook *71*, 141–148.

Parkinson, K. J.: 1968. Apparatus for the simultaneous measurement of water vapor and carbon dioxide exchanges of single leaves. J. Expt. Bot. *19*, 840–856.

Slatyer, R. O.: 1970. Relationship between plant growth and leaf photosynthesis in C_3 and C_4 species of *Atriplex*. Planta *93*, 175–189.

Whittingham, C. P., Coombs, J., Marker, A. F. H.: 1969. The role of glycolate in photosynthetic carbon fixation. *In* Biochemistry of chloroplasts (ed. T. W. Goodwin), vol. 2, pp. 155–173. New York: Academic Press.

5

Optimal Leaf Form

S. ELWYNN TAYLOR

Introduction

The size of leaves typical for specific climates has been studied for many years, and several investigators have considered the "leaf size class" as an indicator of climatic conditions (Raunkiaer, 1934). Bailey and Sinnott (1916) concluded that the form and size of leaves were more a result of environment than of genetic history, although the latter was certainly an influence. Benson *et al.* (1967) reported ecotypic differentiation of leaf form with respect to slope exposure for a hybrid population of *Quercus douglassii* × *Q. turbinella.* They suggested that hybrid variability may permit the rapid evolutionary selection of characters best suited for the particular microclimate. They reported that individuals found on the northeast slope had leaves of significantly larger dimension than did those growing on the more arid southwest slope (Fig. 5.1). It is generally considered that the reduction of leaf size in arid areas has the effect of conserving water, but quantitative evidence of the effects of leaf size has been available only recently.

The effects of leaf size are inseparably coupled with other characteristics of the leaf and the environment. Proper evaluation of the significance of any characteristic must consider all environmental and biological factors. The analysis must include the primary meteorological and edaphic parameters: solar and thermal radiation, air temperature, atmospheric vapor pressure, air speed, atmospheric gas concentrations (CO_2, O_2), and availability of soil moisture. Biological parameters include absorptivity to radiation; stomatal and mesophyll resistance to uptake or loss of carbon dioxide, oxygen, water vapor, and other gases; size, shape, and orientation of the organism; and the temperature range critical to survival. When the biological and environmental parameters are known, one can properly evaluate the biological responses to the environment and thereby determine the significance of variations in the individual characteristics or parameters. The biological responses considered in this paper are leaf temperature, transpiration rate, net photosynthesis, and the ratio of photosynthesis to transpiration.

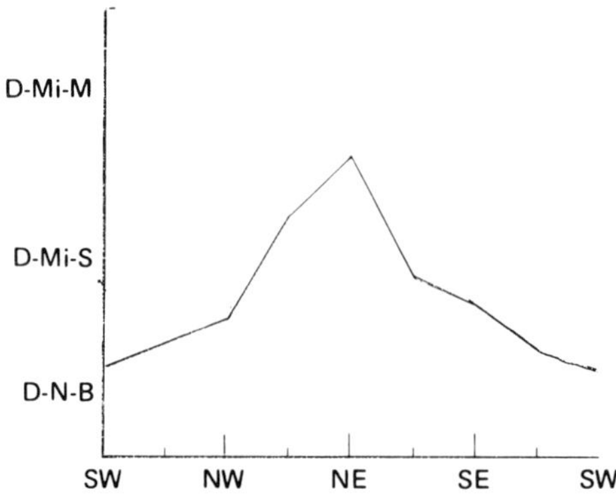

Fig. 5.1. Leaf-dimension class for hybrid *Quercus douglassi* × *Q. turbinella* with slope direction. (Adapted from Benson *et al.*, 1967.)

Leaf Size and Dimension Classification

Leaf dimension directly affects the energy exchange at the leaf because the thickness of the boundary layer, which limits exchange of heat and diffusion of water vapor, depends on air speed and leaf size. The leaf size is best expressed as "characteristic dimension" or "effective leaf width" (Parkhurst *et al.*, 1968; Taylor and Gates, 1970). The effective leaf width is the downwind leaf width that has convective heat transfer equal to a flat rectangular plate having actual dimensions equal to the effective or characteristic dimensions. For any specified leaf shape, a constant can be determined such that maximum leaf width times that constant yields characteristic dimension. The constant is calculated according to the expression

$$R_D = \frac{D}{D_{\max}} \tag{1}$$

where $D_{\max}$ is the maximum leaf width, R_D is the constant for the leaf form, and D is the characteristic dimension defined as

$$D = \left(\frac{\sum_{i=0}^{W} D_i \, \Delta W_i}{\sum_{i=0}^{W} \sqrt{D_i} \, \Delta W_i} \right)^2 \tag{2}$$

where D_i is the length of an increment in the direction of air flow and W is the length of the leaf perpendicular to air flow. The leaf is divided into i increments in the W direction.

Constants for several leaf shapes are presented in Table 5.1 for wind flow across the leaf at right angles to the midvein. Evaluation of leaf dimension to air flow along the vein can be made in a similar manner. Constants for other leaf shapes were given by Taylor and Gates (1970), as adapted from Parkhurst *et al.* (1968).

Table 5.1 Dimension Constants for Leaves Collected in the Canal Zone[a]

R_D	0.791	0.772	0.700	0.607	0.784	0.699	0.766	0.762	0.769	0.661
R_W	0.648	0.668	0.585	0.640	0.647	0.575	0.833	0.664	0.721	0.664

[a] The characteristic leaf dimension is found by multiplying the maximum width by R_D. The characteristic length is found by multiplying the maximum length by R_W.

Leaf size is often classified, according to Raunkiaer (1934), on the basis of surface area. However, for energy-exchange evaluations, leaves should be classified according to "characteristic dimension." Some have found it desirable to express the characteristic dimension using Raunkiaer's classifications as a basis (Brunig, 1970). A dimension classification scheme compatible with the Raunkiaer size classification was formulated according to Eq. (2) for leaves of the basic "elliptic" shape given by Raunkiaer. The characteristic dimension for leaves of each size suggested by Raunkiaer is given in Table 5.2.

Raunkiaer considered the leaflets of compound leaves as individual leaves. Further, deeply lobed leaves were not considered in the size class distribution with entire leaves; i.e., a size class distribution for entire leaves, one for lobed leaves, and one for compound leaves is made for the vegetation of a region by the investigator. All three groups are considered together by use of the "dimension class." Hence only one leaf class distribution is given, rather than three.

It should be noted, however, that the classifying of leaf dimensions should be broken down according to the energy environment typical for the leaves; i.e., leaves considered together should come from similar energy environments, such as deep shade, semishade, exposed to full sun, terrain slope and aspect, and time of solar exposure. Also, moisture regimes should be separated, for example, dry hillside from moist valley environments.

The leaf dimension classification system utilizes the basic nomenclature of Raunkiaer to describe the dimension grouping of leaves. Each group is divided into three subgroups, as suggested by Raunkiaer (1934). The classes are referred to as small–medium–big and are more satisfactory to the author's needs than the addition of another major class, as has been suggested by several investigators (Cooper, 1922; Webb, 1959).

The foliar physiognomy can have a considerable influence on the "leaf dimension," so that a leaf with an entire margin may be larger in dimension than a deeply lobed leaf that has considerably greater surface area (Fig. 5.2). Deep lobing or foliar pinnation can reduce the leaf dimension without changing the leaf size appreciably. Taylor and Sexton (1972) demonstrated that tattering of the leaves in Musaceae effectively reduced the dimension to one better suited for their climate (Fig. 5.3). It must be noted, however, that very fine pinnation may not be effective

Table 5.2 Raunkiaer Leaf-Size Class and Leaf-Dimension Class[a]

Size Class[b]	Leaf Area One Side (cm^2)	Characteristic Dimension Width × 0.742 = D[c] (cm)	Dimension Class
Leptophyll	0–0.25	0–0.33	*D*-leptophyll
S	0–0.056	0–0.16	D–Le–S
M	0.056–0.12	0.16–0.24	D–Le–M
B	0.12–0.25	0.24–0.33	D–Le–B
Nanophyll	0.25–2.25	0.33–0.93	*D*-nanophyll
S	0.25–0.52	0.33–0.47	D–N–S
M	0.52–1.08	0.47–0.68	D–N–M
B	1.08–2.25	0.68–0.93	D–N–B
Microphyll	2.25–20.25	0.93–2.75	*D*-microphyll
S	2.25–4.68	0.93–1.32	D–Mi–S
M	4.68–9.74	1.32–1.80	D–Mi–M
B	9.74–20.25	1.80–2.75	D–Mi–B
Mesophyll	20.25–182.25	2.75–7.38	*D*-mesophyll
S	20.25–42.09	2.75–3.8	D–Ms–S
M	42.09–87.68	3.8–5.3	D–Ms–M
B	87.68–182.25	5.3–7.38	D–Ms–B
Macrophyll	182.25–1640.25	7.38–22.26	*D*-macrophyll
S	182.25–378.82	7.38–10.8	D–Ma–S
M	378.82–789.13	10.8–15.2	D–Ma–M
B	789.13–1640.25	15.2–22.26	D–Ma–B
Megaphyll	1640.25–*x*	22.26–*x*	*D*-megaphyll
S	1640.25–3409.31	22.26–31.5	D–Mg–S
M	3409.31–7102.11	31.5–43	D–Mg–M
B	7102.11–*x*	43–*x*	D–Mg–B

[a] Leaf-dimension classification is directly derived from the leaf-size classification by Raunkiaer (1934) for elliptic leaf form. The elliptic form does not constitute an ellipse which has D = width × 0.87, but the dimension is found as D = width × 0.742. Each class is divided into three groups: small, medium, and big, as suggested by Raunkiaer, with the areas for each division chosen by the author as consistent with the original class size divisions.

[b] S, M, B (small, medium, big) are class divisions suggested by Raunkiaer (1934) but divided (values chosen) by me.

[c] The characteristic dimension for Raunkiaer's leaf outlines is width × 0.7420, after Parkhurst *et al.* (1968).

in reducing dimension since the elements might share a common boundary layer (Parkhurst and Loucks, 1972).

A study of bracken fern (*Pteridium aquilinum*) conducted by the author with Hyrum Johnson at The University of Michigan's Biological Station in August 1968 showed that, for air speeds of 10–300 cm s^{-1}, the final or third pinnation did indeed have a boundary layer in common with the second pinnation. Utilizing the energy-exchange equations (Gates *et al.*, 1968) [see Eq. (3)] to solve for dimension, we found that the first pinnation was the fundamental unit of energy exchange. Calculations of energy exchange based on the second pinnation caused an error of +83 percent in the calculated amount of energy released by convection. Calculations utilizing the final pinnation produced errors of +210 percent. All measure-

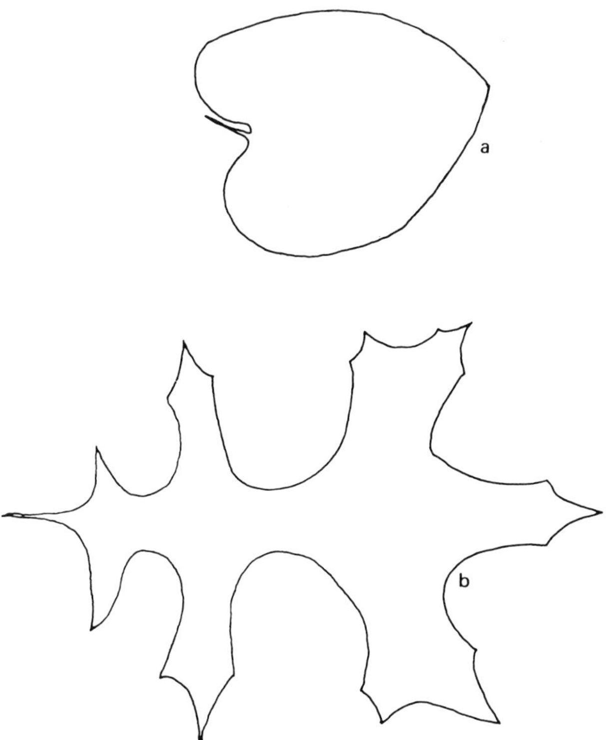

Fig. 5.2. The leaf of *Aristolochia durior* (a) is smaller in area (small mesophyll) than the *Quercus palustris* leaf (mesophyll); (b) yet, because of the leaf shape, the former has the larger characteristic dimension. The *Q. palustris* is a medium microphyll in dimension, whereas *A. durior* has the dimension of small mesophyll.

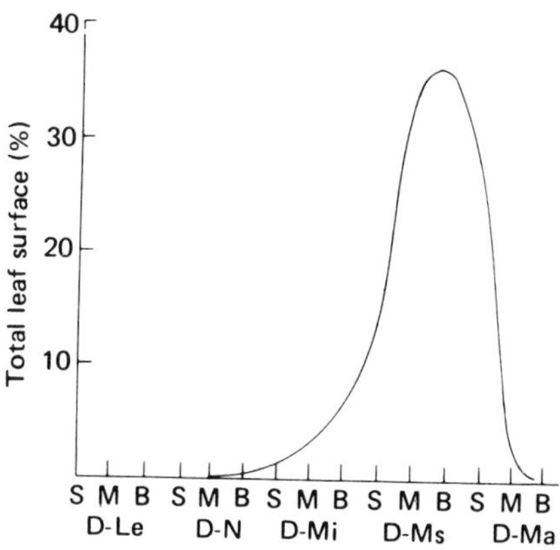

Fig. 5.3. Leaves from the exposed portion of the Barro Colorado Island Forest canopy, Canal Zone. Approximately 80 percent of the total leaf surface has dimensions in the range 4–11 cm, including the banana relatives, whose very large leaves are reduced in dimension by tearing.

ments were made for the natural sunlit environment using potometer-mounted fronds.

The leaf form typical for geographical regions is best expressed as a dimension class distribution, especially in cases where leaf size is limited by the physical environment. For example, Taylor and Sexton (1972) calculated that no leaves larger than a "big-dimension-mesophyll" (7.4 cm) would be anticipated in exposed areas during the dry season at Barro Colorado Island, Canal Zone, unless a continuous water supply were available. The leaves of several Musaceae species have dimension greater than this, but as mentioned above, leaf tearing had the effect of reducing the dimension to within the specified class. A frequency distribution of leaf size and dimension was made for exposed leaves on B.C.I. during December 1970 as the wet season was nearing an end (Fig. 5.3). The analysis included all exposed leaves of randomly chosen plots (9 m^2) in an old clearing on the north side of the island. The mean dimension and the upper limit to dimension for leaves in the specified climate were predicted by the energy-budget and gas-exchange analysis. Commonly, leaves of greatly different surface areas taken from similar environments have similar characteristic dimension. Such leaves are of equivalent dimension insofar as energy-exchange parameters are considered.

Climate Classification

Classification of leaf climates becomes a necessity if finite observations of leaf form and environment are desired. Gates (1968) defined leaf climates for purposes of convenience and discussion. He included air temperature, humidity, wind, and solar radiation in the classification. Taylor (1971) utilized a climate classification based on that of Gates but defined somewhat finer divisions. The classification (Table 5.3) included the solar and thermal energy absorbed by the leaf and also qualitative observations of soil-moisture availability and canopy condition. No range of values was assigned the latter two parameters because of insufficient quantitative observations. Values for realistic natural ranges should be defined as understanding permits.

The climate for an individual leaf is almost a unique condition for that leaf. It can be expected that one would need several climates to describe the environment of a tree or forest canopy.

The proposed "plant climate classification," together with the leaf form classification presented herein, provide a reasonable matrix for the analysis of plant–climate relationships.

Environmental Limitations to Leaf Form

The environmental conditions together with the biological characteristics of the leaf interact to determine, among other parameters, the leaf temperature. Small leaves are close to air temperature since their narrow width allows increased

Table 5.3 Leaf-Climate Classification[a]

Parameter	Designation	Range	Designation	Range	Designation	Range
Air temperature (T_a)	Hot	50–30°C	Warm	29–15°C	Cool	14–0°C
Humidity (rh)	Dry	0–40%	Moist	41–70%	Wet	71–100%
Sunlight (Epp)[b]	Bright	1.26–0.83 × 10^6 ergs cm^{-2} s^{-1}	Hazey	0.83–0.56 × 10^6 ergs cm^{-2} s^{-1}	Cloudy	0.56–0.28 × 10^6 ergs cm^{-2} s^{-1}
Wind (V)	Windy	> 100 cm s^{-1}	Moderate	100–10 cm s^{-1}	Still	< 10 cm s^{-1}
Absorbed radiation (Q_{abs})	High	1.12–0.70 × 10^6 ergs cm^{-2} s^{-1}	Moderate	0.70–0.49 × 10^6 ergs cm^{-2} s^{-1}	Low	0.49–0.21 × 10^6 ergs cm^{-2} s^{-1}
Soil	Dry		Moist		Wet	
Canopy (during observations)	Open		Partial shade		Closed	

[a] The factors of the environment utilized in energy- and gas-exchange analysis are placed in categories that can be used to designate the immediate climate for an organism. A few other factors, considered significant, are included for convenience.

[b] When the solar insolation is less than 0.28 × 10^6 ergs cm^{-2} s^{-1}, the designation "dark" is used.

interaction of the air with the leaf surface; i.e., the magnitude of the boundary layer is related to the characteristic dimension of a leaf. The energy-budget equation describes the dimension effects explicitly. The form of the energy-budget equation presented by Gates *et al.* (1968) is

$$Q_{abs} = \sigma\varepsilon T_L^4 + k_1(V/D)^{1/2}(T_L - T_a) + L\frac{{}_s\rho_L(T_L) - (\text{rh})_s\rho_a(T_a)}{r_i + k_2(W^{0.2}D^{0.35}/V^{0.55})} \quad (3)$$

where Q_{abs} = total absorbed radiation (ergs cm^{-2} s^{-1})
T_L = leaf temperature (°K)
T_a = air temperature (°K)
V = air velocity (cm s^{-1})
σ = Stefan–Boltzmann constant (5.67 × 10^{-5} ergs cm^{-2} s^{-1} $°K^{-4}$)
ε = leaf emissivity (0.94 − 1.0) (as determined by Idso *et al.*, 1969)
rh = relative humidity (0 − 1.0)
L = latent heat of evaporation of water (ergs g^{-1})
r_i = resistance of the leaf to diffusion of water vapor (s cm^{-1})
D = characteristic leaf dimension (cm)
W = width perpendicular to D (cm)
${}_s\rho_L(T_L)$ = saturation density of water vapor at leaf temperature (g cm^{-3})
${}_s\rho_a(T_a)$ = saturation density of water vapor at air temperature (g cm^{-3})
k_1 = 1.13 × 10^4 when W is 5 cm or less and 6.98 × 10^3 when greater than 5 cm
k_2 = 1.56 for W of 5 cm or less and 2.10 when W is greater than 5 cm

The temperature of the leaf is determined by the simultaneous interaction of energy absorbed, air temperature, wind, leaf size, and leaf diffusion resistance. The energy-budget equation was solved to produce a nomogram (Fig. 5.4) showing leaf temperatures as affected by leaf size (D) and leaf resistance to the diffusion of water vapor (r_i) for fair-weather conditions typical of July and August in northern Michigan (warm, moist air, moderate wind, bright sun, and moderate to high Q_{abs}). The temperature of the leaf increases with increased leaf dimension, and the temperature rises with increasing leaf resistance. Large leaves are not expected to sustain injuriously high temperatures in this environment, even when resistance to water loss is high. Small leaves will be close to air temperature unless resistance is quite high.

Leaf temperature affects chemical processes, and indeed extremes of temperature affect the survival of the leaf. The extent of temperature effect varies for individual plants and according to the history of the plant (Yarwood, 1961; Lange, 1965, 1967). The effects of leaf temperature on biochemical processes must be individually known, but the effect of leaf temperature on transpiration rate is clearly defined for the general case by the energy-budget expression [Eq. (3)]. The only significant exception is that of extremely halophytic plants and the case of undefined boundary-layer conditions, such as those caused by extreme pubescence or leaf cupping.

Transpiration rate as a function of dimension (D) and leaf resistance (r_i) was calculated from the energy-budget equation and is shown in Fig. 5.5, which was produced from Eq. (3) using the same environmental conditions described for Fig. 5.4. The transpiration rate is controlled by leaf resistance to the diffusion of water vapor rather than by leaf dimension. In still air the effect of dimension on water loss is more pronounced because of the greater thickness of the leaf boundary layer. Increased leaf dimension results in a greater boundary layer, which impedes water loss. However, in this case, the greater boundary-layer thickness similarly limits the convective transfer of heat, resulting in higher leaf temperatures (Fig. 5.4), and thereby increasing the water-vapor-pressure difference between the leaf and

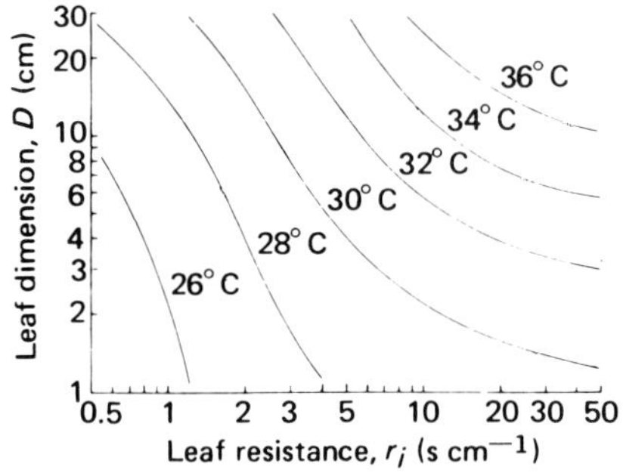

Fig. 5.4. Isolines representing leaf temperature as influenced by leaf dimension and leaf resistance to the diffusion of water vapor for a warm day, moist air, moderate wind, bright sun, and moderate to high radiation absorbed by the leaf.

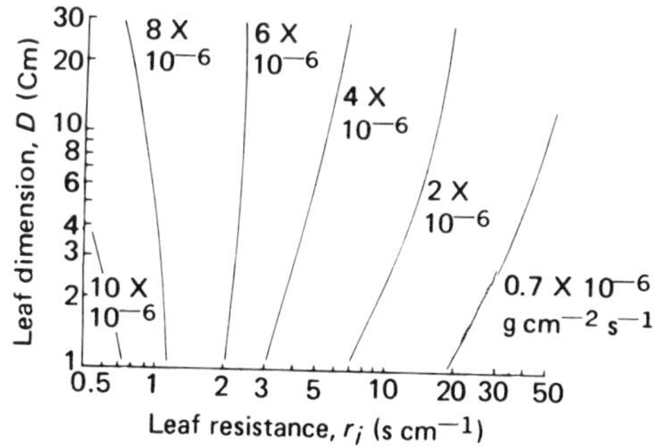

Fig. 5.5. Influence of leaf dimension and resistance on transpiration rate for environmental conditions identical to Fig. 5.4.

the free air. The relative balance of increased vapor pressure at the leaf and additional boundary-layer resistance to the transfer of water vapor results in the limited effect of dimension on transpiration. The transpiration rate may be increased, decreased, or unaffected by leaf dimension depending on the interaction of other energy-exchange parameters. The energy-budget equation is effective as a predictive model to describe biological response to various environmental conditions.

Dimension of the leaf has almost no effect on transpiration for a leaf resistance of 2 s cm^{-1} under the environmental conditions specified for Fig. 5.5. The transpiration rate increases with increased dimension at higher resistance values, and the rate is decreased by increased leaf dimension for low resistances.

Leaf temperature and transpiration rate nomograms may be produced for any environment and can serve as a guide to the limitations on leaf form imposed by the environment. The temperature nomogram is used to determine potential zones of thermal danger to the leaf. The transpiration nomogram delimits the conditions, resulting in excessive transpiration (I have observed that maximum transpiration rates are normally less than 8×10^{-6} g H_2O cm^{-2} s^{-1}).

Photosynthesis and Water-Use Efficiency

Leaf temperature and transpiration as influenced by leaf characters are important aspects of species success. Additionally, the effects of leaf form on photosynthesis and water-use efficiency may profitably be considered. The photosynthesis model developed by D. M. Gates and associates was used to describe effects of leaf tattering on production and water-use efficiency (Taylor and Sexton, 1972).

According to Gates *et al.* (1969), the photosynthesis equation may be expressed in quadratic form as

$$P = \frac{(r'P_m + K + \rho_a') - [(r'P_m + K + \rho_a')^2 - 4r'\rho_a'P_m]^{1/2}}{2r'} \tag{4}$$

where P = rate of CO_2 exchange between the air and the leaf (g CO_2 cm^{-2} s^{-1})
ρ'_a = atmospheric density of carbon dioxide
r' = resistance to diffusion of carbon dioxide from the atmosphere to the chloroplast
K = Michaelis rate constant for the reaction
P_m = maximum carbon dioxide exchange rate possible for given light and temperature conditions (g CO_2 cm^{-2} s^{-1})

The effects of light intensity, leaf temperature, and mesophyll thickness on net photosynthesis must be known individually. Trends of photosynthesis can be described from general approximations of the above biochemically related parameters. Examples of parameter approximation for the temperature and light dependence of photosynthesis and for leaf thickness are developed elsewhere (Taylor, 1971; Taylor and Sexton, 1972).

A photosynthesis nomogram was generated for leaves with optimum photosynthesis near 30°C (Fig. 5.6). The environmental conditions were identical to those used in the above nomograms. Large dimension and low resistance affect the greatest net photosynthesis.

Water-use efficiency or the ratio of carbon dioxide fixation to transpiration is of importance to modern plant management, although I do not consider it the most significant factor of natural leaf adaptation. Cohen (1970) theorized that the most successful species is one that photosynthesizes at the maximum rate when water is available with no specific measures toward water economy that would limit photosynthesis, and is capable of surviving dry periods, although not necessarily being productive. I have found that natural plant communities tend to exist between the extremes of maximum water economy and maximum net photosynthesis (as discussed below).

The water-use-efficiency nomogram (Fig. 5.7) is produced from Figs. 5.5 and 5.6. Small leaves with moderate resistances exhibit greatest water-use efficiency. The zone of maximum efficiency, however, has only one-sixth the potential for net

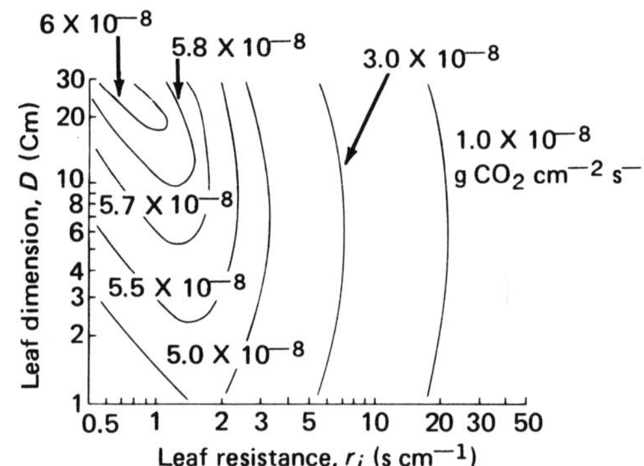

Fig. 5.6. Net photosynthesis is maximum for large leaves with low resistance (see the text for biochemical description) in the environment specified for Figs. 5.4 through 5.8.

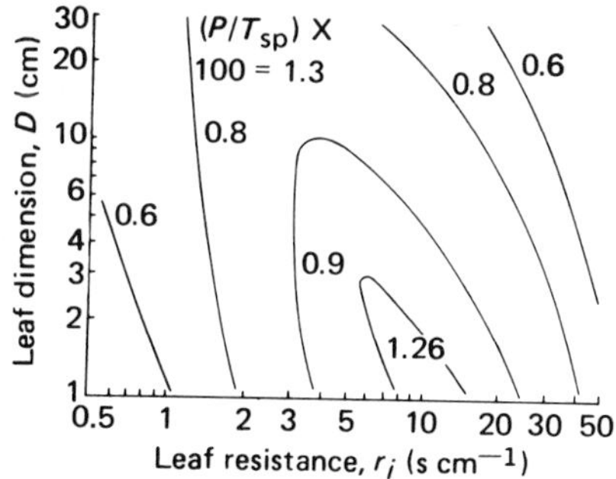

Fig. 5.7. Water-use efficiency expressed as the net weight of carbon dioxide fixed per weight of water expended times 10^2. Small leaves with moderately high resistance have greatest efficiency.

photosynthesis. Unless a plant were free from competition by other vegetation, this low productivity would not be expected to be competitive for space or available water. The possibility exists that a leaf might have such high water-use efficiency that it has little or no productivity since carbon dioxide uptake is reduced, together with decreased water loss.

Natural Communities

A composite nomogram of Figs. 5.5 through 5.7 depicts the zones of maximum photosynthesis and greatest water-use efficiency, together with transpiration values (Fig. 5.8). Actual leaf data taken from Taylor (1971) are plotted on the nomogram. The data represent the dominant vegetation of the study area and show a trend of

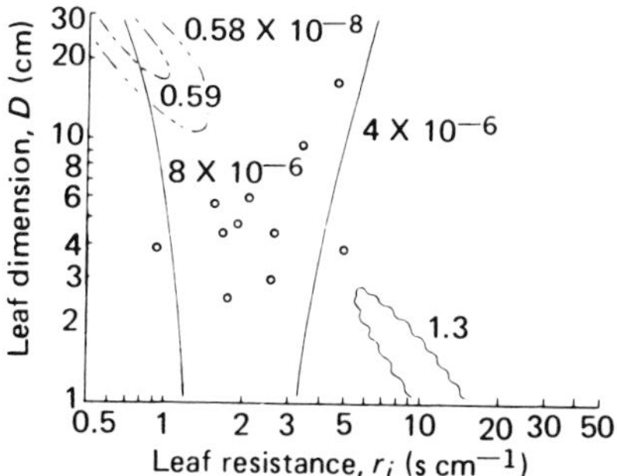

Fig. 5.8. Actual data from the University of Michigan's Biological Station area (July and August 1968) are depicted together with essential elements of Figs. 5.4 through 5.7 to produce a composite nomogram displaying transpiration (solid line), net photosynthesis (dotted-and-dashed line), and water-use efficiency (wavy line) for the actual leaves of each major species of the region (circles). (From Taylor, 1971.)

leaf dimension toward mesophyll–macrophyll class. Most individuals observed exhibited characters midway between greatest water-use efficiency and maximum photosynthesis.

The vegetation of a zone characterized by the climate specified for Figs. 5.4 through 5.7 has, according to the nomograms, no severe environmental limitations to leaf form. Leaves may be very small or very large without severely affecting productivity or being exposed to thermal danger. The tendency of the vegetation to develop toward a characteristic leaf dimension may be considered a response to structural economics or a tendency toward the optimal form that provides the greatest competitive advantage to the species. Harsh environments do, however, restrict leaf form to those characters suitable for survival.

Desert plants, unless capable of enduring leaf temperatures normally considered excessive, must be supplied with copious quantities of water, have small leaf dimension (Fig. 5.9), or be otherwise adapted to arid, harsh conditions with such mechanisms as seasonal dimorphism or defoliation. The nomogram for desert conditions (hot, dry air, moderate wind, bright sun, and high Q_{abs}) shows that the small leaves with very low resistance have optimum net photosynthesis. However, leaves are not expected to function at this maximum because of the excessively great transpiration rate in this energy regime. The leaf exposed to the climate specified for Fig. 5.9 is expected to exhibit resistance greater than 6 s cm^{-1} to maintain water loss within reasonable limits. Leaf temperatures in excess of 50°C are indicated for leaves with dimension greater than 1 cm unless resistance can be maintained at the lower limit of approximately 6 s cm^{-1}. Water-use efficiency is maximum for leaves of small dimension and resistance near 6 s cm^{-1}. The nomogram appears to define the suitable leaf dimension and resistance, but numerous exceptions are noted (Gates *et al.*, 1968) in cases of tolerance to plant temperatures above 50°C.

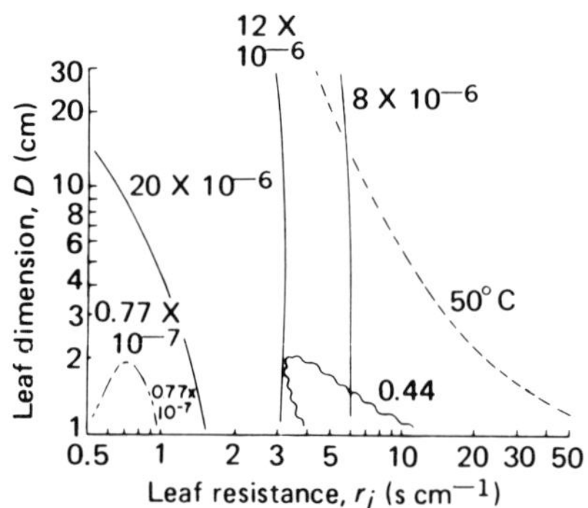

Fig. 5.9. Nomogram typical of desert conditions (hot, dry air; moderate wind; high Q_{abs}; and bright sun). The leaf responses to the environment shown are leaf temperature (dashed line) of 50°C (cooler temperatures occur at lower resistances and/or smaller dimension), transpiration rate (solid line), net photosynthesis (dotted-and-dashed line), which is at maximum for small leaves with low resistance, and the water-use efficiency ratio (wavy line), which is greatest for small leaves with resistance near 7 s cm^{-1}.

Biological Variables

Leaf form and resistance to the diffusion of water vapor are used herein as the primary biological parameters affecting leaf temperature, transpiration, and productivity. Numerous other biological parameters can be evaluated by use of the method discussed above. Leaf orientation, coloration (absorptivity to incident radiation), mesophyll thickness, integument, and the extent to which the surface is convex or concave are significant or potentially significant physical parameters of the leaf. The productivity of the leaf is also affected by numerous biochemical parameters related to the kinetics of photosynthesis and respiration (Gates *et al.*, 1972).

Orientation of the leaf blade significantly affects the amount of radiation absorbed. Numerous species exhibit differential sun–shade leaf orientation, and many have the facility to adjust leaf orientation throughout the diurnal cycle. Orientation with respect to insolation directly affects leaf temperature and light absorbed at the photosynthetic sites. Taylor observed that the variable orientation in *Erythrina berteroana* and *Cercis canadensis* served more as a mechanism for leaf-temperature control than as a water-conservation adaptation. Several other biological parameters were similarly evaluated (Taylor, 1971).

The significance of each biological parameter as it affects water usage, leaf temperature, and net photosynthesis can be evaluated by energy-budget methods.

Conclusions

The model presented in this paper can be used to evaluate the significance of leaf form as it affects water economy, productivity, and leaf temperature. The method has predictive value for agronomy and studies of natural communities. Potentially, the method can be beneficial for defining past climatic conditions from fossil leaf-dimension evidence.

The prime value of the model is its predictive capability, whereby the optimal leaf forms can be defined for any climate. The nature of optimal form is not fully understood because it is not altogether obvious to which effect natural selection tends. It is the task of the informed naturalist to discover this factor. I would encourage other investigators to define "optimal conditions" for themselves whether they wish to do so for maximum productivity or greatest water-use economy.

References

Bailey, I. W., Sinnott, E. W.: 1916. The climatic distribution of certain types of angiosperm leaves. Am. J. Bot. *3*, 24–39.

Benson, L., Phillips, E. A., Wilder, P. A., *et al.*: 1967. Evolutionary sorting of characters in a hybrid swarm: I. Direction of slope. Am. J. Bot. *59*, 1017–1026.

Brunig, E. F.: 1970. Stand structure, physiognomy and environmental factors in some lowland forests in Sarawak. Trop. Ecol. *11*, 26–43.

Cohen, D.: 1970. The expected efficiency of water utilization in plants under different competitive and selection regimes. Israel. J. Bot. *19*, 50–54.

Cooper, W. S.: 1922. The broad-sclerophyll vegetation of California: an ecological study of the chaparral and its related communities. Carnegie Inst. Wash. Publ. *319*, 1–24.

Gates, D. M.: 1968. Energy exchange in the biosphere. *In* Functioning of terrestrial ecosystems at the primary producing level (ed. F. E. Eckhardt), Proc. Copenhagen Symp., pp. 33–43. Paris: UNESCO.

———, Alderfer, R., Taylor, S. E.: 1968. Leaf temperature of desert plants. Science *159*, 994–995.

———, Johnson, H. B., Yocum, C. S., Lommen, P. W.: 1972. Geophysical factors affecting plant productivity. *In* Theoretical foundations of the photosynthetic productivity, pp. 406–419. Moscow: Nauka.

Idso, S. B., Jackson, R. D., Ehrler, W. L., Mitchell, S. T.: 1969. A method for determination of infrared emittance of leaves. Ecology *50*, 899–902.

Lange, O. L.: 1965. The heat resistance of plants, its determination and variability. Method. Plant Eco-physiol., Proc. Montpellier Symp., pp. 399–405. Paris: UNESCO.

———: 1967. Investigations on the variability of heat resistance in plants. *In* The cell and environmental temperature (ed. A. S. Troshin), pp. 131–141. Oxford: Pergamon Press.

Parkhurst, D. F., Loucks, D. L.: 1972. Optimal leaf size in relation to environment. J. Ecol. *60*, 505–537.

———, Duncan, P. R., Gates, D. M., Kreith, F.: 1968. Wind-tunnel modelling of convection of heat between air and broad leaves of plants. Agric. Meteor. *5*, 33–47.

Raunkiaer, C.: 1934. The life forms of plants and plant geography. New York: Oxford Univ. Press.

Taylor, S. E.: 1971. Ecological implications of leaf morphology considered from the standpoint of energy relations and productivity. Ann Arbor, Mich.: Univ. Michigan. Ph.D. diss.

———, Gates, D. M.: 1970. Some field methods for obtaining meaningful leaf diffusion resistances and transpiration rates. Oecologia Plantarum *5*, 105–113.

———, Sexton, O. J.: 1972. Some implications of leaf tearing in Musaceae. Ecology *53*, 143–149.

Webb, L. J.: 1959. A physiognomic classification of Australian rain-forests. J. Ecol. *47*, 551–570.

Yarwood, C. E.: 1961. Acquired tolerance of leaves to heat. Science *134*, 941–942.

6

Aspects of Predicting Gross Photosynthesis (Net Photosynthesis Plus Light and Dark Respiration) for an Energy-Metabolic Balance in the Plant

RANDALL S. ALBERTE, JOHN D. HESKETH, AND DONALD N. BAKER

Introduction

An important goal of most photosynthetic models is to predict photosynthetic energy available for plant metabolism. Earlier workers studying quantum efficiency of conversion of light to photosynthetic energy encountered considerable philosophical and experimental difficulties, which have not yet been resolved (Rabinowitch, 1945; Kok, 1972). Oxygen or carbon dioxide exchange of intact organs is typically used as a measure of photosynthetic production of energy. However, difficulties are created by consumption or generation of oxygen, carbon dioxide, and energy by nonphotosynthetic metabolism within a plant organ. Experimentally, the assumption is and has been that light respiration is equivalent to the dark respiration of a photosynthetic system shortly after the light has been shut off. The frustrating question for many years has been: Is light respiration different from dark respiration of previously irradiated plant organs?

We shall discuss the following points in relation to this question:

1. Sources and sinks for O_2, CO_2, and energy in the cell, the leaf, or the plant.
2. Methods for predicting the photosynthetic reduction of CO_2 in plant communities.
3. Methods for predicting respiration in the light and the dark.

The objective of this review will be to summarize as simply as possible information concerning light respiration for an audience of generalists modeling photosynthesis or plant growth. We will not document every statement; rather, a few relevant reviews and some pertinent papers will be cited for further reading.

Intracellular Energy and Carbon Dioxide Exchange

Using the terminology of Oaks and Bidwell (1970), two types of intracellular compartments can be defined. The most familiar types are the tight compartments. These are the discrete, membrane-bounded organelles of the cell. The other types are the loose compartments. These are not structurally defined, but rather represent cytoplasmic[1] loci of specific intermediary metabolites and metabolic events (Oaks and Bidwell, 1970).

A summary of tight intracellular compartments, their structural characteristics, metabolic functions, and energy relationships is given in Table 6.1. There are rather specific associations and interactions between these compartments. For instance, specific associations of glyoxisomes and chloroplasts are found in most photosynthetic cells. The chloroplast–glyoxisome complexes are relatively immobile and localized primarily adjacent to the plasmalemma. They are bathed in streaming cytoplasm, which contains the loose compartments.

Since the nature and types of exchange between these compartments are quite complex, resistance determinations are difficult to obtain, although they are important to photosynthetic models. For example, a resistance is often hypothesized between glyoxisomes and chloroplasts. This resistance is made up of the bounding membrane resistances, the organellular sap (i.e., chloroplast stroma) resistances, and perhaps more importantly, the biochemical resistances. Unfortunately, the magnitudes of biochemical resistances are the most difficult to assess (Raven, 1970). Within this framework intracellular CO_2 flux and energy exchange will be considered.

The intracellular sources of CO_2 in the light include the mitochondria, glyoxisomes (the site of so-called photorespiration), and the cytoplasmic loose compartments involved in pentose[2] and hexose oxidations. In addition, there are large pools of CO_2 in the cytoplasm of extracellular origin. Therefore, it appears that there are at least three, but probably more, sources of CO_2 for fixation by chloroplasts. Another potential CO_2 source may arise from intermediary metabolism within the chloroplast itself (Givan and Leech, 1971). In addition to chloroplastic CO_2 fixation, several nonphotosynthetic carboxylations (i.e., associated with amino acid metabolism) also contribute an important component to the plant's carbon balance.

The CO_2 from the various sources is subject to different degrees of resistance in its path to the chloroplast. For example, diffusion barriers to CO_2 exchange in the chloroplast–glyoxisome complexes are the respective membranes of the two organelles. The CO_2 arising from mitochondria or cytoplasmic compartments are subject to cytoplasmic resistances in addition to membrane resistances. Interesting, in this respect, is the suggested facilitation of photosynthate transport out of or

[1] Loose compartments are not confined to the cytoplasms but are found as well in the mobile phases of tight compartments. However, we shall restrict our considerations of loose compartments to those with cytoplasmic loci.

[2] The CO_2 contribution of the oxidative pentose phosphate shunt in the light is not fully clear.

Table 6.1 Structural Characteristics, Metabolic Functions, and Energy Relationships of Plant Intracellular Compartments[a]

Intracellular Compartments	NADH, NADPH	ATP	CO_2	O_2	Function
Cytoplasm	+	+	+	0	Contains various metabolic compartments, including: glycolysis and oxidative pentose phosphate shunt; streaming transport of ions and intermediary metabolites, CO_2, O_2; PEP carboxylase (C_4 carbon pathway) and carbonic anhydrase
Endoplasmic reticulum	?	−	−	0	Partitions cytoplasm; backbone for polyribosomes; functions in membrane synthesis
Mitochondria	+	−	+	0	*Mobile phase:* Krebs citric acid cycle; NADH; fatty acid metabolism; amino acid synthesis
	−	+	0	−	*Stationary phase* (membrane bound): cytochrome electron transport chain; oxidative phosphorylation generating ATP
Chloroplast	−	−	−	0	*Mobile phase:* carbon reduction (C_3 and C_4 pathways; reductive pentose phosphate shunt; intermediary metabolism (amino acids, fatty acids, lipids, etc.); starch synthesis and storage
	+	+	0	+	*Stationary phase* (membrane bound): light reactions—photosystem I and II; photophosphorylation; splitting of water
Glyoxisome	−	−	+	−	Glycolate metabolism associated with photosynthesis; site of photorespiration; amino acid synthesis; associated with C_3 carbon pathway
Peroxisome	−	−	?	−	Glycolate metabolism associated with fatty acid and lipid synthesis
Ribosome	?	−	0	0	Protein synthesis; associated with rRNA, tRNA, and mRNA; found in chloroplasts and mitochondria as well as in the cytoplasm
Lysosome or spherosome	?	?	?	?	Degradation of cellular components, perhaps associated with differentiation; contain hydrolytic enzymes
Golgi apparatus	?(−)	?(−)	0	0	Cell wall and membrane synthesis
Amyloplast	0	?(−)	0	0	Starch storage

[a] +, source; −, sink; 0, neither a sink nor a source. (For further details, see Clowes and Juniper, 1968; Pridham, 1968.)

intermediates into the chloroplast through a membranous network called the peripheral reticulum. The peripheral reticulum is found primarily in chloroplasts possessing the C_4 pathway for carbon reduction, and probably contributes to the higher photosynthetic efficiency of these chloroplasts by allowing for rapid removal of photosynthetic end products (Rosado-Albeiro *et al.*, 1968; Laetsch, 1970; Gracen *et al.*, 1972).

Probably more important than these physical diffusion barriers are the biochemical barriers that hinder CO_2 flux between compartments. These resistances probably account for the noted greater intracellular resistances for gas-phase flux (Raven, 1970). More especially, to maintain a favorable diffusion gradient toward and into the chloroplast, it is essential that the metabolic pools of CO_2 in the chloroplast and the surrounding cytoplasm be kept low. This can be accomplished by lowering the biochemical resistance to CO_2 fixation (i.e., increase the V_0 for the carboxylating enzymes), or by increasing the number or kinds of carboxylation sites. In C_4 plants, at least, this has been partially accomplished by increasing the kinds of chloroplastic carboxylation sites. Further, it appears that some C_4 plants have in addition to chloroplastic carboxylation sites an active cytoplasmic site for CO_2 fixation (Coombs and Baldry, 1972). This carboxylation site is capable of maintaining low levels of cytoplasmic CO_2, which in turn favor CO_2 influx. There is some evidence that carbonic anhydrase may also serve an important role in the control of favorable CO_2 gradients for photosynthesis (Grahm and Reed, 1971; Poincelot, 1972). The specific nature of these reactions will be discussed in more detail later.

Consideration of the photosynthetic reassimilation of light-produced respiratory CO_2 in addition to photosynthetic CO_2 exchange is important not only to examinations of carbon balance, but also to those of cellular energy relationships in plants. The cellular energy relationships of photosynthetic cells in the light and in the dark will be examined to try to integrate energy sinks and sources with photosynthetic and respiratory metabolism.

In the dark, reducing power (NADPH or NADH) and ATP are generated by mitochondria and the cytoplasmic glycolytic and oxidative pentose phosphate shunt compartments. These energy supplies are used by the synthetic machinery for growth, maintenance, and storage. In addition, these pools are a source of energy to drive such specific metabolic events as active transport of ions and metabolites, and to support nitrate metabolism (Beevers and Hageman, 1972). Respiratory energy is also important in carbon metabolism in Crassulacean (CAM) plants whose conversion of organic acids to simple sugars in the dark is energy-dependent (Hatch *et al.*, 1971). It is believed that the chloroplast is capable of producing a sufficient amount of ATP (via photophosphorylation) to support not only photosynthetic carbon reduction, but also chloroplastic syntheses of small molecules (i.e., fatty acids, amino acids, etc.), and for transport to cytoplasmic energy sinks in quantities in excess of 20 percent of the total ATP generated in the light (Krause, 1971; Raven, 1972a). There is, though, some discrepancy in the literature concerning the stoichiometry of photophosphorylation and carbon reduction (Hall and Evans, 1972; Kok, 1972; Schurmann *et al.*, 1972; Heber, 1973). Photosynthetically generated reducing power (NADPH), on the other hand, is not available to the cytoplasm (see, further, Raven, 1972a), although there is evidence that reducing equivalents from organic acids are available to the cytoplasm. The light-produced reductant pool is utilized for various chloroplastic syntheses (Givan and Leech, 1971; Lex and Stewart, 1973; Beevers and Hageman, 1972).

Glycolate metabolism associated with photosynthesis carbon cycling, so-called photorespiration, cannot be ignored in energy considerations of photosynthetic cells. Photorespiration places a heavy demand on chloroplastic energy sources as well as on photosynthetic carbon balance (Hatch *et al.*, 1971; Ludwig and Canvin, 1971; Zelitch, 1971; Raven, 1972b).

The contribution of respiratory energy in the light to the cellular energy resources is not fully clear at present. There is support for the notion that mitochondrial oxidative phosphorylation is little if at all inhibited by photophosphorylation (Reid, 1970; Raven, 1972a). If so, then there is a very large potential ATP source available in the light of photosynthetic cells (see also Kok, 1972). Clearly, under photosynthetic conditions the cytoplasm is a major sink for both mitochondrial and chloroplastic ATP, although the reducing power remains compartmentalized. Reducing power from oxidative pentose metabolism and glycolysis is probably the major source of reducing power to the cytoplasm.

Thus respiratory and photosynthetic ATP as well as CO_2 are available to the cytoplasm in the light. The CO_2 and the energy can be recycled efficiently despite the various resistances between the major cellular compartments, most especially the chloroplast, mitochondria, glyoxisome, and cytoplasm. It appears that at least carbon recycling has a higher quantum efficiency than direct photophosphorylation in systems where glycolate metabolism is minimal as in C_4 plants (Raven, 1970, 1972a, 1972b). Therefore, the complex metabolic and energy interactions of the various cellular compartments with their respective resistances constitute a significant force in plant metabolic–energy balance. These interactions and resistances demand consideration in models of plant growth and productivity.

Intercellular Energy and Carbon Dioxide Exchange

Detailed discussions of energy and CO_2 exchange between cells and organs are available (Raven, 1970, 1972a, 1972b; Ludlow and Jarvis, 1971; D'Aoust and Canvin, 1973); therefore, considerations here will be limited to the more general aspects of intercellular and interorgan exchange. Almost all types of exchange between cells (that is, of metabolites and ions) require energy. Exchange between organs apparently is also energy-dependent. Phloem companion cells and sieve elements are rich in ATP-generating mitochondria, and in some plants plastids are found as well (MacRobbie, 1971; Gilder and Cronshaw, 1973; Behnke, 1973).

The form of CO_2 exchange between cells and organs is primarily gaseous CO_2, bicarbonate, or some other small organic molecule (i.e., triose phosphates, organic acids, sucrose, etc.). The form of energy exchanged, on the other hand, is not as ATP or reducing power. It is, instead, in the form of bond energy of organic molecules, which serve as the substrates for respiratory metabolism in the light and in the dark. Thus, in most cases all the energy exchange between cells and organs is indirect. An exception to this may exist in cells which have specialized structures called plasmadesmata, which function in cell-to-cell exchange (Buvat, 1969).

The contribution of given cells or organs to the energy–metabolic balance of the plant is further complicated by cells and tissues that show division of metabolic work. For example, it is impossible to evaluate the respiratory role of the roots in the plant's energy balance without considering the nature and sources of substrate, namely, the photosynthetic tissue, for that energy production. Or, what is the role of stem photosynthesis in the carbon balance and productivity of a given plant (Adams *et al.*, 1967; Adams and Strain, 1969)? Similarly, it is equally difficult to evaluate the contribution of a given cell type to the energy and metabolic relationships of another cell type.

Plants possessing the C_4 pathway for carbon reduction provide a good system in which to examine some aspects of these problems. The vascular tissue of C_4 plants is typically surrounded by photosynthetic cells rich in chloroplasts. These chloroplasts probably function in reassimilation of respiratory as well as photosynthetically produced CO_2 and bound carbon; C_3 plants do not have the benefits of such cellular and organellular associations. In C_3 plants it is likely that the vascular CO_2 diffuses into the leaf air spaces and contributes to some of the gas-exchange aspects of the photorespiration phenomena characteristic of these plants.

At the biochemical level, the photosynthetic light and dark reactions differ in the two different cell types of C_4 leaves. The mesophyll cells of most C_4 plants probably contain the majority of the PEP carboxylase which fixes CO_2 into organic acids, whereas the bundle sheath cells are deficient in this enzyme and are the major site of C_3 carbon reduction and starch synthesis (Hatch *et al.*, 1971). Further, these cells show low activity of enzymes associated with glycolate metabolism characteristic of plants with little or no photorespiration.

It is hypothesized that the early products of photosynthesis in these plants are translocated quickly from the mesophyll cells to the bundle sheath cells. The translocation process may be facilitated by the peripheral reticulum characteristic of C_4 chloroplasts, although not exclusively associated with C_4 chloroplasts. In the mesophyll chloroplasts, the CO_2 is fixed by PEP carboxylase via the following reaction:

$$C_3 + CO_2 \rightarrow C_4$$

It is likely that fixation is analogous in Crassulacean (CAM) plants, except that there are temporal differences in fixation (Hatch *et al.*, 1971).

The C_4 product is then shuttled to the bundle sheath chloroplasts, where it is decarboxylated. The CO_2 generated is then refixed via the Calvin cycle typical of exclusively C_3 plants as follows:

$$C_5 + CO_2 \rightarrow 2C_3$$
RuDP carboxylase

and/or

$$C_5 + O_2 \rightarrow C_3 + C_2$$
RuDP carboxylase

The C_3 product in either reaction can be utilized for the formation of starch, cellulose, or other complex carbohydrates, as well as for respiratory metabolism leading to the production of ATP and reducing power. The C_2 product, on the other hand, is believed to be the major substrate for photorespiration (at least in C_3 plants). It is transported out of the chloroplast to either the mitochondrion or the glyoxisome, which contain enzymes associated with C_2 or glycolate metabolism. If utilized by the glyoxisome, then by a series of reactions, one of which is light-dependent, CO_2 is generated, and perhaps glycine and serine are also produced, at the expense of reducing power and ATP. If utilized by the mitochondrial compartment, then the C_2 component may enter into fat metabolism, amino acid metabolism via the glyoxylate shunt of the Krebs citric acid cycle, or into regeneration of carbohydrate via reverse glycolysis. In either case, the metabolism of the C_2 product requires energy, but consumes O_2 and is light-dependent only in the former case, hence the name "photorespiration." If C_4 plants do have photorespiratory metabolism (although it is not easily detected by standard methods that measure CO_2 since it is probably rapidly refixed), the energy relationships of such metabolism would be quite significant to the energy–metabolic balance in such plants. These parameters must be considered to evaluate reasonably photosynthesis and plant growth for modeling purposes. Further, in terms of carbon balance, the lack of detectable photorespiratory release of CO_2 presents a more favorable picture of photosynthetic efficiency than that found in photorespiratory plants. However, in terms of energy balance, the C_4 plant is likely not more energetically efficient. First, it is more expensive to fix CO_2 via the C_4 pathway in addition to the C_3 pathway; and second, if photorespiratory metabolism is present although there is no apparent release, then this metabolism places a fairly large energy demand on the system, which has not been considered. Clearly, such a demand must be considered in models of photosynthesis and plant growth.

Evaluations of the differences in the light reactions of C_3 and C_4 plants, sun and shade leaves of the same plant, and dimorphic chloroplasts of the same organ are necessary for estimations of the energy–metabolic balance in plants as well. As the light reactions in photosynthesis are currently envisaged (Boardman, 1968; Arnon, 1971), there are two light-dependent reactions, photosystem I and II, in series. These reactions involve the transduction of light energy to electrochemical energy resulting in the splitting of water, the evolution of O_2, and coupled electron flow leading to the production of ATP and reducing power. The efficiencies and the interrelationships of these reactions, too complex to detail here, depend on many factors, of which environmental ones are prime (Kok, 1972). Let it suffice to give just a few examples to point out some of the problems.

The bundle sheath chloroplast of some monocotyledonous C_4 plants are deficient in grana (stacked photosynthetic lamelae) and believed to be deficient in photosystem II activity (i.e., water splitting and production of NADPH). This deficiency has dramatic effects on the relationship between the two photosystems as well as on the energy production efficiency of such chloroplasts. How do these cells support carbon fixation with limited availability of reducing power?

Sun leaves typically have smaller photosynthetic unit sizes (i.e., less chlorophyll for light harvesting per energy transduction site) than shade leaves, making them very much less efficient in light utilization at low light intensities. What is the consequence of these differences on the energy budget and carbon balance of the plant? In this respect it is important to realize that in general photosynthetic unit size is not a static characteristic of a given chloroplast type or species. Instead, it is a very flexible component that responds dramatically to environmental modulation. This property of the photosynthetic lamellae responsible for energy production necessitates examinations of environmental factors such as light intensity and temperature on photosynthesis, the diurnal fluctuations of photosynthesis, and the seasonal changes in photosynthetic capacity as it relates not only to environmental factors but also to the life cycle of the plant. There is little doubt that the efficiency of light capture and its subsequent transduction to chemical energy for plant growth and development is fundamental to any considerations that attempt to model the energy–metabolic interactions of photosynthesis or plant growth.

Methods for Estimating Photorespiration

Ludlow and Jarvis (1971) list various methods for estimating photorespiration:

1. CO_2 efflux into CO_2-free air, with diffusion theory to correct for reassimilation; $^{14}CO_2$ into the leaf versus CO_2 influx.
2. Short-term influx of $^{14}CO_2$ into the leaf versus CO_2 influx.
3. CO_2 influx upon removal of O_2 from the air versus CO_2 influx in air.
4. Diffusion resistance theory, the above, and photosynthetic response to CO_2 concentration.

Method 1, following CO_2 efflux, requires extrapolation to normal CO_2 conditions, plus corrections for reassimilation based on theory. Method 2 requires corrections from 4, inasmuch as diffusion theory predicts inhibition of influx of CO_2 to chloroplasts from outside the leaf because of distances between respiratory sources and chloroplasts. Method 3 needs to be corrected for the inhibition of photosynthesis by oxygen (e.g., $C_5 \rightarrow C_3 + C_2$, no net fixation of CO_2), as well as the effect listed under method 2. Method 4 involves many assumptions that oversimplify processes discussed in earlier sections.

It is generally concluded that photorespiration cannot be estimated accurately; Ludlow and Wilson (1972) report two-fold differences in estimates of photorespiration between method 1 and methods 3 and 4.

Methods for Measuring Photosynthetic Energy Available for Plant Metabolism

There are now four standard methods for estimating photosynthate per plant, listed in Table 6.2.

The chamber method involves measurement of respiration after a period of photosynthesis. Dark respiration is a function of photosynthesis (see below),

Table 6.2 Methods for estimating Plant Photosynthate Production with the Corresponding Respiratory and Synthesis Corrections

Method	Respiratory Correction[a]	Synthesis Corrections[a]
1. Chamber (CO_2 exchange)	Measure light, then dark	Hourly
2. Flux (CO_2 profiles)	Measure day and night, organ by organ	Hourly
3. Leaf model, canopy model	Measure light, then dark	Hourly
4. Dry-weight sampling	Develop a respiration model	Daily or weekly

[a] Corrections are for changes in protein, starch, or fat synthesis.

provided that photosynthesis has been proceeding at a fairly constant rate for a period of time.

Often roots and soil are excluded from the chamber system; therefore, the respiration data are only good for generating a gross photosynthetic rate.

These two measurements of net photosynthesis and subsequent dark respiration must be made together for a range of environmental and physiological (stage of growth, acclimation, and water stress effects) conditions.

The flux method does not account for plant and soil respiration; thus respiration must be determined independent of the method. Vertical flux rates are more difficult to determine than chamber exchange rates, but once systematic errors in flux determinations are resolved and respiration is accounted for, the method may well be used to calibrate more extensive information from models based on chamber data or leaf-canopy characteristics.

The leaf method requires a vegetation:light-interception model, with consideration of leaf physiology (acclimation, age, nutrition, water stress, etc.). Nonleaf photosynthesis should also be accounted for (Adams *et al.*, 1967; Adams and Strain, 1969). The dry-weight method requires an estimate of the carbon equivalent of a unit of dry matter, as well as respiration data.

In the estimation of gross photosynthesis, it is assumed that the plant model contains an independent respiration subsystem, based upon "paper biochemistry" and synthetic behavior or the relative proportions of photosynthate being converted to fat, protein, carbohydrate, or cellulose. Shifts in synthetic behavior greatly affect estimates by these methods. Shifts do occur over the course of a day, as well as during the life cycle of the plant. Such shifts must be monitored.

Water stress during the day greatly complicates the application of "paper biochemistry" to the problem (Kozlowski, 1972). Cell expansion and RNA synthesis are suppressed at high water potentials (-3 to -4 bars). In comparison, photosynthesis is suppressed at -8 to -12 bars. However, translocation and incorporation of carbohydrate into storage organs are inhibited at much lower water potentials than photosynthesis. Obviously different kinds of synthesis are affected differently by water stress. At the moment, actual diurnal measurements of respiration are needed to check estimates from "paper biochemistry." The effects of wind on respiration also need to be accounted for and explained (Todd *et al.*, 1972).

Problems in Measuring Dark Respiration

In 1876 and 1881, Borodin reported that dark respiration, based on CO_2 efflux and O_2 influx, was greater for leaves with an immediate history of light exposure or photosynthetic activity than for leaves previously in the dark (see p. 491, Weintraub, 1944; pp. 561–571 in a chapter on photosynthesis and respiration by Rabinowitch, 1955). This phenomenon was confirmed over and over and aspects of the previous history of the leaf were studied in great detail (Weintraub, 1944). Rabinowitch (1945) also defines and uses the terms "photorespiration" and "light respiration" as effectively as more recent writers.

After the lights are shut off, most C_3 plants exhibit a burst of CO_2, whereas the respiratory efflux of CO_2 from C_4 leaves builds up slowly over several minutes. This phenomenon may be associated with changes in intermediates of the carbon pathways associated with sudden darkness and stoppage of photosynthesis, or with changes in stomatal resistance (Ludlow and Jarvis, 1971). In any case, respiratory efflux of CO_2 seems to be meaningful only after reaching a steady state, and may possibly require a stomatal correction.

Conclusions

Thus far it is impossible to measure directly light respiration or energy production from photophosphorylation independent of carbon reduction. It is not known if part of the light respiration can be ignored in a carbon-energy budget (if it does not interfere with measurements of gas exchange). At least three photosynthate carbon pathways vary in activity in different kinds of cells and in different environments. The activity of the light reactions that synthesize energy for metabolic use varies among photosynthetic cells.

The objective of a model is to summarize information. We propose that review papers associated with process models be generated from time to time, with discussions of how relevant problems might be handled. We do not mean to be critical of process modeling; rather, the models can be useful as educational and logical tools, if care is taken to convey to a general audience, often including non-biologists, as simplified a description as possible of the required assumptions.

Also, the future direction of modeling a subsystem should depend upon careful consideration of experimental and logical problems, such as we present here for the photosynthetic subsystem. As many methods as possible for measuring a process should be made in the same system to cross-check results. We hope that these discussions of process models will soon evolve textbooks for teaching biology at the undergraduate level.

References

Adams, M. S., Strain, B. R.: 1969. Seasonal photosynthetic rates in stems of *Cercidium floridum* Benth. Photosynthetica *3*, 55–62.

———, Strain, B. R., Ting, I. P.: 1967. Photosynthesis in chlorophyllous stem tissue and leaves of *Cercidium floridum*: Accumulation and distribution of ^{14}C from $^{14}CO_2$. Plant Physiol. *42*, 1797–1799.

Arnon, D. I.: 1971. The light reactions of photosynthesis. Proc. Natl. Acad. Sci. *68*, 2883–2892.

Beevers, L., Hageman, R. H.: 1972. The role of light in nitrate metabolism. *In* Photophysiology: current topics in photobiology and photochemistry (ed. A. C. Giese), vol. 7, pp. 85–114. New York: Academic Press.

Behnke, H.-D.: 1973. Plastids in sieve elements and their companion cells. Investigations on monocotyledons, with special reference to *Smilax* and *Tradescantia*. Planta *110*, 321–328.

Boardman, N. K.: 1968. The photochemical systems of photosynthesis. *In* Advances in enzymology (ed. F. F. Nord), vol. 30, pp. 1–79. New York: Wiley-Interscience.

Buvat, R.: 1969. Plant cells: an introduction to plant protoplasm. New York: McGraw-Hill.

Clowes, F. A. L., Juniper, B. E.: 1968. Plant cells. Oxford: Blackwell.

Coombs, J., Baldry, C. W.: 1972. C-4 pathway in *Pennisetum purpureum*. Nature *238*, 268–270.

D'Aoust, A. L., Canvin, D. T.: 1973. Effect of oxygen concentration on the rates of photosynthesis and photorespiration of some higher plants. Can. J. Bot. *51*, 457–464.

Gilder, J., Cronshaw, J.: 1973. Adenosine triphosphatase in the phloem of *Cucurbita*. Planta *110*, 189–204.

Givan, C. V., Leech, R. M.: 1971. Biochemical autonomy of higher plant chloroplasts and their synthesis of small molecules. Biol. Rev. *46*, 409–428.

Gracen, V. E., Jr., Hilliard, J. H., Brown, R. H., West, S. H.: 1972. Peripheral reticulum in chloroplasts of plants differing in CO_2 fixation pathways and photorespiration. Planta *107*, 189–204.

Grahm, D., Reed, M. L.: 1971. Carbonic anhydrase and the regulation of photosynthesis. Nature *231*, 81–83.

Hall, D. O., Evans, M. C. W.: 1972. Photosynthetic photophosphorylation in chloroplasts. Sub-cell. Biochem. *1*, 197–206.

Hatch, M. D., Osmond, C. B., Slatyer, R. O. (eds.): 1971. Photosynthesis and photorespiration. New York: Wiley-Interscience.

Heber, U.: 1973. Stoichiometry of reduction and phosphorylation during illumination of intact chloroplasts. Biochim. Biophys. Acta *305*, 140–152.

Kok, B.: 1972. Efficiency of photosynthesis. *In* Horizons of bioenergetics (eds. A. San Pietro, H. Gest), pp. 153–170. New York: Academic Press.

Kozlowski, T. T. (ed.): 1972. Water deficits and plant growth, vol. 3. New York: Academic Press.

Krause, G. H.: 1971. Indirekter ATP-transport zwischen Chloroplasten und Zytoplasma während der Photosynthesis. Z. Pflanzenphysiol. *65*, 13–23.

Laetsch, W. M.: 1970. Chloroplast structural relationships in leaves of C_4 plants. *In* Advances in C_4 photosynthesis and photorespiration in plants (ed. R. O. Slatyer). Canberra: Austral. Acad. Sci.

Lex, M., Stewart, W. D. P.: 1973. Algal nitrogenase, reductant pools and photosystem I activity. Biochem. Biophys. Acta *292*, 436–443.

Ludlow, M. M., Jarvis, P. G.: 1971. Methods for measuring photorespiration in leaves. *In* Plant photosynthetic production, manual of methods (eds. Z. Sestak, J. Catsky, P. G. Jarvis), pp. 294–315. The Hague: Dr. W. Junk, N.V. Publ.

———, Wilson, G. L.: 1972. Photosynthesis of tropical pasture plants. IV. Basis and consequences of differences between grasses and legumes. Austral. J. Biol. Sci. *25*, 1133–1145.

Ludwig, L. J., Canvin, D. T.: 1971. The rate of photorespiration during photosynthesis and the relationship of the substrate of light respiration to the products of photosynthesis in sunflower leaves. Plant Physiol. *48*, 712–719.

MacRobbie, E. A. C.: 1971. Phloem translocation. Facts and mechanisms: a comparative study. Biol. Rev. *46*, 429–482.

Oaks, A., Bidwell, R. G. S.: 1970. Compartmentation of intermediary metabolites. Ann. Rev. Plant Physiol. *21*, 43–66.

Poincelot, R. P.: 1972. Intracellular distribution of carbonic anhydrase in spinach leaves. Biochem. Biophys. Acta *258*, 637–642.

Pridham, J. B. (ed.): 1968. Plant cell organelles. New York: Academic Press.

Rabinowitch, E. I.: 1945. Photosynthesis and related processes. New York: Wiley-Interscience.

Raven, J. A.: 1970. Exogenous inorganic carbon sources in plant photosynthesis. Biol. Rev. *45*, 167–221.

———: 1972a. Endogenous inorganic carbon sources in plant photosynthesis. I. Occurrence of the dark respiratory pathways in illuminated green cells. New Phytol. *71*, 227–247.

———: 1972b. Endogenous inorganic carbon sources in plant photosynthesis. II. Comparison of total CO_2 production in the light with measured CO_2 evolution in the light. New Phytol. *71*, 955–1014.

Reid, A.: 1970. Energetic aspects of the interaction between photosynthesis and respiration. *In* Prediction and measurement of photosynthetic productivity, pp. 231–246. Wageningen: Pudoc.

Rosado-Albeiro, J., Weir, E., Stocking, C. R.: 1968. Continuity of the chloroplast membrane systems in *Zea mays* L. Plant Physiol. *43*, 1325–1329.

Schurmann, P., Buchanan, B. B., Arnon, D. I.: 1972. Role of cyclic photophosphorylation in photosynthetic carbon dioxide assimilation by isolated chloroplasts. Biochim. Biophys. Acta *267*, 111–124.

Todd, G. W., Chadwick, D. L., Tsai, S.: 1972. Effect of wind on plant respiration. Physiol. Plantarum *27*, 324–346.

Weintraub, R. L.: 1944. Radiation and plant respiration. Bot. Rev. *10*, 383–459.

Zelitch, I.: 1971. Photosynthesis, photorespiration, and plant productivity. New York: Academic Press.

PART II

Extreme Climate and Plant Productivity

Introduction

David M. Gates

As has been evident for many years, our understanding of plant response to environmental factors will be advanced more rapidly by the study of plants growing in extreme environments than the study of plants growing in moderate environments. This attitude has been strongly supported by the discovery of plants with the C_4 or dicarboxylic acid or beta carboxylation pathway of photosynthesis as compared with the Calvin cycle C_3 plants with the conventional reductive pentose phosphate pathway. The C_4 pathway metabolism appears to give plants a photosynthetic advantage under conditions of high light intensity, high temperature, and limited water supply, because it avoids the inhibitory effect of oxygen on net CO_2 uptake. Experiments with plants that have one or the other kind of metabolic pathway are necessary to determine if various intrinsic biochemical properties of rate-limiting enzymes exist which produce an adaptive advantage to warm, dry habitats. It now appears that C_4 plants contain the normal Calvin cycle pathway in the bundle sheath cells and the dicarboxylic acid system in the mesophyll cells. This information adds credence to the suggestion that the ultimate resolution of many fundamental ecological questions concerning plants will by necessity be achieved at the biochemical level when combined with the appropriate biophysical mechanisms and models.

To understand which plant parameters and properties must be included in a theoretical model that properly represents the manner in which a plant truly functions, one must have sufficient field data and information. These data must include information concerning plant anatomy, morphology, physiology, and ecology; about the daily and seasonal variation of certain plant characteristics, particularly photosynthetic capacity, compensation point, heat resistance, stomate opening, and water potential; and about the growth and development of the plant and the extent to which various environmental factors influence photosynthetic capacity, photosynthetic temperature optimum, stomate size and degree of opening, size of leaf, absorptivity of leaf, etc. These are extremely complicated problems, but their complexity must not prevent us from gathering the pertinent data and formulating theoretical models.

Hyrum Johnson discusses some of the important plant properties that affect photosynthesis and water loss. He points out that "periods of rapid carbon assimilation are invariably periods of high transpiration." Two quantities are of particular interest when considering the effectiveness of a plant in its response to various environmental factors. The one is a photosynthetic efficiency defined as the ratio P/P_m, where P is the actual photosynthetic rate and P_m is the maximum photosynthetic rate that would occur if it were limited only by stomatal diffusion. The other is the transpiration ratio defined as the ratio E/P, where E is the transpiration rate and P is the photosynthetic rate, as before. Johnson discusses the C_3 and C_4 metabolic pathways and then describes in considerable detail the behavior of plants with a Crassulacean acid metabolism (CAM). These plants are characteristically perennials with succulent or thick photosynthetic organs.

The paper by Lange, Schulze, Kappen, Buschbom, and Evenari reveals several very important relationships between various plant properties and environmental conditions. This paper demonstrates magnificently the importance of gathering field data with enormous care and precision. The extremely important discovery concerning the influence of the relative humidity of the air on stomate opening and hence on the uptake of carbon dioxide is described in this paper. The diurnal response of this effect is strong and significant. The data contained in this paper were obtained under well-controlled and very well described conditions.

Plants, of course, grow in communities that form more or less compact canopies in response to available light, moisture, carbon dioxide, and nutrients. Although a great deal of our effort in biophysical ecology has been devoted to understanding the autecology of single leaves, it is equally important that we understand the functional behavior of leaves throughout a plant canopy. Ronald Alderfer obtained his Ph.D. degree at Washington University while working with our biophysical ecology group at the Missouri Botanical Garden. At that time he worked on the problem of temperature and transpiration rates of leaves in plant canopies and their utilization of sunlight. He has continued this research at the University of Chicago in association with some of his students there. The paper by Alderfer gives us insight into the problems of productivity in plant canopies.

Methods of photosynthetic measurement are discussed by Boyd Strain. He points out some of the precautions that must be taken to get accurate, valid measurements. The environmental variables that must be accurately measured with time include radiation, air temperature, wind, humidity, carbon dioxide concentration, and soil-water potential. When plants are placed in metabolic chambers for photosynthetic measurements, the same environmental variables must be carefully determined. Strain describes methods of measurement and their application to woody perennials in the field. He reports maximum photosynthetic rates for many species of desert plants and concludes that evergreen species have lower net photosynthetic rates than do more mesophytic deciduous plants. Maximum net photosynthesis usually occurred early in the day. Strain reports valuable field measurements. One wonders whether the measured maximal photosynthetic rates for these plants are absolute maxima or only relative maxima. Certainly, as

Strain points out, the maximum photosynthetic rate a plant can achieve depends upon the time of year, its age, and its history of growth.

In the final paper in this section, Lloyd Dunn describes field data for the photosynthetic rates of sclerophyllic plants in Mediterranean climates and shows that stomatal control in response to environmental factors is extremely significant. Stomatal closure is suggested as a dominant factor limiting CO_2 uptake by a leaf under conditions of severe drought. In particular, the influence of temperature on CO_2 uptake and leaf resistance is discussed with regard to field data from warm conditions. Once again, it is essential to combine theory with precise laboratory and field measurements for an understanding of phenomena as complex as these.

7

Gas-Exchange Strategies in Desert Plants

HYRUM B. JOHNSON

Introduction

A close correspondence between the rates of transpiration and photosynthesis in higher land plants has been routinely observed over the past 30 years (Heath, 1969). A ready explanation for this correspondence lies in the existence of a common diffusion path for the water vapor as it passes from the inside of the plant to the atmosphere in transpiration, and for the carbon dioxide, which passes from the atmosphere to the inside of the plant, where it is fixed in photosynthesis. The correlation between these two processes is not absolute, however, since carbon dioxide must also diffuse through a liquid phase after it reaches the walls and cytoplasm of the internal photosynthetic cells as well as meet the limitations of the biochemical steps involved in carbon fixation. Transpiration, then, is primarily a biophysical process, whereas photosynthesis is a combination of biophysical and biochemical events. The degree to which the two processes correspond may then depend in large part on the relative magnitude of the limitation imposed by the additional diffusive pathway and the biochemical component of photosynthesis. The interrelationships among the biophysical and biochemical events involved will be considered further below.

The degree to which the transpiration and photosynthetic processes correspond is of great significance to all land plants, especially those that inhabit the deserts of the world. The primary evolutionary hurdle encountered by plants in moving from the water to dry land must have been the problem of how to carry out CO_2 assimilation on a *sustained* basis in a desiccating environment, since any tissues porous enough to permit a ready inward diffusion of atmospheric CO_2 would also permit the evaporation of water from the moist surfaces of the cells containing the assimilatory apparatus. The functional solution to this problem must involve essentially all the morphological and anatomical structures of land plants.

Although plant structures at all levels of detail should be viewed in terms of the environments in which they have developed, one should be cautious about assigning specific functions to given structures on the basis of supposed environmental limitations. Suspected relationships based at first on teleological interpretations should be substantiated by biophysical and biochemical evidence before they are fully accepted. This practice has not always been followed, especially not with plants from arid environments, and as a result considerable confusion still seems to exist about the relationships between the gas-exchange processes and the anatomical and morphological structures peculiar to the plants of such environments. Much of the confusion stems from work done around the turn of the century, when a school of academic endeavor, which might best be designated "Ecological Anatomy," developed. The practitioners of this school were careful observers and noted that various kinds of habitats were occupied by plants exhibiting distinctive assemblages of anatomical characteristics. Their work culminated in the classic volumes *Plant-geography on a Physiological Basis* by Schimper (1903) and *Physiological Plant Anatomy* by Haberlandt (1914). A principal theme of these early workers was the deduction of function for anatomical and morphological structures. Thus the syndrome of structures associated with desert environments (e.g., small thick leaves, thick cuticles, sunken stomata, ample pubescence, wax coverings, and the succulent habit) were ascribed roles in water economy. The economy was thought to come about through the establishment and maintenance of low transpiration rates. The associated structures were termed "xeromorphic" and the plants exhibiting them were called "xerophytes." It should not be surprising, then, that the study of ecological anatomy suffered a severe blow when actual data showed the transpiration rates of many xerophytes to be higher than those of the mesophytes (Maximov, 1929). Even though evidence that this is often true has continued to accumulate, the earlier concept became so firmly rooted that even now it frequently appears with little qualification in general textbooks. The existence of plants in desert environments is ample evidence that their structures and functions are compatible with desert conditions. This can mean only that survival in deserts has dimensions other than the water economy as mediated through low transpiration rates, as thought by Schimper. Indeed, as pointed out by Maximov (1931), the consequences of high transpiration rates are more properly viewed in the context of carbon assimilation potential, since plants cannot greatly restrict water loss without affecting photosynthetic activity. Periods of rapid carbon assimilation are invariably periods of high transpiration. As indicated above, this relationship stems from a common diffusion path for the gases involved in the two processes.

The work of Brown and Escomb (1900) and that of Gaastra (1959) provide the framework for describing the biophysical components of this relationship in terms of diffusion gradients and the diffusion resistances acting along these gradients. The resulting formulation is based on Fick's law of diffusion and an analog of Ohm's law, so that for water moving out of the plant by transpiration,

$$E = \frac{\Delta[H_2O]}{r} \tag{1}$$

where E = transpiration rate ($g\ cm^{-2}\ s^{-1}$)
$\Delta[H_2O]$ = water-vapor-density difference between the immediate vicinity of the evaporative surfaces of the plant and the surrounding atmosphere ($g\ cm^{-3}$)
r = total diffusion resistance to water vapor ($s\ cm^{-1}$), which combines the epidermal resistances (stomatal and cuticular) with that of the boundary layer

The value of r is a critical property of the plant and is in large part a function of plant structure. It depends mostly on stomatal and cuticular characteristics and, to a lesser degree, especially in microphyllus desert plants, on leaf size and shape. The effect of wind on the boundary layer can be more or less disregarded here since the resistance to gaseous diffusion offered by the layer is almost always small in relation to the total. I hasten to add, however, that organ size and shape, together with wind speed, are of great significance with respect to the total energy regime of desert plants, and through this latter means play significant roles in water-use considerations (Gates *et al.*, 1968; Taylor and Sexton, 1972). Equation (1) can be solved for r after determining values for E and $\Delta[H_2O]$. E can be measured in a variety of ways and the value of $\Delta[H_2O]$ can be determined by measuring the vapor density of the atmosphere and subtracting this from the vapor density of the internal plant cavities, which by convention are considered to be at saturation.

Carbon dioxide assimilation may be similarly treated, as follows:

$$P = \frac{\Delta[CO_2]}{r'} \tag{2}$$

where P = net uptake of CO_2
$\Delta[CO_2]$ = difference in CO_2 concentration at the site of photosynthesis and that of the atmosphere
r' = total resistance to CO_2 uptake and is comprised of biochemical as well as diffusional parameters

Thus r' corresponds only in part to r.[1] It is helpful to our purpose to partition r' in terms of that portion which corresponds to r, r'_s, and that which does not, r'_m, so that $r' = r'_s + r'_m$, and Eq. (2) becomes

$$P = \frac{\Delta[CO_2]}{r'_s - r'_m} \tag{3}$$

Since the value of P is directly proportional to $\Delta[CO_2]$, which under natural conditions has a maximum value of 300 ppm or approximately 12.5 nm cm^{-3}, there is a maximum value for P at any given resistance. It then follows that as r'_m becomes large with respect to r'_s, carbon assimilation will be low relative to the transpiration rate, since only r'_s is directly related to r, and the resulting water-use efficiency will be

[1] r is related to r'_s through the ratio of the diffusion coefficient of water over the diffusion coefficient of CO_2. At 20°C this amounts to approximately 1.56. The value of r'_s can thus be easily attained by determining r as explained in the text and multiplying the resulting value by 1.56.

poor. It now becomes clear that a plant's capacity to utilize water efficiently is intimately related to the way the resistances in photosynthesis are partitioned. A measure of this efficiency has been proposed by Ting *et al.* (1972), formulated as follows:

$$e = \frac{P}{P_m} \tag{4}$$

where e = carbon dioxide assimilation-efficiency ratio
P = observed CO_2 assimilation rate as limited by all diffusional and biochemical resistances
P_m = maximum rate at which CO_2 assimilation could proceed if it were limited only by r_s' at a given $[CO_2]$.

That is,

$$P_m = \frac{\Delta[CO_2]}{r_s'} \tag{5}$$

To reiterate, it is r_s' that relates directly back to the plant's capacity for transpiration.

In the past, water-use efficiency has often been evaluated in terms of a transpiration ratio (TR) defined as follows:

$$TR = \frac{E}{P} \tag{6}$$

but, as has been pointed out previously (Ting *et al.*, 1972), such a ratio combines properties of the plant with those of the environment so that neither can be separately evaluated. This should be kept clearly in mind whenever the ratio is used, since it can change dramatically with environmental conditions even though the plant characteristics remain constant, and vice versa (Gates *et al.*, 1972; Lommen *et al.*, 1971).

Water-use efficiency for desert plants can be properly evaluated only in an appropriate environmental context. Radiation, air temperature, wind, humidity, and carbon dioxide concentration are the factors that affect transpiration and photosynthesis most directly. These factors have been effectively combined in an energy-exchange framework that embraces the biochemical and the biophysical aspects of both processes (Lommen *et al.*, 1971). The Lommen *et al.* model provides a powerful tool for determining the relative importance of physiological and environmental factors, both of which must be considered when attempting to understand adaptive strategies. Plant adaptation can proceed (1) by selection for specific environmental conditions or (2) by increased physiological efficiency for any given set of conditions.

Photosynthetic Systems in Southern California Deserts

Interest in photosynthetic efficiency has increased sharply during the past few years. It is now generally recognized that there are at least three contrasting systems by which plants may take up CO_2. These are designated as the C_3, C_4, and

CAM pathways. All three are represented in plants of the Southern California deserts. The basic distinction between C_3 and C_4 pathways is the first product of CO_2 assimilation. In C_3 plants $^{14}CO_2$ first appears in the three-carbon compound PGA; in C_4 plants it first appears in the four-carbon compound OAA, which is quickly converted to malate or in some cases aspartate (Downton, 1971a). Schematically, the processes are represented as

$$\text{Rudp} + CO_2 \rightarrow \text{PGA}$$
$$\text{PEP} + CO_2 \rightarrow \text{Malate}$$

for C_3 and C_4, respectively. The CAM pathway is presumed to be essentially the same as the C_4 except that it operates at night, storing vast amounts of malate. Much work has now been done on comparative aspects of these pathways (Hatch *et al.*, 1971). I will consider only those characteristics which appear to be most pertinent to gas-exchange strategies in desert plants.

The recognition of these alternative pathways for CO_2 assimilation renews interest in ecological anatomy, since in essentially all cases the type of pathway can be correctly surmised from gross morphological characteristics. C_4 plants have "Kranz"-type leaf anatomy, a name that refers to the presence of a compact sheath of cells filled with chloroplasts surrounding the vascular tissues of the photosynthetic organs so that the veins appear as small green cylinders under magnification (Laetsch, 1971). CAM is an acronym for Crassulacean acid metabolism, found in succulent plants as represented in the family Crassulaceae, from which the name is derived. Many other plant families have species with the CAM pathway, however (Ting *et al.*, 1972). Essentially all such individuals have either succulent or thickened fibrous photosynthetic organs. Plants having the C_3 pathway are the most common and are distinguished by the lack of C_4 and CAM characteristics.

The relative efficiency of these pathways is currently of great interest (Black, 1971; Björkman, 1972; Downton, 1971a; Evans, 1971; Ting *et al.*, 1972). The most excitement is centered around C_4 plants, which are capable of unusually high photosynthetic rates. Important relevant features that distinguish them from C_3 are (1) the attainment of maximum photosynthesis at higher temperatures and higher light intensities; (2) comparatively low residual resistance [r'_m in Eq. (2)]; and (3) the capacity to maintain positive CO_2 assimilation at very low ambient CO_2 concentrations. The contrast between C_4 and CAM in these respects is less well known and will be considered further below

There has been speculation about the evolutionary position of C_4 and CAM with respect to the more common C_3 plants. It seems obvious that the first two have had a polyphyletic origin (Evans, 1971). Some authors (Black, 1971; Björkman, 1972) suggest a superiority of the C_4 system for desert-type environments and argue that plants with this system will have a competitive advantage (Black *et al.*, 1969). Representatives of all three groups are found in the deserts of southern California. A question of ecological importance, then, is: How are the plants with these pathways inserted into their environment?

The southern California desert area is renowned for its high summer temperatures and intense solar radiation. Winter temperatures are rather moderate. The vegetation has permanent and ephemeral aspects. The permanent plant cover is provided by perennials, primarily shrubs, and various types of succulents. The relative importance of each perennial group varies from place to place. In some areas succulents are almost entirely absent, whereas in others they may comprise up to 80 percent of the fresh weight biomass. The ephemeral aspect is provided by a host of annuals that can conveniently be divided into two species groups: those that respond to precipitation falling in the hot summer period (June through September), summer annuals; and those which respond to precipitation falling during the cooler season (November through April), winter–spring annuals.

Species responding to summer precipitation are limited in number. Those most commonly observed in 1970, following August and September rains, were:

Pectis paposa	*Euphorbia setiloba*
Allionia incarnata	*Euphorbia* spp.
Boerhavia erecta	*Amaranthus fimbriatus*
Bouteloua barbata	*Bouteloua aristidoides*

All have Kranz-type leaf anatomy and therefore presumably the C_4 pathway.

Favorable moisture conditions during the cool season of 1972–1973 produced an abundant growth of winter–spring annuals. The number of species was much greater than in the summer group. Representatives of the following genera were commonly observed:

Erodium (two species)	*Plantago* (one species)
Gilia (several species)	*Anisocoma* (one species)
Phacelia (several species)	*Oenothera* (several species)
Eschscholzia (one species)	*Cryptantha* (several species)
Malacothrix (two species)	*Plagiobothrys* (several species)
Descurania (one species)	*Mentzelia* (several species)
Chaenactis (three species)	*Namma* (one species)
Lupinus (one species)	*Abronia* (one species)
Festuca (one species)	

None of the species examined from this group exhibited the Kranz-type leaf anatomy and, on the basis of the presently known taxonomic distribution of this feature (Welkie and Caldwell, 1970; Downton, 1971b), none of them should be expected to. Many other less abundant taxa belonging to this group could be listed. All are evidently C_3's. Different gas-exchange strategies are obviously employed by the two phenological types. Further evidence for this is provided by the anatomy of two species that bridge the gaps between the two major groups. *Erodium circutarium* and *Salsola iberica*, both annuals introduced from the old world and now widely naturalized, differ from each other and from most of the natives in their phenology. *E. circutarium* may germinate with early fall rains along with some summer annuals. It persists, however, and makes its major growth in the cool winter and spring (Went, 1948). It has C_3-type anatomy. *S. iberica* germinates in late spring and continues to grow through the early hot summer period. It has C_4 anatomy.

Perennial herbs are not abundant in the desert, but they too show differentiation with respect to the C_3 and C_4 pathways. Some of those which grow most vigorously during the summer, as exemplified by *Euphorbia polycarpa*, *E. albomarginater*, *Hilaria rigida*, *Sporobolus airoides*, and *Tridens pulchellus*, show the C_4-type Kranz anatomy, whereas those that grow most vigorously during the cooler period of the year, e.g., *Mirabilis bigelovii*, *Brandigea bigelovii* and *Sphaeralcea ambigua*, show the C_3 structure. The separation here is not so clear cut as in the annuals, however, since some species with C_3 anatomy, such as *Curcurbita palmata*, *Datura meteloides* and *Proboscida altheaefolia*, thrive throughout the summer period.

The functional response of the perennial shrubs in relation to phenology is ambiguous. Many of them, C_3 as well as C_4, are able to become metabolically active with high temperatures whenever water is available. The most abundant shrubs of the desert have C_3 anatomy. Major representatives of this group are as follows:

Larrea divarticata	*Ephedera nevedensis*
Ambrosia dumosa	*Salizaria mexicana*
Hymenoclea salsola	*Thamnosma montana*
Lycium andersonii	*Brickellia* spp.
Dalea spp.	*Encelia farinosa*
Acacia gregii	*Acamptopappus sphaerocephalus*
Tetradymia spp.	*Cercidium floridum*
Salvia spp.	*Hyptic emoryii*

Shrubs with C_4 anatomy seem to be restricted to the genus *Atriplex* of the family Chenopodiaceae. Members of this genus (i.e., *A. polycarpus*, *A. canescens*, and *A. lentiformis*) are found mostly along drainage ways and in valley bottoms, where there is some salt accumulation and where soil water is undoubtedly available for extended periods after precipitation, owing to the accumulation and concentration of water from outlying areas. Growth is vigorous in the summertime when moisture becomes available. *Tidestromia oblogifolia*, a subshrub belonging to the family Amaranthaceae, is also a C_4 plant and grows in valley bottoms and along disturbed roadways. In some respects it is more like the herbaceous perennials and annuals than other shrubs. Its contribution to the total vegetation is rather minor. It is of special interest, however, since it has been observed to show very high photosynthetic rates, 58 mg dm^{-2} h^{-1} at extreme temperatures, 46°C (Björkman *et al.*, 1972).

Shrubs occupying well-drained hillsides and bajadas lack C_4 anatomy almost entirely, even though some of them, such as *Larrea diverticata* and *Ambrosia dumosa*, grow vigorously on hot summer days. Some species, such as *Acacia gregii*, *Petalonyx thuberi*, and *Chilopsis linearis*, initiate new growth only after hot weather arrives.

Desert plants that exhibit the CAM pathway are essentially all perennials with succulent or thick fibrous photosynthetic organs. The important families are the Cactaceae, Agavaceae, and Liliaceae. The following species are important components of the desert vegetation:

Opuntia bigelovii	*Agave desertii*
Opuntia acanthocarpa	*Yucca brevifolia*
Opuntia basilaris	*Yucca schidigera*
Opuntia echinocarpa	*Echinocactus acanthoides*
Opuntia ramossisima	*Echinocereus engelmani*

Carbon Dioxide Assimilation Efficiency and Transpiration Ratios

Carbon dioxide assimilation efficiency and transpiration ratios [Eqs. (1)–(6)] have been estimated for summer and winter annuals in terms of the kind of environment they are known to occupy (Table 7.1). The environmental and physiological conditions on which the e and TR values are based appear in Table 7.2. Even though direct observations have not been made on the species listed in the two groups, enough data have accumulated in the literature on C_3 and C_4 plants to establish that the values used are reasonable. The conditions given for the C_4 plants are those reported by Björkman (1972) for *Tidestromia obligifolia*.

Contrasting strategies for water-use efficiency are evident in the two classes of annuals. It appears that the C_3's have selected favorable environments for water use, whereas the C_4's have developed more efficient physiological systems to cope with more severe growing conditions. It seems of special interest that the physiologically less efficient C_3's are still able to maintain the more favorable transpiration ratio, owing to the environment they have selected. If the C_4's in their environment had the same r'_m characteristics as the typical C_3 plant, the TR value would be over 1500, an increase of nearly 1000. On the other hand, if a C_3 plant had the same e value as the C_4, its TR value would improve only from 225 to 77, a change of 148. It thus appears that the selection pressure for physiologically efficient systems is much greater for plants growing in dry, hot environments than for those growing under dry, cool conditions.

The desert shrubs able to respond to moisture in the hot summer must use water extravagantly if they possess typical C_3 characteristics. One is almost led to suggest the possibility of an intermediate or alternative photosynthetic pathway to explain the remarkable success of some "C_3" species, such as *Patalonyx thurberi*, under very hot temperatures. Shrubs that grow under cool conditions should show water-use characteristics comparable to the winter annuals, whereas the C_4 species

Table 7.1 Carbon Dioxide Assimilation Efficiency and Transpiration Ratios of Summer and Winter Annuals in Appropriate Environments Assuming Realistic Photosynthetic Characteristics

Photosynthetic Type and Season of Growth	Carbon Dioxide Assimilation Efficiency, e^a	Transpiration Ratio, TR^b
C_4, summer annuals	0.89	560
C_3, winter–spring annuals	0.31	225

[a] Eq. (4).
[b] Eq. (6), based on volume.

Table 7.2 Environmental and Photosynthetic Characteristics Used in Determining Carbon Assimilation Efficiency and Transpiration Ratios Presented in Tables 7.1 and 7.3

Photosynthetic Type and Season of Growth	Temp. (°C)[a]	Rel. Hum.	r_s (s cm^{-1})	Transpiration Rate (g dm^{-2} h^{-1})	r'_s (s cm^{-1})	P (mg dm^{-2} h^{-1})	P_m (mg dm^{-2} h^{-1})
C_4, summer annuals	46	—	2	12.60	3.1	58	65
C_3, winter annuals	20	0.40	2	1.78	3.1	20	65
CAM (*Opuntia basilaris*)							
Winter	8	0.23	28	0.08	45	4.2	4.5
Summer	26	0.24	21	0.31	43	5.0	6.0

[a] Leaf temperature and air temperature.

of *Atriplex* should be expected to show efficiencies similar to those which have been observed for *Tidestromia.*

Gas-Exchange Strategies in CAM

Gas-exchange efficiencies for CAM species are less well known. A few reports summarized by Ting *et al.* (1972) indicate the possibility of low *TR* ratios relative to C_3 species. In recent studies of our own with three species of *Opuntia*, water-use evaluations have been possible, and so the strategies associated with these species as representatives of the CAM system will be given special emphasis. The three species studied, *O. basilaris*, *O. bigelovii*, and *O. acanthocarpa*, showed CO_2 uptake in response to moisture both winter and summer. But CO_2 uptake in CAM should not be categorically assumed to be equivalent to photosynthesis, since CO_2 is taken up primarily at night by combining with PEP, provided from stored carbohydrate, to form malate. In the light the malate is decarboxilated to form pyruvate and CO_2, which is then available for photosynthesis. Some workers have reported CO_2 uptake by CAM species in light as well as in darkness under laboratory conditions (Kluge and Fischer, 1967). In our work with *Opuntia* spp. under field conditions, our observations indicate that essentially all net CO_2 uptake occurs during the night and very early morning hours. This makes it convenient to measure the amount of CO_2 assimilated over intervals of time, since, if we assume a stoichiometric relationship between CO_2 uptake and increased acidification, the amount of CO_2 assimilated can be quantitatively determined by titration of the accumulated acid. This has been done on a more or less routine basis over the last 2 years under all kinds of environmental conditions.

The photosynthetic stem joints, even when severed from the parent plant, continue to function in CO_2 assimilation and transpiration throughout the dark period with little change. Stomatal resistance appears to be little affected for a period of several hours. Thus resistance to water vapor, *r*, can be determined gravimetrically as acidification proceeds. Table 7.3 gives the carbon dioxide assimilation efficiencies and transpiration ratios for *Opuntia basilaris* under two very different field conditions, as determined in the above manner.

Table 7.3 Carbon Dioxide Assimilation Efficiency and Transpiration Ratios of *Opuntia basilaris* under Favorable Moisture Conditions for Winter and Summer Conditions

Season	Photosynthetic Efficiency e^{a}	Transpiration Ratio TR^{b}
Winter	0.92	45
Summer	0.85	153

[a] Eq. (4).
[b] Eq. (6), based on volume.

The total acid accumulations at the ends of the dark periods for the two seasons showed little difference and was 38.4 μEq cm^{-2} in January and 37.6 μEq cm^{-2} in July. The carbon assimilation efficiency was relatively high in both instances, but the transpiration ratio was more than three times as great for the July conditions. When the values in Table 7.3 are compared with those in Table 7.1, it appears that the CAM's are more efficient than the C_3's or C_4's with respect to actual water use (*TR*) under both winter and summer conditions. The great advantage over C_4 in the summer, as shown by the more favorable *TR* values, is a direct consequence of nighttime gas exchange as regulated by the stomata and the CO_2 storage system. If CO_2 assimilation were to proceed under typical daytime temperature and radiation conditions, instead of at night, the associated transpiration rate would be expected to increase over three times. Thus the inverted pattern of gas-exchange characteristic of CAM's must be considered a major strategy for water-use efficiency. The evolution of this strategy is a form of environmental selection coupled with the development of an appropriate physiological system that could store CO_2 in the dark and yet use it for photosynthetic energy accumulation in the light.

Photosynthesis in CAM

A possible drawback of the CO_2 assimilation process as it operates in CAM may occur at the point of deacidification, where the CO_2 stored in malate is made available for photosynthesis. The release of CO_2 during this stage may proceed more

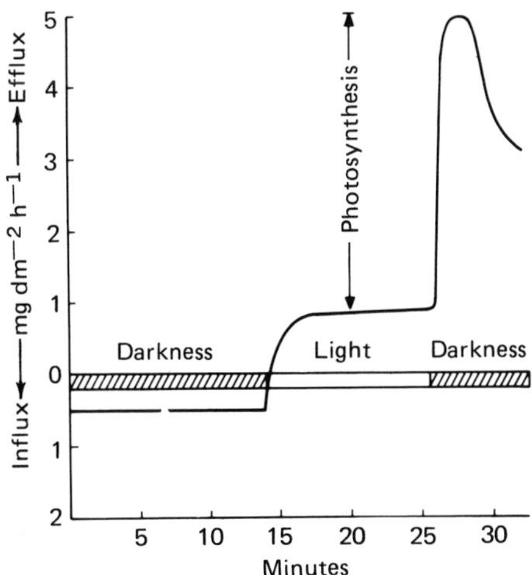

Fig. 7.1. Carbon dioxide exchange for *Opuntia basilaris* in response to a plus *e* of light after a night of CO_2 assimilation.

rapidly than it can be photosynthesized. Thus all the CO_2 stored in the dark may not be incorporated into high-energy compounds in the light. The rate of photosynthesis (Fig. 7.1) was estimated as will be explained. A severed *Opuntia basilaris* stem joint that had been actively assimilating CO_2 through the night was placed in a small, closed photosynthetic chamber system near the end of its dark period, and the CO_2 flux was measured with an IRGA during subsequent light and dark cycles. At the beginning the joint was actively assimilating CO_2 but at a low rate: 0.5 mg dm^{-2} h^{-1} (a few hours earlier CO_2 uptake had been measured at a rate of 5 mg dm^{-2} h^{-1}). Immediately upon illumination, CO_2 evolution began. The lights were turned off after 12 min and a burst of CO_2 efflux was observed. The difference in CO_2 release rates in light and darkness after deacidification has been initiated is considered an approximate measure of real photosynthesis, which in this case was just over 4 mg dm^{-2} h^{-1}. The treatment was repeated after $\frac{1}{2}$ h of light, and the rate was observed to have increased to 9 mg dm^{-2} h^{-1}. The increased rate with time after illumination is related to stomatal closure, which increases the internal CO_2 concentration by trapping the released CO_2 inside the joint. Diffusion resistances have commonly been observed to increase rapidly after sunrise. It is likely also that the deacidification process is speeded up after a period of illumination.

Diffusion Resistance

The view held by Schimper (1903) that plants in arid environments must have low transpiration potentials if they are to survive is not entirely without merit. For, as was pointed out earlier with respect to Eq. (3), as r'_s becomes large with respect to r'_m, the correspondence between transpiration and photosynthesis

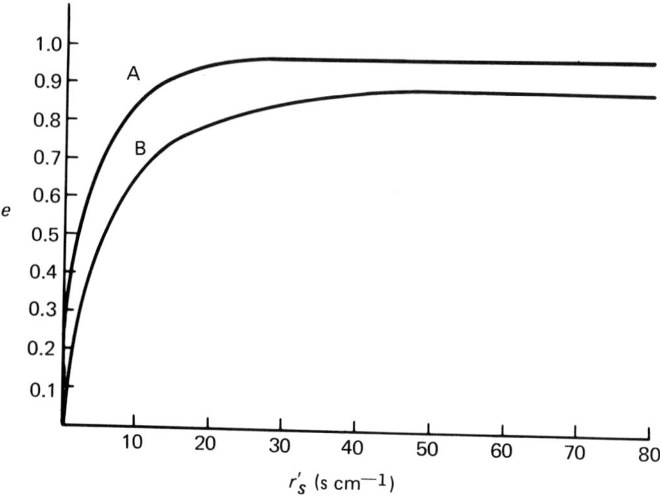

Fig. 7.2. CO_2 assimilation efficiency (*e*) as a function of stomatal diffusion resistance (r'_s) to CO_2 with a constant r'_m of 2 s cm^{-1} (A) and 5 s cm^{-1} (B).

approaches its optimum value and depends more and more on the conditions of the physical environment. Thus over a certain range of resistances the efficiency will improve markedly with stomatal closure (Fig. 7.2). This has significance for CAM's, which most generally have high diffusion resistances relative to C_3's and C_4's (Ting *et al.*, 1972). Stomatal resistances in *Opuntia* may operate at levels nearly equal to the cuticular resistance of more mesophytic plants. Cuticular resistance for the three species of *Opuntia* being studied has been measured to be on the order of 1000 s cm^{-1}. It thus appears fairly certain that any resistances less than 100 s cm^{-1} for mature joints indicate some degree of stomatal opening. Studies on CO_2 uptake also indicate that this is the case. Minimum resistances measured in the field by gravimetric means are close to 10 s cm^{-1}. The operating ranges of most joints are commonly higher than this, ranging above 20 s cm^{-1}. This is in contrast to *r* values of 2–5 s cm^{-1} reported for C_3's and C_4's. Such high resistances are associated with low stomatal densities, which appear to be a general feature of CAM's (Ting *et al.*, 1972). High stomatal diffusion resistances may also account partly for the apparent lack of temperature sensitivity indicated by the similar maximum acidification values for winter and summer nights, since, if CO_2 uptake is controlled principally by stomatal resistance, the effect of temperature will be minimized (Gates *et al.*, 1972).

Carbon Conservation

Throughout much of the year soil moisture in the desert environment is lacking and the stomata of CAM plants remain closed both day and night. Under such conditions transpiration is very low and atmospheric CO_2 uptake is nil. Still,

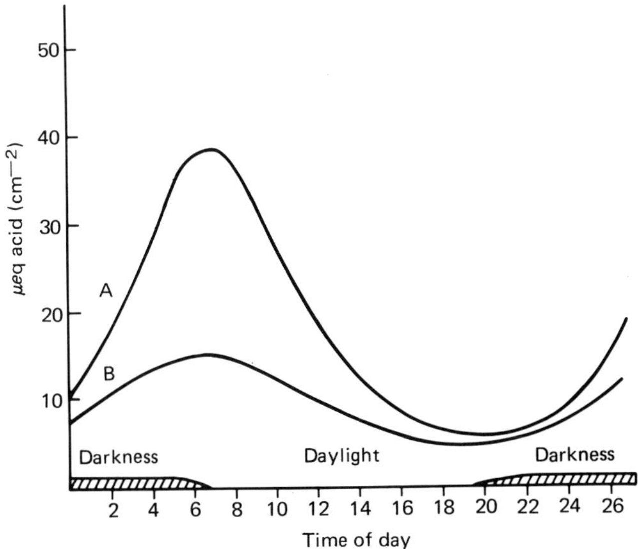

Fig. 7.3. Generalized diurnal acid fluctuation curves for *Opuntia* spp. Under low (A) and high (B) water stress conditions. CO_2 is taken up from the atmosphere under low water stress but not under high water stress.

however, the acid level inside the stem joints of all three species of *Opuntia* studied fluctuates up and down in a diurnal pattern (Fig. 7.3). The magnitude of the fluctuations is greatly reduced in comparison with that which occurs when atmospheric CO_2 is being actively assimilated. The dampened fluctuation is viewed as representing the recycling of respiratory CO_2 through acidification and photosynthesis. That the observed nighttime acidification is actually due to the fixation of respiratory CO_2 is illustrated by the data in Table 7.4, which gives the effects of temperature on severed stem joints of *Opuntia basilaris* having closed stomata. In all cases the degree of acidification produced during the dark period was approximately twice as great at night temperatures of 32°C as for night temperatures of 15°C. Other experiments (Szarek *et al.*, 1973) on respiration have shown that indeed the rate of O_2 uptake at 32°C is double that at 15°C. Energy fixation is thus

Table 7.4 Effect of Two Night Temperatures on the Dark Acidification in *O. basilaris* Stem Tissue

Night temp. (°C)	Increase in acidity after 12 h of darkness (μ Eq/cm^{-2})				
	1	**2**	**3**	**4**	**Av.**
32	8.9	8.7	7.7	10.7	9.0
15	5.5	4.1	2.4	5.3	4.3

able to continue on a daily basis even though no new structural carbon is added. Plants persisting in this condition for extended periods of time are not dormant, but are maintaining a fairly active metabolic existence. This enables them to respond quickly to moisture when it becomes available. In the summer of 1972, 12 h after a 6-mm rain, stems of *O. basilaris* that had been inactive for months were found to be actively transpiring and assimilating CO_2. Within 3 days they were operating at full capacity. Perennial shrubs and annuals in the area did not show any obvious responses to this meager shower.

The strategy of carbon conservation and a sustained energy fixation through daily photosynthesis, then, appears to be that of holding on and surviving adverse periods in a vigorous enough condition to be able to respond quickly to moisture when it becomes available and which in the desert is extremely transient. The conservation of respiratory carbon may be of significance in plants other than CAM's. Any available reservoir of CO_2 will be tapped by the photosynthetic process. Once CO_2 enters a desert plant, there should be advantages in retaining it, since water had to be expended in its acquisition.

An experiment that suggests one aspect of such a strategy was carried out with young fruiting pods of the desert shrub *Isomerus arborea*. The internal atmosphere of the pod was sampled and its photosynthetic rate was measured in a chamber

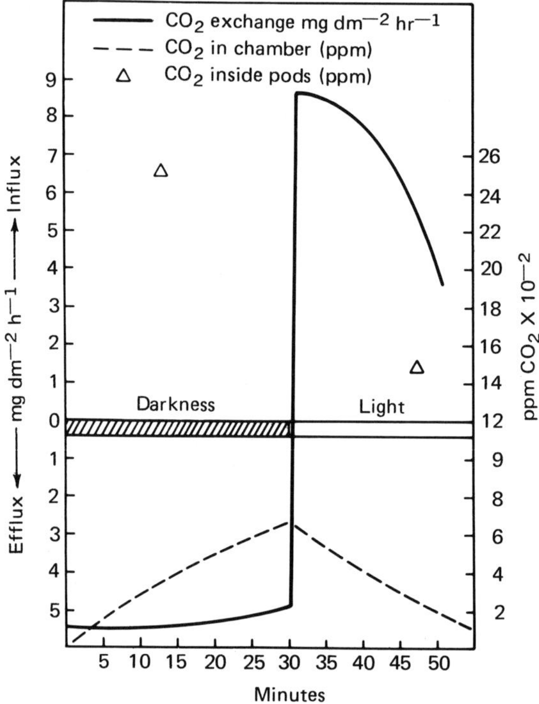

Fig. 7.4. Carbon dioxide exchange in the fruit pods of *Isomeric arboria* in a closed system, showing CO_2 influx in light and efflux in darkness and the changing CO_2 concentration inside the chamber and pods.

(Fig. 7.4). The results clearly indicate that much of the CO_2 produced by the rapidly respiring developing seeds was reassimilated by the photosynthetic outer wall of the pod.

Conclusions

The evolution of gas-exchange strategies in desert plants has proceeded in two directions, one involving physiological refinements and the other phenological adjustments. Both directions may lead to increased water-use efficiency. Physiological refinements are especially evident in species with the C_4 and CAM pathways. Plants that exhibit the characteristics of these pathways show carbon dioxide assimilation efficiencies (*e* values) approaching the theoretical maximum. This means that r'_m is small and that CO_2 assimilation rates are highly dependent on r'_s. CAM species have magnified this dependence further by developing the capacity for maintaining comparatively high minimum r'_s values. Another significant refinement of the CAM system is the mechanism for carbon conservation, wherein respiratory CO_2 is recycled, thus decreasing the plant's dependence on external CO_2, which can be acquired only at the expense of transpiration. Phenological adjustments that favor water-use efficiency are evident in annual C_3's and CAM's. The less physiologically efficient C_3's are able to maintain favorable *TR*'s by growing only during the cooler part of the year. The phenological adjustment of CAM's has a much shorter time base, since it is fitted to the diurnal cycle. By carrying on gas exchange at night, transpiration is greatly reduced over what it would be under daytime conditions.

Acknowledgments

I would like to thank Terry Yonkers for his assistance in all aspects of the field and laboratory work, and Irwin Ting and Stan Szarek for helpful discussions.

This research was supported in part by the National Science Foundation Grant GB15886 through the US/IBP Desert Biome.

References

Björkman, O., Pearcy, R. W., Harrison, T. A., Mooney, H.: 1972. Photosynthetic adaptation to high temperatures: a field study in Death Valley, California. Science *175*, 786–789.

Black, C. C.: 1971. Ecological implications of dividing plants into groups with distinct photosynthetic production capacities. Adv. Ecol. Res. *7*, 87–114.

———, Chen, T. M., Brown, R. H.: 1969. Biochemical basis for plant competition. Weed Sci. *17*, 338–344.

Brown, H., Escombe, F.: 1900. Static diffusion of gases and liquids in relation to the assimilation of carbon and translocation in plants. Phil. Trans. Roy. Soc. London, Ser. B *193*, 233–291.

Downton, W. J. S.: 1971a. Adaptive and evolutionary aspects of C_4 photosynthesis. *In* Photosynthesis and photorespiration (eds. M. D. Hatch, C. B. Osmond, R. O. Slatyer). New York: Wiley.

———: 1971b. Check list of C_4 species. *In* Photosynthesis and photorespiration (eds. M. D. Hatch, C. B. Osmond, R. O. Slatyer). New York: Wiley.

Evans, L. T.: 1971. Evolutionary, adaptive and environmental aspects of the photosynthetic pathways: assessment. *In* Photosynthesis and photorespiration (eds. M. D. Hatch, C. B. Osmond, R. O. Slatyer). New York: Wiley.

Gaastra, P.: 1959. Photosynthesis of crop plants as influenced by light, carbon dioxide, temperature and stomatal diffusion resistance. Med. Ed. Landboushogeschool Wageningen *59*(11).

Gates, D. M., Alderfer, R., Taylor, S. E.: 1968. Leaf temperatures of desert plants. Science *159*, 994–995.

———, Johnson, H. B., Yocum, C. S., Lommen, P. W.: 1972. Geophysical factors affecting plant productivity. *In* Theoretical foundations of the photosynthetic productivity, pp. 406–419. Moscow: Nauka.

Haberlandt, G.: 1914. Physiological plant anatomy. London: Macmillan.

Hatch, M. D., Osmond, C. B., Slatyer, R. O. (eds.): 1971. Photosynthesis and photorespiration. New York: Wiley.

Heath, O. V. S.: 1969. The physiological aspects of photosynthesis. Stanford, Calif.: Stanford Univ. Press.

Kluge, M., Fischer, K.: 1967. Über Zusammenhänge zwischen den CO_2-Austausch und der Abgabe von Wassendampf durch *Bryophyllum daigremontianum*. Berg. Planta *77*, 212–223.

Laetsch, W. M.: 1971. Chloroplast structural relationship in C_4 plants. *In* Photosynthesis and photorespiration (eds. M. D. Hatch, C. R. Osmond, R. O. Slatyer). New York: Wiley.

Lommen, P. W., Schwintzer, C. R., Yocum, C. S., Gates, D. M.: 1971. A model describing photosynthesis in terms of gas diffusion and enzyme kinetics. Planta *98*, 195–220.

Maximov, N. A.: 1929. The plant in relation to water. London: MacMillan.

———: 1931. The physiological significance of the xeromorphic structure of plants. J. Ecol. *19*, 273–282.

Schimper, A. F. W.: 1903. Plant-geography upon a physiological basis. (Eng. trans. W. R. Fisher). Weinheim: H. R. Engelmann (New York: Oxford Univ. Press).

Szarek, S. R., Johnson, H. B., Ting, I. P.: 1973. Drought adaptation in *Opuntia basilaris*: significance of recycling carbon through Crassulacean acid metabolism. Plant Physiol. *52*, 539–541.

Taylor, S. E., Sexton, O. J.: 1972. Some implications of leaf tearing in *Musaceae*. Ecology *53*, 43–149.

Ting, I. P., Johnson, H. B., Szarek, S.: 1972. Net CO_2 fixation in crassulacean acid metabolism plants. *In* Net carbon dioxide assimilation in higher plants (ed. C. C. Black), pp. 26–53. Raleigh, N.C.: Cotton, Inc.

Welkie, G. W., Caldwell, M.: 1970. Leaf anatomy of species in some dicotyledon families as related to C_3 and C_4 pathways of carbon fixation. Can. J. Bot. *48*, 2135–2146.

Went, F. W.: 1948. Ecology of desert plants. I. Observations on germination in the Joshua Tree National Monument, California. Ecology *29*, 242–253.

8

Photosynthesis of Desert Plants As Influenced by Internal and External Factors

OTTO L. LANGE, ERNST–D. SCHULZE, LUDGER KAPPEN,
UWE BUSCHBOM, AND MICHAEL EVENARI

Introduction

One of the main aims of current research in physiological ecology is to investigate and causally interpret the photosynthetic productivity of plants and of plant communities in different habitats. Efforts have been made to develop mathematical models to compute the net fixation of CO_2 by plants from meteorological parameters and to predict their productivity (e.g., Lommen *et al.*, 1971; Chapter 4, by Hall and Björkman). One basic requirement for the realization of such models is a detailed knowledge of the functional relationships between the photosynthetic efficiency of a plant and the external conditions characteristic for its particular habitat. Special attention must be paid to the responses of the different morphological types, considering the variability of their physiological state and their capacity for regulative adaptations. The more sophisticated models recently proposed have made apparent large gaps in our knowledge about the influence of important internal and external factors on the CO_2 exchange of plants. Therefore, during our work on productivity and water relations of desert plants, we first focused our interest on a functional analysis of the photosynthetic responses of the plants in their natural habitat, using two approaches. On the one hand, we investigated net photosynthesis and transpiration in relation to the natural conditions in which the different plant types were growing. On the other, we analyzed the importance of single environmental factors by carrying out experiments under artificially changed conditions in the field.

The investigations were conducted in the Negev desert in Israel during the dry season in 1967 (see Lange *et al.*, 1969) and during the whole vegetation season in 1971.

Plants and Methods

The Negev desert, which has a mean annual precipitation of about 80 mm, is located at the borderline between the Irano-Turanian and the Saharo-Arabian phytogeographical region. The natural vegetation (see Zohary, 1962) is characterized by a thin cover of dwarf-shrub formations (Fig. 8.1) which belong to three different plant communities (Orshan, 1973; Zohary, 1953; Zohary and Orshan,

Fig. 8.1. View of the central Negev vegetation near the ancient town of Avdat (background). The north-facing foreground is covered with an *Artemisietum*, the loessial plain of the valley floor with a *Hammadetum*, and the opposing south slope with a *Zygophylletum*. The edge of the runoff farm Avdat can be seen to the right. In the center is the mobile laboratory, connected with electric leads and tubings to the installations for measurement of different plants in the natural vegetation and in the farm.

1954). The widespread associations of the *Artemision herbae-albae* alliance are found primarily in the north-facing slopes with *Artemisia herba-alba* Asso, *Reaumuria negevensis* Zohary et Danin, and *Noaea mucronata* (Forssk.) Aschers. et Schweinf. as dominant chamaephytes. The loess-covered flood plains are inhabited by communities with *Hammada scoparia* (Pomel.) Iljin as dominant plants. In contrast to this, associations with *Zygophyllum dumosum* Boiss. grow on extreme south-facing slopes and plateaus of the stone desert. Most of these plants show typical phenomenological changes of the anatomical and morphological structure of their photosynthetic organs (Orshan, 1963; 1973). *Z. dumosum* sheds all its succulent leaflets during the dry season, and the petioles alone remain photosynthetically active. *A. herba-alba* also shows a considerable reduction in its photosynthesizing organs under dry conditions: the large pinnate winter leaves are replaced by small scaly summer leaves. *R. negevensis* has heavily salt-encrusted brachyblasts during the dry season. In *H. scoparia*, the green stem cortex functions as the photosynthetic tissue. During the dry season, part of this cortex is shed *Salsola inermis* Forssk. is one of the few summer active annuals of the Negev, having its main growing time during the dry season (Evenari *et al.*, 1971; Evenari, 1937–1938).

Fig. 8.2. New type of climatized gas-exchange chamber in operation in the field. The acrylic plastic housing, with the humidity sensors placed next to it, is connected to the heat-exchanger assembly by flexible tubes. The reference air temperature is taken near an adjacent plant using a ventilated radiation shield.

Net photosynthesis and transpiration together with all the important meteorological parameters were measured on nonirrigated and experimentally irrigated samples of these typical wild desert species. In addition to this, investigations were carried out with cultivated species (*Prunus armeniaca* L., *Vitis vinifera* L.) at the runoff farm Avdat (Evenari *et al.*, 1971) and with continuously irrigated plants of *Datura metel* L. and *Citrullus colocynthis* (L.) Schrad.

The measurements were made with a mobile laboratory (Fig. 8.1) equipped with peltier-cooled temperature- and humidity-controlled cuvettes (Fig. 8.2). The cuvettes were installed on single twigs or whole plants in their natural habitat either matching the natural temperature and humidity conditions or by using the chambers as small portable phytotrons to keep all but one factor constant (see Lange and Schulze, 1971; Koch *et al.*, 1971; Schulze *et al.*, 1972d; Schulze 1972). The layout of the measuring system is shown in Fig. 8.3. Carbon dioxide uptake or

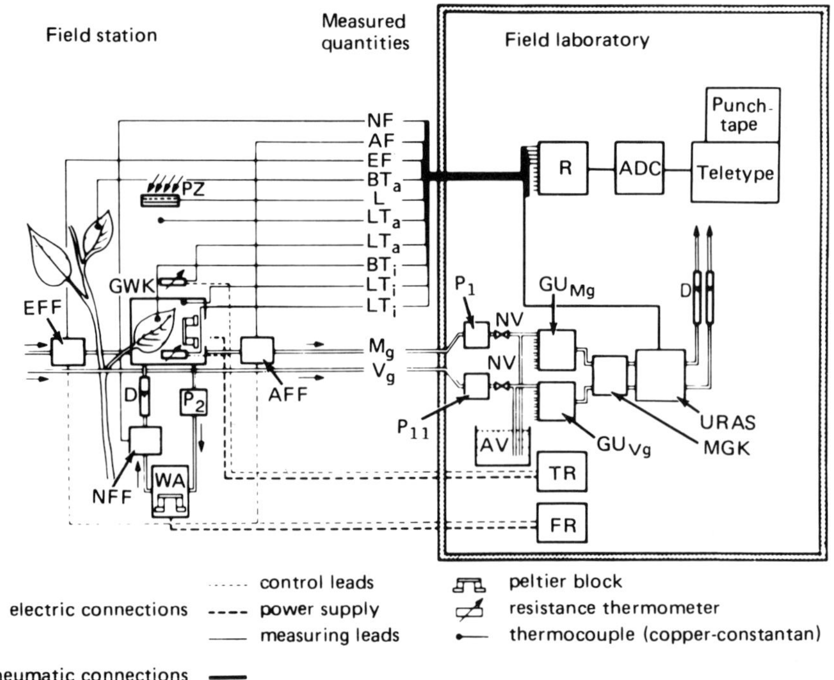

Fig. 8.3. Layout of the measuring system: ADC, analog digital converter; AFF, humidity sensor measuring the humidity, AF, in the air outlet of the chamber; AV, shutoff valve; BTa and BTi, leaf temperature outside and inside the chamber; D, flow meter; EFF, humidity sensor measuring the humidity, EF, in the air inlet of the chamber; FR, humidity controller; GU, gas changeover switch; GWK, gas-exchange chamber; L, light intensity; LTa and LTi, air temperature outside and inside the chamber; MGK, air cooler; Mg, measuring air; NFF, humidity sensor measuring the humidity, NF, in the circulating air stream of the humidity control system; NV, needle valve; P, pump; PZ, photo cell; R, compensation point recorder; TE, thermocouple; TR, temperature controller; URAS, infrared gas analyzer; Vg, reference air; WA, water-vapor dew-point trap. (With changes, after Koch *et al.*, 1971, Fig. 2.)

evolution of the enclosed plant material was measured in an open system as the difference between the inlet and outlet CO_2 concentration of an air stream passing through the chamber, using a differential infrared gas analyzer. Transpiration measurements were obtained by a compensation system that simultaneously also controlled the air humidity inside the chamber. Light intensity, leaf temperature, air temperature, and air humidity inside and outside the cuvette were recorded together with the CO_2 exchange and the transpiration of the plant, all by the same recorder. Ten chambers were connected in sequence with their recorders to a teleprinter that produces a punch tape for computer processing.

The water relations of the plants were investigated by leaf water-content determinations using the beta-ray absorption method (Buschbom, 1970) inside the plant chambers. In addition to the osmotic potential, we measured the water potential of the experimental plants, using the pressure bomb (Boyer, 1967).

CO_2 Exchange in Relation to External and Internal Conditions

The light dependency of net photosynthesis under favorable temperature and humidity conditions is shown for several wild and cultivated plants in Fig. 8.4 (Schulze *et al.*, 1972b). The measurements were carried out in the middle of the dry season. When photosynthesis is based on dry weight, three types of response can be recognized. *C. colocynthis* and *D. metel*, which were artificially and continuously irrigated and for which soil water was never a limiting factor, show the highest photosynthetic rates. The nonirrigated wild desert plants have rates of about one-tenth of the irrigated ones, and the cultivated apricot and grapevine are intermediate. This order changes when net photosynthesis is based on the surface area of the photosynthesizing organs. But the greatest change in order occurs when net photosynthesis is based on the chlorophyll content. *S. inermis* and *N. mucronata* are now in the order of magnitude of *C. colocynthis* and *D. metel*. *H. scoparia* show higher values than the apricot. *R. negevensis*, *Z. dumosum*, and *A. herba-alba* remain at the lowest level. The results show that all species reach their light saturation only at the very high values of 60–80 klx. This may be considered an adaptation to the high light intensities typical of deserts (see also Björkman *et al.*, 1972). The absolute efficiency of the photosynthetic mechanisms, if expressed by relating photosynthesis to chlorophyll content, is as high for some wild plants as that of the irrigated ones. Obviously the desert plants can keep up this high efficiency only because their photosynthetic apparatus is protected against damaging water loss by xeromorphic structures with high dry weight and relatively small surface area.

Such high rates of net photosynthesis are achieved under optimum light, temperature, and humidity conditions, and occur only rarely during the course of a desert day. The water deficit and the climatic conditions soon cause a decline in CO_2 uptake, the magnitude of which depends on the time of year and the type of photosynthetic reaction of the individual species (Schulze *et al.*, 1972c; Stocker, 1960, 1970, 1971, 1972). As shown in Fig. 8.5, in April *H. scoparia* has a single-peaked daily course of photosynthesis with highest rates at noon, whereas *A.*

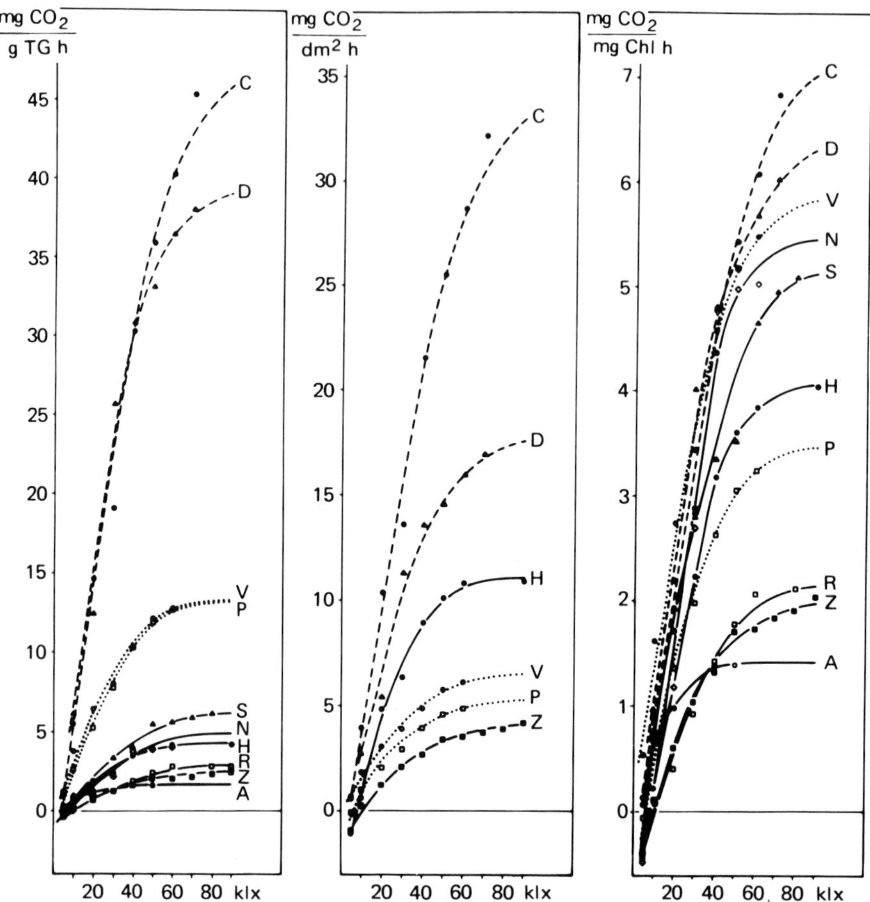

Fig. 8.4. Light dependency of net photosynthesis as related to gram dry weight (left), dm^2 surface area (middle), and milligram chlorophyll content (right) of the photosynthesizing organs. Ordinate: CO_2 exchange; abscissa: light intensity. *Artemisia herba-alba* (A), *Citrullus colocynthis* (C), *Datura metel* (D), *Hammada scoparia* (H), *Noaea mucronata* (N), *Prunus armeniaca* (P), *Reaumuria negevensis* (R), *Salsola inermis* (S), *Vitis vinifera* (V), *Zygophyllum dumosum* (Z). (After Schulze *et al.*, 1972b, Fig. 2.)

herba-alba and *Z. dumosum* show a slight depression of CO_2 uptake in the afternoon. In June, photosynthesis is lower and both species have a pronounced two-peaked daily curve with a higher rate of photosynthesis in the morning, a deep depression falling down to the compensation point around midday, and a second peak of CO_2 uptake in the afternoon. This midday depression is only moderate in July in *H. scoparia*, but is much more severe in September. At that time of year *A. herba-alba* and *Z. dumosum* have only a low rate of net photosynthesis in the morning hours and a respiratory loss during the rest of the day. A pronounced noon depression of CO_2 uptake is exhibited throughout the summer by the cultivated apricot trees (Fig. 8.6). Net photosynthesis is high in the early morning, drops to

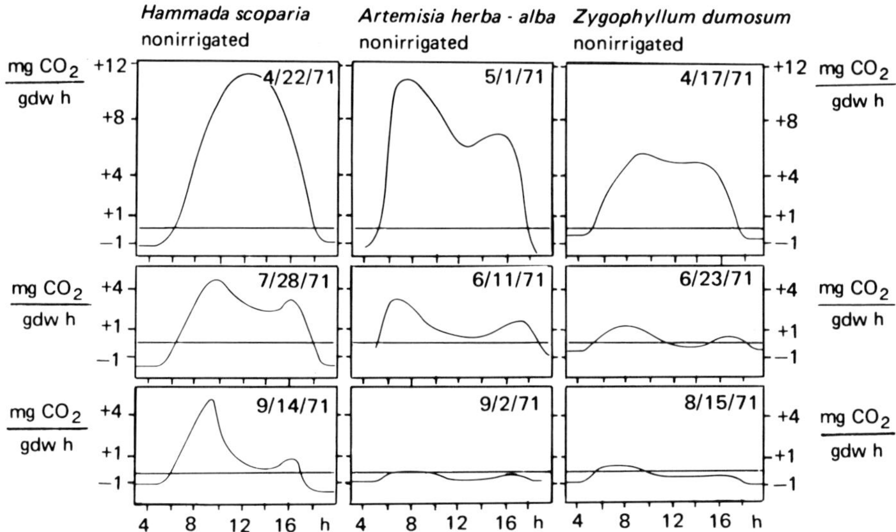

Fig. 8.5. Diurnal course of net photosynthesis of natural-growing *Hammada scoparia*, *Artemisia herba-alba*, and *Zygophyllum dumosum* in the different seasons. Ordinate: CO_2 exchange per gram dry weight; abscissa: time of day.

the compensation point during the midday hours, and increases somewhat again in the late afternoon.

The intensity and the extent of the midday depression have considerable influence on the net carbon gain, i.e., the primary productivity of the desert plants. Because of the importance of the two-peaked curve of photosynthesis under arid conditions, it is necessary to investigate its cause and regulating mechanisms to gain an understanding of the biomass production and to learn its functional background for modeling purposes.

The "classic" interpretation for the causes of the two-peaked photosynthesis curve is that, owing to a higher water loss in the morning, the leaves reach a water deficit, which results in a hydroactive closure of the stomata. This certainly could be correct, but then it becomes difficult to explain why the stomata remain closed during the day at a limited water loss and why they open again in the late afternoon when the water potentials are still low. In addition to the internal control of the stomatal opening system via the water potential of the entire leaf, there must be an effective external component of stomatal regulation. For this control system, temperature and air humidity would have to be considered as important factors.

The temperature has several effects on the photosynthetic gain of plants. On the one hand, it influences the enzymatic photosynthetic processes, and the relation between photosynthetic CO_2 fixation and respiration. A high net photosynthetic capacity even under high temperatures of the leaf tissue seems to be typical for native desert plants (Adams and Strain, 1968; Schulze *et al.*, 1972b; Björkman *et al.*, 1972). An example is given in Fig. 8.7 for *H. scoparia* during the middle of the dry season. In this plant photosynthesis attains a temperature optimum of 37.7°C and

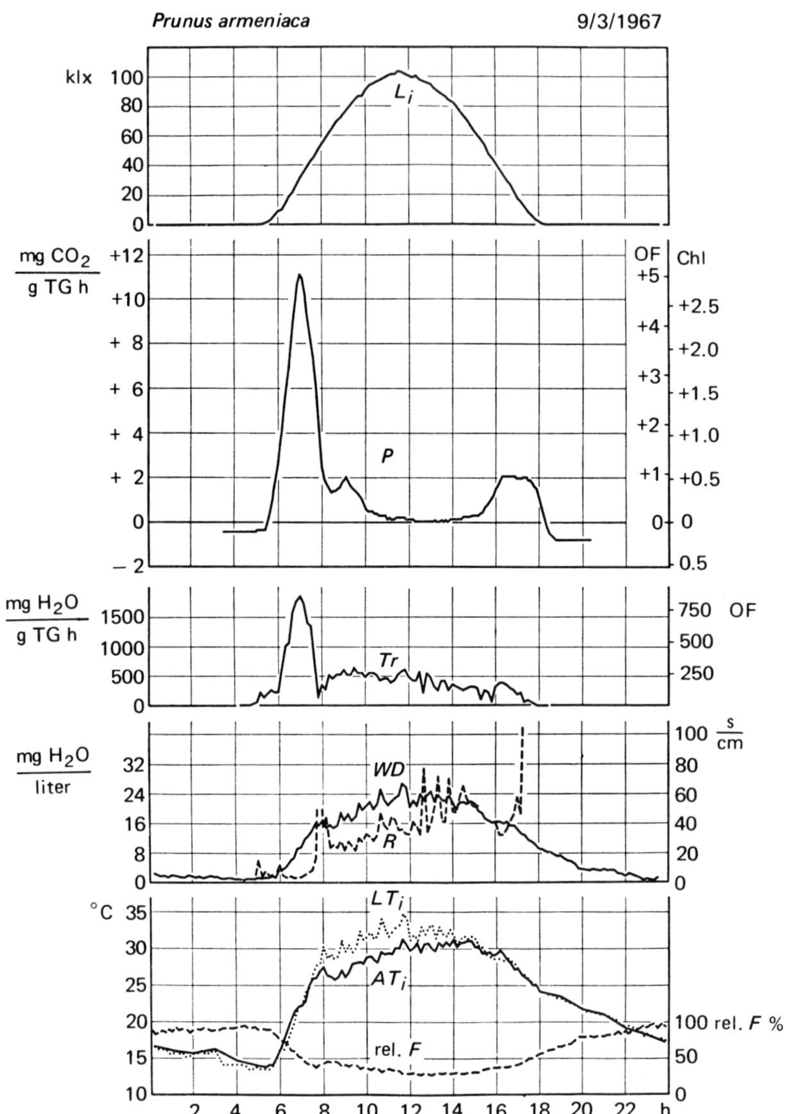

Fig. 8.6. Diurnal course of CO_2 exchange of *Prunus armeniaca*. From top to bottom: Light intensity (L_i, klx); net photosynthesis (*P*, in relation to g dry weight TG, mg chlorophyll content Chl, and surface area OF); transpiration (*Tr*, in relation to g dry weight TG, and surface area OF); water-vapor difference between leaf mesophyll and surrounding air (*WD*, mg H_2O liter^{-1}); total diffusion resistance to water vapor of the leaves (R, in s cm^{-1}); air temperature (AT_i), leaf temperature (LT_i) and relative humidity (rel. *F*) inside the chamber. (After Schulze *et al.*, 1972c, Fig. 6.)

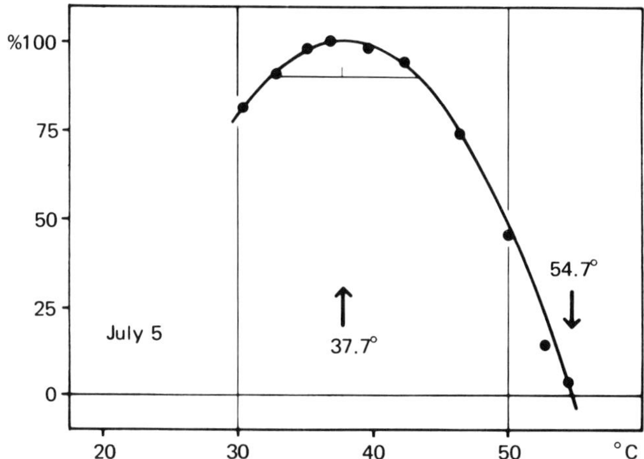

Fig. 8.7. Temperature dependency of net photosynthesis of *Hammada scoparia*. Ordinate: % net photosynthesis as related to the maximal rate; abscissa: temperature. Arrows indicate the optimum temperature and the upper temperature compensation point.

reaches the upper temperature compensation point at 54.7°C. Note that the temperature dependence of the net photosynthesis does not remain constant but changes throughout the year (Mooney and West, 1964; Strain and Chase, 1966; Caldwell, 1972). Figure 8.8 demonstrates this effect for *H. scoparia*. In spring the temperature optimum of net photosynthesis and the upper temperature compensation point of the CO_2 exchange lie at relatively low values. Later in the season the temperature dependence of net photosynthesis shifts to higher temperatures, with both temperature optimum and temperature compensation point reaching maximum values in midsummer, and decreasing again in late summer and fall. The amplitude of this seasonal shift exceeds 10°C.

On the other hand, temperature also influences the photosynthetic CO_2 uptake of the plants through its effect on stomatal movement, which is different at a low and at a high water stress (Schulze *et al.*, 1973). Figure 8.9 shows results of a field experiment in which temperature was changed experimentally but the water-vapor-concentration difference between the leaf and the surrounding air was kept constant. At a low water stress in response to the temperature increase, the total diffusion resistance for water vapor of *Z. dumosum* leaves was lowered and the transpiration rate increased. A subsequent decrease in temperature resulted in an increase in diffusion resistance. The same result was found by Hall and Kaufmann (this volume). They also demonstrated the great importance of a constant water-vapor-concentration difference for the results of such experiments. With increasing water stress on a given site, the opening response of the stomata to temperature becomes increasingly less and the reaction can even reverse. In *R. negevensis*, which had a minimal water potential of −87 bars during the day, an increase of leaf temperature resulted in increased diffusion resistance and a decreased transpiration rate (Fig. 8.10). From the ecological point of view, this change in stomatal reaction

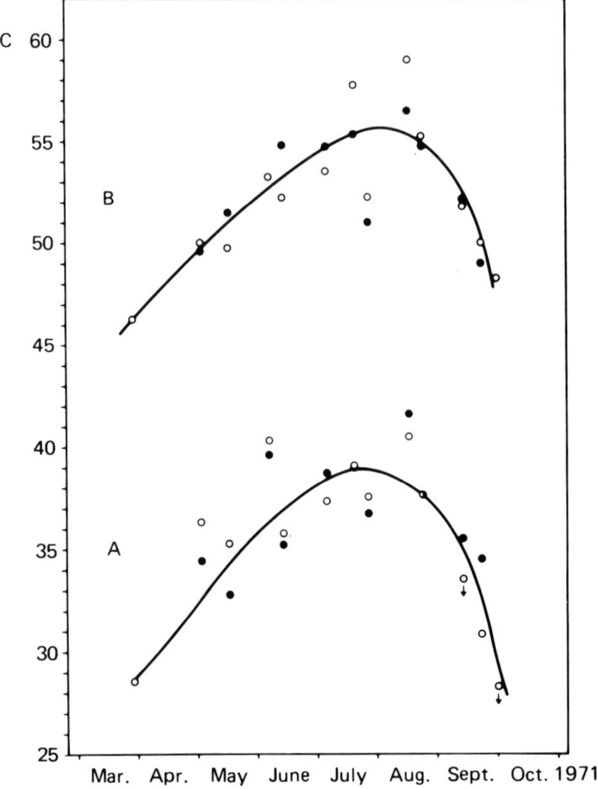

Fig. 8.8. Seasonal course of the temperature optimum of net photosynthesis (A), and of the upper temperature compensation point of CO_2 exchange (B) for *Hammada scoparia*. Abscissa: time of year; ordinate: tissue temperature. Dots: watered plants; circles: nonwatered plants under natural water stress; arrows: actual values might be lower.

is a very expedient behavior: at low water stress the increasing transpiration due to stomatal opening leads to increasing cooling, which results in a more favorable temperature range for the various reactions in the leaf tissue (Lange, 1959; Lange and Lange, 1963). Under water stress, which also induces an increased heat resistance of the cytoplasm in the leaves (Hammouda and Lange, 1962; Kappen, 1966), the restriction of water loss is of greater significance for survival of the plant. The dependence of the temperature-controlled stomatal response on plant water stress differs from one species to another. The slope of the change in resistance caused by change in temperature ($\Delta R/\Delta T$)—which is positive for stomatal closure and negative to stomatal opening—is plotted in relation to the plant water potential of the experimental plants during the vegetation period in Fig. 8.11. All species exhibited stomatal opening with increasing temperature at high plant-water potentials. At a threshold water stress, different for each species, the opening response was reversed to a closing reaction.

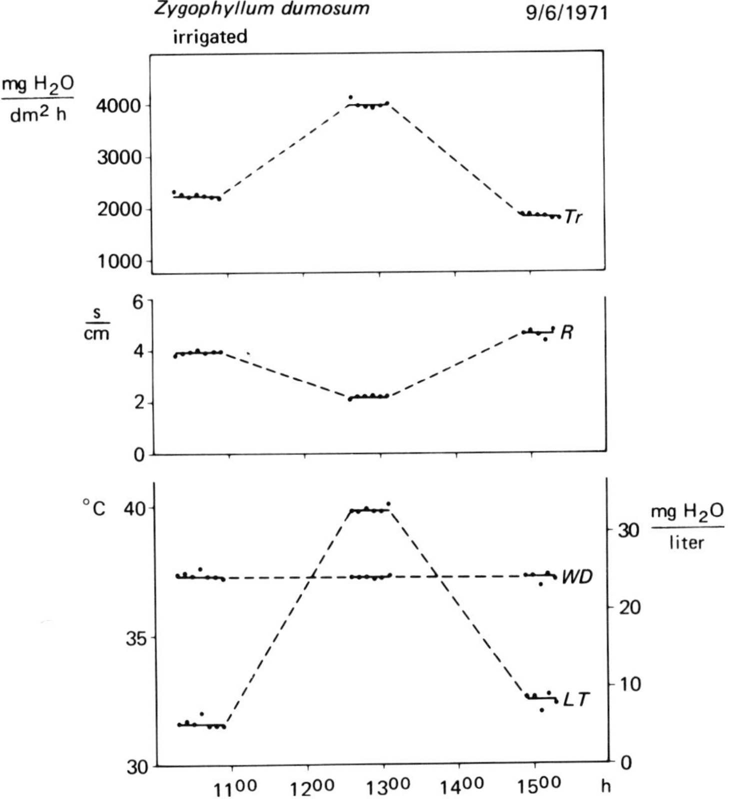

Fig. 8.9. Change in transpiration (*Tr*) and total diffusion resistance (*R*) with changes in leaf temperature (*LT*) for an irrigated *Zygophyllum dumosum* plant. The water-vapor-concentration difference between the leaf and the surrounding air (*WD*) was held constant. Abscissa: time of day. (After Schulze *et al.*, 1973, Fig. 1.)

Experiments on the effect of air humidity lead to the question of whether stomatal opening and closing reactions can be caused directly by a changing water-vapor-concentration difference between the leaf and the air rather than by changes in the water potential of the leaf tissue. Basic information regarding this question was obtained in experiments with isolated epiderms (Lange *et al.*, 1971), the outer surface of which was exposed to air with different water-vapor contents. Figure 8.12 shows that the stomata repeatedly open and close in response to the humidity conditions of the air above them, even though the inner surface was exposed to a constant high air humidity. Even stomata that are very close to each other react independently from each other to the humidity conditions in the air of the outer side of the epidermis (Fig. 8.13). Stomata exposed to dry air close, whereas those exposed to moist air open. Thus it becomes evident that the stomata have the ability to function as humidity sensors for the ambient air, measuring the difference in the water potential of the air inside and outside the leaf. The mechanism of this system possibly works through the peristomatal transpiration (Maercker, 1965).

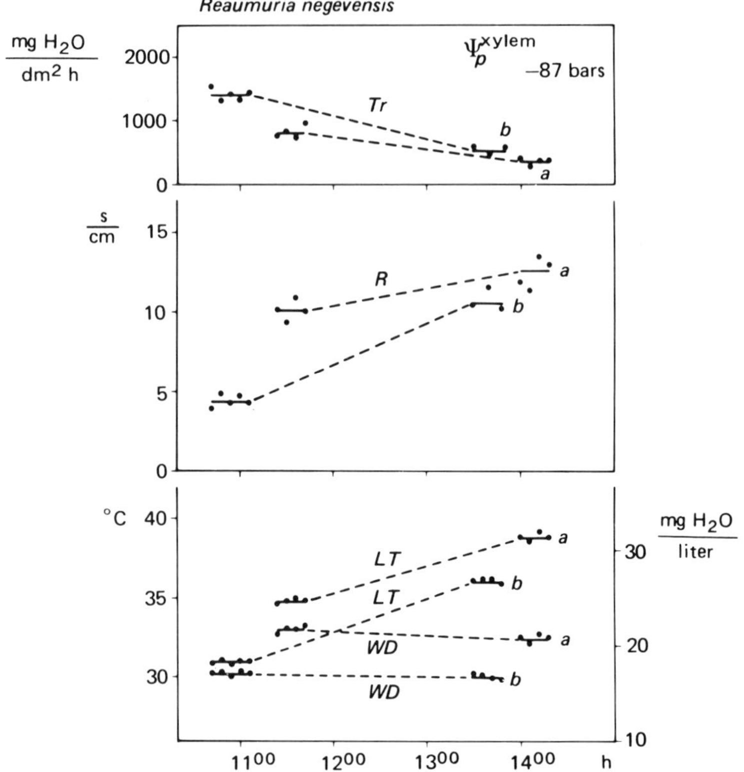

Fig. 8.10. Change in transpiration (*Tr*) and total diffusion resistance (*R*) with changes in leaf temperature (*LT*) for a large (a) and a small (b) water-vapor-concentration difference (*WD*) in *Reaumuria negevensis* under high water stress. Abscissa: time of day. (After Schulze *et al.*, 1973, Fig. 5.)

The same reaction was also found with intact leaves under desert conditions (Schulze *et al.*, 1972a). Figure 8.14 shows the results of an experiment with *P. armeniaca* in which the water-vapor-concentration difference between the leaf and its surrounding air was gradually increased from 10 to 21.5 mg H_2O liter^{-1} at constant leaf temperatures. The total diffusion resistance for water vapor increased simultaneously, and the rate of transpiration and net photosynthesis decreased. From the measurements of the leaf water content by beta-ray absorption, it follows that the stomatal reaction cannot be caused hydroactively by lowering of the water potential of the entire leaf. The water conditions improved simultaneously with the closing of the stomata. With subsequent lowering of the water vapor difference, the stomata reopened, in spite of increased water stress in the leaf. The same experiments were carried out with the xeromorphic photosynthesizing organs of *H. scoparia* and with the succulent leaves of *Z. dumosum* (Fig. 8.15). Hall and Kaufmann made similar findings with sesame, sunflower, and sugar beet. Their results show a linear increase of leaf resistance with *WD*, which turns to an exponential curve at larger values of *WD*. This response can be affected by temperature

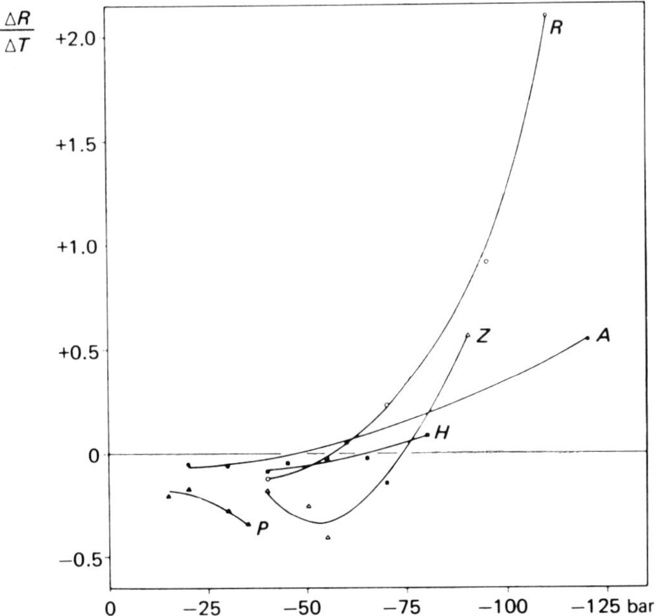

Fig. 8.11. Change in total diffusion resistance per 1-degree increase in leaf temperature ($\Delta R/\Delta T$) as related to the minimal water potential in the experimental plants during the day (bar). Positive values of $\Delta R/\Delta T$ designate stomatal closure, and negative values indicate stomatal opening with increasing temperature. *Artemisia herba-alba* (A), *Hammada scoparia* (H), *Prunus armeniaca* (P), *Reaumuria negevensis* (R), and *Zygophyllum dumosum* (Z). (After Schulze *et al.*, 1973, Fig. 6.)

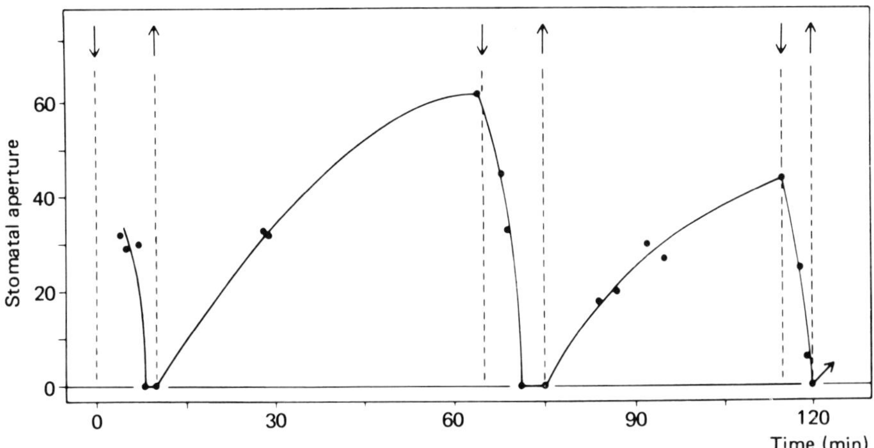

Fig. 8.12. Degrees of aperture of one stoma (*Polypodium*), in terms of relative value, during changes between moist and dry air over the outer side of the epidermis, and at a constant high air humidity at its inner side (↓, beginning; ↑, end of treatment with dry air). (After Lange *et al.*, 1971, Fig. 3.)

Fig. 8.13. Treatment of the outer side of an epidermis (*Polypodium*) with moist and dry air through two faintly visible capillaries (arrows). At a constant high air humidity at the inner side of the epidermis, stomata open in moist air (top, left) and closed in dry air (bottom, right). Magnification ca. 70 times. (After Lange *et al.*, 1971, Fig. 7.)

at the same time. A comparison of irrigated and nonirrigated plants showed that the stomatal reaction in the irrigated plants occurred not only at a higher level of diffusion resistance, but that its increase was also greater than in the irrigated plants. Since an air-humidity-controlled stomatal response avoids excess water loss in the plant before the water potential in the assimilatory organs decreases, it has special ecological significance for plants in arid habitats.

The results show a complicated control mechanism that is apparently of considerable importance for the daily balance of CO_2 exchange, the water relations, and thus for the primary production on a desert site. Endogenous and exogenous controls supplement each other, and the total diffusion resistance needs to be analyzed to ascertain which system ultimately determines the two-peaked assimilation and transpiration curves. From the temperature and humidity experiments, a model was developed that allows the computation of the changes in the leaf's total diffusion resistance to water vapor in relation to leaf temperature and ambient humidity. From this model of a particular plant in a particular phenological stage, a diurnal curve of the changes of the diffusion resistance can be predicted, as far as it depends solely on leaf temperature and humidity. These computed values of diffusion resistance can be compared with the directly observed gas-exchange

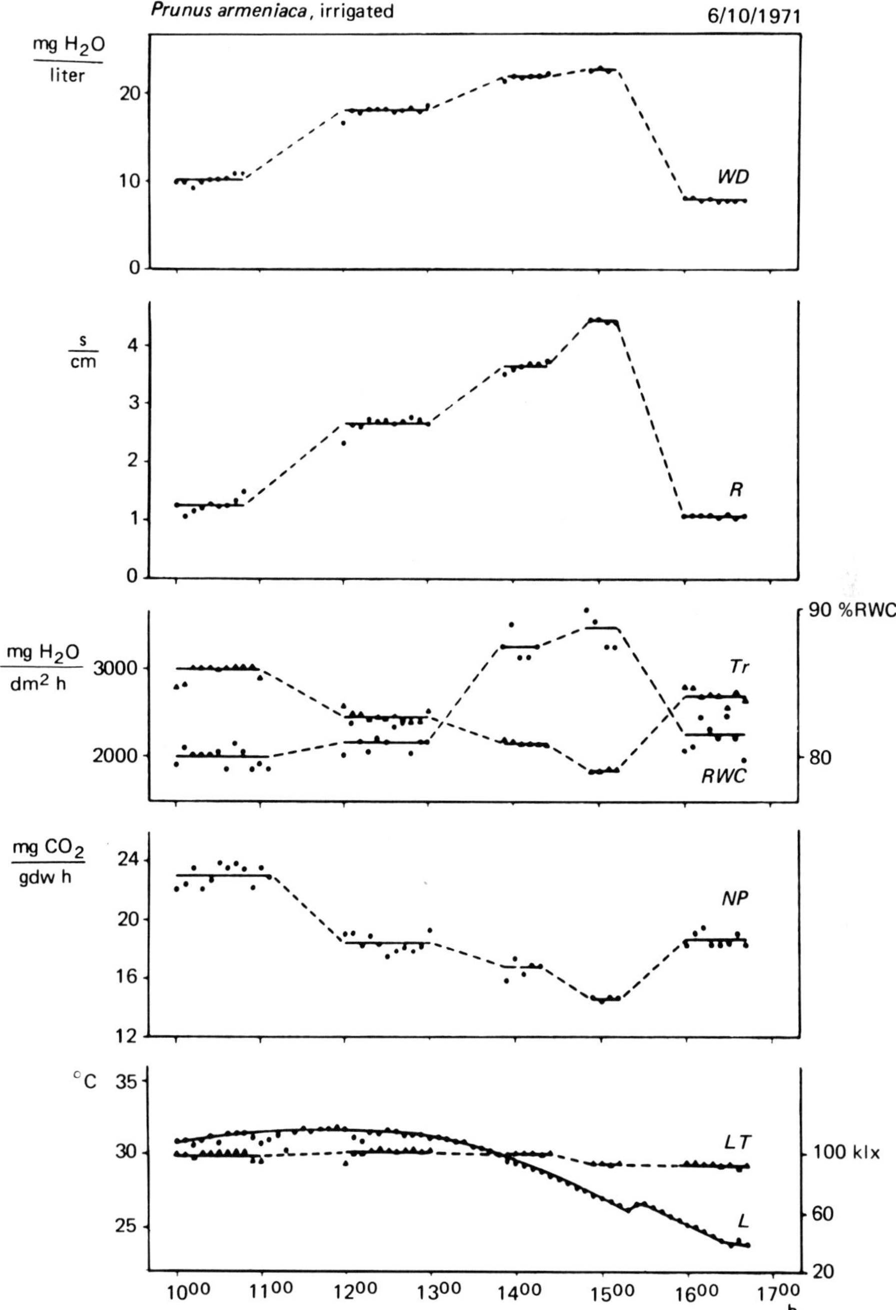

Fig. 8.14. Changes in the water-vapor-concentration difference between the leaf and the surrounding air (*WD*), together with the changes in the total diffusion resistance (*R*), transpiration (*Tr*), net photosynthesis (*NP*), and relative water content (%RWC) of the leaves in the irrigated *Prunus armeniaca*. The leaf temperature (*LT*) is kept constant. The light conditions are natural. Abscissa: time of day. (After Schulze *et al.*, 1972a, Fig. 1.)

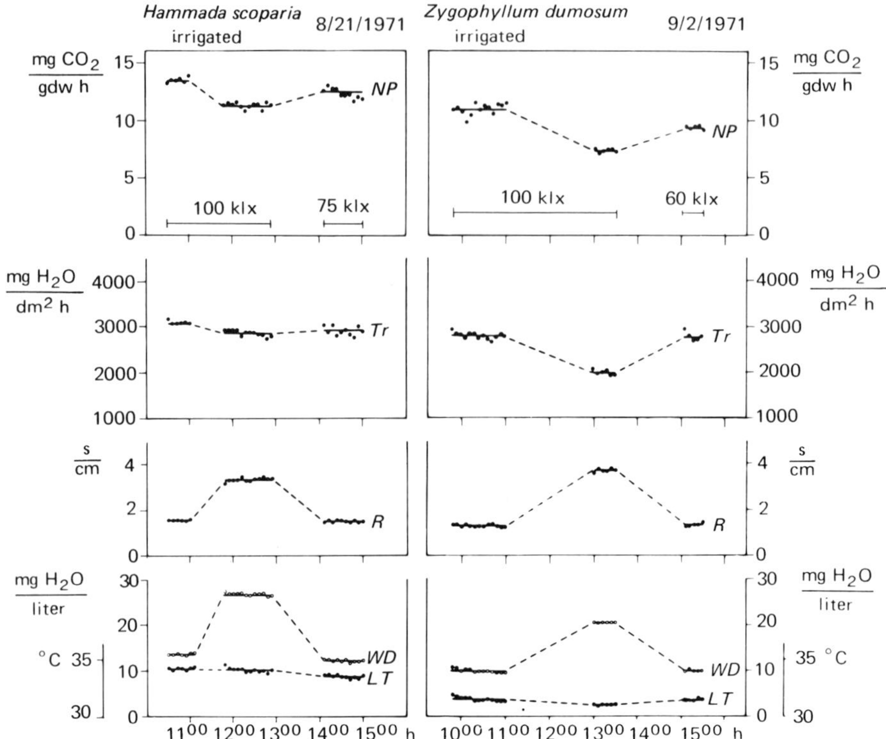

Fig. 8.15. Changes in net photosynthesis (*NP*), transpiration (*Tr*) and total diffusion resistance (*R*) as a reaction to a change in water-vapor-concentration difference (*WD*) between the leaf and the air in the irrigated *Hammada scoparia* and the irrigated *Zygophyllum dumosum*. The leaf temperature is kept constant. Abscissa: Time of day. (After Schulze *et al.*, 1972a, Fig. 5.)

measurements under natural conditions in the field. This will be demonstrated by two examples (see Schulze *et al.*, 1974).

Figure 8.16 summarizes data from a day with high leaf temperature and a pronounced peak in the water-vapor-concentration difference between leaf and air at noon. The change in stomatal aperture of *P. armeniaca* caused by this change in water-vapor difference is computed from the experimental data (Δr_{WD}). It shows a curve parallel to the course of water-vapor difference with a distinct peak at noon. At these water conditions of the apricot, the leaf temperature causes stomatal opening, reducing the humidity-induced stomatal closure. This reduction (Δr_{LT}) is subtracted from the first curve as shown by the hatched area. The resultant ($\sum \Delta r$) is a change in resistance induced by air humidity and temperature. The predicted diffusion resistance (r_p) is compared (half-hour intervals) with the observed values of R (r_0). The predicted values are almost identical with the observed ones for the first part of the day. There is a small deviation only in the afternoon,

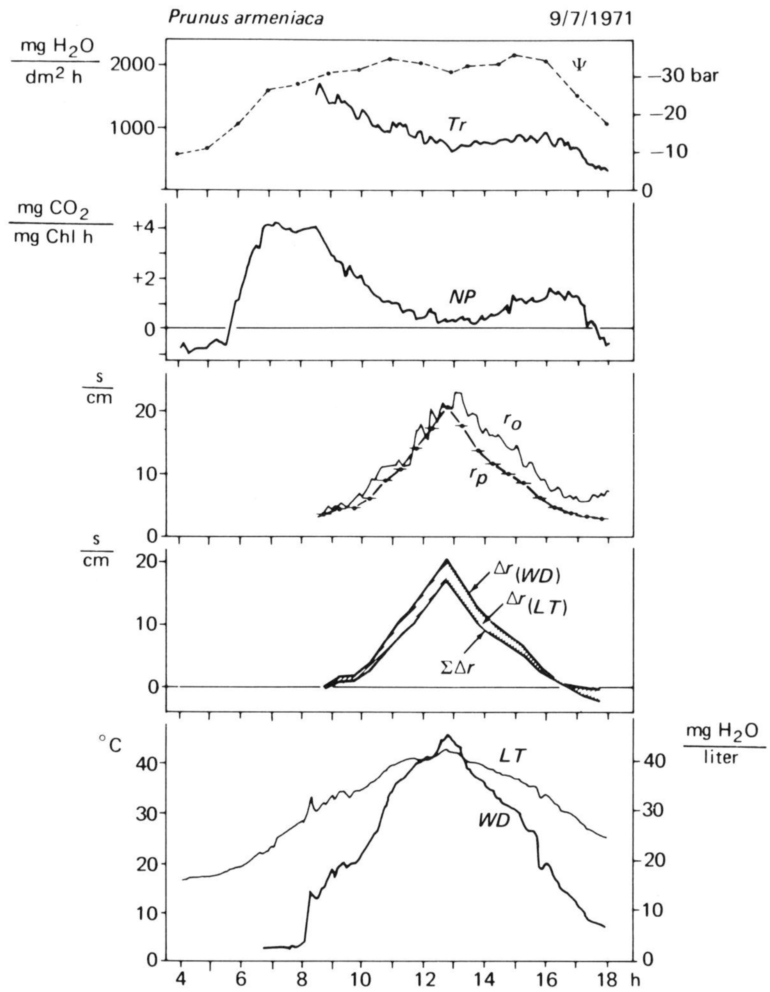

Fig. 8.16. Diurnal course of the negative hydrostatic pressure in the xylem (ψ), transpiration (Tr), net photosynthesis (NP), observed values of total diffusion resistance (r_o), predicted values of total diffusion resistance (r_p), change in diffusion resistance caused by changes in the water-vapor-concentration difference between the leaf and the surrounding air ($\Delta r_{(WD)}$), change in diffusion resistance caused by changes in leaf temperature ($\Delta r_{(LT)}$), total change in diffusion resistance caused by changes in leaf temperature and water-vapor-concentration difference ($\sum \Delta r$), leaf temperature (LT), and water-vapor-concentration difference between the leaf and the surrounding air (WD). Abscissa: time of day.

indicating the additional internal controlling mechanism for that part of the day. It is still clearly demonstrated, however, that the exogenous control of stomatal movement leads to the large noon depression of CO_2 uptake. The water potential of the leaf is stabilized during the day, even showing a slight increase at noon. The second peak of CO_2 intake is again primarily controlled by the change of the external climatic conditions. The stomata open at decreasing water-vapor differences, in spite of the low values of water potential and even in spite of its slight decrease. The significance of the climate-controlled stomatal movement becomes even more evident in comparison to a moist day (Fig. 8.17). For that day, the humidity-controlled stomatal closure is almost compensated by the temperature-induced stomatal opening. The predicted and observed resistance values are low throughout most of the day and increase only slightly in the later afternoon. This makes a high rate of photosynthesis possible throughout the day with only a gradual decrease in the afternoon. The high transpiration loss leads to a decrease in water potential, which reaches values as low as on the dry day.

Under the same plant water stress (similar maximum water potential in the morning and minimum values during the day), two distinctly different daily courses of net photosynthesis are consequently possible. There is a high carbon gain on the moist day and only a small gain on the dry day. These results demonstrate the great

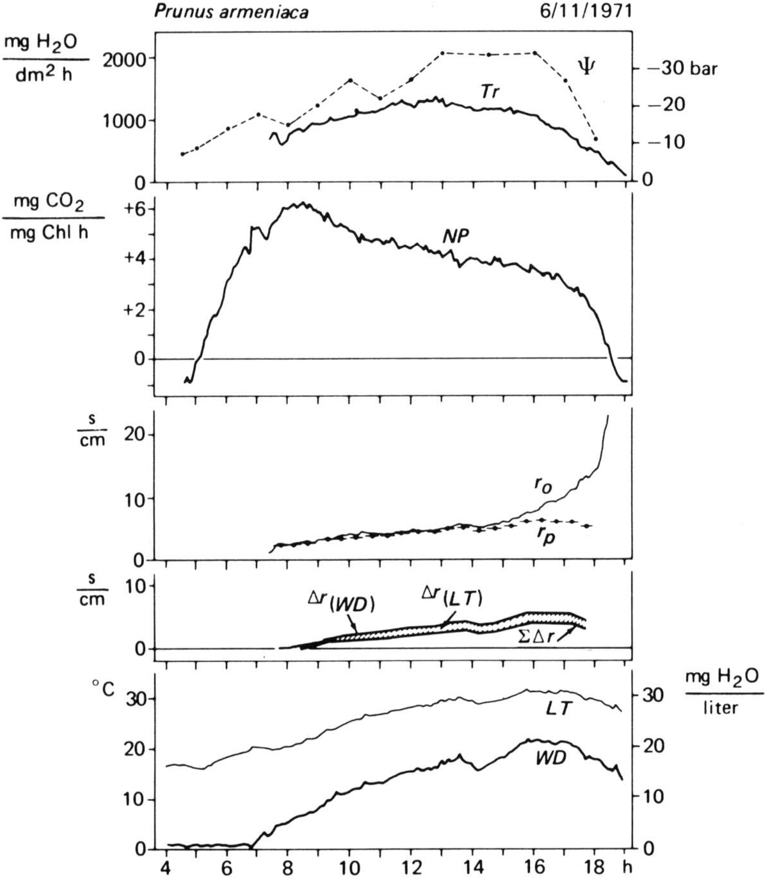

Fig. 8.17. Same as Fig. 8.16.

influence of the external climatic conditions on the CO_2 balance for the dry and hot desert climate. As shown in Fig. 8.5, the daily courses of CO_2 uptake are to a great extent the result of this complicated control mechanism. Internal and phenomenological factors probably have the greatest effect on the range within which the diffusion resistance is modulated by external control. They also determine the maximum rates of CO_2 uptake in the morning, which decreases throughout the season (Fig. 8.4; Oechel *et al.*, 1972; Hellmuth, 1968; Slatyer, 1957). But Moore *et al.* (1972) found a controlled transpiration rate at −115 bars, even under extreme water stress, and Kappen *et al.* (1972) showed that *A. herba-alba* is still photosynthetic active at −163 bars negative hydrostatic pressure in the xylem and an osmotic potential in its leaves of −92 bars.

The Daily Gain of Photosynthesis

Preliminary results of the annual change of the daily gain of net photosynthesis per gram dry weight together with the minimal daily water potential are given in Fig. 8.18. There is a great difference in the annual change between *H. scoparia* and *Z. dumosum*. *H. scoparia* has high rates of CO_2 uptake throughout most of the season. The decrease in autumn is not due to an impaired water balance. The water potential is the same in June at a high daily rate of CO_2 uptake and in September at a low one. This autumn depression is caused by the prevailing temperature and humidity conditions at that time of year and by the increasing dry weight of the xeromorphic structures in the photosynthesizing stems during the year. It is obvious that days of little and of high CO_2 gain follow each other closely. Light is not limiting under these conditions. These differences are mainly regulated by the external temperature and humidity conditions and not by water stress. The average daily CO_2 gain is only 48 percent of the highest daily rate. The minimal daily rate is just 11 percent of the maximum.

The seasonal change of the photosynthetic production in *Z. dumosum* is, however, closely related to the seasonal decrease in the water potential of the plant. High rates of CO_2 uptake continue only until June. Later in the year the gain of CO_2 uptake is low, finally becoming even negative in value. Since *Z. dumosum* sheds about 90 percent of its photosynthesizing organs at decreasing water stress (Zohary and Orshan, 1954; Orshan, 1963), the respiratory loss of the remaining leaves at the end of the year has only small influence on the total annual CO_2 gain of the plant. The main proportion of the total annual gain is fixed before leaf-shedding in June, during the period of high photosynthetic activity. However, during this time the large effect of the temperature- and humidity-controlled stomatal reactions is also obvious from the variation in the CO_2 gain from one day to the next. At a constant daily solar radiation, it varies from 10.8 mg CO_2 gdw^{-1} d^{-1} to 71.1 mg CO_2 gdw^{-1} d^{-1}. The average rate is only 45 percent of the maximum daily CO_2 uptake.

The reduction of the daily CO_2 gain through stomatal regulation by the

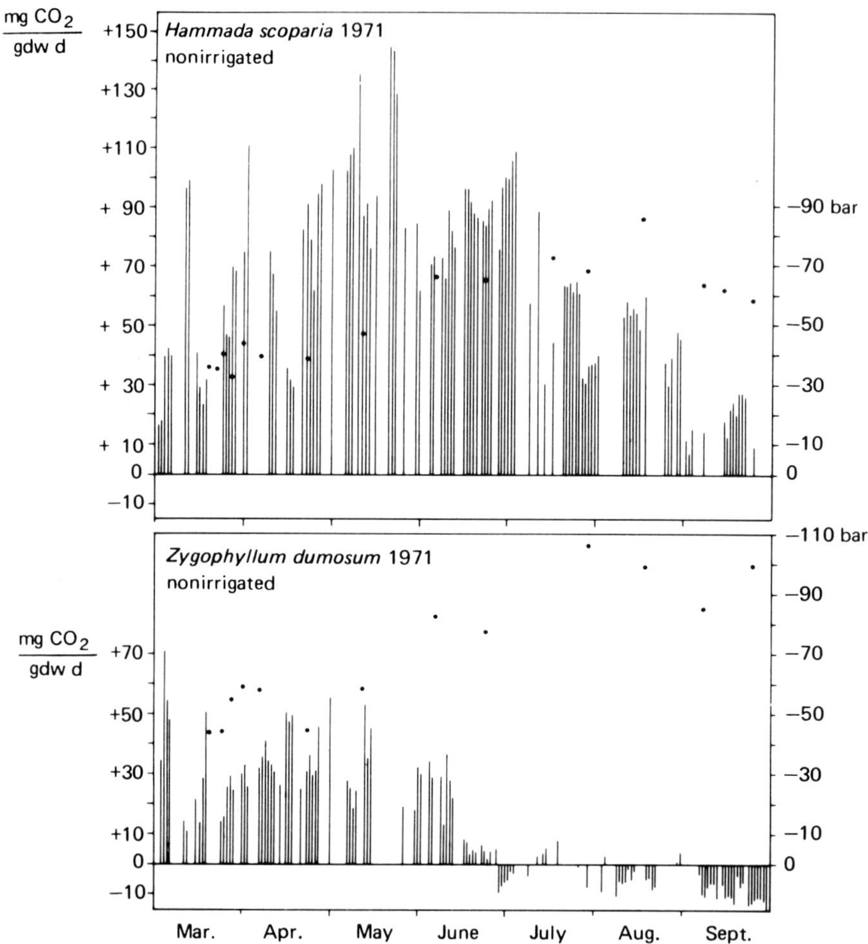

Fig. 8.18. Annual course of the daily gain of net photosynthesis (vertical lines) together with the minimal daily water potential (dots) for *Hammada scoparia* and *Zygophyllum dumosum*. Abscissa: time of year.

external climate is in a certain sense "disadvantageous" for the primary production. However, it is surely necessary for the water balance of the plants (Evenari and Richter, 1937–1938; Evenari *et al.*, 1971; Orshan, 1973). Under these stress conditions, this mechanism allows a stabilization of plant water relations for a long time, which again is the basis for any primary production. These plants could not survive without this stomatal regulation, nor without the ability of changing leaf structure and the biomass reduction. In these habitats, CO_2 balance and water balance are connected decisively for plant existence. Both are determined by the external conditions and the specific functional responses of the plant, which may vary from one species to another. Consideration of these facts will be the base for development of photosynthetic models of desert plants.

Acknowledgment

We thank H. Wiebe (Utah State University, Logan) for reading the manuscript of this article.

References

Adams, M. S., Strain, B. R.: 1968. Photosynthesis in stems and leaves of *Cercidium floridum*: spring and summer diurnal field response and relation to temperature. Oecologia Plantarum *3*, 285–297.

Björkman, O., Pearcy, R. W., Harrison, A. T., Mooney, H. A.: 1972. Photosynthetic adaptation to high temperatures: a field study in Death Valley, California. Science *175*, 786–789.

Boyer, J. S.: 1967. Leaf water potential measured with a pressure chamber. Plant Physiol. *42*, 133–137.

Buschbom, U.: 1970. Zur Methodik kontinuierlicher Wassergehalt-Bestimmungen an Blättern mittels β-Strahlenabsorption. Planta *95*, 146–166.

Caldwell, M. M.: 1972. Adaptability and productivity of species possessing C_3 and C_4 photosynthesis in a cool desert environment. *In* Ecophysiological foundation of ecosystems productivity in arid zone, pp. 27–29. Leningrad: Nauka.

Evenari, M.: 1937–38. The physiological anatomy of the transpiratory organs and the conducting systems of certain plants of the wilderness of Judaea. J. Linnean Soc. London *51*, 389–407.

———, Richter, R.: 1937–38. Physiological-ecological investigations in the wilderness of Judaea. J. Linnean Soc. London *51*, 333–351.

———, Shanan, L., Tadmor, N.: 1971. The Negev: the challenge of a desert. Cambridge, Mass.: Harvard Univ. Press.

Hammouda, M., Lange, O. L.: 1962. Zur Hitzeresistenz der Blätter höherer Pflanzen in Abhängigkeit von ihrem Wassergehalt. Naturwissenschaften *49*, 500.

Hellmuth, E. O.: 1968. Eco-physiological studies on plants in arid and semi-arid regions in western Australia. I. Autecology of *Rhagodia baccata* (Labill.) Moq. J. Ecol. *56*, 319–344.

Kappen, L.: 1966. Der Einfluß des Wassergehaltes auf die Widerstandsfähigkeit von Pflanzen gegenüber hohen und tiefen Temperaturen, untersucht an Blättern einiger Farne und von *Ramonda myconi*. Flora, Abt. B *156*, 427–445.

———, Lange, O. L., Schulze, E.-D., Evenari, M., Buschbom, U.: 1972. Extreme water stress and photosynthetic activity of the desert plant *Artemisia herba-alba* Asso. Oecologia *10*, 177–182.

Koch, W., Lange, O. L., Schulze, E.-D.: 1971. Ecophysiological investigations on wild and cultivated plants in the Negev Desert. I. Methods: A mobile laboratory for measuring carbon dioxide and water vapour exchange. Oecologia *8*, 296–309.

Lange, O. L.: 1959. Untersuchungen über Wärmehaushalt und Hitzeresistenz mauretanischer Wüsten- und Savannenpflanzen. Flora *147*, 596–651.

———, Lange, R.: 1963. Untersuchungen über Blattemperaturen, Transpiration und Hitzeresistenz an Pflanzen mediterraner Standorte (Costa Brava, Spanien). Flora *153*, 387–425.

———, Schulze, E.-D.: 1971. Measurement of CO_2 gas exchange and transpiration in the beech (*Fagus silvatica* L.). *In* Ecological studies (ed. H. Ellenberg), vol. 2, pp. 16–28. New York: Springer-Verlag.

———, Koch, W., Schulze, E.-D.: 1969. CO_2-Gaswechsel und Wasserhaushalt von Pflanzen in der Negev-Wüste am Ende der Trockenzeit. Ber. Dtsch. Botan. Ges. *82*, 39–61.

———, Lösch, R., Schulze, E.-D., Kappen, L.: 1971. Responses of stomata to changes in humidity. Planta *100*, 76–86.

Lommen, P. W., Schwintzer, C. R., Yocum, C. S., Gates, D. M.: 1971. A model describing photosynthesis in terms of gas diffusion and enzyme kinetics. Planta *98*, 195–220.

Maercker, U.: 1965. Zur Kenntnis der Transpiration der Schließzellen. Protoplasma *60*, 61–78.

Mooney, H. A., West, M.: 1964. Photosynthetic acclimation of plants of diverse origin. Am. J. Bot. *51*, 825–827.

Moore, R. T., White, R. S., Caldwell, M. M.: 1972. Transpiration of *Atriplex confertifolia* and *Eurotia lanata* in relation to soil, plant, and atmospheric moisture stress. Can. J. Bot. *50*, 2411–2418.

Oechel, W. C., Strain, B. R., Odening, W. R.: 1972. Photosynthetic rates of a desert shrub *Larrea divaricata* Cav. under field conditions. Photosynthetica *6*, 183–188.

Orshan, G.: 1963. Seasonal dimorphism of desert and Mediterranean chamaephytes and their significance as a factor in their water economy. *In* The water relations of plants (eds. A. J. Rutter, F. H. Whitehead), pp. 206–222. Oxford: Blackwell.

———: 1973. Morphological and physiological plasticity in relation to drought. *In* Wildland shrubs—their biology and utilization (eds. C. M. McKell, J. P. Blaisdell, J. R. Goodin), USAD Forest Service Gen. Tech. Rep. INT-1, pp. 245–254. Ogden, Utah: Intermountain Forest and Range Experiment Station.

Schulze, E.-D.: 1972. A new type of climatized gas exchange chamber for net photosynthesis and transpiration measurements in the field. Oecologia *10*, 243–251.

———, Lange, O. L., Buschbom, U., Kappen, L., Evenari, M.: 1972a. Stomatal responses to changes in humidity in plants growing in the desert. Planta *108*, 259–270.

———, Lange, O. L., Koch, W.: 1972b. Ökophysiologische Untersuchungen an Wild- und Kulturpflanzen der Negev-Wüste. II. Die Wirkung der Außenfaktoren auf CO_2-Gaswechsel und Transpiration am Ende der Trockenzeit. Oecologia *8*, 334–355.

———, Lange, O. L., Koch, W.: 1972c. Ökophysiologische Untersuchungen an Wild- und Kulturpflanzen der Negev-Wüste. III. Tagesläufe von Nettophotosynthese und Transpiration am Ende der Trockenzeit. Oecologia *9*, 317–340.

———, Lange, O. L., Lembke, G.: 1972d. A digital registration system for net photosynthesis and transpiration measurements in the field and an associated analysis of errors. Oecologia *10*, 151–166.

———, Lange, O. L., Kappen, L., Buschbom, U., Evenari, M.: 1973. Stomatal responses to changes in temperature at increasing water stress. Planta *110*, 29–42.

———, Lange, O. L., Evenari, M., Kappen, L., Buschbom, U.: 1974. The role of air humidity and leaf temperature in controlling stomatal resistance of *Prunus armeniaca* L. under desert conditions. I. A simulation of the daily course of stomatal resistance. Oecologia (in press).

Slatyer, R. O.: 1957. The influence of progressive increases in total soil moisture stress on transpiration, growth, and internal water relationships of plants. Austral. J. Biol. Sci. *10*, 320–336.

Stocker, O.: 1960. Die photosynthetischen Leistungen der Steppen- und Wüstenpflanzen. *In* Handbuch der Pflanzenphysiologie (ed. W. Ruhland), vol. 5, pp. 460–491. New York: Springer-Verlag.

———: 1970. Der Wasser- und Photosynthesehaushalt von Wüstenpflanzen der mauretanischen Sahara. I. Regengrüne und immergrüne Bäume. Flora *159*, 539–572.

———: 1971. Der Wasser- und Photosynthesehaushalt von Wüstenpflanzen der mauretanischen Sahara. II. Wechselgrüne, Rutenzweig- und stammsukkulente Bäume. Flora *160*, 445–494.

———: 1972. Der Wasser- und Photosynthesehaushalt von Wüstenpflanzen der mauretanischen Sahara. III. Kleinsträucher, Stauden und Gräser. Flora *161*, 46–110.

Strain, B. R., Chase, V. C.: 1966. Effect of past and prevailing temperatures on the carbon dioxide exchange capacities of some woody desert perennials. Ecology *47*, 1043–1045.

Zohary, M.: 1953. Ecological studies on the vegetation of the near eastern deserts. III. Vegetation map of the central southern Negev. Palestine J. Bot. *6*, 27–36.

———: 1962. Plant life of Palestine. New York: Ronald Press.

———, Orshan, G.: 1954. Ecological studies in the vegetation of the near eastern deserts. V. The Zygophylletum dumosi and its hydroecology in the Negev of Israel. Vegetatio *5–6*, 340–350.

9

Field Measurements of Carbon Dioxide Exchange in Some Woody Perennials

BOYD R. STRAIN

Introduction

The objectives and problems of making measurements of net carbon dioxide exchange under natural field conditions have been discussed by several investigators, including Talling (1961), Larcher (1963), Lange, Schulze and Koch (1970), and Mooney (1972). In addition, several workers have reviewed reports of field measurements and have provided lists of maximum values of net photosynthesis under normal field conditions (Stocker, 1935; Rabinowitch, 1951; Verduin, 1953; Spector, 1956; Verduin *et al.*, 1959; Altman and Dittmer, 1964; Larcher, 1969; Wolf, 1969; Šesták *et al.*, 1971).

As recently discussed by Mooney (1972), however, many of the rates determined prior to the development of plant chambers (cuvettes) with environmental control (Bosian, 1953, 1955) are questionable. In fact, temperature control was not generally available in field studies until the early 1960s. We described one of the first temperature- and turbulence-controlled field-chamber designs to be extensively used (Mooney *et al.*, 1964). Lange (1962) had earlier described the "Klapp-Kuvette," but, except for Eckardt (1968), that design has not been widely adopted.

Although chamber vapor-pressure control was finally developed for routine use by Lange and his coworkers (1969), many measurements are still being conducted without vapor-pressure control. None of the measurements to be reported in this paper was obtained under vapor-pressure control, partly because all the data presented in this paper were collected prior to 1970 except for that shown for *Pinus taeda*.

Several investigators have considered the carbon dioxide exchange characteristics of woody plants (Kramer, 1958; Kozlowski and Keller, 1966; Larcher, 1969). Physiological and ecological problems associated with the perennial growth habit and water relations associated with the resistances of woody vascular tissue are examples of the characteristics of woody plants receiving attention. In trees, foliage is often distributed over vertical gradients several meters long. Thus microenvironmental conditions may vary considerably and require particular attention in physiological–ecological analyses of large trees.

The specific objectives of this paper are to present values for maximum net photosynthesis rates of some woody plants, most of which have not previously been published; to relate the values to selected environmental and phenological factors; and to discuss further the use of maximal values of net photosynthesis.

Methods

The technical procedures we have followed through the years will be briefly described here. All these procedures are discussed in detail in the recent manual of methods edited by Šesták *et al.* (1971). The data were collected over a period of 11 years (1962–1973). Consequently, technical procedures and instrumentation vary. In general, the measurements were begun in early morning and were continued throughout the daylight hours following the ambient environment. Some of the measurements, however, were short term and made at various times of day. In those measurements, time of day is not available. Following the period of photosynthesis measurements, a dark hood was placed over the chamber, the temperature in the chamber adjusted to 25°C, and carbon dioxide exchange monitored until equilibrium was attained. This rate is reported as 25°C dark respiration, where available. The data presented here were all obtained with the system shown in Fig. 9.1 (Strain, 1969) or a modification of it. The basic system is designed to accommodate 1–3 plant chambers, whose temperature control is obtained by circulating temperature-controlled water from the refrigerated water bath through an internally mounted radiator. In the chambers used since 1970, a 300-W nichrome resistance coil mounted between the radiator and the chamber space is used to back-heat the chamber air to the desired temperature. In other words, the air leaves the radiator core 1–2°C below the required temperature and is heated to the desired temperature. Chamber temperatures can be set and maintained where desired (± 0.5°C), or the ambient air temperatures can be tracked by means of thermistor-controlled devices. If desired, leaf temperatures may be monitored by thermocouples and their mean used to determine chamber temperature. Heat exchange and air circulation within the chamber are regulated with an electric fan.

Ambient air is drawn from a source upwind of the field site with the air inlet placed well above the vegetation canopy. The air stream is metered to the test chambers and the reference cell of the IRGA through rotameter-type flow meters. If changing CO_2 concentrations are encountered due to industrial or natural contamination, a large-volume container (>200 liters) placed on the inlet line will decrease errors caused by sudden and frequent changes.

From 1962–1964 sensitivity was low because we were using a Beckman 15-A infrared gas analyzer (IRGA) with a closed reference cell. The analyzer was cali-

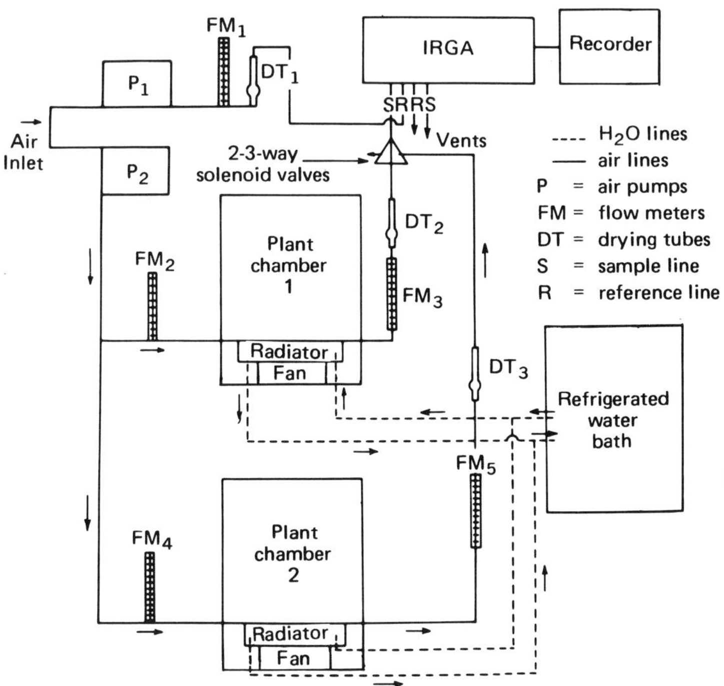

Fig. 9.1. Basic configuration of the system used since 1964. Various modifications are explained in the text. (From Strain, 1969.)

brated with untested, commercially supplied calibration gases on a 0–600 ppm CO_2 span. Since 1964, only differential analyzers with open reference cells have been used in these measurements, calibrating on a 0–100 ppm CO_2 span by using gases from the U.S. Bureau of Standards and other procedures to check values supplied from commercial sources.

If vapor-pressure control is being attempted, a refrigerated condenser is placed in the air line before each chamber to set the dew point of the air entering the chambers (Koch *et al.*, 1968). A measurement of the dew point or the temperature and relative humidity of the air leaving the chamber will provide an estimate of transpiration if compared to the incoming vapor-pressure and air-flow rate.

At this point I would like to discuss briefly a practical difficulty in measuring transpiration in a chamber which uses an internally mounted heat exchanger. The investigator must accurately measure the temperature of the coldest surface within the chamber for comparison with the effluent dew point. If any temperature in the system drops below the dew-point temperature of the downstream measurement, transpiration estimates will be invalid. Various solutions for this problem exist (Chapter 2 in Šesták *et al.*, 1971), but they are generally complicated and to date have rarely been applied. The practical solution is to monitor the various temperatures and avoid condensation in the chamber by adjusting flow rates or incoming dew point. The researcher must be aware, however, of possible measurement errors and loss of sensitivity associated with changing flow rates (Slatyer, personal communication).

The CO_2 content of the air entering and leaving the chamber is determined. The differential between the air supply and the chamber effluent is a function of the rate of flow into the chamber and the rate of net carbon dioxide exchange of the plant sample. The differential should be kept small (< 10–30 ppm CO_2) to minimize the effect of abnormal CO_2 concentrations. The null-point procedures developed by Koller and Samish (1964) and Koller (p. 81 in Šesták *et al.*, 1971) solves the above problem, but requires more complicated control equipment than is generally available.

Results and Discussion

Desert Shrubs and Trees

Table 9.1 presents selected measurements of desert shrubs during the period March 1963–July 1969 made with the mobile laboratory previously described (Strain, 1965, 1969). The data source column indicates whether the data have been previously published and gives citations if available.

The species are arranged in Table 9.1 according to growth form as indicated below:

Xerophytic, evergreen:	*Larrea divaricata*
	Simmondsia chinensis
	Isomeris arborea
Mesophytic, drought deciduous:	*Encelia farinosa*
	Hymenoclea salsola
	Fouquiera splendens
Phreatophytic, winter deciduous:	*Chilopsis linearis*
	Prosopis juliflora
	Acacia greggii

Cercidium floridum has characteristics of all three categories (Adams, 1968). The leaves of *Cercidium floridum* are mesophytic in structure and often die and fall from the plants during the summer droughts. If rain is not received until late in the summer, when daylength is becoming shorter, new foliage will not appear until the following spring. The chlorophyllous stem cortex, however, is photosynthetic and thus is like an evergreen in its year-round productivity. *Fouquiera splendens* also has photosynthetic bark (Mooney and Strain, 1964), but the rates observed in March 1963 were much lower than those observed in *Cercidium.*

Two important points may be realized by comparing the results shown in Table 9.1. Desert plants tend to reach maximum net photosynthesis early in the day and, in general, evergreen species have lower net photosynthesis rates than do the more mesophytic deciduous plants.

Table 9.1 Maximum Net Photosynthesis of Desert Shrubs Under Natural Field Conditions

Species	Data Source[a]	Date	Time (PST)	Net Photo. Rate[b]	Temp. (°C)	Radiation ergs × 10^5	Radiation fc × 10^3	25°C Dark Resp.[b]	Water Pot. (atm.)	Comments
Larrea divaricata	1	3/18/66	1130	4.8	20	7.0		0.6		Varner Road, Calif.
	1	3/19/66	0900	11.6	16	5.3		1.1		Varner Road, Calif.
	1	4/28/66	1647	8.8	19		11	1.1		Varner Road, Calif.
	1	7/13/66	0908	6.1	35	7.8		0.4		Deep Canyon, Calif.
	2	7/20/66	0900	7.2	35	7.4				Varner Road, Calif.
	1	8/02/66	0830	5.2	39	6.0		0.3		Deep Canyon, Calif.
	1	8/03/66	0735	5.7	34	5.2		0.6		Deep Canyon, Calif.
	3	9/08/66	0800	9.3	34	5.9		0.5		Varner Road, Calif.
	1	10/29/66	0930	4.0	30		10	1.3		Deep Canyon, Calif.
	1	12/21/66	0915	6.2	14		10		−47	Tucson, Arizona
	1	12/30/66	0930	5.5	12		10			Deep Canyon, Calif.
	2	1/27/67	0910	5.3	18	7.4		1.0		Deep Canyon, Calif.
	1	5/02/67	0800	4.2	24	7.2				Deep Canyon, Calif., sample 1
	1	5/03/67	0730	3.4	24	7.2				Deep Canyon, Calif., sample 1
	1	5/03/67	1630	7.1	27	4.5				Deep Canyon, Calif., sample 2
	1	5/04/67	0800	8.3	30	6.6			−34	Deep Canyon, Calif., sample 2
	1	6/29/67	0720	9.2	28		12			Deep Canyon, Calif.
	1	7/19/67	0710	13.3	26	5.0				Deep Canyon, Calif.
	4,5	7/09/68	0800	3.0	24		9		−55	Deep Canyon, Calif.
	4,5	9/19/68	0900	2.0	34		12		−62	Deep Canyon, Calif.
	4,5	10/12/68	0900	2.0	25		12		−62	Deep Canyon, Calif., morning
	4,5	10/12/68	1530	2.0	29		12		−67	Deep Canyon, Calif., afternoon
	4,5	11/23/68	1100	2.7	26		12		−55	Deep Canyon, Calif.
	4,5	2/02/69	1330	7.0	18		12		−34	Deep Canyon, Calif.
	4,5	4/08/69	0800	5.0	20		9		−44	Deep Canyon, Calif.
	4,5	5/20/69	0700	8.0	20		5		−35	Deep Canyon, Calif.
	4,5	6/08/69	0700	4.0	28		4		−50	Deep Canyon, Calif.

(*continued*)

Table 9.1—*continued*

Species	Data Source[a]	Date	Time (PST)	Net Photo. Rate	Temp. (°C)	Radiation: ergs × 10^5	Radiation: fc × 10^3	25°C Dark Resp.[b]	Water Pot. (atm.)	Comments
Simmondsia chinensis	6	7/20/67	0915	5.7	33		12	0.2	−35	Morongo Canyon, Calif., male
	6	7/21/67	1000	4.1	35		12	0.5	−41	29 Palms, Calif., male
	6	7/24/67	0915	9.5	24		12	2.0	−29	San Diego, Calif., male
	6	8/07/67	1000	2.2	34		12	0.2	−55	Tucson, Arizona, male
	6	8/07/67	1030	3.5	29		12	0.2	−60	Tucson, Arizona, female
Isomeris arborea	1	7/21/66	0830	10.1	37	7.2		1.7		Morongo Canyon, Calif.
Hymenoclea salsola	2	7/14/66	0730	7.9	29	7.4		1.2		Deep Canyon, Calif.
	2	1/28/67	0900	15.2	19	7.4		2.3		Deep Canyon, Calif.
	1	5/03/67	1000	14.9	27	6.6		2.1	−10	Deep Canyon, Calif.
	1	6/28/67	0855	6.9	35		12	1.3		Deep Canyon, Calif.
	1	7/18/67	0912	9.2	34	5.9		1.6		Deep Canyon, Calif.
Fouquieria splendens	7	3/ /63		5.2	20		5	1.1		Deep Canyon, Calif., bark and leaves
	7	3/ /63		6.5	20		5	0.9		Deep Canyon, Calif., bark and leaves
	7	3/ /63		0.4	20		5	1.0		Deep Canyon, Calif., bark only
	7	3/ /63		0.2	20		5	0.8		Deep Canyon, Calif., bark only
Encelia farinosa	2	7/22/66	0800	2.8	30	7.4		1.2		Deep Canyon, Calif.
	1	10/28/66	1000	4.8	30		12		−35	Deep Canyon, Calif.
	2	12/29/66	0945	15.3	16	7.4	12	2.6	−22	Deep Canyon, Calif.
	1	2/26/67	1015	15.0	22		12		−36	Deep Canyon, Calif.
	1	5/02/67	0912	6.7	18		12		−40	Deep Canyon, Calif.
	1	6/27/67	0800	11.8	41		12		−27	Deep Canyon, Calif.
	1	9/07/67	0800	37.5	35		12		−11	Deep Canyon, Calif.
	8,9	10/28/67		3.5	30		10		−30	Deep Canyon, Calif.
	8,9	12/29/67		12.5	15		12		−26	Deep Canyon, Calif.

	8,9	2/26/68		13.6	23		12		−34	Deep Canyon, Calif.
	8,9	5/02/68		3.7	24		12		−40	Deep Canyon, Calif.
	8,9	6/27/68		+6.7	47		12		−36	Deep Canyon, Calif., loss of CO_2
	8,9	9/07/68		35.0	35		12		−14	Deep Canyon, Calif.
	8,9	12/01/67		25.7	12		8		−5	Riverside, Calif.
	8,9	1/13/68		41.5	23		12		−17	Riverside, Calif.
	8,9	4/03/68		14.4	16		12		−18	Riverside, Calif.
	8,9	5/25/68		12.3	28		12		−23	Riverside, Calif.
	8,9	7/26/68		6.0	38		12		−38	Riverside, Calif.
Chilopsis linearis	1	6/25/66	1000	14.7	30	7.7				Burns Canyon, Calif.
	1	7/13/66		12.9	30			0.7		Deep Canyon, Calif.
Prosopis juliflora	3	7/20/66	0830	14.3	35	7.0		1.4		Varner Road, Calif.
	3	9/07/66	0945	7.1	37	6.9		0.6		Varner Road, Calif.
	3	9/08/66	0845	6.3	35	5.9				Varner Road, Calif.
Acacia greggii	2	7/15/66	0700	14.2	31	7.4		0.7		Deep Canyon, Calif., 1966 leaves
	1	8/04/66	0810	11.2	34	5.0		1.4		Deep Canyon, Calif., 1966 leaves
	2	1/26/67	1100	14.7	26	7.4		0.8		Deep Canyon, Calif., 1966 leaves
	1	5/04/67	0900	6.1	28	6.6			−22	Deep Canyon, Calif., 1967 leaves
	1	6/29/67	0600	17.2	34	6.8		0.8		Deep Canyon, Calif., 1967 leaves
	1	7/17/67	1010	29.1	34	6.5		1.3		Deep Canyon, Calif., 1967 leaves
Cercidium floridum	10	4/06/66	0845	9.0	23	7.0				Deep Canyon, Calif., bark and leaves
	10	4/06/66	0845	1.1	24	7.0				Deep Canyon, Calif., bark only
	10,11	8/39/66	0840	2.3	25	6.2				Deep Canyon, Calif., bark only
	10,11	8/31/66	1400	3.1	44	7.8				Deep Canyon, Calif., bark only
	10,11	6/28/67	0755	7.2	35		12			Deep Canyon, Calif., bark and leaves
	10,11	7/19/67	0600	15.0	20	6.0			−25	Deep Canyon, Calif., leaves only
	10,11	7/19/67	0730	6.0	30	7.8			−30	Deep Canyon, Calif., bark and leaves
	1	7/10/68	1168	2.3	42		12			Deep Canyon, Calif., bark and leaves

[a] 1, Strain, previously unpublished; 2, Strain, 1969; 3, Strain, 1970; 4, Oechel, 1970; 5, Oechel *et al.*, 1972; 6, Al-Ani *et al.*, 1972; 7, Mooney and Strain, 1964; 8, Cunningham, 1968; 9, Cunningham and Strain, 1969; 10, Adams, 1968; 11, Adams and Strain, 1968.

[b] Rates are all mg CO_2 g leaf dry weight^{-1} h^{-1}.

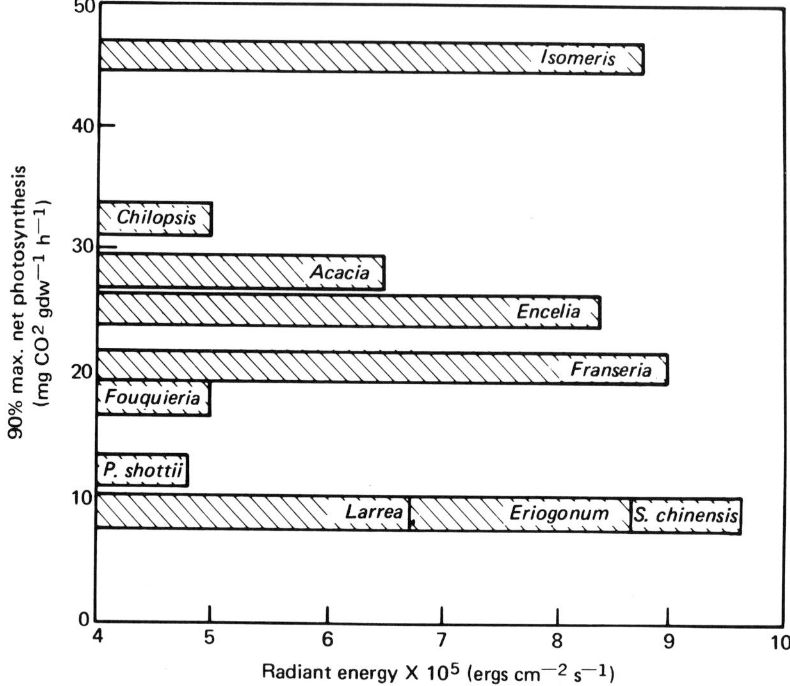

Fig. 9.2. Effect of radiant energy flux on net photosynthesis of potted plants grown in a standard greenhouse. Complete identifications for the various species may be found in Table 9.2. Radiant energy was provided by a tungsten lamp (GE cool beam 300 PAR/MF) and was measured with a YSI-Kettering Radiometer, Model 65.

Table 9.2 Effect of Radiant Energy Flux on Maximum Net Photosynthesis Rates of Some Woody Desert Perennials Grown in Pots in a Standard Greenhouse[a]

Species	Chamber Temp. (°C)	Radiant Energy (*ergs* cm^{-2} s^{-1} $\times$ 10^5)	Photosynthesis (mg CO_2 gdw^{-1} h^{-1})
Isomeris arborea	30	10.0	50.8
Chilopsis linearis	25	10.0	35.6
Acacia greggii	30	10.0	31.1
Encelia farinosa	25	10.0	28.1
Franseria dumosa	25	10.0	22.2
Fouquieria splendens	20	10.0	20.0
Peucephyllum shottii	25	9.4	13.1
Eriogonum fasciculatum	20	9.4	9.3
Simmondsia chinensis	30	10.0	9.2
Larrea divaricata	25	10.0	9.2

[a] Light curves were measured at the temperature of maximum net photosynthesis for the respective species; 10 $\times$ 10^5 ergs cm^{-2} s^{-1} was the maximum energy applied.

The potential for this characteristic is clearly shown in Fig. 9.2, which presents data obtained from potted plants grown under common greenhouse conditions. Each species was measured at its own temperature optimum. The vertical line following each generic name indicates the radiant energy flux required to reach 90 percent maximum net photosynthesis. The maximum values obtained in the experiment are shown in Table 9.2. Only *Isomeris arborea* does not follow the trends observed in Table 9.1. Although *Isomeris arborea* is foliated year round on the Colorado Desert, and consequently must be classified as evergreen in growth form, the species is physiologically drought-deciduous, as shown experimentally by Tobiessen (1970). Presumably the plants only survive on the desert in habitats with sufficient soil moisture to allow them to remain foliated. If this is true, the species may more appropriately be classified as a phreatophyte. Unfortunately, no information exists on the extent or the distribution of the root system in *Isomeris arborea*.

All the data shown in Table 9.1 were taken from all-day tracking runs. The magnitudes of net photosynthesis understandably vary with season and phenological status of the respective plants. Table 9.3 presents relative net photosynthesis rates for nine tracking runs made from July 13, 1966, to June 29, 1967. The rates shown

Table 9.3 Relative Net Photosynthesis Rates of *Larrea divaricata*[a]

Time (PST)	1966						1967		
	7/13	7/20	8/2	8/3	9/8	10/29	1/27	5/4	6/29
0630		88			88			79	—
0700		81			81			79	94
0730		88		100	81		44	100	100
0800	72	90		78	100		58	98	62
0830	92	99	100	78	96	70	82	59	65
0900	100	100	89	73	89	75	100	50	3
0930	92	95	100	82	49	100	92	69	6
1000	90	94	94	64	30	75	70	76	10
1030	85	100	76	63	23	53	87	73	6
1100	90	87	74	55	22	70	—	86	27
1130	72	90	53	44	20	60	87	86	30
1200	47	87	41	51	25	54	80	82	13
1230	38	78	50	44	17	75	56	83	20
1300	20	78	39	38	16	70	56	80	23
1330	20	68	48	40	27	75	50	80	27
1400	46	68	32	47	34	83	56	83	16
1430	20	25	0	46	50	91	54	74	20
1500	28	25		43	47	91	53	65	16
1530	26	50		6	60	83	38	84	20
1600	26	25		41	50	83	31	70	22
1630	26	43		35	65	83	38	42	27
1700	28	46		23	53	41		2	16
1730	26	41		24	42	0			10

[a] See Fig. 9.1 for absolute values and environmental conditions.

for the various days and times are percentages of the maximum observed on the respective days. It is clear from this arrangement of the data that maximum rates are observed in the early morning. It is also clear that there are great differences in the daily trends of net photosynthesis. Space does not allow a complete analysis of the significance of this phenomenon here; suffice it to say that single observations of maximum rates of net photosynthesis are rather limited in their usefulness. In addition, it is clear that time of day and year, as well as environmental parameters, must be considered in the interpretation of maximum photosynthesis rates.

Southern Pine: Pinus taeda

Table 9.4 presents maximum net photosynthesis rates of *Pinus taeda* observed on 19 tracking days from June 22, 1972, to July 7, 1973. Measurements have been made on the same leaves from the time of their production in April 1972 to the present. *Pinus taeda* is an evergreen conifer which retains leaves for two summer seasons. Most of the leaves produced in April 1972 will fall from the trees in November 1973.

As expected, net photosynthesis declines with decreasing water potential, radiation, and extreme temperature changes. It is interesting to note that maximum

Table 9.4 Maximum Net Photosynthesis of *Pinus taeda* Leaves[a]

Date	Time (EST)	Net Photosynthesis (mg CO_2 dm^{-2} h^{-1})	Temp. (°C)	Radiation (μEq m^{-2} s^{-1})	Water Potential (atm.)
6/22/72	1200	6.3	22	2500	−12
7/7/72	1000	5.9	22	2380	−16
7/14/72	1100	2.9	30	2350	−16
9/7/72	1400	2.5	26	2480	−10
9/8/72	1300	3.7	25	2500	−10
9/29/72	1300	2.5	26	850	−8
10/13/72	1100	2.8	23	1150	−10
10/27/72	0900	2.3	10	425	−8
11/10/72	1100	3.3	13	1200	−12
11/28/72	1400	2.7	15	625	−10
12/28/72	1100	3.2	12	500	−10
1/26/73	1400	3.0	13	875	−10
2/28/73	1000	4.0	10	1750	−9
3/23/73	1100	5.5	13	1825	−11
3/27/73	1100	5.5	13	2178	−9
4/17/73	1100	7.0	20	1050	−11
5/1/73	1000	8.0	24	2020	−15
7/2/73	1342	6.4	30	2025	−12

[a] The foliage measured was produced in April 1972 in the top of a 16-year-old plantation canopy. The same needles were measured throughout the study. Surface areas were determined by length on surface regression relationships previously determined by destructive sampling.

net photosynthesis values were generally obtained in midday, from 10 A.M. to 2 P.M. This is in contrast to the trends observed in the desert. It is also seen that the midday values tend to remain closer to the maximum values. In other words, a maximum value in the relatively moderate environment of the southeastern United States is more reflective of diurnal photosynthetic activity than in the extreme environment of the desert.

It is well known that tissue which develops in low radiation (shade tissue) has different photosynthetic activity per energy unit than high radiation (sun tissue) (Kramer and Kozlowski, 1960). It is also known that this condition obtains in forest canopies. The leaves at the canopy top have the characteristics of sun leaves, and those in the bottom have shade leaf characteristics (Helmers, 1943; Bourdeau and Laverick, 1958). In addition, it has been shown that age of tissue may affect photosynthetic activity (Freeland, 1952). McLaughlin (1967) confirmed these observations for *Pinus taeda* in a laboratory study using severed branches brought in from field-grown trees. Our field data (Fig. 9.3) further confirm this. Figure 9.3 also indicates a significant change in physiological activity during the season.

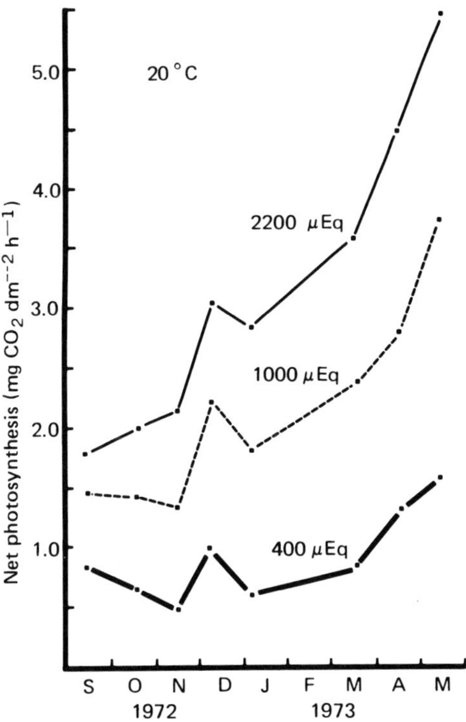

Fig. 9.3. Effect of radiant energy flux on net photosynthesis of field-grown 16-year-old *Pinus taeda* foliage. The same needles were measured throughout the study. Radiant energy from a tungsten lamp (GE cool beam 300 PAR/MF) was varied by screen and was measured by Lambda Quantum Sensors, Model LI-190SR.

Conclusions

Single observations of maximum net photosynthetic activity have only limited utility. Investigators and reviewers who devise such lists should follow the example set by Larcher (1969) and provide rates for as many categories of tissue and environmental factors as possible. This is particularly important now that we have begun to construct predictive models of plant and ecosystem behavior.

Acknowledgment

I am grateful to Kenneth O. Higginbotham for his permission to use data from his dissertation research on loblolly pine.

This research was supported in part by the Eastern Deciduous Forest Biome, US/IBP, funded by the National Science Foundation under Interagency Agreement AG-199, 40-193-69 with the Atomic Energy Commission, Oak Ridge National Laboratory, and in part by grants GB-4146, GB-7230, and GB-17357 from the National Science Foundation. This paper is Contribution No. 107 from the Eastern Deciduous Forest Biome, US/IBP.

References

Adams, M. S.: 1968. The significance of photosynthetic carbon dioxide fixation in the chlorophyllous stems of *Cercidium floridum* Benth. Riverside, Calif.: Univ. California. Ph.D. diss.

———, Strain, B. R.: 1968. Photosynthesis in stems and leaves of *Cercidium floridum*: spring and summer diurnal field response and relation to temperature. Oecologia Plantarum *3*, 285–297.

Al-Ani, H. A., Strain, B. R., Mooney, H. A.: 1972. The physiological ecology of diverse populations of the desert shrub *Simmondsia chinensis*. J. Ecol. *60*, 41–57.

Altman, P. L., Dittmer, D. S. (eds.): 1964. Biology data book: photosynthesis, maximum rates, natural conditions. Washington, D.C.: Fed. Am. Soc. Exptl. Biol.

Bosian, G.: 1953. Über die Vollautomatisierung der CO_2-Assimilationsbestimmung. Ber. Deutsch Bot. Ges. *66*, 35–36.

———: 1955. Über die Vollautomatisierung der CO_2-Assimilations-bestimmung und zur Methodik des Küvettenklimas. Planta *45*, 470–492.

Bourdeau, P. E., Laverick, M. L.: 1958. Tolerance and photosynthetic adaptability to light intensity in white pine, red pine, hemlock, and *Ailanthus* seedlings. Forensic Sci. *4*, 196–207.

Cunningham, G. L.: 1968. The ecological significance of seasonal leaf variability in a desert shrub. Los Angeles, Calif.: Univ. California. Ph.D. diss.

———, Strain, B. R.: 1969. Ecological significance of seasonal leaf variability in a desert shrub. Ecology *50*, 400–408.

Eckardt, F. E.: 1968. Techniques de mesure de la photosynthèse sur le terrain, basées sur l'emploi d'enceintes climatisées. *In* Functioning of terrestrial ecosystems at the primary production level (ed. F. E. Eckhardt), pp. 289–319. Paris: UNESCO.

Freeland, R. O.: 1952. Affect of age of leaves on photosynthesis. Plant. Physiol. *27*, 685–690.

Helmers, A. E.: 1943. Ecological anatomy of ponderosa pine needles. Am. Midland Naturalist *29*, 55–71.

Koch, W., Klein, E., Walz, H.: 1968. Neuartige Gaswechsel-Messanlange für Pflanzen in Laboratorium und Freiland. Siemens-Z. *42*, 392–404.

Koller, D., Samish, Y.: 1964. A null-point compensating system for simultaneous and continuous measurement of photosynthesis and transpiration by controlled gas-stream analysis. Bot. Gaz. *125*, 81–88.

Kozlowski, T. T., Keller, T.: 1966. Food relations of woody plants. Bot. Rev. *32*, 293–382.

Kramer, P. J.: 1958. Photosynthesis of trees as affected by their environment. *In* The physiology of forest trees (ed. K. V. Thimann), pp. 157–186. New York: Ronald Press.

———, Kozlowski, T. T.: 1960. Physiology of trees. New York: McGraw-Hill.

Lange, O. L.: 1962. Eine "Klapp-Küvette" zur CO_2-Gaswechselregistrierung an Blättern von Freilandpflanzen mit dem URAS. Ber. Deutsch Bot. Ges. *75*, 41–50.

———, Koch, W., Schulze, E. D.: 1969. CO_2-Gaswechsel und Wasserhaushalt von Pflanzen in der Negev-Wüste am Ende der Trockenzeit. Ber. Deutsch Bot. Ges. *82*, 39–61.

———, Schulze, E. D., Koch, W.: 1970. Evaluation of photosynthesis measurements taken in the field. *In* Prediction and measurement of photosynthetic productivity, pp. 339–352. Wegeninger: Centre for Agricultural Publishing and Documentation.

Larcher, W.: 1963. Die Leistungsfähigkeit der CO_2-Assimilation höherer Pflanzen unter Laborbedingungen und am natürlichen Standort. Mitt. florist.-soz. Ange. *10*, 20–33.

———: 1969. The effect of environmental and physiological variables on the carbon dioxide gas exchange of trees. Photosynthetica *3*, 167–198.

McLaughlin, S. B.: 1967. Variations in needle morphology and anatomy as related to variations in photosynthetic potential within loblolly pine (*Pinus taeda* L.) canopies. Blacksburg, Va.: Va. Polytechnic Inst. M.S. thesis.

Mooney, H. A.: 1972. Carbon dioxide exchange of plants in natural environments. Bot. Rev. *38*, 455–469.

———, Strain, B. R.: 1964. Bark photosynthesis in ocotillo. Madroño *17*, 230–233.

———, Wright, R. D., Strain, B. R.: 1964. The gas exchange capacity of plants in relation to vegetation zonation in the White Mountains of California. Am. Midland Naturalist *72*, 281–297.

Oechel, W. C.: 1970. The utilization of seasonally produced ^{14}C-labelled photosynthates in the growth of the desert shrub, *Larrea divaricata* Cav. Riverside, Calif.: Univ. California. Ph.D. diss.

———, Strain, B. R., Odening, W. R.: 1972. Photosynthetic rates of a desert shrub, *Larrea divaricata* Cav. under field conditions. Photosynthetica *6*, 183–188.

Rabinowitch, E. I.: 1951. Photosynthesis and related processes, 2 vols. New York: Wiley-Interscience.

Šesták, Z., Čatský, J. Jarvis, P.: 1971. Plant photosynthetic production: manual of methods. The Hague: Dr. W. Junk, N.V. Publ.

Spector, W. S. (ed.): 1956. Apparent maximum rates of photosynthesis: natural conditions. *In* Handbook of biological data, p. 250. Wright Air Develop. Rept. 56–273.

Stocker, O.: 1935. Assimilation und Atmung west javanischer Tropenbäume. Planta *24*, 402–445.

Strain, B. R.: 1965. Another mobile laboratory. Bull. Ecol. Soc. Am. *46*, 190.

———: 1969. Seasonal adaptations in photosynthesis and respiration in four desert shrubs growing *in situ*. Ecology *50*, 511–513.

———: 1970. Field measurements of tissue water potential and carbon dioxide exchange in the desert shrubs *Prosopis juliflora* and *Larrea divaricata*. Photosynthetica *4*, 118–122.

Talling, J. F.: 1961. Photosynthesis under natural conditions. Ann. Rev. Plant Physiol. *12*, 133–154.

Tobiessen, P. L.: 1970. Temperature and drought stress adaptations of desert and coastal populations of *Isomeris arborea*. Durham, N.C.: Duke Univ. Ph.D. diss.

Verduin, J.: 1953. A table of photosynthetic rates under optimal, near natural conditions. Am. J. Bot. *40*, 675–679.

———, Whitwer, E. E., Cowell, B. C.: 1959. Maximal photosynthetic rates in nature. Science *130*, 268–269.

Wolf, F. T.: 1969. Plants with high rates of photosynthesis. Biologist *51*, 147–155.

10

Environmental Stresses and Inherent Limitations Affecting CO_2 Exchange in Evergreen Sclerophylls in Mediterranean Climates

E. Lloyd Dunn

Introduction

For several years we have been investigating the adaptive physiological basis for convergent evolution of the evergreen sclerophyllous shrub form in mediterranean-type climates. The objectives of this discussion are twofold. First, it is necessary to summarize briefly the results of the field studies designed to measure the effects of environmental stresses characteristic of mediterranean-type climates on the seasonal metabolic activity of evergreen shrubs growing under natural conditions in California and Chile. Second, the results of laboratory studies intended to analyze the inherent limitations of the photosynthetic capacity in these species, and the basis of some of the observed field responses, is reported.

Field Results

California

The vegetation, methods, and results have previously been described in some detail (Mooney *et al.*, 1970; Mooney and Dunn, 1970a, 1970b; Dunn, 1970). Briefly, the seasonal changes in the diurnal course of photosynthesis were measured in temperature-controlled cuvettes at monthly intervals throughout the year. The temperature dependence of light-saturated photosynthesis and dark respiration was also determined monthly. Concurrently, tissue water status as percent relative water content and tissue water potential was measured.

Some of the important climatic variables at the study site before and during the period of study are summarized in Fig. 10.1. In addition to showing a general picture of the mediterranean-type climatic pattern (hot, dry summers, and cool, wet winters), Fig. 10.1 also shows the difference in rainfall and soil moisture before and during the two summers of the study, 1966–1967. From July 1965 to June 1966 rainfall was 57.8 cm (83.4 percent of the annual mean), with much of this rain coming in late fall. By September 1966 the soil profile was dry down to 350 cm. In contrast, before the summer of 1967, there had been 114.2 cm of rainfall (164.8 percent of the annual mean), with much of this rain coming in spring storms. As a result, water was still available at 200 cm in August 1967, with the soil moisture at 350 cm, almost at field capacity, into September.

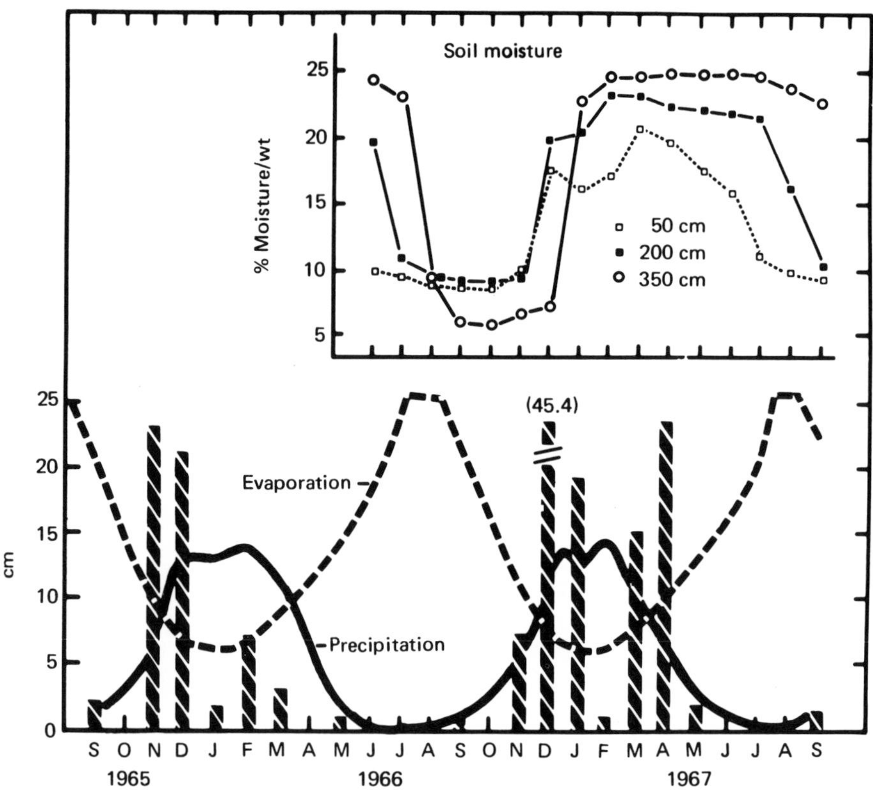

Fig. 10.1. Summary of the seasonal pattern of precipitation, evaporation, and soil moisture content at Tanbark Flat Station, San Dimas Experimental Forest, California. Twenty-seven-year mean monthly precipitation (solid line) and 20-year mean monthly evaporation (dashed line) are from Sinclair (1958). Monthly precipitation for the study period (diagonal bars) is from U.S. Weather Bureau records. Rainfall of 45.4 cm is indicated for December 1966. Soil-moisture content (percent water/gram dry weight of soil) at three depths at monthly intervals is shown above.

The seasonal changes in photosynthesis and tissue water stress are shown for one representative California species in Figs. 10.2–10.5. One important feature is the reduction in CO_2 uptake through the summer drought period (July to September 1966, Figs. 10.2 and 10.3), the recovery in rates after the winter rains (February 1967, Fig. 10.4), and the recovery of photosynthesis during the second, less severe

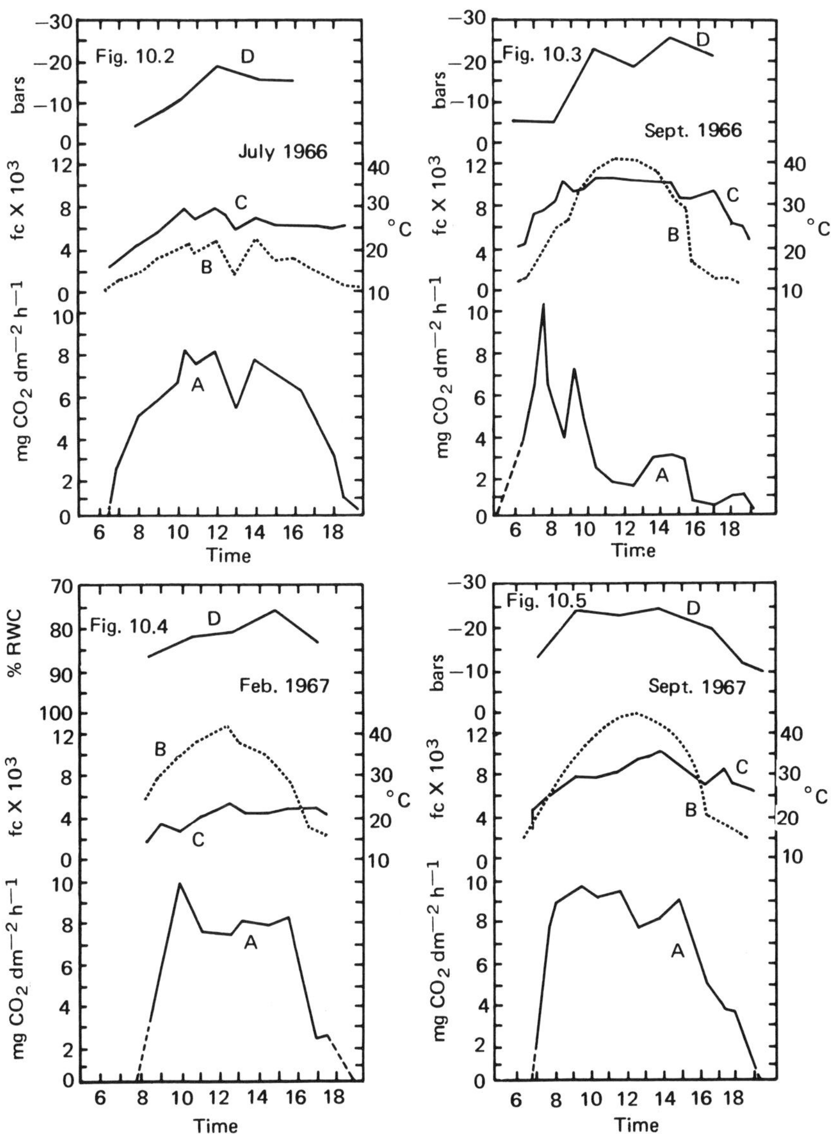

Fig. 10.2–10.5. Diurnal course of net photosynthesis (A) and tissue water stress (D) in relation to light intensity (B) and chamber air temperature (C) for *Prunus ilicifolia* on July 29, 1966 (Fig. 10.2); Sept. 21, 1966 (Fig. 10.3); Feb. 26, 1967 (Fig. 10.4); and Sept. 13, 1967 (Fig. 10.5).

summer (September 1967, Fig. 10.5). The other important feature is the pattern of diurnal tissue water stress in relation to photosynthesis. The maximum reduction of photosynthesis is not correlated with the maximum level of water stress; instead, the greatest reduction of photosynthesis occurs as tissue water status improves. These results suggest very strongly that stomatal control limits CO_2 uptake rather than that drought affects photosynthetic metabolism directly.

This same general reduction and recovery in photosynthesis was seen in the temperature-dependence experiments. Representative examples of these results are shown in Fig. 10.6 for *Rhus ovata.* Net photosynthesis increased with increasing chamber temperature up to some optimum and then declined fairly rapidly as the temperature was increased further. Thus some of the reduction in the daily CO_2 uptake was due to the high air temperatures, which were above the temperature for maximum net photosynthesis.

Also, a general reduction in CO_2 uptake at all temperatures and a decreasing temperature dependence of net photosynthesis (a flattening of the temperature-photosynthesis response curve) was found between certain observations. These changes occurred very rapidly between the morning and afternoon determinations on the same day (Fig. 10.6, Aug.). They also occurred as a trend through the drought period.

After the soil water had been renewed by the winter rains, the rates of net photosynthesis recovered to some extent (February 1967, Fig. 10.6), but not to the pre-drought levels. However, after new leaf tissue was produced in the late spring, net photosynthesis rates were back to the same or higher levels than the previous June and July, with the same general temperature-dependence characteristics. Even with the improved soil-moisture situation the second summer, however, the

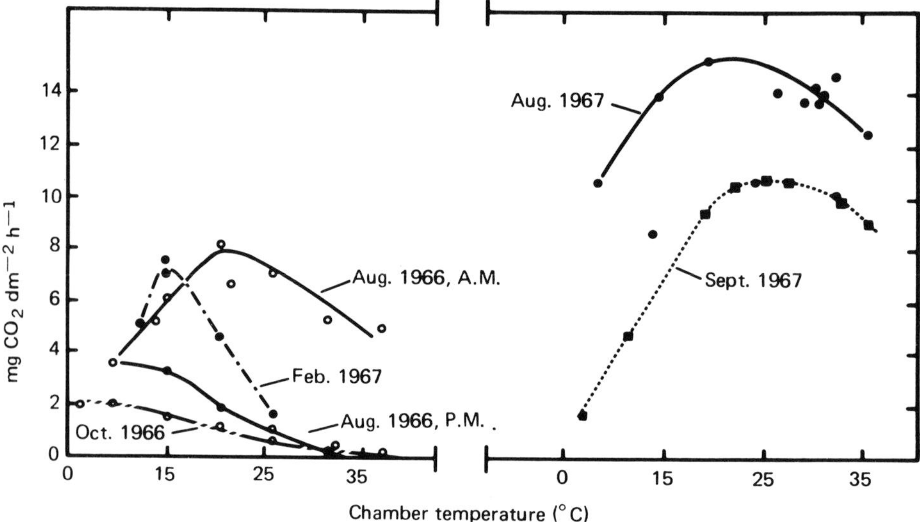

Fig. 10.6. Seasonal and diurnal changes in the relationship between net photosynthesis and chamber air temperature for *Rhus ovata.*

same trend toward a seasonal reduction in photosynthetic capacity had become apparent by September 13, 1967 (Fig. 10.6).

To summarize briefly the California field results, a progressive reduction of photosynthesis was observed as the summer drought became more severe in 1966. These rates recovered to some extent after the winter rains. During the summer of 1967, under less severe drought conditions, these effects were less pronounced. In all cases it appeared that the reduction in photosynthesis could be accounted for by stomatal control and not necessarily by direct drought effects on metabolism.

Chile

Similar field studies were conducted in Chile under as near equivalent conditions as possible during the Southern Hemisphere summer and fall of 1967–1968 (see Dunn, 1970). The greatest difference between the two sites was the amount of precipitation received before the summer drought periods. During the rainy season in 1967, prior to the beginning of this study in December 1967, only 17.2 cm of rain had been recorded, compared to a 60-year mean annual precipitation of 36.95 cm. Thus the rainfall before the field measurements was only 47 percent of the long-term average compared to 57.8 cm of precipitation, or 83 percent of average before the first summer's measurements in California in 1966 and 114.2 cm or 165 percent of the yearly average before the second summer's measurements in 1967 in California.

Examples of the seasonal changes in the daily course of net photosynthesis and tissue water stress in *Lithraea caustica* are shown in Fig. 10.7 for the beginning of the summer and 6 weeks later near the height of the drought (Fig. 10.8). At the first measurement period, *L. caustica* showed an active photosynthetic capacity, responding to the fluctuations in light intensity and chamber temperature. Tissue water potential decreased rapidly in the morning to -25 bars, then recovered slightly, followed by a steady, slow decrease through the afternoon until a couple of hours before sunset, when water potential increased to the level observed early in the morning.

On a clear day in late February, a definite midday reduction in CO_2 uptake was observed in *Lithraea*, with photosynthetic rates in general lower than in the previous month. Tissue water potential showed a slow, steady decrease, reaching a minimum at approximately the same time as the maximum reduction in net photosynthesis. As tissue water potential began to recover in the afternoon, net photosynthesis also increased for a couple of hours, until the light intensity decreased. The maximum water stress (minimum water potential) was greater during this observation period, also, but still only reached -34 bars.

During the same period in the middle of the summer, *Quillaya saponaria* showed quite a different response (Fig. 10.9). Net photosynthesis was very low all day, with slightly higher rates early in the morning and late in the afternoon. Water potential had recovered to only -32 bars during the night and had dropped to

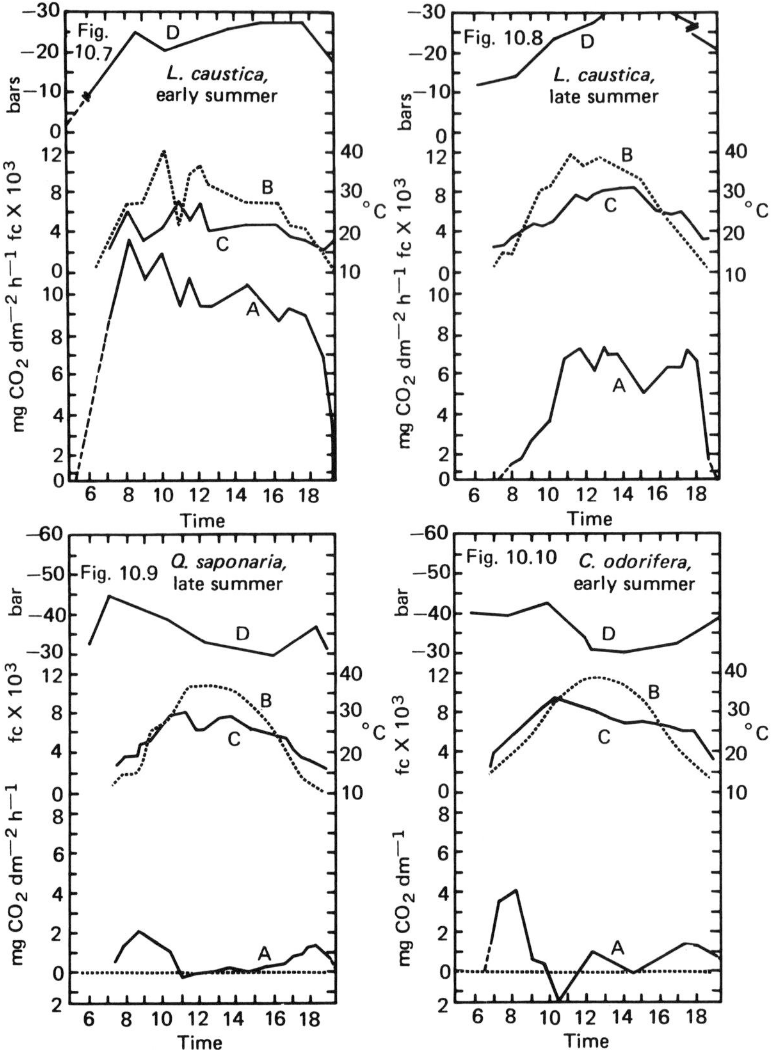

Fig. 10.7–10.10. Diurnal course of net photosynthesis (A) and tissue water stress (D) in relation to light intensity (B) and chamber air temperature (C) for *L. caustica* on Jan. 18, 1968 (Fig. 10.7); *L. caustica* of Feb. 29, 1968 (Fig. 10.8); *Q. saponaria* on March 12, 1968 (Fig. 10.9); and *C. odorifera* on Jan. 23, 1968 (Fig. 10.10).

−45 bars by 7:00 A.M. During the day, water potential increased slowly until late in the afternoon, when it decreased slightly, at the same time that photosynthesis showed the second increase. This increase in tissue water status was in the opposite direction of the environmental stresses leading to more evaporation, i.e., increasing radiation and temperature.

Similar diurnal patterns were observed for *Colliguaya odorifera* during all three observation periods. Figure 10.10 illustrates these daily trends even as early as

January. As for the previous example, higher photosynthesis rates were observed early in the morning when water potential was lower, and decreased during the day as water potential increased (water stress decreased). Again, the changes in tissue water status were contrary to the changes in environmental stresses.

The experiments to determine the temperature dependence of net photosynthesis showed the same low photosynthetic activity as the ambient tracking experiments. These experiments also showed a reduction of photosynthetic capacity at moderate temperatures with net CO_2 release increasing with temperature above 25°C. Figure 10.11 shows the seasonal change in the temperature-dependence characteristics of *Quillaya saponaria*. These results were typical of all the species except *L. caustica*, which maintained net CO_2 uptake up to temperatures around 35°C.

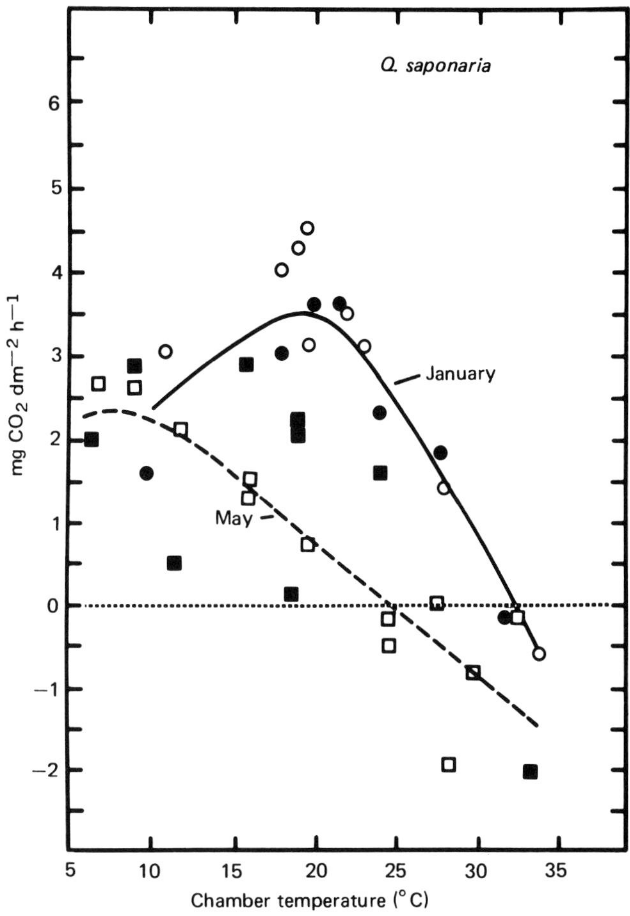

Fig. 10.11 Seasonal changes in the relationship between net photosynthesis and chamber air temperature for *Q. saponaria* on January 30, 1968 (circles and solid line) and May 6, 1968 (squares and dashed line). The solid and open symbols are duplicate branches on the same dates.

The artificially watered shrubs provided very interesting comparisons to those under natural conditions. Figure 10.12 shows the photosynthesis–temperature relationship in both *L. caustica* and *C. odorifera* under natural conditions and with irrigation at the beginning of the summer. The measurements of maximum and minimum plant water stress for these two conditions are also shown.

As a result of the added water, *L. caustica* showed only a small increase in net photosynthesis rate, with water stress remaining at about the same level. In striking contrast, *C. odorifera* under natural conditions had a very low photosynthesis rate with a flat photosynthesis–temperature response curve. With artificial watering, maximum net photosynthesis rates equal to the watered *L. caustica* were observed. Also, a "normal" temperature dependence of photosynthesis rate was evident. The changes in tissue water stress were also evident.

To summarize briefly the Chile field results, a reduction in photosynthetic activity was observed in all species. *L. caustica* showed a photosynthetic activity comparable to that observed in the California species and showed a comparable level of tissue water stress to that observed in the California species. The other three species, however, had much lower photosynthesis rates at the beginning of the study, and these rates dropped even further as the summer drought progressed. Also, these species had much greater levels of tissue water stress than either *L.*

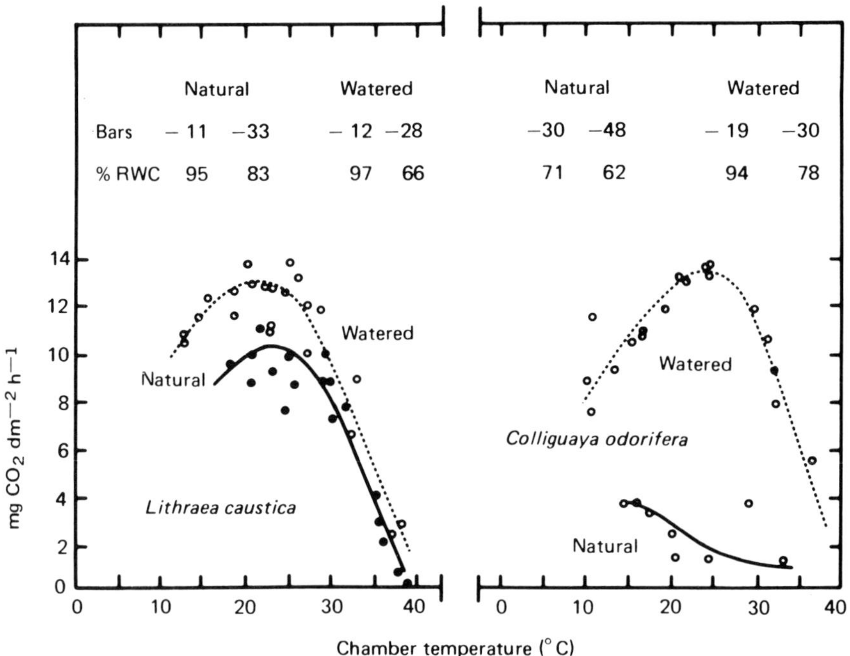

Fig. 10.12. Relationship between net photosynthesis and chamber air temperature for *L. caustica* and *C. odorifera* under natural and artificially watered conditions in early summer in Chile. Values above are maximum and minimum tissue water stress as water potential (bars) and percent relative water content (%RWC).

caustica or the California shrubs. However, with artificial irrigation, *C. odorifera* showed much reduced levels of tissue water stress and a photosynthetic capacity equal to that observed in *L. caustica* and the California species in natural but less severe environmental conditions.

Laboratory Studies and Discussion

The field results from both California and Chile demonstrated the effects of the mediterranean climatic pattern on the capacity of typical evergreen sclerophylls to maintain photosynthesis. With only a few exceptions under severe drought conditions, stomatal closure was suggested as a dominant factor limiting CO_2 uptake. Since precise field measurements of stomatal resistance were not made, it was not possible to evaluate separately stomatal closure and metabolic or other limitations responsible for the observed reductions in net photosynthesis. Therefore, laboratory experiments were conducted to provide quantitative estimates of stomatal resistance simultaneously with CO_2 exchange. Once the effects of stomatal resistance could be determined, it was possible to determine quantitatively some of the other internal factors limiting the rates of CO_2 uptake.

Theory and Calculations

The definition of specific internal resistances to CO_2 uptake depends on the methods used to determine them. One of the first clear statements of this resistance was Gaastra's (1959) mesophyll resistance, r_m. As a diffusional process, photosynthesis can be described by the equation

$$F = \frac{(CO_2)_{ext} - (CO_2)_{chl}}{r'_a + r'_s + r'_m} \tag{1}$$

where F = measured rate of net photosynthesis
$(CO_2)_{ext}$ = concentration of CO_2 in the bulk air
$(CO_2)_{chl}$ = concentration at the chloroplasts
r'_a, r'_s = resistances to CO_2 diffusion of the boundary layer and stomates, determined from simultaneous measurements of photosynthesis and transpiration and corrected for the difference in diffusion rates of water vapor and CO_2

Ideally, then, the mesophyll resistance, r'_m, is an analogous resistance to CO_2 transfer from the mesophyll cell wall to the sites of photosynthetic carboxylation. r'_m defined in this way, or by rearranging as in Eq. (1A), includes all the factors limiting the rate of CO_2 uptake other than stomatal and boundary-layer resistances.

$$r'_m = \frac{(CO_2)_{ext} - (CO_2)_{chl}}{F} - (r'_a + r'_s) \tag{1A}$$

To solve Eq. (1A) for r'_m, it has been necessary to make some assumption about $(CO_2)_{chl}$, earlier assumed to be zero (Gaastra, 1959; more recently by Wuenscher and Kozlowski, 1971). Some assumptions also have to be made about rates of respiration in the light to correct net photosynthesis to gross photosynthesis.

These assumptions have been questioned for various reasons by others, who have proposed alternative methods for calculating the mesophyll resistance or equivalent value (Bierhuizen and Slatyer, 1964; Lake, 1967a, 1967b; Samish and Koller, 1968; Troughton and Slatyer, 1969; Slatyer, 1970; McPherson and Slatyer, 1973). This estimate of internal resistance (r_i) depends on the experimental determination of the CO_2 dependence of net photosynthesis as follows:

$$r'_i = \frac{d(CO_2)_{int}}{dF} \tag{2}$$

where dF is the measured change in net photosynthesis for a given change in the intercellular concentration of CO_2, $d(CO_2)_{int}$, determined from the linear portion of the CO_2 dependence of net photosynthesis under light-saturated conditions.

The internal concentration of CO_2 is calculated from the relationship

$$(CO_2)_{int} = (CO_2)_{ext} - (r'_a + r'_s) \times F \tag{3}$$

All these measurements are subject to various sources of error in their determination. These errors have been discussed in much more detail elsewhere (Slatyer, 1971; Jarvis, 1971).

In addition to providing a more precise estimate of r'_i, the experimental determination of the relationship between net photosynthesis and $(CO_2)_{int}$ also provides an estimate of photorespiration and the CO_2 compensation point of photosynthesis. By extrapolating the linear slope to zero $(CO_2)_{int}$, photorespiration (L) is defined as

$$L = -F \qquad \text{where } (CO_2)_{int} \text{ is } 0 \tag{4}$$

or, arithmetically,

$$L = -\frac{\Gamma}{r'_i} \tag{5}$$

where L is photorespiration or CO_2 release in the light, in units of ng CO_2 cm^{-2} s^{-1}. Γ is the CO_2 compensation point derived from the net photosynthesis–$(CO_2)_{int}$ relationship as follows:

$$\Gamma = (CO_2)_{int} \qquad \text{where } F \text{ is } 0 \tag{6}$$

Γ is in units of ng CO_2 cm^{-3}. This value also has been proposed as a better estimate of $(CO_2)_{chl}$ than zero in Eq. (1). If Γ rather than zero is used in Eq. (1), the calculated mesophyll resistance is equal to the value calculated by Eq. (2) for r'_i.

A more detailed discussion of assumptions and errors involved in determining r'_m or r'_i with Eq. (1) or Eq. (2) has been given elsewhere (Dunn, 1973). Briefly summarizing the comparison of these two estimates of internal resistances to CO_2 uptake, r'_m defined by Eq. (1A) [i.e., assuming $(CO_2)_{chl}$ to be 0] includes several factors in the resistance term. Because of this inclusiveness, it is a less precise term or description of the photosynthetic process. This explains at least partially the differing conclusions reached by several workers about the effects of various environmental variables on the internal resistance term (Troughton and Slatyer, 1969; compared to Gale *et al.*, 1966, and to Wuenscher and Kozlowski, 1971).

Even though r_i' as defined and determined by Eq. (2) (i.e., reciprocal of the photosynthesis–CO_2 curve) is a more precise term and more theoretically sound in terms of a transfer resistance, it still involves some limitation due to metabolic activity. This can be demonstrated fairly easily by doing the same experimental determination of the Ps–CO_2 dependency in a low oxygen environment (1 percent O_2). The absence of oxygen appears to do two things: (1) it eliminates photorespiration, and (2) it eliminates the "oxygen inhibition" of photosynthesis. This results in higher rates of photosynthesis and also a change in the internal resistance, r_i'. Since it is unlikely that a lower oxygen concentration would change the physical diffusion characteristics of a leaf, it follows that some of the r_i' determined in air of 21 percent O_2 is due to metabolic processes changed by the low oxygen concentration.

The value of r_i' in low oxygen is a minimum measurable internal resistance. This can also be interpreted as the maximum resistance in a leaf that can be attributed solely to physical diffusion resistance. This value still includes some metabolic component, which is reduced to a minimum under these conditions. The relative contributions of the metabolic component and the physical diffusive resistance in this value cannot be evaluated from this analysis.

Laboratory Results

Two values for the internal resistance to CO_2 uptake have been defined and discussed in relation to some of the errors and assumptions inherent in each. The experimental apparatus and procedures have been described previously (Dunn, 1970). The following discussion will give an example of the type of analysis using resistance values to understand the interaction between environmental variables and inherent plant parameters in determining photosynthesis rates.

The results from a representative example of the experiments to determine the CO_2 dependence of net photosynthesis at ambient and low (ca. 1 percent) oxygen concentrations at 25°C are shown in Fig. 10.13 for *Heteromeles arbutifolia.* The changes in leaf resistance ($r_l = r_a + r_s$) with changing CO_2 concentration are also shown. The effects of the changes in r_l' have been accounted for, however, since the photosynthesis rate is expressed relative to $(CO_2)_{int}$.

Several important points are evident from this type of analysis. First, net photosynthesis was linearly related to $(CO_2)_{int}$ over the entire range used in this study (ambient concentrations and below). Thus some factor relating to the rate of CO_2 supply or carboxylation was determining the rate of photosynthesis. The reciprocal of the slope of the photosynthesis–CO_2 relationship is a quantitative measure of these limiting factors, the mesophyll or internal resistance, r_i'.

In low oxygen, much higher rates of net photosynthesis were observed at the same $(CO_2)_{int}$. The F–CO_2 relationship was still linear in low oxygen, but the slope of this relationship and thus the r_i' were different.

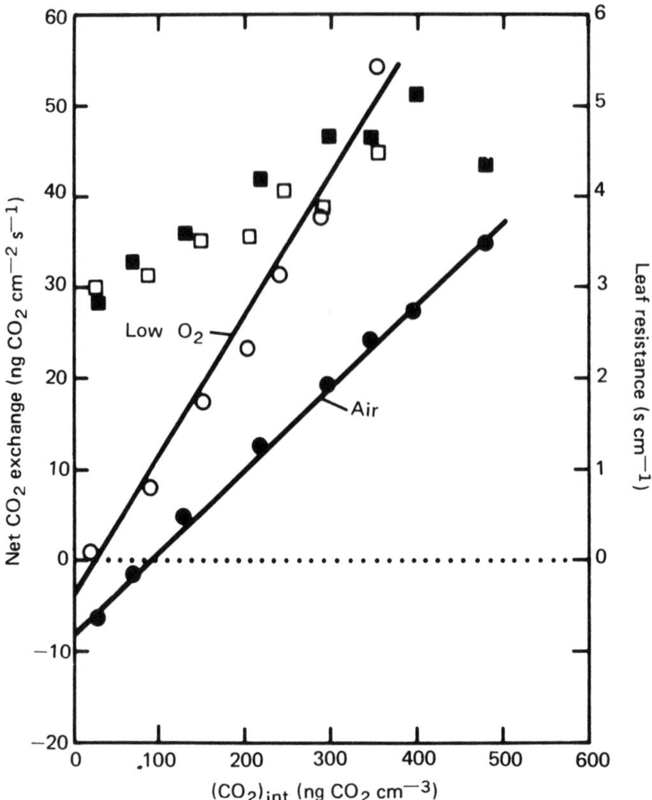

Fig. 10.13. CO_2 dependence of net photosynthesis (circles) and leaf diffusive resistance to water-vapor transfer (squares) under normal air (21 percent O_2, solid symbols) and low oxygen (1 percent O_2, open symbols) concentrations.

A CO_2 compensation point of 91 ng CO_2 cm^{-3} (approximately 50 ppm CO_2) was observed in air for this experiment [the *X*-intercept, see Eq. (6)]. However, in low oxygen, the CO_2 compensation point was very much reduced. This compensation point represents the balance between photosynthesis and photorespiration, thus indicating a significant level of respiration occurring in the light. This interpretation is reinforced by the calculated rate of photorespiration [the *Y*-intercept, Eq. (4)].

Table 10.1 summarizes several parameters determined from the photosynthesis–CO_2 relationships for two California species, *Heteromeles arbutifolia* and *Rhus ovata*, and two Chilean species, *Kageneckia oblonga* and *Lithraea caustica*. Several important characteristics relative to the photosynthetic potential of evergreen sclerophylls from these two different mediterranean climates are evident from these data.

First, there was a significant internal resistance to CO_2 uptake determined from Eq. (2) in air. These values ranged from 6.9 to 11.9 s cm^{-1} in young, fully expanded

Table 10.1 Summary of Various Characteristics and Limitations of the Photosynthetic Capacity of Sclerophylls[a]

	$r'_a + r'_s$ (s cm^{-1})	r'_i in air (s cm^{-1})	r'_i in low O_2 (s cm^{-1})	Γ (ng CO_2 cm^{-3})	L (ng CO_2 cm^{-2} s^{-1})
Heteromeles arbutifolia					
New[b]	4.7	11.9	6.5	85	7.03
Old[c]	7.8	32.6	17.3	146	4.53
Rhus ovata					
New[b]	4.4	9.0	6.9	104	12.44
Old[c]	10.7	26.4	8.0	43	1.67
Kageneckia oblonga					
New[b]	3.9	6.9	3.9	108	16.00
Lithraea caustica					
New[b]	5.3	9.3	4.1	131	15.06
Means of new tissue	4.6	9.3	5.35	108	4.55

[a] All values determined at 25°C.
[b] Young, fully expanded to mature leaf tissue.
[c] Leaf tissue approximately 1 year old.

tissue. Two branches older than 1 year had much higher values, ranging from 26.4 to 32.6 s cm^{-1}.

Also, all these species showed substantial rates of photorespiration, as calculated from Eq. (4). The values for the CO_2 compensation point also reflected the high respiratory activity. The enhancement of net photosynthesis observed in low oxygen compared to air also reflected the photorespiration process, which was inhibited by the low oxygen. However, some of the enhanced rate of photosynthesis in low oxygen is a result of an "oxygen inhibition" of the rate of photosynthesis at normal O_2 concentrations in air.

The minimum leaf resistance (r'_l) values were similar for all four species. The highest minimum leaf resistances were observed in the two samples of old tissue. These values were about half the values of the internal resistance; thus, under optimum conditions of water supply, the stomatal resistance component would provide only one-third of the total limitation to photosynthesis; the metabolic and other internal components would provide two-thirds of the total resistance to CO_2 uptake.

The attempts to separate and evaluate the different internal factors limiting photosynthesis are summarized in Table 10.2. The total mesophyll resistance to CO_2 uptake was calculated from Eq. (1), assuming $(CO_2)_{chl}$ to be zero. The magnitude of this value depends on the $(CO_2)_{int}$ at which the photosynthesis rate used for its calculation was observed (see, e.g, Bierhuizen and Slatyer, 1964; Whiteman and Koller, 1967; Dunn, 1973). Since differences in leaf resistance at the same external concentration of CO_2 result in changes in $(CO_2)_{int}$, this value is also sensitive to the value of r'_l during its determination. To reduce these effects to a minimum, total r'_m was calculated from the rate of photosynthesis corrected to a

Table 10.2 Comparison of the Various Components of the Internal Resistance to CO_2 Transfer in Sclerophylls

	r'_i in air ($s\ cm^{-1}$)	Total r'_m ($s\ cm^{-1}$)	Resp. (%)	O_2 inhib. (%)	Transfer Resistance (%)
Heteromeles arbutifolia					
New	11.9	13.4	11.2	40.3	48.5
Old	32.6	44.6	26.9	34.3	38.8
Rhus ovata					
New	9.0	11.2	19.6	18.8	61.6
Old	26.4	28.7	8.0	64.1	27.9
Kageneckia oblonga					
New	6.9	8.6	19.8	34.9	45.3
Lithraea caustica					
New	9.3	12.7	26.6	40.9	32.3
Means of new tissue	9.3	11.5	18.7	38.9	42.4

standard $(CO_2)_{int}$ of 300 ppm by extrapolation of the F–$(CO_2)_{int}$ relationship. In effect, this calculation takes the limitation to the rate of net photosynthesis due to photorespiration, which is reflected in the proportion of Γ to the CO_2 available for net photosynthesis [i.e., $(CO_2)_{int} - \Gamma$], and includes it in the resistance value. Total r'_m is a function of the relative rate of photorespiration and photosynthesis at the $(CO_2)_{int}$ at which it is calculated.

The difference between total r'_m and r'_i [Eq. (2)] in air was considered to be a value equivalent to the limitation resulting from photorespiration and is expressed as a percentage of the total r'_m. However, r'_i in air still includes some metabolic component as well as any physical diffusion limitation to CO_2 uptake. This metabolic component can be reduced to a minimum by determining the internal resistance from the photosynthesis–CO_2 relationship in a low oxygen atmosphere, thus eliminating both photorespiration and the "oxygen inhibition" of net photosynthesis (Björkman, 1966; Forrester *et al.*, 1966a, 1966b). Figure 10.13 exemplifies the changes in net photosynthesis in low oxygen in one experiment. These changes are summarized for all four species in Table 10.1. The difference between the values of r'_i determined in air and low oxygen was considered to be the "oxygen inhibition" and is expressed in Table 10.2 as a percentage of the total r'_m.

The r'_i value determined in low oxygen provides an estimate of the maximum possible physical diffusion resistance inherent in the leaf structure of these evergreen sclerophylls. The percentage contribution of this value to the total r'_m is given in Table 10.2. r'_i in low O_2 still includes a metabolic component, which is reduced to a minimum in low-oxygen conditions. The relative contributions of the metabolic component and the physical diffusive resistance in this value cannot be evaluated from the present analysis.

Several important aspects of the photosynthetic capacity of evergreen sclerophylls are evident from the data in Table 10.2. Metabolic components account for approximately 58 percent of the observed limitation to photosynthesis not attributable to stomatal resistance at 25°C and adequate water supply. These metabolic limitations can be separated into a photorespiration component of about 19 percent of an "oxygen inhibition" of about 40 percent. The remaining 42 percent of the total r'_m is a maximum estimate of an actual physical diffusion resistance with some of this value actually due to metabolic aspects of the CO_2-dependent carboxylation reactions.

Thus a considerable proportion of the resistance to CO_2 uptake other than stomates is determined by metabolic components and should be susceptible to severe environmental stresses. From the type of analysis presented here, it would be possible to determine the point of action and the severity of these stresses. Some possible effects of environmental stresses on internal resistances have been discussed in more detail elsewhere (Dunn, 1970).

Table 10.3 compares some of the resistance components limiting photosynthesis in a wide range of species. The evergreen sclerophylls in this study showed substantial internal resistance values, even with adequate water (from 6.9 to 11.9 s cm^{-1}) The samples of old tissues had values approximately three times greater than the younger tissue. These lower values were still several times higher than those found in mesophytic species such as cotton (4 s cm^{-1}, Troughton and Slatyer, 1969) and sunflower (2.4 s cm^{-1}, Holmgren *et al.*, 1965) but were equal to values for some woody species reported by Holmgren *et al.* (1965). However, mesophyll resistance values determined in low oxygen were approximately 50 percent lower than the values in air. Since there is no reason to suspect that low oxygen concentration would decrease physical resistance to diffusion (Gauhl, 1970; Gauhl and Björkman, 1969), the observed decrease must be attributed to an effect of oxygen concentration on the metabolic processes limiting photosynthesis even while there is a linear dependence of photosynthesis on CO_2 concentration (i.e., CO_2 is limiting). Thus physical resistances are not the only limiting factors to photosynthesis, as

Table 10.3 Comparison of Resistances to CO_2 Uptake in Various Species

	r'_s (s cm^{-1})	r'_i *in air* (s cm^{-1})	r'_i in low O_2 (s cm^{-1})
Evergreen sclerophylls[a]	4.6	9.3	5.4
Populus tremula[b]	3.9	7.5	—
Betula verrucosa[b]	2.1	6.1	—
Quercus robur[b]	12.1	10.1	—
Acer platanoides[b]	13.5	8.0	—
Helianthus annuus[b]	0.72	2.4	—
Cotton[c]	—	4.2	2.9
Atriplex hastata[d] (C_3)	0.71	3.0	2.3
Atriplex spongiosa[d] (C_4)	1.97	0.92	0.92
Tropical grasses[e]	2.5	0.74	0.93
Tropical legumes[e]	1.4	3.31	2.32

[a] Means of new tissue only.
[b] Holmgren *et al.*, 1965.
[c] Troughton and Slatyer, 1969.
[d] Slatyer, 1970.
[e] Ludlow and Jarvis, 1971.

has been suggested (Gaastra, 1959; Holmgren *et al.*, 1965), over this range of CO_2 concentration.

These lower values of r_i' under low oxygen are much closer to comparably determined values for more mesophytic species, with only the oldest tissue having much higher values (8 to 17 s cm^{-1}).

The mesophyll resistance determined under low oxygen still involves limitations due to the metabolic activity of the carboxylation reactions of photosynthesis as well as actual physical resistances to the transport of CO_2 from the intercellular spaces to the chloroplasts (Troughton and Slatyer, 1969; Slatyer, 1970). Several characteristic features of these evergreen sclerophylls, such as thick cell walls, dense, compact mesophyll cells, and possible internal suberization of cell walls (Scott, 1950) suggest that the physical resistances may contribute a large proportion to this minimum r_i' value.

There is still a metabolic or biochemical component inherent in the internal resistance, even when determined under low-oxygen conditions. This component is associated with the amount and activity of the carboxylating and associated enzymes (Troughton and Slatyer, 1969; Osmond *et al.*, 1969; Downton and Slatyer, 1972). Recently several investigators have shown a close correlation between levels of these enzymes and rates of photosynthesis at light saturation, thus suggesting the importance of this factor in determining the photosynthetic capacity of plants (Björkman, 1968a, 1968b; Wareing *et al.*, 1968; Gauhl and Björkman, 1969; Medina, 1971). Woolhouse (1967–1968) has also shown an increased mesophyll resistance associated with a decrease in the activity of the carboxylating enzymes during senescence.

The consequence of a high mesophyll resistance on relative water use efficiency has been discussed elsewhere (Holmgren *et al.*, 1965; Wuenscher and Kozlowski, 1971; Dunn, 1970). Usually interpretations are based on the ratio of the relative diffusion resistances to H_2O and CO_2 exchange. The ratio of the total resistance to each process determines the relative efficiency of water utilization per unit of CO_2 exchange. It is apparent from an analysis of the relevant resistances that photosynthesis is proportionately less affected by changes in leaf resistance than is transpiration, all other factors remaining constant.

However, under natural conditions, the other factors that affect the absolute water-use efficiency of photosynthesis do not remain constant. These other factors, such as the water-vapor gradient between the leaf and air and the CO_2 gradient between the ambient air and the intercellular spaces, are influenced by environmental conditions. The external concentration of CO_2 remains fairly stable compared to the changes in the water-vapor concentration. The opposite changes in these two driving forces occur with increasing temperature. The saturation vapor pressure doubles for every 10°C increase in temperature, thus increasing the vapor-pressure deficit between the leaf and the surrounding air, while the CO_2 compensation point in plant species exhibiting photorespiration tends to increase with temperature (by a factor of approximately 1.8 for every 10°C increase, according to Heath, 1969), thus reducing the external driving force for photosynthesis.

Examples of these changes were seen in some experiments to determine the temperature dependence of net photosynthesis and leaf resistance. The changes in these values as well as changes in the water-vapor-pressure deficit between leaf and air are shown in one such experiment for *Heteromeles arbutifolia* in Fig. 10.14 and Table 10.4. Leaf resistance increased rapidly with increasing temperatures above 25°C; net photosynthesis decreased over the same temperature range.

Table 10.4 compares the changes in diffusive resistances, and concentration gradients determining the rates of photosynthesis and transpiration, from 25 to 37°C in this experiment. The r_i' value was calculated from the CO_2-dependence experiment at 25°C. This was assumed to remain constant over this temperature range for the purpose of this analysis (see, e.g., Troughton and Slatyer, 1969; Downton and Slatyer, 1972). The CO_2 concentration gradient and compensation point at 25°C were determined from the CO_2-dependence experiment with the same branch. These two values for 37°C were calculated from the observed rate of photosynthesis, the measured changes in leaf resistance, and the assumed constant value or r_i'.

From these data it is apparent that the total diffusive resistance to CO_2 uptake increased proportionately less than for water-vapor loss (55 versus 185 percent

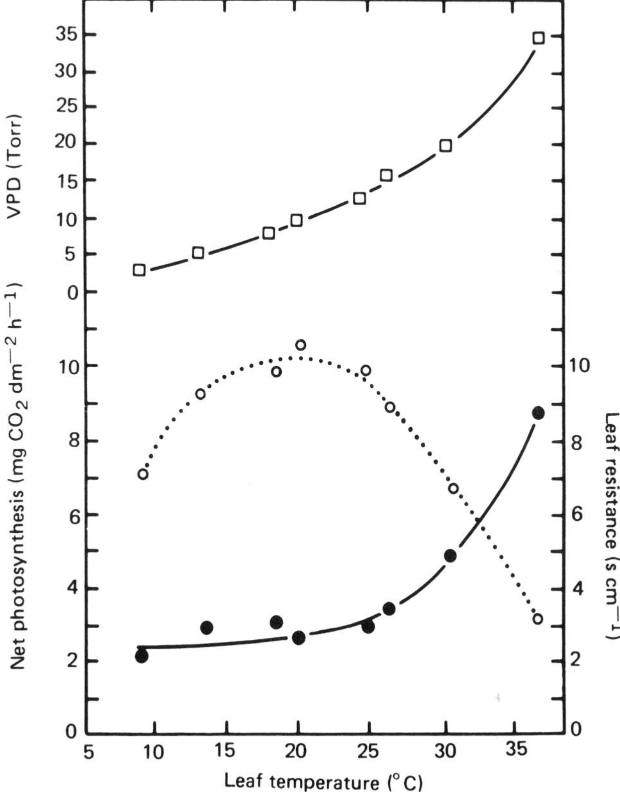

Fig. 10.14. Temperature dependence of net photosynthesis (open circles and dotted line), leaf diffusive resistance to water-vapor transfer (solid circles and line), and vapor-pressure deficit (open squares) in *H. arbutifolia*.

Table 10.4 Effects of Temperature on the Factors Determining the Water-Use Efficiency of Photosynthesis in a Representative Evergreen Sclerophyll, *Heteromeles arbutifolia*[a]

	25°C	37°C	Change from 25 to 37°C[b]
Net photosynthesis ($mg\ CO_2\ dm^{-2}\ h^{-1}$)	9.92	3.19	68% decrease
Leaf diffusive resistance to CO_2 ($s\ cm^{-1}$)	4.9	14.0	185% increase
Internal resistance ($s\ cm^{-1}$)	11.6	[11.6]	—
$r'_l + r'_i = R'_{total}$ ($s\ cm^{-1}$)	16.5	[25.6]	55% increase
CO_2 concentration gradient (ppm CO_2)	250	[125]	50% decrease
CO_2 compensation point (ppm CO_2)	50	[175]	250% increase
Transpiration ($mg\ CO_2\ dm^{-2}\ h^{-1} \times 10^3$)	1.45	1.32	9% decrease
Leaf resistance to water vapor r_l ($s\ cm^{-1}$)	3.0	8.7	185% increase
Water-vapor-concentration gradient ($mg\ H_2O\ cm^{-3} \times 10^{-2}$)	1.22	3.15	158% increase
Transpiration/photosynthesis ratio	146.4	413.2	182% increase

[a] Calculated values based on an assumed constant r'_i in brackets.
[b] (Value at 37°C–value at 25°C)/value at 25°C multiplied by 100.

increase). However, the water-vapor-concentration gradient for water loss increased 158 percent, while the CO_2 gradient for CO_2 uptake decreased 50 percent. Thus even though transpiration was slightly less at the higher temperature, net photosynthesis was reduced even more as a result of the increased leaf resistance and the decreased CO_2 gradient from the air to the chloroplast. Both of these changes resulted in an unfavorable change in the T/P ratio or a decreased efficiency of water utilization per unit of CO_2 fixed.

A similar analysis from 25°C toward cooler temperatures would show the opposite changes, with the vapor-pressure deficit decreasing at lower temperature, while the CO_2 gradient would tend to increase as internal respiratory activity decreased and the CO_2 compensation point was reduced. Changes in this direction would improve the water-use efficiency to some point where photosynthesis was limited by low temperature.

This response to high temperatures, and other features (Dunn, 1970), emphasize some of the inherent limitation of these evergreen sclerophylls (which possess the C_3 photosynthetic pathway) in relation to environmental stress. Increasing photorespiration with increasing temperature, in addition to a substantial diffusion resistance component, reduce the capacity for CO_2 uptake at high temperatures.

The relative significance of the leaf diffusive resistance compared to the mesophyll resistance and other internal components in determining CO_2 exchange rates at any point in time depends to a large extent on the response of the stomates to environmental conditions at some earlier time. Stomatal response to obvious environmental factors such as light, temperature, and plant water status are well documented (see Meidner and Mansfield, 1965, for review). Recently stomatal responses to changes in CO_2 concentration have received considerable attention (Raschke, 1965, 1966).

The results from the CO_2 dependence experiments showed that stomates in these sclerophylls were responsive to changes in CO_2 concentration, with lower leaf resistances at the lower concentrations. These results also showed higher leaf resistances in the species or individual plants with higher mesophyll resistances. This same trend was evident from the data of Holmgren *et al.* (1965, Table 1). These authors, however, concluded that differences in leaf resistance were the dominant factor accounting for differences in maximum photosynthesis rates between species, even though in most of the species investigated, the mesophyll resistance was equal to or higher than the minimum leaf resistance.

This apparent contradiction in their data and conclusions emphasizes the important point that leaf resistance is a dynamic component, capable of changing over a large range very rapidly in response to environmental stimuli. Therefore, a "minimum" r_l may reflect past environmental factors as well as present conditions. To assess fully the stomatal restrictions to CO_2 uptake, more information is needed about the range of leaf-resistance values as well as the external and internal factors that affect this character. Before meaningful models or predictions of photosynthetic responses are possible in natural conditions in which drought is an important environmental limitation, a better understanding of stomatal responses is necessary.

An example from recent field studies in the chaparral with improved equipment capabilities (Morrow, 1971) has shown the extreme importance of changes in leaf

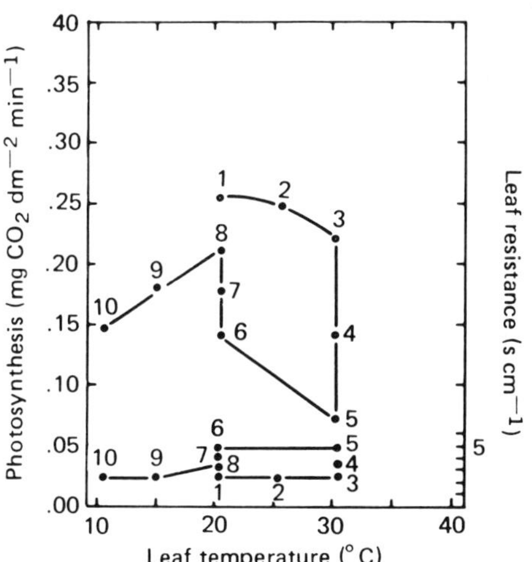

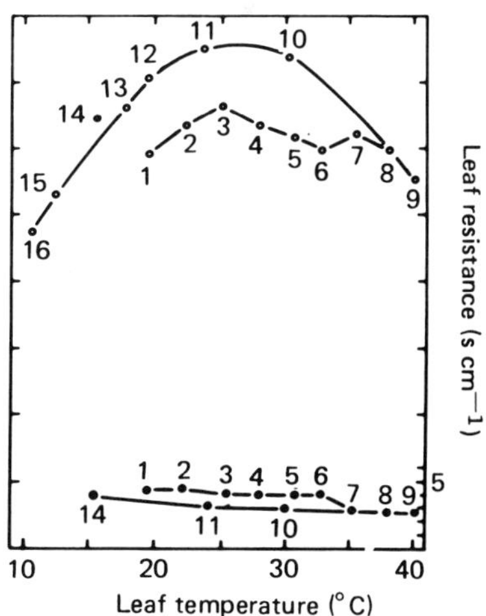

Fig. 10.15. Stomatal resistance to water-vapor transfer (solid circles) and net photosynthesis (open circles) as a function of high temperature in artificially watered plants of *Arbutus menzeisii*, in February 1970 (left) and *H. arbutifolia* in September 1970 (right). Data points are numbered chronologically. Thirty to 90 minutes elapsed between successive points. (From Morrow, 1971.)

resistance affecting CO_2 exchange rates. Examples of the temperature-dependence curves of photosynthesis and leaf diffusion resistance for two evergreen sclerophylls are shown in Fig. 10.15 (from Morrow, 1971, Fig. 11). The points are numbered and plotted sequentially in the order in which they were taken.

Figure 10.15a shows a decrease of photosynthesis and an increase of leaf resistance with increasing temperature. As temperature was reduced back to 25°C, photosynthesis recovered as leaf resistance decreased again. A similar but less complex relation between leaf resistance and photosynthesis is shown in Fig. 10.15b. These results are similar to the response observed in the field in this study before simultaneous measurements of CO_2 and water-vapor exchange could be made. These results emphasize again the importance of measuring the effects of leaf resistance on gas exchange.

Summary and Conclusions

As a result of the environmental pattern characteristic of mediterranean-type climates, several stresses are imposed on the capacity of an evergreen plant to maintain a positive CO_2-fixation rate. Typical evergreen sclerophylls from California and Chile, two representative of this climate type, responded to these stresses in the following ways.

High temperatures much above 25°C caused a reduction in net photosynthesis rates. This reduction appeared to be a result of stomatal closure and probably increased photorespiration. High evaporative demands and progressively depleted soil moisture resulted in internal water deficits. Severe water deficits were delayed or prevented by effective stomatal closure and cuticular restriction of water loss. Where soil moisture was limited, however, owing to insufficient precipitation (or a limited root system), internal water stress increased to a sublethal stage at which the evergreen leaf was a liability from an energy standpoint, losing more carbon through respiratory activity than could be fixed by photosynthesis.

Any estimate of internal resistance to CO_2 uptake requires measurements of stomatal and boundary-layer resistances to gaseous diffusion. Stomatal resistance is extremely important in determining photosynthesis rates and can hardly be overemphasized. To understand the effects of various environmental factors on rates of photosynthesis, the effects of these factors on stomatal resistances must be known.

Once external diffusion resistances were known, estimates of a comparable internal or "mesophyll" resistance were made. The best estimate of r_i' is derived from the reciprocal of the slope of the linear portion of the CO_2 dependence of net photosynthesis. This experimental procedure also provides estimates of photorespiration and the CO_2 compensation point of photosynthesis. The determination of r_i' in low O_2 permits a further separation of some of the metabolic and physical components inherent in the r_i' value.

The separation of the photosynthetic process into some of its components expressed as resistances provides a mechanism of analyzing the effects of environmental stresses on photosynthesis. The various resistance values provide an excellent means of characterizing and comparing the photosynthetic potential of different species under different environmental conditions.

In addition to restrictions by a thick cuticle and stomatal responses (i.e., high leaf-diffusive resistances), evergreen sclerophylls had significant internal resistances to CO_2 uptake: the mesophyll resistance. This factor definitely has a substantial metabolic component and probably another substantial component due to the leaf structure necessary to withstand severe drought stress.

Acknowledgments

This research was supported by National Science Foundation grants GB-5223 and GB-8184 to H. A. Mooney, Stanford University, and by an NSF Graduate Fellowship and a General Biological Supply House (Turtox) Scholarship to E. L. Dunn. The help of S. B. Dunn, A. T. Harrison, J. A. Martinez, H. A. Mooney, P. A. Morrow, F. M. Shropshire, L. C. Song, Jr., and F. A. Stevens was greatly appreciated.

References

Bierhuizen, J. F., Slatyer, R. O.: 1964. Photosynthesis of cotton leaves under a range of environmental conditions in relation to internal and external diffusive resistances. Austral. J. Biol. Sci. *17*, 348–359.

Björkman, O.: 1966. The effect of oxygen concentration on photosynthesis in higher plants. Physiol. Plantarum *19*, 618–633.

——— 1968a. Carboxydismutase activity in shade-adapted and sun-adapted species of higher plants. Physiol. Plantarum *21*, 1–10.

——— 1968b. Further studies on differentiation of photosynthetic properties in sun and shade ecotypes of *Solidago virgaurea*. Physiol. Plantarum *21*, 84–99.

Downton, J., Slatyer, R. O.: 1972. Temperature dependence of photosynthesis in cotton. Plant Physiol. *50*, 518–522.

Dunn, E. L.: 1970. Seasonal patterns of carbon dioxide metabolism in evergreen sclerophylls in California and Chile. Los Angeles: Univ. California. Ph.D. diss.

———: 1973. Estimates of internal resistance to CO_2 uptake-effects of photorespiration. *In* Terrestrial primary production: Proc. Interbiome Workshop on Gaseous Exchange Methodology (eds. B. E. Dinger, W. F. Harris), pp. 109–127. Oak Ridge Natl. Lab.

Forrester, M. L., Krotkov, G., Nelson, C. D.: 1966a. Effect of oxygen on photosynthesis, photorespiration, and respiration in detached leaves. I. Soybean. Plant Physiol. *41*, 422–427.

———, Krotkov, G., Nelson, C. D.: 1966b. Effect of oxygen on photosynthesis, photorespiration, and respiration in detached leaves. II. Corn and other monocotyledons. Plant Physiol. *41*, 428–431.

Gaastra, P.: 1959. Photosynthesis of crop plants as influenced by light, carbon dioxide, temperature, and stomatal diffusion resistance. Mededelingen Landbouwhogesch *59*, 1–68.

Gale, J., Kohl, H. C., Hagan, R. M.: 1966. Mesophyll and stomatal resistances affecting photosynthesis under varying conditions of soil, water, and evaporation demand. Israel J. Bot. *15*, 64–71.

Gauhl, E.: 1970. Leaf factors affecting the rate of light saturated photosynthesis in ecotypes of *Solanum dulcamara*. Carnegie Inst. Wash. Yearbook *68*, 633–636.

———, Björkman, O.: 1969. Simultaneous measurements on the effect of oxygen concentration on water vapor and carbon dioxide exchange in leaves. Planta *88*, 187–191.

Heath, O. V. S.: 1969. The physiological aspects of photosynthesis. Stanford, Calif.: Stanford Univ. Press.

Holmgren, P., Jarvis, P. G., Jarvis, M. S.: 1965. Resistance to carbon dioxide and water vapor transfer in leaves of different plant species. Physiol. Plantarum *18*, 557–573.

Jarvis, P. G.: 1971. The estimation of resistances to carbon dioxide transfer. *In* Plant photosynthetic production: manual of methods (ed. Z. Šesták, J. Čatský, P. G. Jarvis), pp. 566–631. The Hague: Dr. W. Junk, N. V. Publ.

Lake, J. V.: 1967a. Respiration of leaves during photosynthesis. I. Estimates from an electrical analogue. Austral. J. Biol. Sci. *20*, 487–483.

———: 1967b. Respiration of leaves during photosynthesis. II. Effects on the estimation of mesophyll resistance. Austral. J. Biol. Sci. *20*, 495–499.

Ludlow, M. M., Jarvis, P. G.: 1971. Methods for measuring photorespiration in leaves. *In* Plant photosynthetic production: manual of methods (eds. Z. Šesták, J. Čatský, P. G. Jarvis), pp. 294–315. The Hague: Dr. W. Junk, N. V. Publ.

McPherson, H. G., Slatyer, R. O.: 1973. Mechanisms regulating photosynthesis in *Pennisetum typhoides*. Austral. J. Biol. Sci. *26*, 329–339.

Medina, E.: 1971. Relationships between nitrogen level, photosynthetic capacity, and carboxydismutase activity in *Atriplex patula* leaves. Carnegie Inst. Wash. Yearbook *69*, 655–662.

Meidner, H., Mansfield, T. A.: 1965. Stomatal responses to illumination. Biol. Rev. *40*, 483–509.

Mooney, H. A., Dunn, E. L.: 1970a. Convergent evolution of mediterranean-climate evergreen sclerophyll shrubs. Evolution *24*, 292–303.

———, Dunn, E. L.: 1970b. Photosynthetic systems of mediterranean-climate shrubs and trees of California and Chile. Am. Naturalist *104*, 447–453.

———, Dunn, E. L., Shropshire, F., Song, L. C., Jr.: 1970. Vegetation comparisons between the mediterranean climatic areas of California and Chile. Flora *159*, 480–496.

Morrow, P. A.: 1971. The eco-physiology of drought adaptation of two mediterranean climate evergreens. Stanford, Calif.: Stanford Univ. Ph.D. diss.

Osmond, C. B., Troughton, J. H., Goodchild, D. J.: 1969. Physiological biochemical and structural studies of photosynthesis and photorespiration in two species of *Atriplex*. Z. Pflanzenphysiol. *61*, 218–237.

Raschke, K.: 1965. Die Stomata als Glieder eines schwingungsfähigen CO_2-Regelsystems. Experimenteller Nachweis an *Zea mays*. Z. Naturforsch. *20*, 1261–1270.

———: 1966. Die Reaktionen des CO_2-Regelsystems in den Schliesszellen von *Zea mays* auf weisses Licht. Planta *68*, 111–140.

Samish, Y., Koller, D.: 1968. Estimation of photorespiration of green plants and of their mesophyll resistance to CO_2 uptake. Ann. Bot. *32*, 687–694.

Scott, F. M.: 1950. Internal suberization of tissues. Bot. Gaz. *111*, 378–394.

Sinclair, J. D., Hamilton, E. L., Waite, M. N.: 1958. A guide to the San Dimas Experimental Forest. U.S. Forest Service, Calif. Forest and Range Expt. Sta. Misc. Paper *11*.

Slatyer, R. O.: 1970. Comparative photosynthesis, growth and transpiration of two species of *Atriplex*. Planta *93*, 175–189.

———: 1971. Effect of errors in measuring leaf temperature and ambient gas concentration on calculated resistances to CO_2 and water vapor exchange in plant leaves. Plant Physiol. *47*, 269–274.

Troughton, J. H., Slatyer, R. O.: 1969. Plant water status, leaf temperature, and the calculated mesophyll resistance to carbon dioxide of cotton leaves. Austral. J. Biol. Sci. *22*, 815–827.

Wareing, P. F., Khalifa, M. M., Treharne, K. J.: 1968. Rate-limiting processes in photosynthesis at saturating light intensities. Nature *220*, 453–457.

Whiteman, D., Koller, D.: 1967. Interactions of carbon dioxide concentration, light intensity, and temperature on plant resistances to water vapor and carbon dioxide diffusion. New Phytol. *66*, 463–473.

Woolhouse, H. W.: 1967–1968. Leaf age and mesophyll resistance as factors in the rate of photosynthesis. Hilger J. *11* (1), 7–12.

Wuenscher, J. E., Kozlowski, T. T.: 1971. Relationship of gas exchange resistance to tree-seedling ecology. Ecology *52*, 1016–1023.

PART III

Water Transport and Environmental Control of Diffusion

Introduction

David M. Gates

Water is a critical environmental factor for plants and greatly affects net photosynthesis, productivity, and growth. The role of water is complex since it enters all aspects of the plant and its environment. The water potential of the soil and plant tissue is important for physiological activity, and the water status of the atmosphere is significant in its influence on the rate of loss of water from the plant surface.

There has been much debate by plant physiologists about the physiological importance of transpiration to a plant. The loss of water from a plant is generally through the stomates and only minimally through the cuticular surface. Transpiration may be purely the fortuitous consequence of taking in carbon dioxide for assimilation. Whether or not this is true may not be particularly important in itself. However, the rate of transpiration does affect the water potential of leaf cells and their physiological "health." The rate of transpiration affects the energy budget of a leaf and its temperature. There are many circumstances when the energy consumed by transpiration has relatively little influence on leaf temperature (less than 2°C). There are some conditions, however, particularly those of high radiation intensity, high air temperature, and dry air, when the evaporation of water has notable influence on leaf temperature (as much as 10–20°C) and prevents denaturation of leaf tissue. Such strong temperature differences notably influence photosynthetic and respiration rates. The rate of water loss from a plant is proportional to the difference in water-vapor pressure inside and outside of the leaf and inversely proportional to the resistance of the pathway. The resistance is strongly controlled by the stomates.

It has been known for many years that light intensity affects the stomatal apertures of a leaf. The influence of temperature on stomatal aperture is less clearly understood because of its intimate involvement with the photosynthetic and respiration rates of the guard cells and their feedback effect on stomatal apertures. The effect of light and temperature on stomates is discussed in the following articles. A very significant new idea is described by A. E. Hall and M. R. Kaufmann: a negative feedback effect between the water-vapor pressure of the air, the stomate aperture, and the transpiration rate. This concept must be included in advanced models of leaf response to environmental factors. Most of the analytical models derived to date have been essentially static or steady state. It is increasingly clear that time-dependent models must be formulated and that stomate apertures affecting gas diffusion must be considered as dynamic; furthermore, the model must include the appropriate feedback mechanisms that occur within the cells of the leaf.

James O'Leary discusses transpiration rates of whole plants grown in growth chambers under very controlled conditions. Each environmental factor affecting growth and transpiration rate was explored in great detail, including air temperature, light intensity, relative humidity, carbon dioxide concentration, and wind speed. Once again the great significance of stomatal aperture on transpiration rate and CO_2 uptake is evident. John Tenhunen and I explore in further detail the effect of environmental factors on stomatal apertures and show how these effects are included in the theoretical model of whole-leaf photosynthesis. Then we apply this analysis to an evaluation of photosynthetic efficiency for milkweed plants. Once again the importance of theoretical analysis combined with laboratory and field measurements is demonstrated by this discussion.

The last two papers in this section deal with the interaction of radiation with plants. Ronald Alderfer explores the combined effects on the photosynthetic capacity of the plant canopy of the physiological age of the leaf and the radiation flux reaching the leaf. He believes that there are significant relationships to be discovered between the physical environment of a developing leaf and the rate at which specific proteins are synthesized in it. Richard Lee, W. G. Hutson, and S. C. Hill write about the difficulty of revegetation of disturbed lands, which, because of dark soil and low reflectivity, create a hot, harsh environment for seedling plants. This is a problem of energy exchange, environmental temperature, and plant growth.

Today the demand for energy and mineral resources by our industrialized society is causing massive excavation of coal, oil shale, iron, copper, and other minerals from the earth. Often these must be obtained by strip mining, a process that leaves the earth's green surface relatively lifeless and unproductive. Modern society realizes that mankind will survive in good health only if a productive, relatively pollution-free environment can be maintained. The restoration of the land surface to a productive state by green vegetation is essential. A thorough understanding of the ecology of plants in harsh environments will make it possible for man to improve his management of his environment and, in particular, to revegetate vast areas of land surface denuded through mining, grazing, and other activities.

11

Regulation of Water Transport in the Soil–Plant–Atmosphere Continuum

Anthony E. Hall and Merril R. Kaufmann

Introduction

The regulation of water loss from plants has particular significance to the water use, photosynthetic rates, water-use efficiencies, and development of water stress in plants. These features strongly influence plant performance where the evaporative demand is not low and/or where water supplies in the root zone are limiting. Surprisingly, there are important conflicts of opinion concerning the systems by which water loss from plants is regulated. These conflicts must be resolved if we are to develop effective mathematical models of water transport in the soil–plant–atmosphere continuum.

It is widely assumed that, under otherwise constant environments, transpiration responds linearly to the vapor-pressure gradient between leaf and air until critical levels of water stress are reached in the leaf, whereupon hydroactive closure of stomata take place which reduces the transpiration rate. This system cannot totally eliminate water stress, because its function depends on the development of a critical level of stress, which may, according to our observations, be close to the wilting point with some mesomorphic leaves.

Lange *et al.* (1971) have presented evidence for an additional, more sensitive control mechanism that involves a dependence of stomatal aperture on the vapor-pressure gradient between the leaf and the air. Some of this information is in Chapter 8 by Lange *et al.* Evidence for the influence of the vapor-pressure gradient on leaf resistance is also presented in this volume by O'Leary. A direct negative feedback between transpiration and stomatal aperture would be capable of restricting water loss so that it would prevent the further development of water stress in the subepidermal tissues. Schulze *et al.* (1972) demonstrated, with both xerophytic species and a mesophytic species, that increases in leaf resistance accompanying increases in evaporative demand reduced the level of water stress in leaves. This supports the claim that stomata may respond to humidity independent of the traditional hydroactive mechanism and illustrates the ability of the system to prevent increases in leaf water stress. Schulze *et al.* (1972) commented that all species did not react in this manner.

This issue is, however, controversial for the following reasons. Meidner and Mansfield (1968) stated in a review that stomatal behavior is comparatively unaffected by changes in the relative humidity of the air. Stomatal responses to humidity would be important adaptive features in regions of high evaporative demand, yet Whiteman and Koller (1967) reported that stomata did not respond to humidity changes with *Atriplex halimus* or *Kochia indica*. Similarly, *Tidestromia oblongifolia* had low stomatal resistances even when subjected to very high evaporative demands (Björkman *et al.*, 1972). Other workers have observed increases in leaf resistance or decreases in net photosynthesis as evaporative demands were increased but have attributed the effects to different causes. Bierhuizen and Slatyer (1964) observed decreases in CO_2 uptake by cotton at high vapor-pressure gradients, but attributed it to partial drying of the internal evaporating surfaces of the leaves. Whiteman and Koller (1964) reported decreases in CO_2 uptake and increases in leaf resistance to water-vapor transport with increases in vapor-pressure gradient in *Pinus halepensis*. They suggested that these responses were due to progressive dehydration of the mesophyll, leaf water stress, and "incipient drying." Drake *et al.* (1970) observed increases in stomatal resistance in dry air compared with moist air using *Xanthium strumarium* and invoked the classical negative feedback mechanism, involving moisture stress in the bulk leaf tissue caused by the high transpiration rates. None of these alternative explanations was substantiated. Also, the extent to which hydroactive stomatal closure contributes to the regulation of transpiration in well-watered plants of certain species may be questioned. It has been observed that, with optimal soil water, the water potential of leaves may not be influenced by large changes in transpiration rate with a number of species (Barrs, 1970; Camacho *et al.*, 1974).

The relationship between stomatal resistance and temperature is also important, owing to its effect on the energy balance of the leaf, net photosynthesis, and water-use efficiencies. As pointed out by Schulze *et al.* (1973), there are conflicts in the literature concerning the influence of temperature on stomatal resistance, and these conflicts may be partially due to variations in vapor-pressure gradients during the experiments that influenced stomatal aperture independently from temperature.

The influence of the humidity gradient on transpiration, leaf resistance, and photosynthesis was studied at different leaf temperatures with different species to attempt to resolve these conflicts of opinion.

Materials and Methods

Plant Material

Helianthus annuus L. cv. Mammoth Russian was chosen because earlier work by Whiteman and Koller (1967) had indicated that its stomata do not respond to humidity. *Sesamum indicum* L. cv. Glauca was chosen because earlier work had indicated that transpiration rates were much lower than those of sunflower when both species were well watered and subjected to similar high evaporative demands. Also, both of these species exhibited fairly constant, low levels of leaf water stress regardless of large changes in transpiration rate when their root zones were irrigated frequently (Camacho *et al.*, 1974). *Beta vulgaris* L. cv. USH8 was chosen because of its tendency to exhibit temporary wilting when subjected to high evaporative demands in the field. The plants were grown in potting mix (University of California Mix) in a greenhouse with moderate to high light intensities at maximum daytime temperatures of 32°C and minimum nighttime temperatures of 16°C. Young plants with four to six true leaves were studied and less than 50 percent of the total leaf area was removed when the intact shoot was placed in the measurement chamber. The leaves were arranged to minimize shading and held in place by nylon filaments. The plants were carefully irrigated every 2 h during gas-exchange measurements.

Measurements

A steady-state, gas-exchange system was used because it is the most precise and accurate method, of those currently available, for the simultaneous measurement of transpiration, leaf resistance, and net photosynthesis. Systematic errors could produce experimental artifacts with this method; therefore, the system is described in detail.

Leaf resistance to water-vapor transfer (r_l in s cm^{-1}) was determined by indirect procedures from the transpirational flux per unit projected area of leaf (T in μg cm^{-2} s^{-1}) and the absolute humidity difference ($q_{\text{leaf}} - q_{\text{air}}$) in μg cm^{-3}:

$$r_l = \frac{q_{\text{leaf}} - q_{\text{air}}}{T}$$

q_{leaf} was obtained from leaf temperatures by assuming saturation with respect to water vapor at the evaporating surfaces within the leaf. Leaf temperature was measured with a T-shaped thermocouple clamp of 36-gauge copper and constantan (Gale *et al.*, 1970). Gaseous fluxes were determined using an open system with a ventilated, controlled-environment, plant chamber (Hoeffer Scientific Instruments, San Francisco). Inlet humidities were controlled by passing air through a sintered-glass wash bottle containing water and a condenser in a controlled-temperature bath. Exit humidity (also equivalent to the humidity outside the leaf) was measured with a dew-point hygrometer (Cambridge Systems, Inc., Model 880, with amplified output to a potentiometric recorder). Carbon dioxide concentrations were measured differentially with an infrared gas analyzer (Beckman Model 315), and gaseous flow rates were measured with 250-mm-tube flow meters (Matheson Gas Products). Two 500-W quartz-iodine lamps in heat-filtered fixtures (Dicrolite Company, Inc.) provided radiation through a 6-cm water filter. It was assumed that the boundary-layer resistance to water-vapor transfer, which is included in the leaf resistance term used here, would be small and reasonably constant for individual plants, owing to the high rates of turbulent ventilation in the chamber.

Procedures

A plant was placed in the chamber in the morning and allowed to equilibrate at the required light intensity and temperature for 2 or more hours at the lowest humidity gradient possible without condensation of water vapor within the system. Humidity gradients between the leaf and the air were progressively increased, as indicated by the individual points on the figures, by both lowering the inlet humidity and increasing the flow rate through the chamber. Humidity gradients across the leaf were small, owing to the high-capacity fan in the chamber. Measurements were taken when steady-state gaseous exchanges were attained; this took from 15 to 90 min, depending upon the species and the climate. The time to equilibration was longest when the largest changes in leaf resistance occurred. When temperature-response curves were conducted, the flow rate and inlet humidity were adjusted to achieve the required humidity gradient between the leaf and air. Leaf temperatures were maintained better than $\pm 0.1°C$ in any one position of the clamp on the leaf.

Laboratory air from a basement compressor was passed through an activated charcoal filter and a mixing vessel. Consequently, the carbon dioxide concentration external to the leaf varied somewhat from day to day and during specific experiments. For example, the ranges in [CO_2] external to the leaf for the three sesame plants of Fig. 11.6 were 342 ± 6, 307 ± 25, and 328 ± 29 ppm; yet the responses of leaf resistance to the humidity gradient were very similar. The [CO_2]'s inside the leaf were calculated from [CO_2] external, CO_2 flux rates, and leaf resistances

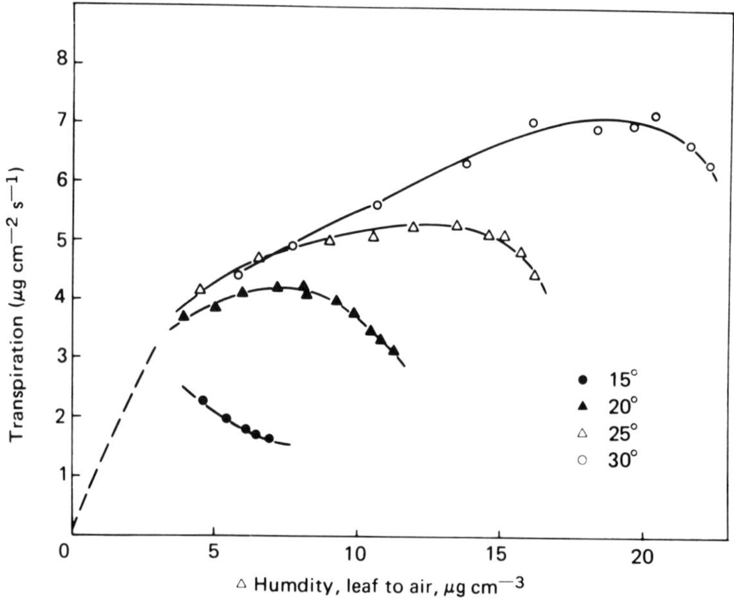

Fig. 11.1. Influence of the abolute humidity gradient between the leaf and the ambient air, and leaf temperature (°C) on transpiration rates by sesame in normal air and PAR of 7.7×10^4 ergs cm^{-2} s^{-1}.

assuming a ratio of diffusivities of 1.61 (Šesták *et al.*, 1971). In some cases, as the leaf resistance increased the [CO_2] inside the leaf decreased, and in a few cases the [CO_2] inside the leaf remained constant as the leaf resistance increased. In all cases the changes in leaf resistance were consistently related to changes in humidity. At no time were the variations in leaf resistance consistently related to changes in [CO_2] internal or [CO_2] external. It was assumed, therefore, that the variations in [CO_2]'s had little influence on the leaf-resistance responses reported here. Zelitch (1969) has suggested that CO_2 concentration does not have an important role in the normal opening and closing of stomata. As Dunn reports elsewhere in this volume, leaf resistance may be correlated with internal CO_2 concentration in some situations. The authors consider that species may differ with respect to the influence of internal CO_2 concentrations on stomatal aperture, but we suggest that humidity gradients and light-intensity interactions should also be considered when the effects on stomata of normal and lower concentrations of CO_2 are reevaluated.

Results

Influence of the Humidity Gradient and Leaf Temperatures on Gaseous Exchanges by Sesame

Transpiration rates were determined as a function of the humidity gradient leaf to air with sesame at different temperatures and photosynthetically active radiation (PAR, 400–700 nm) of 7.7×10^4 ergs $cm^{-2} s^{-1}$ (Fig. 11.1). One curve was conducted each day starting at 15°C and progressing to 30°C, using the same plant. Another plant was examined, using the temperature sequence 30°C, 20°C, 30°C. Both plants produced the same results, indicating that acclimation was not a significant factor. Transpiration did not respond linearly to the humidity gradient but achieved optimum rates that increased with temperature. Another way of expressing these data is that leaf resistances increased as the humidity gradient increased, with more drastic responses occurring at lower temperatures (Fig. 11.2). This increase in leaf resistance in response to increases in the humidity gradient regulated the upper limits of transpiration but interacted with temperatures, permitting more transpiration at higher temperatures (Fig. 11.1).

Net photosynthesis was measured concurrently. The rates were corrected to the average CO_2 concentration of 300 ppm by assuming a linear response to CO_2 concentration and CO_2 compensation points that increased with temperature. Net photosynthesis decreased as the humidity gradient increased in a manner that would be expected if the increases in leaf resistance were due to stomatal closure (Fig. 11.3). Apparently the penalty for reducing water loss is reduced net photosynthesis in these conditions.

The nature of the compromise between controlling water loss and maintaining net photosynthesis was assessed by computing the transpiration ratio (=g water

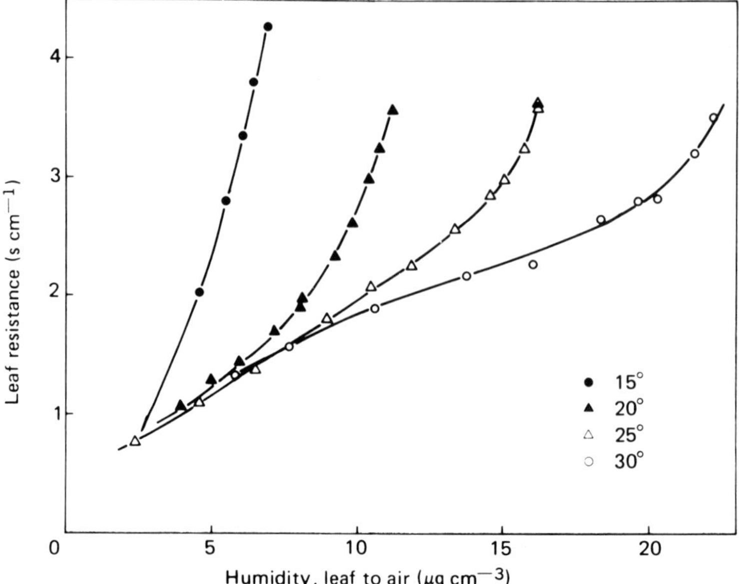

Fig. 11.2. Influence of the absolute humidity gradient between the leaf and the ambient air, and leaf temperature (°C) on the resistance to water-vapor transfer of sesame.

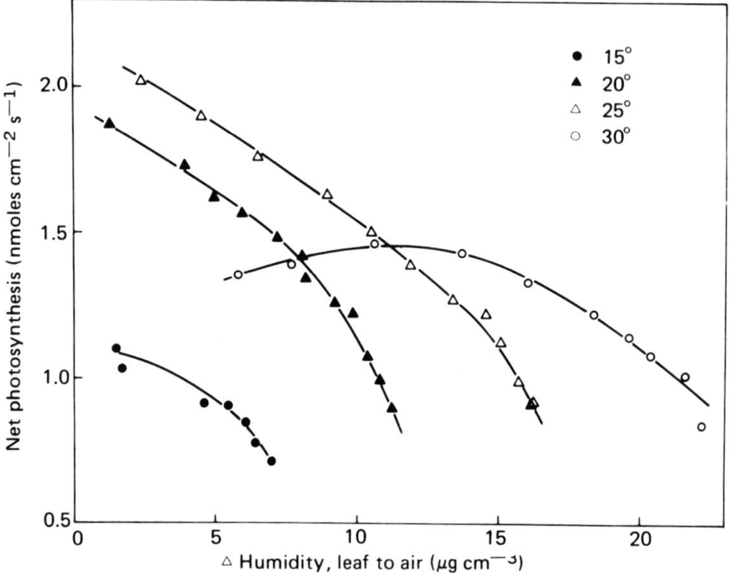

Fig. 11.3. Influence of the absolute humidity gradient between the leaf and the ambient air, and leaf temperature (°C) on net photosynthesis by sesame in normal air and PAR of 7.7 × 10^4 ergs cm^{-2} s^{-1}.

lost/g CO_2 taken up). The influence of the humidity gradient on the transpiration ratio was remarkably similar at all temperatures (Fig. 11.4). Differences in stomatal responses to humidity at different temperatures were partially balanced by differences in intrinsic photosynthetic capabilities. The dashed line describes the theoretical response if stomata were to remain open as the humidity gradient increased (Bierhuizen and Slatyer, 1965). Apparently, stomatal closure at low external humidities substantially improves water-use efficiency with this plant in these conditions, in comparison with the theoretical response.

Another consequence of the responses presented in Fig. 11.2 is that the influence of temperature on both leaf resistance and net photosynthesis may be highly dependent upon the humidity gradient between the leaf and the air. This was tested by simultaneous measurement of leaf resistance and net photosynthesis while leaf temperature was increased in increments of 2°C with PAR = 7.7×10^4 ergs cm^{-2} s^{-1}. On the first day the absolute humidity of the air surrounding the leaf was kept constant at 11 $\mu g\ cm^{-3}$ as the leaf temperature was raised, and a progressive increase in leaf resistance was observed (Fig. 11.5). On the second day, with the same sesame plant, the humidity difference between the leaf and the air was maintained at a low constant level of 3 $\mu g\ cm^{-3}$. This resulted in a tendency for the leaf resistance to become smaller at higher temperatures (Fig. 11.5). The net photosynthetic rates were corrected for variations in CO_2 concentration to representative values at 300 ppm CO_2. The optimum temperature for net photosynthesis and the maximum rate of net photosynthesis were higher when the humidity

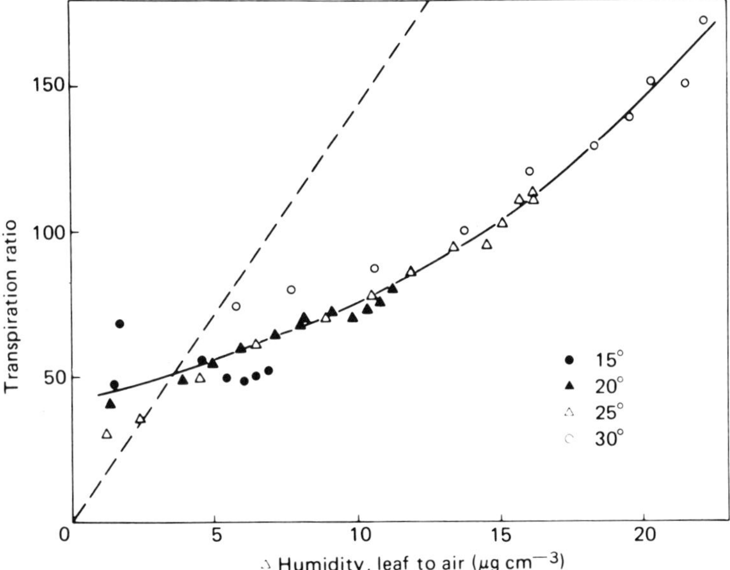

Fig. 11.4. Influence of the absolute humidity gradient between the leaf and the ambient air, and leaf temperature (°C) on the transpiration ratio (g water lost/g CO_2 taken up) of sesame in normal air and PAR of 7.7×10^4 ergs cm^{-2} s^{-1}.

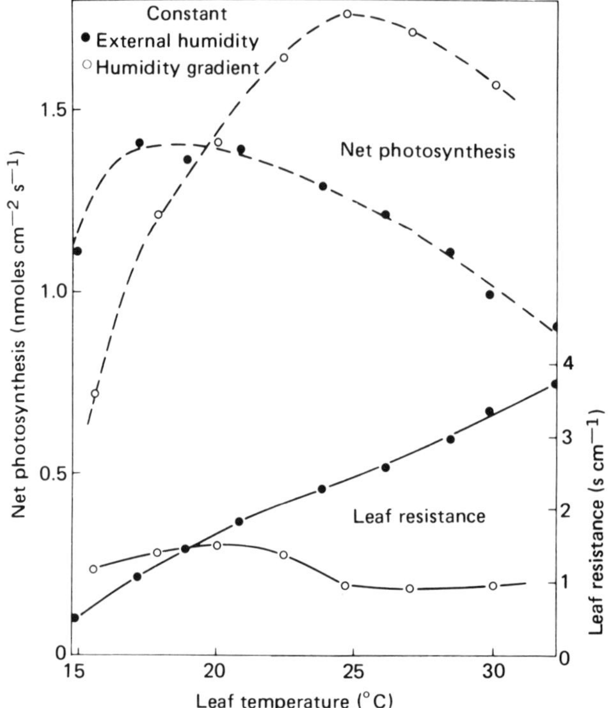

Fig. 11.5. Influence of leaf temperature on net photosynthesis and the resistance to water-vapor transfer in normal air and PAR of 7.7×10^4 ergs cm^{-2} s^{-1} by sesame. At a constant absolute humidity difference between the leaf and ambient air of 3 μg cm^{-3} (circles) and at a constant, ambient, absolute humidity of 11 μg cm^{-3} (dots).

gradient was maintained at a low constant level. The curves exhibited crossovers in both leaf resistance and net photosynthesis at similar temperatures, indicating that the variations in photosynthesis from day to day were associated with variations in leaf resistance from day to day.

Humidity and Temperature Effects on Gaseous Exchanges by Sesame, Sunflower, and Sugar Beet

Gaseous exchanges were measured in response to progressive increases in humidity gradient at 20 and 30°C, taking 2 days per plant. Higher irradiances were used to approach light saturation of net photosynthesis in sunflower and sugar beet (PAR = 1.8×10^5 ergs cm^{-2} s^{-1}). The influences of the humidity gradient on leaf resistances of three representatives of each species are presented for leaf temperatures of 30°C (Fig. 11.6) and 20°C (Fig. 11.7). At the higher temperature, the humidity gradient had little influence on leaf resistance with sunflower and sugar beet, but the response of sesame was similar to that observed before (Fig. 11.2). At 20°C all species tested exhibited increased leaf resistances in response to

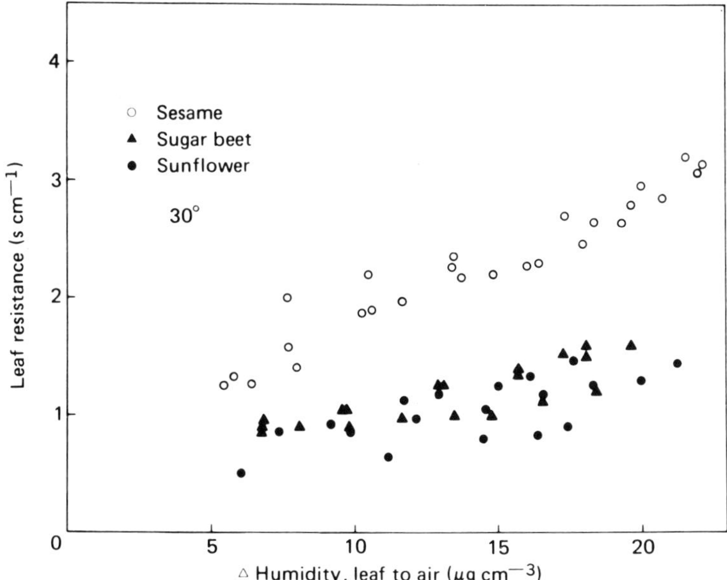

Fig. 11.6. Influence of the absolute humidity gradient between the leaf and the ambient air on the resistance to water-vapor transfer of sesame (circles), sunflower (dots), and sugar beet (triangles) plants in normal air, PAR of 1.8×10^5 ergs cm^{-2} s^{-1} and leaf temperatures of 30°C.

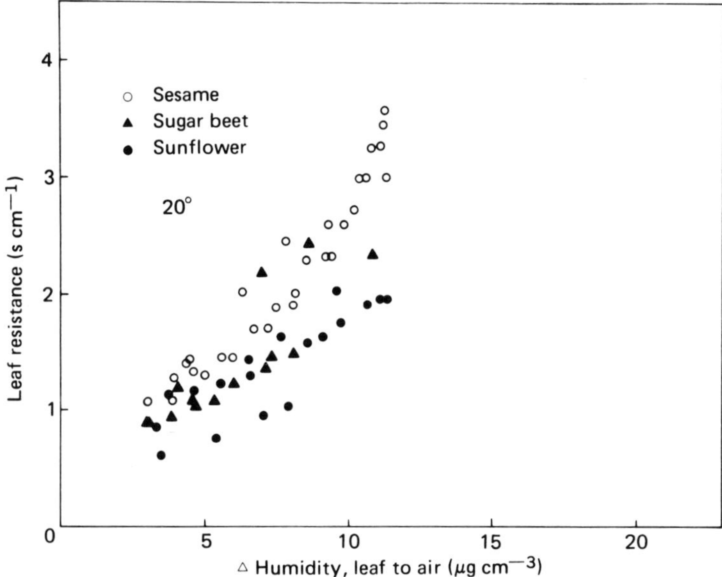

Fig. 11.7. Influence of the absolute humidity gradient between the leaf and the ambient air on the resistance to water-vapor transfer of sesame (circles), sunflower (dots), and sugar beet (triangles) plants in normal air, PAR 1.8×10^5 ergs cm^{-2} s^{-1} and leaf temperatures of 20°C.

increases in humidity gradient, but the sunflower still tended to be less responsive than the sesame. The impact of these differences in leaf resistance on transpiration may be observed in Fig. 11.8. At 30°C and large humidity gradients, the transpiration rate per unit projected leaf area of sunflower was more than twice that of sesame. At 20°C the transpiration rates were still higher with sunflower than with sesame, but the differences between them were not as pronounced as at the higher temperature.

Photosynthetic responses to humidity were determined at the same time (Fig. 11.9). Rates of net photosynthesis were adjusted for variations in CO_2 concentration (the range of variation for the total experiment was 302–348 ppm) to representative values at 330 ppm. The influence of the humidity gradient on net photosynthesis was more pronounced with both species at the lower temperature than at 30°C, and net photosynthesis was higher with sunflower than with sesame. Negative correlations were apparent between net photosynthesis and leaf resistance except at 30°C and moderate humidity gradients. This was also observed in an earlier experiment (Figs. 11.2 and 11.3).

The impact on the transpiration/net photosynthesis compromise of these differences in stomatal response to humidity was evaluated. The transpiration ratios (g water loss/g CO_2 taken up) were examined for sunflower and sesame in relation to the humidity gradient (Fig. 11.10). Both species exhibited improved water-use efficiency at high humidity gradients in comparison with the theoretical response, but the sesame was superior in this respect.

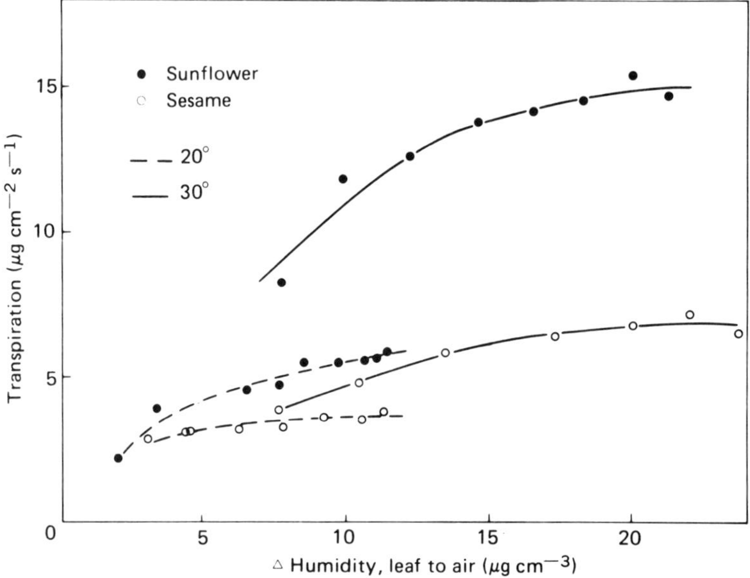

Fig. 11.8. Influence of the absolute humidity gradient between the leaf and the ambient air on transpiration rate per unit leaf area by sesame (circles) and sunflower (dots) plants at 20 and 30°C.

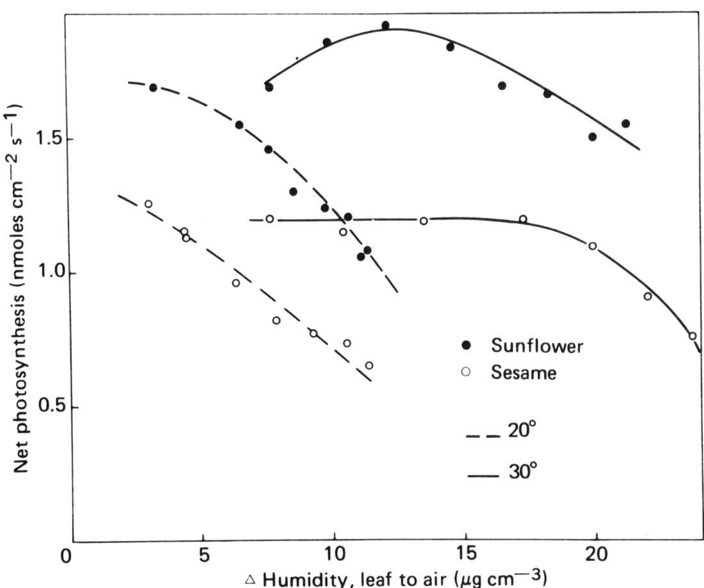

Fig. 11.9. Influence of the absolute humidity gradient between the leaf and the ambient air on net photosynthesis per unit leaf area of sesame (circles) and sunflower (dots) plants in normal air, PAR of 1.8 × 10^5 ergs cm^{-2} s^{-1}, and leaf temperatures of 20 and 30°C.

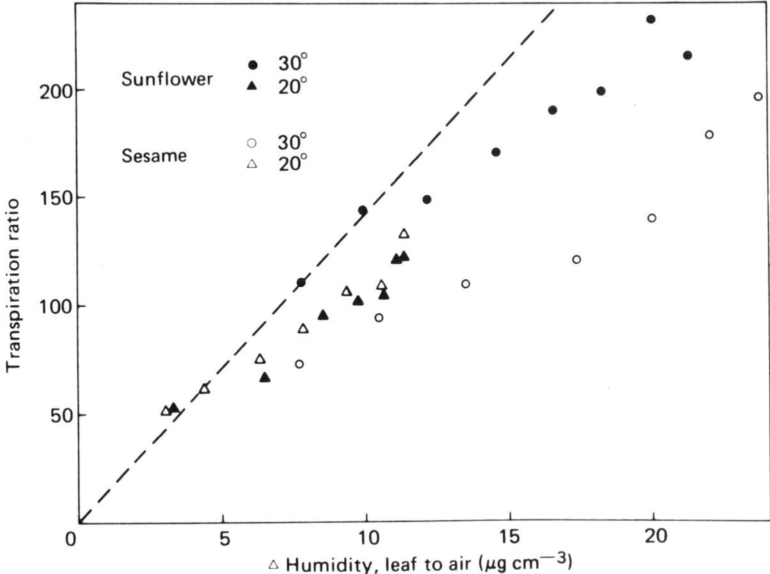

Fig. 11.10. Influence of the absolute humidity gradient between the leaf and the ambient air on the transpiration ratio (g water loss/g CO_2 taken up) of sesame at 20°C (open triangles) and 30°C (circles) and sunflower at 20°C (solid triangles) and 30°C (dots). Normal air and PAR of 1.8 × 10^5 ergs cm^{-2} s^{-1}.

Discussion

Mechanism by Which Water Transport Was Regulated

The similarities between the effect of the humidity gradient on transpiration and net photosynthesis and the evidence of Lange *et al.* (1971) suggest that the stomata are the controlling agent rather than "incipient drying" of the cell walls inside the leaves. Changes in the cuticular pathway could not be directly responsible for these effects, because changes in cuticular resistance from 10 s cm^{-1} to 10,000 s cm^{-1} would only increase the total resistance 10 percent with a stomatal resistance of 1 s cm^{-1}. Also, it is possible that these stomatal responses are not primarily due to changes in water status of the leaf as a whole. Leaf-water potentials of sesame and sunflower remained between -1 and -4 bars and did not become more negative as transpiration was increased to rates in excess of those used in this study (Camacho *et al.*, 1974). Also, Schulze *et al.* (1972) observed increases in leaf-water content when leaf resistances increased in response to decreases in ambient humidity.

The mechanism by which stomata respond to the gradient in absolute humidity between the leaf and the ambient air has not been established, but Lange *et al.* (1971) have proposed a feasible explanation based upon the concept of "peristomatal transpiration." They suggested that direct cuticular loss of water from the guard cells may be more pronounced than the direct cuticular loss of water from other epidermal cells. Consequently, as humidity gradients increase, water loss from the guard cells would increase more than would the water loss from the other epidermal cells. Differences in turgor pressure would then develop between the guard cells and the other epidermal cells, causing stomatal closure. However, Pallaghy (1972) has demonstrated that thallium localization may not be a valid method for demonstrating the existence of "peristomatal transpiration." Localization studies with lead chelate do, however, support the concept of "peristomatal transpiration" (Tanton and Crowdy, 1972).

Adaptive Significance of This Regulatory Mechanism for Water Movement in Plants

Stomatal response to humidity and temperature could be coordinated with the efficiency of the water-transport system in the plant so that the development of water stress in the subepidermal tissues of the leaf would be minimized within the constraints set by differential needs for maintaining net photosynthesis, evaporative cooling, and an optimal level of water-use efficiency. The relative importance of these competing requirements would be set by many environmental and plant factors. Examples are presented to illustrate this complex concept.

The sugar beet is native to coastal environments with low evaporative demands; when subjected to high evaporative demands at high temperatures, it exhibits temporary wilting even when soil moisture levels are high. The small response of its stomata to increases in humidity gradient at high temperatures (Fig. 11.6) could be partially responsible for this wilting tendency, together with contributions from its water-transport system that may not be as efficient as that of sunflower. Studies by Tal (1966) and Barrs (1970) with "wilty mutants" of tomato support this concept of a *balance* between the regulation of water transport by stomatal response to humidity and the efficiency of water transport to the leaves. Tal suggested that the wilting tendency of the mutants was mainly due to reduced regulation of water loss by the stomata of the mutants. His grafting experiments had indicated that the stem and root systems of the mutants had, at most, only a small effect on water supply to the leaves in comparison with the parent material. The data of Barrs indicate that the stomata of the parent tomato responded to humidity, whereas the stomata of the mutant, "sitiens," did not. The mutant had higher transpiration rates than the parent only at relative humidities less than 75 percent. Leaf-water potentials of the parent tomato did not change significantly at relative humidities less than 75 percent, whereas the leaves of the mutant did exhibit greater water stress. This indicates that stomatal closure in the parent was not caused by leaf-water stress in these conditions, but that it was preventing the development of water stress in the parent tomato. Barrs (1970) presents a somewhat different interpretation of these data. The data of Tal (1966) also indicate that other aspects of stomatal regulation, such as the dependence of stomatal aperture on light, are also defective in some of the mutants.

Sunflower appears to have evolved a system that gives priority to the maintenance of photosynthetic rates when evaporative demands increase at high temperatures (Fig. 11.9). Consequently, it also must have an efficient water-transport system, as suggested by the data presented by Camacho *et al.* (1974).

Sesame appears to give more priority to the efficient use of water (Fig. 11.4) than to the maintenance of photosynthetic rates (Fig. 11.3). The more dramatic control of water loss by sesame (Fig. 11.8) should also confer an advantage in comparison with sunflower when soil moisture is limited and evaporative demands are high. In this event the interaction of water stress with stomatal responses to temperature as described by Schulze *et al.* (1973) becomes particularly relevant together with differences in hydroactive stomatal regulation (Boyer, 1970).

The possible significance of the tendency for stomatal responses to humidity to be less pronounced at high temperatures (Figs. 11.2, 11.6, and 11.7) should be evaluated. This may be related to the change in value of evaporative cooling at low and high temperatures to broad-leaved plants such as these. For example, in the experiment of Fig. 11.6 temperatures of sunflower leaves were as much as 3 to 4°C lower than air temperatures at large humidity gradients, whereas sugar beet leaves were only 2°C lower and sesame leaves were close to air temperature. At lower leaf temperatures (Fig. 11.7) all the species exhibited leaf temperatures that were about 1°C higher than air temperature at large humidity gradients. The advantages of higher leaf temperatures at low air temperatures and lower leaf temperatures at high air temperatures are obvious. However, these effects would be smaller with smaller leaves where leaf temperatures are closer to air temperatures. Temperature interactions with the stomatal response to humidity also provide a system through which solutions may have evolved to the complex problem of the compromise involved in controlling water loss while simultaneously maintaining photosynthetic rates at adequate levels (Fig. 11.4). The influences of leaf water status on stomatal responses to humidity (Schulze *et al.*, 1972) and to temperature (Schulze *et al.*, 1973) are also important mechanisms by which the compromise could be achieved. Ultimately the compromise would depend upon the interactions of many plant and environmental factors.

Physiological Significance

The results presented in Figs. 11.2 and 11.5 may provide answers to current controversies concerning the responses of stomata and net photosynthesis to temperature. When the humidity gradient is not controlled, any stomatal response is possible. If humidity gradients are allowed to increase with increasing temperatures (due to constant ambient, absolute humidities or to condensation in the measurement system) stomata will tend to close at high temperatures. If the humidity gradient between the leaf and the air is kept constant, then either the stomatal resistance will remain constant or it may decrease with increasing temperature, depending upon the magnitude of the humidity gradient and the species. There are many examples of these different types of stomatal response to temperature in the literature (Heath and Orchard, 1957; Hofstra and Hesketh, 1969; Drake *et al.*, 1970; Hall and Loomis, 1972). It is also likely that extremely high temperatures will result in stomatal closure regardless of the humidity gradient due to the general occurrence of metabolic lesions.

The same arguments also apply to photosynthetic responses of leaves to temperature, of which there are a bewildering array of different types in the literature. One example will be given. Kriedemann (1968) published photosynthetic responses to temperature by citrus in humid and dry air which are very similar to the responses of Fig. 11.5. Camacho (unpublished data), working in Hall's laboratory, has observed that the stomata of citrus also respond to the humidity gradient between the leaf and the air.

Conflicts concerning the presence or absence of direct effects of the humidity gradient on stomata may be resolved by the observations that species differ in their response to humidity; and that stomatal responses to the humidity gradient are greater at lower leaf temperatures. It is likely that many other interactions between stomata and their environment must be elucidated before the regulation of water transport in the soil–plant–atmosphere continuum is adequately understood.

Acknowledgments

The authors thank Saul Camacho for his contribution to their studies, both through his research, which provided supporting evidence, and through the stimulating discussions he provoked. The work reported here was partially supported by National Science Foundation Grant GB-7621 to M. R. Kaufmann.

References

Barrs, H. D.: 1970. Controlled environment studies of the effect of variable atmospheric water stress on photosynthesis, transpiration and water status of *Zea mays* L. and other species. *In* Proc. Symp. Plant Response to Climatic Factors, pp. 249–258. Uppsala, Sweden: UNESCO.

Bierhuizen, J. F., Slatyer, R. O.: 1964. Photosynthesis of cotton leaves under a range of environmental conditions in relation to internal and external diffusive resistances. Austral. J. Biol. Sci. *17*, 348–359.

———, Slatyer, R. O.: 1965. Effect of atmospheric concentration of water vapor and CO_2 in determining transpiration–photosynthesis relationships of cotton leaves. Agric. Meteorol. *2*, 259–270.

Björkman, O., Pearcy, R. W., Harrison, A. T., Mooney, H.: 1972. Photosynthetic adaptation to high temperatures: a field study in Death Valley, California. Science *175*, 786–789.

Boyer, J. S.: 1970. Differing sensitivity of photosynthesis to low leaf water potentials in corn and soybean. Plant Physiol. *46*, 236–239.

Camacho, S. E., Hall, A. E., Kaufmann, M. R.: 1974. Efficiency and regulation of water transport in some woody and herbaceous species. Plant Physiol. In press.

Drake, B. G., Raschke, K., Salisbury, F. B.: 1970. Temperatures and transpiration resistances of *Xanthium* leaves as affected by air temperature, humidity, and wind speed. Plant Physiol. *46*, 324–330.

Gale, J., Manes, A., Poljakoff-Mayber, A.: 1970. A rapidly equilibrating thermocouple contact thermometer for measurement of leaf-surface temperatures. Ecology *51*, 521–525.

Hall, A. E., Loomis, R. S.: 1972. Photosynthesis and respiration by healthy and beet yellows virus-infected sugar beets (*Beta vulgaris* L.). Crop Sci. *12*, 556–572.

Heath, O. V. S., Orchard, B.: 1957. Midday closure of stomata. Temperature effects on the minimum intercellular space carbon dioxide concentration. Nature *180*, 180–181.

Hofstra, G., Hesketh, J. D.: 1969. The effect of temperature on stomatal aperture in different species. Can. J. Bot. *47*, 1307–1310.

Kriedemann, P. E.: 1968. Some photosynthetic characteristics of citrus leaves. Austral. J. Biol. Sci. *21*, 895–905.

Lange, O. L., Lösch, R., Schulze, E. D., Kappen, L.: 1971. Responses of stomata to changes in humidity. Planta *100*, 76–86.

Meidner, H., Mansfield, T. A.: 1968. Physiology of stomata. New York: McGraw-Hill.

Pallaghy, C. K.: 1972. Localization of thallium in stomata is independent of transpiration. Austral. J. Biol. Sci. *25*, 415–417.

Schulze, E. D., Lange, O. L., Buschbom, U., Kappen, L., Evenari, M.: 1972. Stomatal responses to changes in humidity in plants growing in the desert. Planta *108*, 259–270.

———, Lange, O. L., Kappen, L., Buschbom, U., Evenari, M.: 1973. Stomatal responses to changes in temperature at increasing water stress. Planta *110*, 29–42.

Šesták, Z., Čatský, J., Jarvis, P. G.: 1971. Plant photosynthetic production: Manual of methods. The Hague: Dr. W. Junk, N. V. Publ.

Tal, M.: 1966. Abnormal stomatal behavior in wilty mutants of tomato. Plant Physiol. *41*, 1387–1391.

Tanton, T. W., Crowdy, S. H.: 1972. Water pathways in higher plants. III. The transpiration stream within leaves. J. Expt. Bot. *76*, 619–625.

Whiteman, P. C., Koller, D.: 1964. Environmental control of photosynthesis and transpiration in *Pinus halepensis*. Israel J. Bot., *13*, 166–176.

———, Koller, D.: 1967. Species characteristics in whole plant resistances to water vapor and CO_2 diffusion. J. Appl. Ecol. *4*, 363–377.

Zelitch, I.: 1969. Stomatal control. Ann. Rev. Plant Physiol. *20*, 329–250.

12

Environmental Influence on Total Water Consumption by Whole Plants

JAMES W. O'LEARY

Introduction

Transpiration is a process of fundamental importance in the physiology of plants. It is the link through which many physiological processes in the plant are coupled to the environment. Many of the growth responses of plants to such things as light intensity, relative humidity, air temperature, and wind velocity can be explained on the basis of the direct effect of these environmental parameters on transpiration and the subsequent effects on leaf temperature and stomatal aperture. The relationships between these environmental factors and transpiration and the effects on plant temperature, energy exchange, and photosynthesis have largely been determined from studies with single leaves. This is understandable because of the precise control and monitoring of the environment possible within a transparent leaf chamber.

On the other hand, most of the voluminous literature relating total water consumption and growth as affected by the environment has resulted from studies of plant communities. This, too, is understandable because of the great economic interest in the relationship between the total amount of water required to produce a crop on a given area of land and the yield of that crop. Thus we have an abundance of information on evapotranspiration from plant communities under various environmental conditions. We also know a good deal about consumptive water use or water-use efficiency from similar studies of plant communities.

From an ecological standpoint, it is the individual plant that is of major importance. The plant is composed of many leaves, each behaving in accordance with its physiological age and the microenvironment around it, and the plant has to integrate the activities of all those individual leaves. The plant also must compete with other individuals within the community. The sum of all this then determines the response of the individual plant to a specific environment, how successfully it competes, how efficient it is in use of available water, and how well it grows.

The transpiration and total water consumption of individual plants in a community situation have been studied comparatively little. The closest approach has been the use of potted plants, which periodically are removed from the growth chamber or greenhouse bench and weighed, then returned, probably not in the same position relative to the other plants as before. But this does not tell us how the plant behaves in an undisturbed community, i.e, the situation where the leaves expand and position themselves with respect to the plants next to them and with respect to the microenvironments within the community. For plants in this situation, we can ask questions such as the following. What is the total leaf area of each plant? What is the total water loss from each plant in this undisturbed situation? How efficient is each one in water use? How does this influence the growth of each plant? What effect does the macroenvironment have on these relationships for the individual plant?

Over the past few years we have been studying plant growth under carefully controlled environmental conditions in our laboratory. We have tried to grow plants in undisturbed communities and in a closed, controlled environment with a precision approaching that of the leaf chamber as nearly as possible. Our growth chambers are, in effect, large cuvettes. An analysis of transpiration in response to specific environmental parameters was not the primary objective in any of these studies (O'Leary and Knecht, 1971, 1972; Knecht and O'Leary, 1972). However, a sufficient number of parameters was measured in some of these experiments to allow calculation of transpiration rates and analysis of the relationships among transpiration rate, total water consumption, and plant growth under specific environmental conditions. This analysis is provided in the hope that it will contribute some useful information on whole-plant behavior in undisturbed communities, and aid in evaluating extrapolations to whole plants from data on single-leaf and total-community studies.

Materials and Methods

The experimental plant used for all the studies described here was *Phaseolus vulgaris* L. In all cases, seeds were germinated in vermiculite or paper towels (Prisco and O'Leary, 1970) and seedlings transferred to containers of nutrient solution (O'Leary and Prisco, 1970) 7 to 10 days after seeding. Except for the air-velocity experiment, all studies were conducted in ISCO Model E-3 plant growth chambers. The light source in each chamber was composed of 3200 W of Sylvania metal-arc lamps (metal halide vapor) and 800 W of incandescent light. Precise control of dry- and wet-bulb temperatures was achieved by an electronic control system. The air was humidified through evaporation from a free water surface whose temperature was controlled and dehumidified by condensation on a cold coil. In the 95–100 percent relative humidity treatments, water was added to the air stream from an atomizer. Air flow in the chambers was upward through the plants at a velocity of 25–50 cm s^{-1} across the plant zone, as measured with a Hasting hot wire anemometer. Carbon dioxide concentration in the chambers was controlled in some experiments with an electronic control system designed and constructed in our laboratory.

The growing area within each chamber was 86 dm^2. Twenty 1-liter containers were mounted in the tray of each chamber. The chambers were sealed with a Plexiglas panel mounted in the door opening, and individual plant containers were connected by opaque tubing to valves mounted in this panel. Similarly, aeration lines and nutrient solution supply lines ran to each plant container. This arrangement allowed the plant containers to be drained and refilled without disturbing the plants or the environment within the chamber. Thus daily water consumption could be monitored for each plant without moving the plants and changing their orientation with respect to each other and without changing the constant environmental conditions inside the chambers. This was especially important where relative humidity or carbon dioxide concentration was being maintained at levels greatly differing from the ambient levels in the laboratory. The door could be opened and the plants observed, yet the chamber remained closed, because of the transparent Plexiglas panel.

Leaf areas were determined at harvest with an air-flow planimeter. Leaf temperatures were measured with a Barnes PRT-10 infrared thermometer in some cases and with a PRT-4 in other cases.

The air-velocity experiment was conducted in inflated polyethylene-covered chambers (phytocells) which have been described elsewhere (Gensler, 1972). Briefly, they are inflated chambers whose air temperature is controlled by circulating the air within the chamber in a closed loop through a packed column washed-air heat exchanger. The temperature of the water passing through the packed column is controlled, and the sensible heat transfer between the water and the air passing through the column results in good control of air temperature. This also maintains the humidity of the air near saturation. The light source is sunlight. The air velocity through the plant growing zone could be controlled by changing pulley ratios between the fan and the motor driving the fan.

Growth and Water Use

The effects of varying some environmental parameters on growth and water consumption are summarized in Table 12.1. In general, increased growth and increased water consumption go hand in hand, regardless of which environmental factor is varied. Some observations about the effects of specific environmental factors can be made. Relative humidity (rh) had little effect on the final size of the plants after 20 days growth, but at the high rh, the total water consumption was considerably less than at low and medium rh. As a result, the transpiration ratio (TR) was much lower for the plants grown at high rh. Bierhuizen and Slatyer (1965) found the same relationship between the transpiration ratio and humidity in studies on cotton leaves. Furthermore, they used the data collected by Briggs and Shantz

Table 12.1 Growth and Water Use of *Phaseolus vulgaris* L. Plants Under Various Environmental Conditions

Environmental Variables	Total Plant Dry Weight (g)	Total Water Consumption (g)	Transpiration Ratio (g H_2O/g dw)
Relative humidity (%)			
35–40	5.9	2252	387
70–75	6.5	2524	403
95–100	6.4	770	114
Light (fc)			
2000	4.8	1709	354
4000	6.8	2948	437
6000	6.8	2731	378
8000	6.1	2287	379
Air temperature (day/night; °C)			
65/65	5.5	835	142
75/65	7.5	2747	367
90/65	9.5	4811	428
CO_2 concentration (ppm)			
400	11.1	3148	287
800	17.4	4112	239
1200	18.0	5134	288

(1913) to demonstrate that the same relationship holds for a wide range of plants grown in different parts of the world. On this basis, then, it seems safe to conclude that the transpiration ratio decreases with increasing rh, and this is not at the expense of reduced plant growth. Of more interest, however, is that total water consumption was as low as it was at the lowest rh. In spite of the great difference in rh between the low and medium levels, there was very little difference in total water consumption, especially if the difference in plant size is considered.

The response of the plants to light intensity shows the typical plateau response to limiting factors, and the highest light intensity may even have been slightly detrimental. Again water consumption and growth seem reasonably well correlated. It looks as if water consumption was reduced more than growth at the lowest light level, however.

The effects of day temperature on growth and total water consumption seem straightforward. They both increased with increasing temperature, but the water use was reduced more than growth at the lowest level of temperature, just like the response to light. Thus, at less than optimum levels of both light and temperature, the plants grew less, but they were much more efficient in their use of water.

The response of plant growth to atmospheric CO_2 concentration also showed the expected plateau response to limiting factors, with not much difference between 800 and 1200 ppm, but both resulting in much better growth than 400 ppm. Water consumption also increased with increasing CO_2 concentration. While the growth increase was about the same when going from 400 to 1200 ppm, the water-consumption increase was much less at 800 than at 1200 ppm CO_2.

The relationship between water consumption and some of these environmental factors will now be discussed more fully.

Relative Humidity

Twenty plants were placed into each of three growth chambers maintained at 35–40, 70–75, and 95–100 percent rh, respectively. The chambers were programmed to give a cycle of a 12 h day/1-h dusk/10 h night/1 h-dawn, with air temperatures maintained at 24/21/18/21°C, respectively. The light intensity was 4000 fc during the day and 1200 fc during the dawn and dusk periods. After 20 days of growth in the chambers, a 3-h water-consumption measurement was made on five randomly selected plants in each chamber. Those plants were then harvested and leaf areas were measured as well as fresh and dry weights. Transpiration rates were calculated from the water-uptake and leaf-area data for each plant. Also, the water-consumption data for the 24-h period prior to harvest were used to calculate the transpiration per day. Those data are presented in Table 12.2.

Table 12.2 Transpiration Rates of *Phaseolus vulgaris* L. Plants Grown at Three Humidities

Relative Humidity (%)	3-Hour Measurement ($g\ dm^{-2}\ h^{-1}$)	24-Hour Measurement ($g\ dm^{-2}\ d^{-1}$)
35–40	1.7	8.8
70–75	1.3	7.9
95–100	0.4	1.5

The transpiration rate declines with increasing relative humidity as expected, whether calculated from the short-term experiment or the total daily water consumption. To see the true relationship between the atmospheric moisture content and the transpiration rate, we calculate the vapor-pressure gradient from leaf to outside air (Δe). Leaf temperature was measured directly, and the leaf vapor pressure was assumed to be the saturation vapor pressure for that temperature. The average leaf temperature of the upper leaves exposed to direct light was used as the leaf temperature for these calculations. The vapor pressure of the bulk air was calculated from the wet- and dry-bulb temperatures. The transpiration rate is plotted against the vapor-pressure gradient in Fig. 12.1. The points for the high and medium humidities fall on a line that passes through zero as predicted by diffusion theory. The point for the low humidity does not conform, however, indicating that the plants were not transpiring in direct proportion to the driving gradient. If they were, the transpiration rate for the plants in low humidity would be over 6.0 $g\ dm^{-2}\ h^{-1}$ instead of 1.7. This suggests that the plants are exerting some control over the transpiration rate, presumably by reducing stomatal aperture

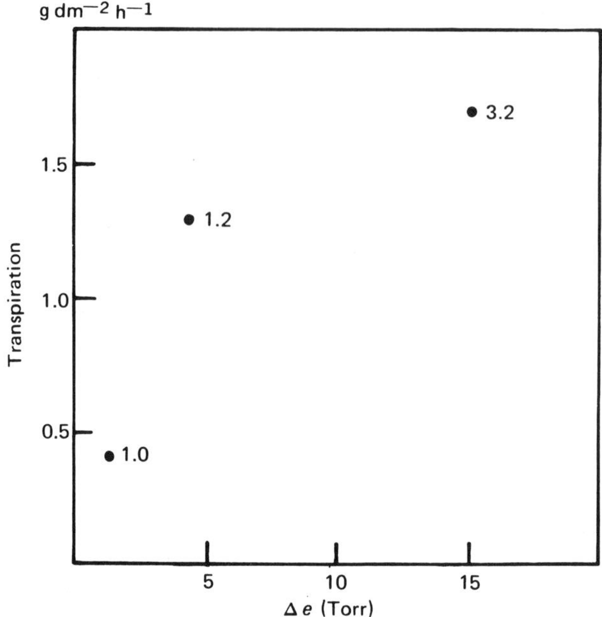

Fig. 12.1. Transpiration rate of *Phaeolus vulgaris* L. plants as affected by the vapor-pressure gradient from leaf to air (Δe). Numbers alongside data points are the relative total resistances for the respective environments.

and increasing the leaf diffusion resistance. Others have also found that plant control over transpiration is exerted when the evaporative demand of the air becomes high. For example, Hoffman *et al.* (1971) found that water loss in cotton at 25 percent rh was only double that at 90 percent rh, even though the vapor-pressure gradient from leaf to air was about six times greater at the lower rh. This was due to reduced stomatal aperture at the lower rh. Similarly, Drake *et al.* (1970) found that transpiration from individual leaves did not increase proportionately with the increase in vapor-pressure gradient between 70 and 10 percent rh, which also was attributed to higher leaf diffusion resistance at the lower humidity.

In the present example, we can estimate the relative differences in plant diffusion resistance at the different humidities by using a generalized transpiration equation of the form

$$T = \frac{C_p - C_a}{\sum r} \tag{1}$$

where T is the transpiration rate (g cm^{-2} s^{-1}) and C_p and C_a are the water-vapor concentrations in g cm^{-3} of the leaf intercellular air spaces and the bulk air, respectively. The denominator is the sum of the diffusion resistances in the water-vapor diffusion pathway. Since the air movement through the plant community was high enough to cause mild leaf flutter, we can assume that the boundary-layer resistance was minor relative to the leaf resistances. In any case, the boundary-layer resistance should have been the same in each chamber, and the differences in the calculated values of the total resistance term should be indicative of differences in average stomatal resistances for the plants in the respective environments. Those relative calculated resistance values are indicated next to the corresponding data points in Fig. 12.1. As can be seen, the total diffusion resistance is higher at the lower humidity; as a result, the transpiration rate does not increase in direct proportion with the increase in vapor-pressure gradient.

When one compares the data for transpiration rates during the short-term experiment with the transpiration rates calculated from the previous day's water consumption (Table 12.2), it is clear that they do not agree. That is, if we were to assume that the plants at 35–40 percent rh transpired at the rate of 1.7 g dm^{-2} h^{-1} for the entire 12 h of the previous light period, they would have lost 20.4 g H_2O/dm^2 leaves/day. However, the direct measurement of the previous day's water consumption shows that those plants lost only 8.8 g H_2O/dm^2 leaves during that time. The same discrepancy holds at each of the other humidities also. The magnitude of this difference between predicted and actual water-loss rates may not be as great as it appears, owing to growth during the 24-h period prior to the short-term transpiration study. However, the amount of growth during this time (i.e., from day 19 to day 20 in the chambers) can explain only a small amount of the difference.

Thus regardless of the magnitude of the difference between calculated and measured daily water use, it seems clear that there is a real difference between the two. The available data permit no more than speculation about the reasons for this difference. It may be significant, however, that the 3-h transpiration experiment was conducted between the fourth and seventh hours of the 12-h light period. It is possible that, through the course of the day, the transpiration rate per plant gradually declines, and the particular period of time selected for our 3-h transpiration experiment corresponded with the period of maximum transpiration intensity during the normal day.

After 6 more days of growth in the chambers, five more plants were harvested from each and the same measurements made. In addition, the net assimilation rate was calculated from the dry-weight and leaf-area data. The data are presented in Table 12.3. Note that the daily transpiration rates are lower when calculated over the 6-day period than over the 1-day or 3-h period, but the same relationships hold as before. The net assimilation rate was about the same in the medium and high humidities in spite of the different transpiration rates. This mirrors the results in

Table 12.3 Transpiration and Net Assimilation Rates for *Phaseolus vulgaris* L. Plants Grown at Three Humidities[a]

Relative Humidity (%)	Transpiration (g dm^{-2} d^{-1})	Net Assimilation Rate (g dm^{-2} d^{-1})
35–40	5.2	0.16
70–75	3.6	0.20
95–100	1.8	0.19

[a] Based on two harvests of five plants each, six days apart, for each treatment.

Table 12.1, where final dry weight was the same at medium and high humidities, but total water consumption was much lower at the high humidity.

The transpiration rate at the lowest humidity again is lower than it would be if it followed the increase in vapor-pressure gradient. In the previous experiment the data supported the suggestion that the reason for this was increased diffusion resistance of the plant. The lower net assimilation rate found at the low humidity further supports that conclusion, and also mirrors the slight reduced growth at low humidity indicated in Table 12.1.

The major conclusion to be drawn from this analysis of the relationships between atmospheric moisture content, transpiration, and growth is that water loss from plants can be reduced considerably without reducing plant growth as long as the method of doing so does not have other effects. For example, if water loss from plants is reduced by lowering the vapor-pressure gradient, growth is not affected. But if water loss is reduced by the plant's increasing its diffusion resistance, there is a price to be paid. The increased diffusion resistance affects other gaseous exchange by the plant, such as CO_2 uptake, and growth reduction is the price.

Carbon Dioxide Concentration

The well-known relationship between CO_2 concentration and stomatal aperture (Meidner and Mansfield, 1968) indicates that there should be a correlation between the CO_2 concentration of the atmosphere and transpiration. The same approach for analyzing this relationship will be applied here as for relative humidity. *Phaseolus vulgaris* L. plants grown in sealed growth chambers with the CO_2 concentration continuously maintained at 400, 800, and 1200 ppm were harvested after 20 days of growth in the chambers. Light intensity was 5000 fc, relative humidity was 70 percent, and air temperature was 24°C in all the chambers. Using the leaf area of the plants at harvest and the water consumption of each plant during the 24 h immediately prior to harvest, we calculated transpiration rates (Table 12.4). As CO_2 was increased from 400 to 800 ppm, the transpiration rate was reduced, suggesting an effect of the increased CO_2 level on stomatal aperture. The calculated diffusion resistances for the plants at 400 and 800 ppm were 6.4 s cm^{-1} and 11.7 s cm^{-1}, respectively, supporting that conclusion. However, it is puzzling why

Table 12.4 Transpiration Rates of *Phaseolus vulgaris* L. Plants Grown at Three Carbon Dioxide Concentrations

CO_2 Concentration (ppm)	Transpiration (g dm^{-2} d^{-1})
400	6.6
800	4.6
1200	6.8

the transpiration rate at 1200 ppm CO_2 should be the same as at 400 ppm. The leaf temperatures were almost identical in the two chambers; the average leaf temperature for each plant was about 26°C. The average leaf temperature in the 800 ppm CO_2 environment was 28°C, reflecting the decreased transpiration rate. In another experiment comparing plant growth and water use at 400 and 1200 ppm CO_2, the same relationship was obtained again. The calculated transpiration rates were 9.7 and 9.6 g dm^{-2} d^{-1}, respectively. Why there should be a reduction of transpiration at 800 ppm CO_2 and not at 1200 ppm CO_2 remains unanswered at this time.

Air Velocity

Plants grown in the naturally lighted phytocells were used for analysis of the effect of air velocity on transpiration in high humidity. The relative humidity was maintained at 95–100 percent to minimize the chances of water stress occurring and confounding the results. Air temperature was 25°C, and the radiation intensity inside the chambers was 1.07 cal cm^{-2} min^{-1}. Air velocity was 75 cm s^{-1} in one case and 225 cm s^{-1} in the other in a horizontal direction. Five plants were randomly harvested from within each chamber at two times 3 days apart, and leaf areas and dry weights were measured. The water consumption per plant over the 3-day period was also measured. The transpiration rates were calculated from these data (Table 12.5). The transpiration rate was higher in the lower-air-velocity environment. When total resistance to diffusion was calculated as before, it was found that diffusion resistance was reduced by the increased air movement (Table 12.5). Stomatal diffusion resistances were measured on several plants in each phytocell, and they were similar in each one. Most of the difference between the two air flows probably is due to the effect on boundary-layer resistances within the community. However, the increased air flow apparently caused enough increase in sensible heat transfer from the leaves to lower significantly the average leaf temperature, and the reduced vapor-pressure gradient more than compensated for the reduced diffusion resistance and kept transpiration rate low. The relative water content (Slatyer, 1967) was 90 percent in the low- and 92 percent in the high-air-

Table 12.5 Transpiration Characteristics of *Phaseolus vulgaris* L. Plants As Affected by Air Velocity[a]

Air Velocity ($cm\ s^{-1}$)	Transpiration ($g\ dm^{-2}\ d^{-1}$)	Total Resistance ($s\ cm^{-1}$)	Average Leaf Temperature (°C)
75	2.8	14.4	30
225	2.4	7.9	27

[a] Based on two harvests of five plants each, 3 days apart, for each treatment.

velocity environments, respectively. At lower atmospheric moisture content, the effects probably would not be the same.

References

Bierhuizen, J. F., Slatyer, R. O.: 1965. Effect of atmospheric concentration of water vapor and CO_2 in determining transpiration–photosynthesis relationships of cotton leaves. Agric. Meteorol. *2*, 259–270.

Briggs, L. J., Shantz, H. L.: 1913. The water requirement of plants. 1. Investigations in the Great Plains. 2. A review of literature. USDA Bur. Plant Ind. Bull. *284*, 1–48; *285*, 1–96.

Drake, B. G., Raschke, K., Salisbury, F. B.: 1970. Temperature and transpiration resistances of *xanthium* leaves as affected by air temperature, humidity, and wind speed. Plant Physiol. *46*, 324–330.

Hoffman, G. J., Rawlins, S. L., Garber, M. S., Cullen, E. M.: 1971. Water relations and growth of cotton as influenced by salinity and relative humidity. Agron. J. *63*, 822–826.

Knecht, G. N., O'Leary, J. W.: 1972. The effect of light intensity on stomate number and density of *Phaseolus vulgaris* L. leaves. Bot. Gaz. *133*, 132–134.

Meidner, H., Mansfield, T. A.: 1968. Physiology of stomata. New York: McGraw-Hill.

O'Leary, J. W., Knecht, G. N.: 1971. The effect of relative humidity on growth, yield, and water consumption of bean plants. J. Am. Soc. Hortic. Sci. *96*, 263–265.

———, Knecht, G. N.: 1972. Salt uptake in plants grown at constant high relative humidity. J. Ariz. Acad. Sci. *7*, 125–128.

———, Prisco, J. T.: 1970. Response of osmotically stressed plants to growth regulators. Adv. Frontiers Plant Sci. *25*, 129–139.

Prisco, J. T., O'Leary, J. W.: 1970. Osmotic and toxic effects of salinity on germination of *Phaseolus vulgaris* L. seeds. Turrialba *20*, 177–184.

Slatyer, R. O.: 1967. Plant water relationships. New York: Academic Press.

13

Light Intensity and Leaf Temperature As Determining Factors in Diffusion Resistance

John D. Tenhunen and D. M. Gates

Introduction

In 1971, Lommen *et al.* proposed a model to describe the interaction of biological and environmental factors governing the gas diffusion and biochemical processes of photosynthetic carbon dioxide fixation in a single leaf. This model, combined with an equation describing the leaf-energy budget, allows calculation of simultaneous rates of transpiration and photosynthesis for a given set of environmental conditions. If this steady-state model is applied to a leaf in its natural environment, one can examine a number of ecologically important relationships for a particular leaf by integrating the established steady-state rates. The amount of photosynthate accumulated, the amount of water used, and the efficiency of water utilization can be estimated for any time period.

The regulation of stomatal aperture has been of great interest to plant scientists for several decades because of the important role of stomata in regulating the transfer of carbon dioxide and water vapor to and from the leaf, respectively. But the complex control of aperture has not yet been adequately described. Recent investigations into the underlying physiology of stomatal opening and closing have demonstrated some important controlling environmental factors (Humble and Raschke, 1971; Raschke, 1966; Raschke, 1970) and have elaborated potentially involved molecular mechanisms (Pallas and Wright, 1973; Willmer *et al.*, 1973). Several workers have attempted to model stomatal functioning with appropriate feedback loops (Cowan, 1972; Penning de Vries, 1972). Nevertheless, our inability to predict accurately stomatal diffusion resistance remains one of the most important obstacles to understanding plant growth in response to environmental factors (Lemon *et al.*, 1971; Stapleton and Meyers, 1971).

The primary effort of Gates and his colleagues at present is to describe steady-state levels of photosynthesis, but in addition a method of treating leaf diffusion resistance is desired that will be suitable for computer simulations integrating photosynthesis and transpiration. The original model (Lommen *et al.*, 1971) for photosynthesis contained only a light dependency for stomatal resistance. Although facts concerning the field behavior of stomata are extensive and sometimes contradictory, we know little about the relative importance of a particular controlling factor. The following experimental study was undertaken to assess the relative importance of light and other environmental factors affecting leaf diffusion resistance in the field. In certain situations, simple description of diffusion resistance appears adequate and can be used in simulating transpiration rates and water use. In other situations, a model of stomatal behavior must undoubtedly approximate the living system more precisely.

Methods

The field work was conducted at the western edge of a clearing in the forest at The University of Michigan Biological Station at Douglas Lake, Michigan, during July and August 1972. A full canopy extended to the west, consisting primarily of maple (*Acer saccharum*), oak (*Quercus rubra*), and aspen (*Populus grandidentata*). Plants near the ground at the edge of the clearing experienced direct solar beam radiation in the morning (with some interference from a building and an isolated tree), and sunlight filtered through the canopy in the afternoon. We chose the common milkweed (*Asclepias syriaca*) as the main experimental plant because of its regular arrangement of horizontal opposite leaves, but we also studied red oak (*Quercus rubra*) leaves.

We monitored the microclimate of selected leaves, continuously recording information on short-wave incident radiant flux (Eppley pyrheliometer), wind velocity (Hastings hot wire anemometer), air temperature (shielded thermocouples), ground temperature (Barnes radiometer), and dew point (EG&G dew-point hygrometer). A diffusion porometer of the type designed by Kanemasu *et al.* (1969) (Lambda Instruments, Lincoln, Nebraska) was used in determining diffusion resistance. The porometer was calibrated on the basis of leaf and porometer temperatures being equal, and although the porometer was shielded from the sun and maintained close to air temperature, we did not try to equilibrate it with the variable leaf temperatures of the large number of leaves measured. Leaf diffusion resistance and leaf temperature (Barnes radiometer) were determined for 10 to 18 selected leaves as a function of time.

We attempted to stress the plants being studied by digging a trench around the plot and constructing a clear plastic canopy 3 m above the plot, depriving it of rainfall for several weeks. The soil moisture as a percentage of the soil dry weight was determined on each day of measurement, and samples were saved for possible later study. *Asclepias* appears to have a very shallow root system in this sandy podzol soil, and we hoped that these simple determinations of soil-moisture content would provide a good measure of water available to the plants.

Measurements were made on four plants of *Asclepias syriaca* on July 24 and 25 and on August 9 and 10 after several weeks under the plastic canopy. One oak seedling was studied in the covered area on August 10. The data collected were analyzed according to the simple model described below, using multiple regression techniques of The University of Michigan's computer statistical programs.

Coefficients were predicted by regression, giving leaf diffusion resistance as a function of light and leaf temperature. The resulting equations were used in simulating the water loss from *Asclepias* leaves for a sunny day at Douglas Lake. We used the energy-budget technique as described by Gates (1968) and made the following measurements as a prerequisite to simulation. The spectral quality of short-wavelength radiation in the clearing was determined with an ISCO spectral radiometer for the direct solar beam during the morning and for the canopy-filtered radiation during the afternoon. Transmittance and reflectance spectra of *Asclepias syriaca* leaves were obtained with a Beckman DK2 spectrophotometer. The mean short-wavelength absorptances to morning and afternoon radiation were calculated from these.

Results and Discussion

The photosynthetic model of a single leaf (Lommen *et al.*, 1971) used Michaelis–Menten kinetics to describe the photosynthetic reactions. We hope that this basis for modeling will lend itself, in the future, to incorporating the more detailed biochemistry of photosynthesis. Utilizing the same concept, an extremely simple formulation of stomatal conductance plus internal mesophyll conductance of a broad leaf (if mesophyll conductance cannot be ignored) was devised in which conductance is a function of two environmental variables known to be of significance. The model included a light effect similar to that of Turner and Begg (1973), and, in addition, a linear effect of temperature was assumed similar to that shown for several species by Wuenscher and Kozlowski (1971). This is elaborated in equation form as

$$\text{conductance} = \frac{\text{conductance}_{\max}}{1 + K/L}$$

Here the maximal conductance is attained at high light intensity (approximately 10 times the light intensity value of K, which is the light intensity at which one-half maximal opening occurs). Since conductance is the reciprocal of resistance, the following relationships are valid:

$$\text{resistance} = \frac{1 + K/L}{C_{\max}}$$

$$= \frac{1}{C_{\max}} + \frac{K}{C_{\max}L}$$

Furthermore, the reciprocal of $C_{\max}$ is the minimum resistance ($R_{\min}$) and $K/C_{\max}$ is a constant (a). Then

$$\text{resistance} = R_{\min} + a\left(\frac{1}{L}\right)$$

Finally, R_{min} is taken to be a linear function of temperature. R_{min} contains any resistance attributable to the mesophyll.

$$R_{min} = b \cdot T + c$$

The model then takes the form

$$R = a\left(\frac{1}{L}\right) + b \cdot T + c$$

where R = leaf diffusion resistance from the cell walls to the leaf surface (s cm^{-1})
a = empirically determined constant (s cal cm^{-3} min^{-1})
L = incident short-wavelength radiant flux (cal cm^{-2} min^{-1})
b = empirically determined constant (s cm^{-1} $°C^{-1}$)
T = leaf temperature (°C)
c = empirically determined constant (s cm^{-1})

Physiologically, one can postulate that both light and temperature manifest their effects via changes in the carbon dioxide concentration in the intercellular air space. Increasing light reduces air-space carbon dioxide concentration by increasing net photosynthesis according to a Michaelis–Menten type of response. Increasing temperature may be expected to increase or decrease the air-space carbon dioxide concentration according to the sum effect of temperature on photosynthesis and respiration. Although we present no evidence that the effects to be described are causally related as suggested here, these postulates do, in fact, provide a plausible basis for the model as outlined above.

The equation above is analogous to a series of Lineweaver–Burk plots (where $1/C$ is plotted versus $1/L$). This type of plot has been shown to be inferior (Dowd and Riggs, 1965) in graphical determination of V_{max} and K_m for an enzymatic reaction, owing to heavy weighting of the least reliable determinations of reaction rate (in this case, the least reliable determinations of conductance when light intensity is low). The following form of the model (Dowd and Riggs, 1965) is favored and was used in the analysis:

$$R \cdot L = a + b \cdot T \cdot L + c \cdot L$$

This equation describes a plane surface and is appropriate for analysis by multiple regression techniques. In terms of Michaelis–Menten responses, this equation describes a series of Michaelis–Menten curves where the C_{max} and K change with temperature.

Stomata on the upper and lower surfaces of a leaf do not respond in the same fashion (Turner and Begg, 1973). The model, when applied to empirical data obtained with a porometer, gives the diffusion resistance of one surface of a leaf as a function of light and temperature. The leaf diffusion resistance is obtained by treating the resistances of upper and lower surfaces as parallel resistances. Then

$$\frac{1}{\text{resistance}_{\text{leaf}}} = \frac{1}{R_{\text{lower surface}}} + \frac{1}{R_{\text{upper surface}}}$$

For *Asclepias*, the resistance of the lower surface is always low, and the resistance of the upper surface is very high (30–75 s cm^{-1}). Consequently, the resistance of the leaf is approximately the same as the resistance of the lower surface. Under this assumption, the equations given below for *Asclepias* are equations for leaf diffusion resistance. Since there are no stomata on the upper surface of *Quercus rubra*, this resistance is also high. The equation to be presented is an equation for the leaf diffusion resistance.

We found, upon initial inspection of the data, that sample variances differed with respect to each other. Since one assumption of regression analysis is an equality of variances, samples were tested with the F_{max} test for heterogeneous variances as outlined by Sokal and Rohlf (1969). All observations taken at one time were treated as a sample. Sample variances were seen to increase significantly in those samples taken at times when the light intensity was very low (less than 0.015 cal $cm^{-2} min^{-1}$). These samples were therefore removed from further consideration. This should not lead to any great error in calculations of water used or photosynthate produced because the equations predicting diffusion resistance are based on observations from times of greatest importance to these processes, i.e., when light intensity is more than negligible with respect to these processes.

The following coefficients were predicted by multiple regression analysis. All coefficients were significant at the 0.01 percent level ($p = 0.0001$) and have the units indicated above.

$$R \cdot L = a + b \cdot T \cdot L + c \cdot L$$

Asclepias syriaca ($n = 236$) $R_{leaf}(L) = 0.044 + 0.027(T)(L) + 0.627(L)$
"wet plot," before drying

Asclepias syriaca ($n = 168$) $R_{leaf}(L) = 0.105 + 0.163(T)(L) - 1.622(L)$
"dry plot," after drying

Quercus rubra ($n = 31$) $R_{leaf}(L) = 0.162 + 0.234(T)(L) - 2.635(L)$

For each multiple regression, the value of r^2 can be derived which is the ratio of the sum of the squares explained by the regression to the total sum of squares. In short, it is a measure of the percentage of the observed variation explained by the model, and is a criterion by which one can assess the usefulness of the model for summarizing the empirical data. The r^2 values for the pooled regressions of resistance on reciprocal light and temperature ($R = a(1/L) + b \cdot T + c$) were about 0.6. Temperature alone explained about 5 percent of the variation, and light alone accounted for 55 to 60 percent of the variation. In this particular habitat, light and temperature are more or less independent (correlation coefficient =0.37). When applied to individual shoots of *Asclepias*, the model performed somewhat better, with the average r^2 value at 0.75. Interaction terms for light and temperature were insignificant in the pooled population equations as derived above. When the soil was dried from its original state, containing 13 g H_2O/100 g dry soil to 9 g H_2O/100 g dry soil, the coefficients predicted for *Asclepias* were significantly different ($p = 0.01$) and the response to temperature changed dramatically.

Resistance in each case increased with increasing temperature. Similar effects of temperature on resistance have been observed previously (Heath and Orchard, 1957) in onion, coffee, and *Pelargonium*, and were attributed to increase in the intercellular-air-space carbon dioxide concentration. The predicted temperature response for *Quercus rubra* is in agreement with results obtained by Wuenscher

Table 13.1 Predicted Light Intensity for One-Half Maximal Conductance (K) and Predicted Minimum Resistance (R_{min}) for Selected Cases of Temperature

	°C	K (cal cm^{-2} min^{-1})[a]	R_{min} (s cm^{-1})
"Wet plot" *Asclepias*$_{leaf}$	15	0.043	1.03
	20	0.037	1.18
	25	0.034	1.30
"Dry plot" *Asclepias*$_{leaf}$	15	0.130	0.82
	20	0.064	1.64
	25	0.042	2.45
Quercus rubra$_{leaf}$	15	0.185	0.87
	20	0.079	2.04
	25	0.051	3.21

[a] Conversion factor: 1 cal cm^{-2} min^{-1} = 6.98×10^5 ergs cm^{-2} s^{-1}.

and Kozlowski (1971) for the same species from transpiration data, i.e., a method independent of temperature in calibration. Thus although it is conceivable that the temperature effect observed might arise from error introduced by not equilibrating porometer and leaf temperature, we believe that the temperature response is real. Future work on the photosynthetic characteristics of these plants should resolve this question.

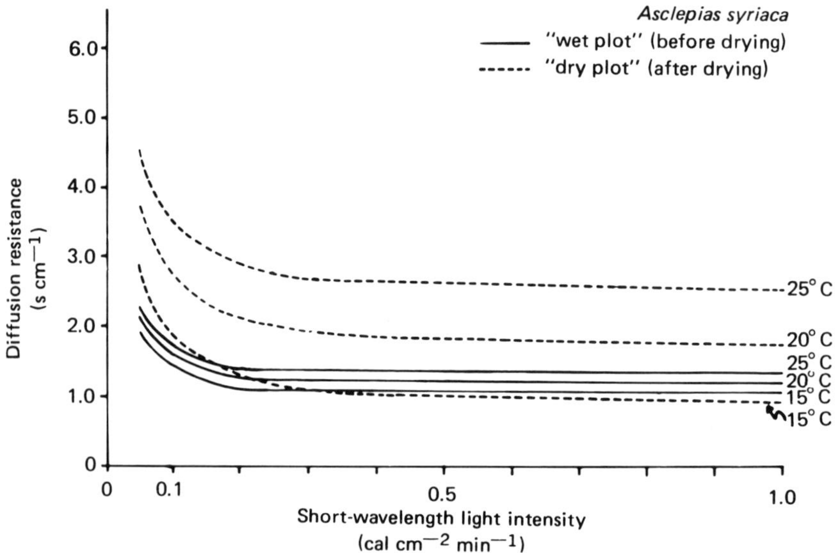

Fig. 13.1. Leaf diffusion resistance of *Asclepias syriaca* as a function of light at three temperatures before and after the drying treatment.

The value of R_{min} and K produced from these equations for the pooled leaf responses are shown in Table 13.1. K (the light intensity of one-half maximal conductance) and R_{min} are given for several temperatures in the range for which the equations are felt to be valid. In all cases, K decreases with increasing temperature while R_{min} increases. Both R_{min} and K tend to increase in *Asclepias* after the soil is dried. Figure 13.1 shows the predicted relationship of leaf diffusion resistance and light intensity for three temperatures for *Asclepias* before and after the drying treatment. Figure 13.2 shows the relationship of leaf diffusion resistance for *Quercus rubra* as a function of light and temperature. The temperature component appears to be quite important in *Quercus rubra.* A Lineweaver–Burk plot of R versus $(1/L)$ is shown in Fig. 13.3 for *Asclepias* and *Quercus* at 20°C and compared with maize and sorghum for Turner and Begg (1973). The K and R_{min} reported for maize and sorghum were higher than those found in this study.

Simulated Water Use

To demonstrate the utility of these equations, water use by *Asclepias* plants growing in the forest clearing was simulated. The amount of water used is dependent on three factors: (1) water-vapor concentration in the air, (2) leaf temperature, and (3) resistance of the diffusion pathway. The first of these can be obtained by monitoring the air temperature and dew point. The second and third factors are interrelated by the equation for internal leaf diffusion resistance that was determined and by the leaf energy-budget equations. Leaf temperature and diffusion resistance

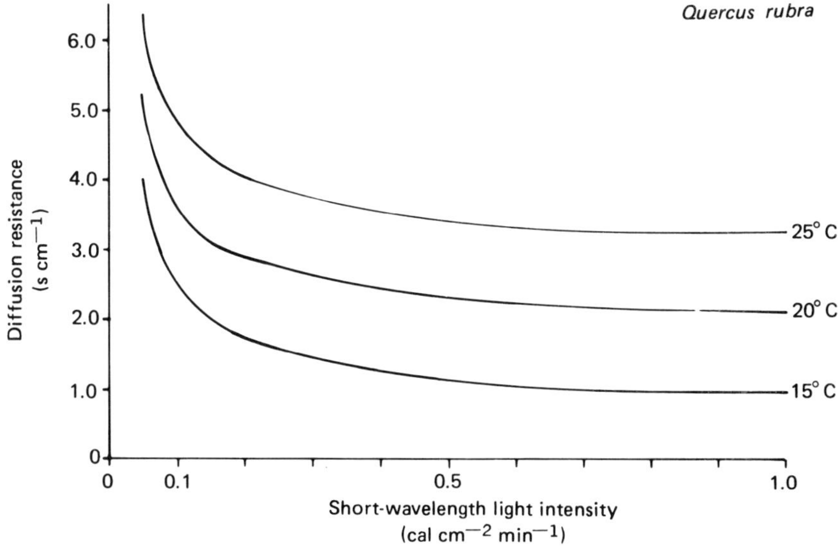

Fig. 13.2. Leaf diffusion resistance of *Quercus rubra* as a function of light at three temperatures.

for different sets of environmental conditions were found by an iterative solution of the two equations. In the process transpiration rate was determined also.

The energy-budget concept is straightforward: the temperature of a leaf in steady-state conditions is such that energy gained by the leaf equals the energy lost. The energy budget of a single leaf is expressed as follows (Gates, 1968):

$$Q_{abs} = \mathrm{RAD} + C + L_v E$$

$$Q_{abs} = \varepsilon\sigma(T + 273)^4 + K_1\left(\frac{V}{D}\right)^{1/2}(T - T_A) + L_v \frac{d(T) - \mathrm{rh}\cdot d(T_A)}{r_s + K_2(B^{0.2}D^{0.35}/V^{0.55})}$$

where
- Q_{abs} = absorbed radiation (ergs cm^{-2} s^{-1})
- RAD = radiation emitted (ergs cm^{-2} s^{-1})
- C = convective heat transfer (ergs cm^{-2} s^{-1})
- L_v = latent heat of vaporization of water (ergs nmole^{-1})
- E = transpiration rate (nmoles cm^{-2} s^{-1})
- ε = emissivity of the leaf (=0.95)
- σ = Stefan–Boltzmann constant = 5.67×10^{-5} ergs cm^{-2} s^{-1} °K^{-4} (T in °C + 273 = T in °K)
- K_1 = empirically determined convection coefficient: $K_1 = 1.13 \times 10^4$ for $B \ll D$ or $B = D \leqq 5$ cm; $K_1 = 7.0 \times 10^3$ for $B \gg D$ or $B = D > 5$ cm
- T = leaf temperature (°C)
- T_A = air temperature (°C)
- V = wind speed (cm s^{-1})
- D = leaf dimension along wind flow (cm)
- B = leaf dimension perpendicular to wind flow (cm)
- $d(T)$, $d(T_A)$ = saturation densities of water vapor at T and T_A, respectively (nmoles cm^3)
- rh = relative humidity expressed as a number between 0 and 1
- r_s = stomatal diffusion resistance to water vapor (s cm^{-1})
- $K_2(B^{0.2}D^{0.35})/V^{0.55}$ = boundary-layer resistance to water vapor (s cm^{-1}), where K_2 is another empirically determined constant = 1.56 for $B \ll D$ or $B = D \leqq 5$ cm, $K_2 = 2.10$ for $B \gg D$ or $B = D > 5$ cm

The radiation absorbed by the plants in the forest clearing was treated as follows. A horizontal leaf in that situation receives short-wavelength radiation from above whose intensity and spectral quality are extremely different in the morning and afternoon, as shown in Fig. 13.4. The leaf also receives thermal radiation from the ground, canopy, and atmosphere. Reflection from the ground was very low and was ignored. The radiation absorbed was approximated by summing (1) a short-wavelength radiation term from above; (2) black-body radiation from the ground at measured ground surface temperature (1 hemisphere); (3) black-body radiation from the canopy, considered as being at air temperature ($\frac{1}{2}$ hemisphere); and (4) long-wavelength radiation from the atmosphere at air temperature [$\frac{1}{2}$ hemisphere according to Swinbank's (1963) empirically derived equation]; and

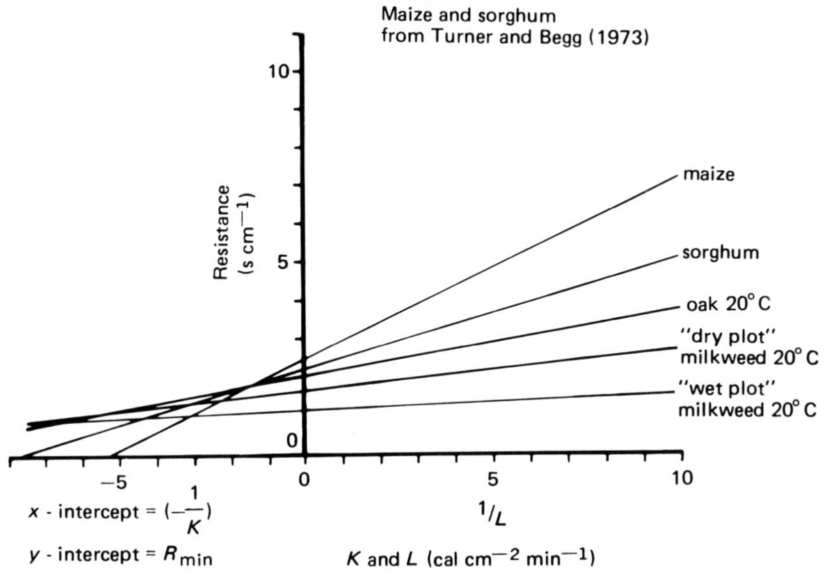

Fig. 13.3. Lineweaver–Burk plot ($R = 1/C$ versus $1/L$) comparing *Asclepias*, *Quercus*, maize, and sorghum.

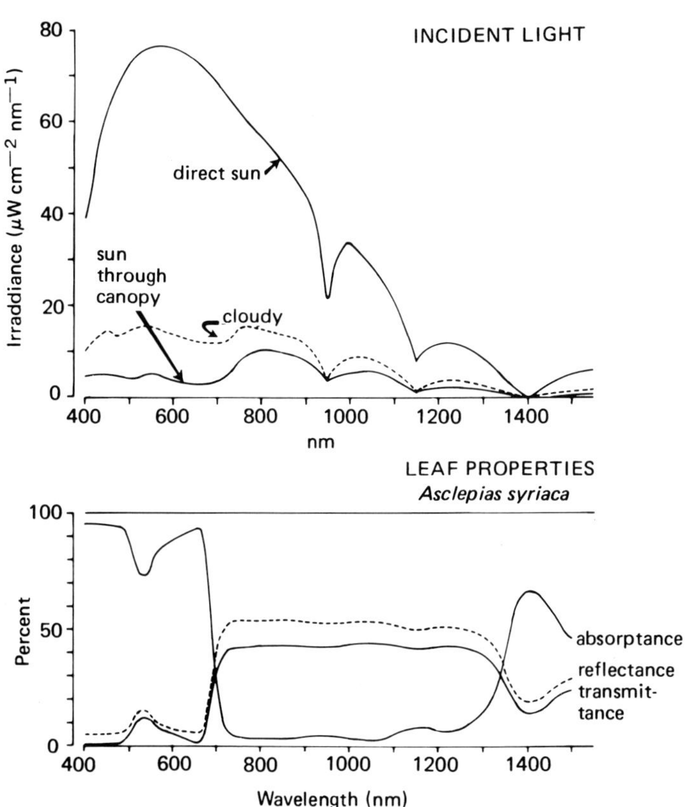

Fig. 13.4. Spectral quality of incident light in the clearing and spectral properties of the *Asclepias syriaca* leaf.

dividing the total by one-half. The result gives the radiation absorbed per unit surface area of leaf:

$$Q_{\text{abs}} = \tfrac{1}{2}\{\alpha_s I_s + \alpha_L \varepsilon_G \sigma (T_G + 273)^4 + \tfrac{1}{2}[\alpha_L \varepsilon_C \sigma (T_C + 273)^4] + \tfrac{1}{2}[\alpha_L S_w (T_A + 273)^6]\}$$

where α_s = mean absorptance of the milkweed leaf to short-wavelength radiation expressed as a number between 0 and 1 dependent on spectral quality of incident light and spectral properties of leaf

α_L = long-wavelength absorptivity = 0.95

I_s = incident short-wavelength radiant flux (ergs $\text{cm}^{-2}\ \text{s}^{-1}$)

ε_G = emissivity of the ground surface = 0.95

ε_C = emissivity of the canopy = 0.95

σ = Stefan–Boltzmann constant = 5.67×10^{-5} ergs $\text{cm}^{-2}\ \text{s}^{-1}\ {}^\circ\text{K}^{-4}$

T_G = temperature of the ground surface (°C)

T_C = temperature of leaves in the canopy = T_A(°C)

T_A = temperature of the air (°C)

S_w = empirically determined constant (Swinbank, 1963) = 5.31×10^{-10}

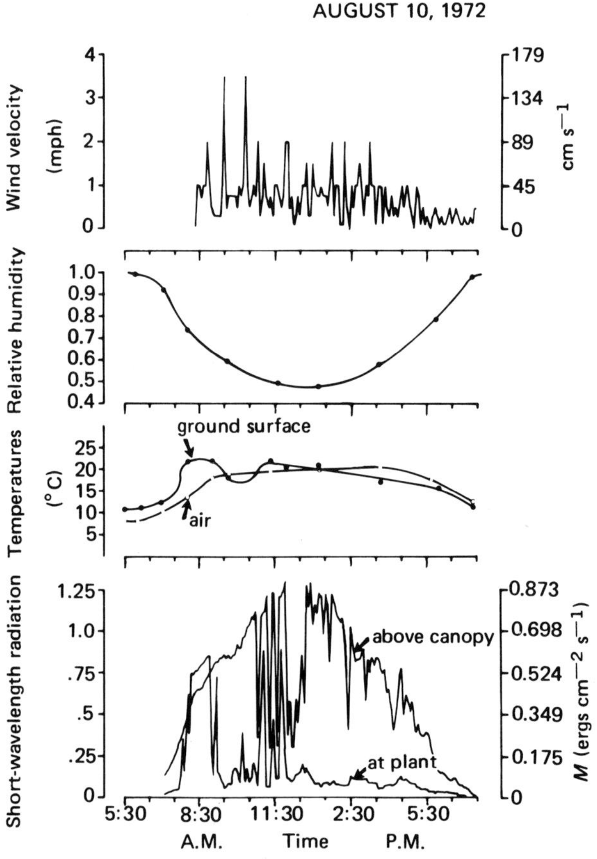

Fig. 13.5. Meteorological data from the clearing on August 10, 1972.

In this equation the long-wavelength absorptivity and the emissivity of ground and canopy were taken as 0.95.

The spectral reflectance, transmittance, and absorptance for an *Asclepias* leaf are shown in Fig. 13.4. The mean absorptance was determined from these data to be 0.48 for direct radiation (morning) and 0.18 for radiation filtered through the canopy (afternoon). In the simulation, the former absorptance was applied (α_s in the equation for Q_{abs}) if the intensity of short-wavelength radiation was greater than 0.15 cal cm^{-2} min^{-1} or 1×10^5 ergs cm^{-2} s^{-1} and the latter when it was less than this. Study of the meteorological data from the clearing, shown in Fig. 13.5, demonstrates that, whenever the radiation was filtered through the canopy, a fairly constant amount of radiation was incident which was less than 0.15 cal cm^{-2} min^{-1}.

Meteorological data from August 10, 1972 (Fig. 13.5), were used in the simulation of water use. Points at different times were read for each variable so that

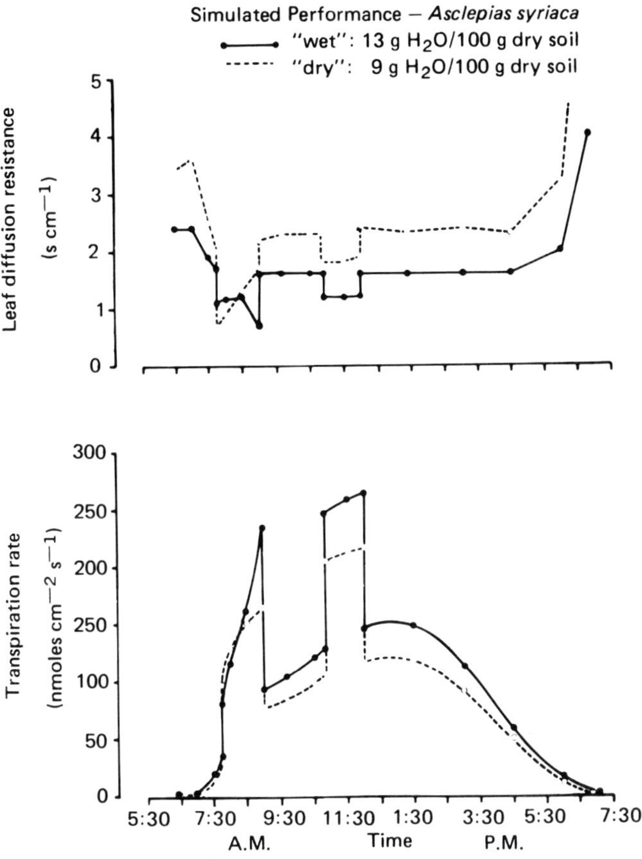

Fig. 13.6. Predicted performance of *Asclepias syriaca* leaves before and after drying treatment when experiencing a day similar to August 10, 1972.

curves were produced as computer input data compatible with all major trends seen in Fig. 13.5. The diffusion-resistance equation and energy-budget equations were solved for a leaf of characteristic dimensions $D = 10.0$ cm and $B = 5.0$ cm at each time that input was provided. The calculated leaf diffusion resistances and transpiration rates (moles cm^{-2} s^{-1}) are shown as a function of time in Fig. 13.6 for an *Asclepias* leaf before and after the drying treatment. The total amount of water used as a function of time during this day (moles cm^{-2}) by each leaf is shown in Table 13.2. The more "conservative strategy" exercised after drying is apparent. Water used after the drying treatment is predicted to be 83 percent of that used before the treatment.

Table 13.2 Amount of Water Used by *Asclepias syriaca* Leaves on a Sunny Day in the Clearing (Simulation) to the Time Indicated from 6:30 A.M.

	Amount of Water Used (nmoles cm^{-2})	
Time of Day	**"Wet-Plot" Leaf**	**"Dry-Plot" Leaf**
6:30 A.M.	0.0	0.0
7:00	0.398×10^{-6}	0.296×10^{-6}
7:30	0.175×10^{-4}	0.148×10^{-4}
7:45	0.405×10^{-4}	0.353×10^{-4}
7:46	0.425×10^{-4}	0.376×10^{-4}
8:00	0.128×10^{-3}	0.135×10^{-3}
8:30	0.377×10^{-3}	0.385×10^{-3}
9:00	0.731×10^{-3}	0.668×10^{-3}
9:01	0.736×10^{-3}	0.672×10^{-3}
9:30	0.910×10^{-3}	0.815×10^{-3}
10:30	0.132×10^{-2}	0.115×10^{-2}
10:48	0.145×10^{-2}	0.126×10^{-2}
10:49	0.146×10^{-2}	0.126×10^{-2}
11:30	0.208×10^{-2}	0.178×10^{-2}
12:00	0.255×10^{-2}	0.216×10^{-2}
12:01 P.M.	0.256×10^{-2}	0.217×10^{-2}
1:30	0.334×10^{-2}	0.280×10^{-2}
3:00	0.404×10^{-2}	0.336×10^{-2}
4:30	0.450×10^{-2}	0.374×10^{-2}
6:00	0.470×10^{-2}	0.391×10^{-2}
7:00	0.473×10^{-2}	0.393×10^{-2}

When this approach is coupled with integration of photosynthesis, it will be possible to investigate in great detail the efficiency of photosynthesis with respect to water use for particular plants in natural situations. A clearer picture should evolve, allowing greater insight into plant ecology and physiology as affected by the fundamental processes of photosynthesis and transpiration.

References

Cowan, I.: 1972. Oscillations in stomatal conductance and plant functioning associated with stomatal conductance: Observations and a model. Planta *106*, 185–219.

Dowd, J. E., Riggs, D.: 1965. A comparison of estimates of Michaelis–Menten kinetic constants from various linear transformations. J. Biol. Chem. *240*, 863–869.

Gates, D. M.: 1968. Transpiration and leaf temperatures. Ann. Rev. Plant Physiol. *19*, 211–238.

Heath, O. V. S., Orchard, B.: 1957. Temperature effects on the minimum intercellular space carbon dioxide concentration. Nature *180*, 180–181.

Humble, G. D., Raschke, K.: 1971. Stomatal opening quantitatively relates to potassium transport. Evidence from electron probe analysis. Plant Physiol. *48*, 447–453.

Kanemasu, E. T., Thurtell, G., Tanner, C.: 1969. Design, calibration and field use of a stomatal diffusion porometer. Plant Physiol. *44*, 881–885.

Lemon, E., Stewart, D., Shawcroft, R.: 1971. The sun's work in a cornfield. Science *174*, 371–378.

Lommen, P., Schwintzer, C., Yocum, C., Gates, D.: 1971. A model describing photosynthesis in terms of gas diffusion and enzyme kinetics. Planta *98*, 195–220.

Pallas, J. E., Jr., Wright, B.: 1973. Organic acid changes in the epidermis of *Vicia faba* and their implication in stomatal movement. Plant Physiol. *51*, 588–590.

Penning de Vries, F. W. T.: 1972. A model for simulating transpiration of leaves with special attention to stomatal functioning. J. Appl. Ecol. *9*, 57–77.

Raschke, K.: 1966. Die Reaktion des CO_2-Regelsystems in den Schliesszellen von *Zea mays* auf Weisses Licht. Planta *68*, 111–140.

———: 1970. Stomatal responses to pressure changes and interruptions in the water supply of detached leaves of *Zea mays* L. Plant Physiol. *45*, 415–423.

Sokal, R., Rohlf, F. J.: 1969. Biometry, p. 370. San Francisco: W. H. Freeman.

Stapleton, H., Meyers, R.: 1971. Modelling subsystems for cotton—the cotton plant simulation. Trans. Am. Soc. Agric. Engs. *14*, 950–953.

Swinbank, W. C.: 1963. Long-wave radiation from clear skies. Quart. J. Roy. Meteorol. Soc. *89*, 339–348.

Turner, N., Begg, J.: 1973. Stomatal behavior and water status of maize, sorghum, and tobacco under field conditions. I. At high soil water potential. Plant Physiol. *51*, 31–36.

Willmer, C., Kanai, R., Pallas, J., Black, C., Jr.: 1973. Detection of high levels of phosphoenolpyruvate carboxylase in leaf epidermal tissue and its significance in stomatal movements. Life Sci. *12*, 151–155.

Wuenscher, J., Kozlowski, T.: 1971. Relationship of gas-exchange resistance to tree-seedling ecology. Ecology *52*, 1016–1023.

14

Photosynthesis in Developing Plant Canopies

RONALD G. ALDERFER

Introduction

All biologically useful energy derives ultimately from that stored in the process of photosynthesis. On land this takes place in leaves and stems of plants arranged in canopies of highly variable structure and density. Parameters of the physical environment that affect photosynthesis (radiation flux density in the waveband 4000–7000 Å; air temperature, total radiation, water-vapor content of the air, and air speed, which together affect leaf temperature; and carbon dioxide content of the air) may vary considerably throughout the total space occupied by a particular plant canopy. Furthermore, the physiological–biochemical parameters of leaf tissue which govern photosynthetic capacity will also vary throughout the canopy, depending on the age and environmental history of component leaves.

In this paper we shall discuss those factors that regulate the photosynthetic rate of plant canopies, including parameters of the physical environment and parameters affecting the physiological–biochemical capacity for photosynthesis in leaf tissue. Our discussion will deal primarily with recent data obtained from soybean plants (*Glycine max* var. Amsoy) raised under controlled environmental conditions. The aim of this work is threefold: (1) to examine the extent of change in the photosynthetic capacity throughout the lifetime of individual leaves making up the canopy; (2) to identify, if possible, the cause of such changes in the photosynthetic capacity; and (3) to investigate the effect of changes in both environmental and physiological–biochemical parameters on the photosynthesis of the entire canopy.

Materials and Methods

Soybean plants (*Glycine max* var. Amsoy) were used in this study because of their rapid growth rates, their general morphology, and the extent of physiological–biochemical information already available for the species. Seeds were obtained from the U.S.D.A. Regional Soybean Laboratory, Urbana, Illinois. The plants were raised from seed in 900-ml plastic containers filled with vermiculite. The plants were watered twice weekly with one-half strength Hoagland's solution and on other days with distilled water. The growth chamber housing them was controlled for an air temperature of 28°C during the day and 22°C during the night. The temperature-control system was such that the daytime air temperature oscillated between plus and minus 2° from set temperature. The nighttime air temperature oscillated with a smaller amplitude. The oscillation had a period of approximately 6 min. Leaf temperature under these growth conditions stayed within 1°C of air temperature. The photoperiod of the growth chamber was 16 h and was followed by a dark period of 8 h in each 24-h cycle. The relative humidity of the air in the growth chamber varied between 60 and 75 percent.

The plants were arranged in rows and columns so that the stems were 10 cm apart in both directions. The photosynthetically active radiation flux density was measured twice a week at each of the developed nodes. A sample size of 10 was used for each node. The sensor used was the Model LI-190SR Quantum Sensor, Lambda Instruments Co., Inc.

When we conducted defoliation experiments, we cut all leaflets developing at nodes above the second at the main stem prior to unfolding. All photosynthesis measurements were made on the central leaflet of the trifoliate leaf. A carbon dioxide concentration of 300 μl liter^{-1} in air and a leaf temperature of 28°C were used for all photosynthetic measurements. The measurements were made with a differential CO_2 gas analyzer (Beckman Instruments, Inc., Model IR 215). The sample chamber had a total volume of approximately 700 cm^3 and the air in it was efficiently mixed with a fan.

The water-vapor content of the air entering and leaving the sample chamber was measured with a dew-point hygrometer (Model 880, EG&G International, Inc.). Incoming dew-point temperature was controlled for 12.5°C. Change in dew point between incoming and outgoing air was used to compute the apparent transpiration rate of the experimental leaf tissue.

Chlorophyll content of sample leaflets was determined spectrophotometrically from acetone extracts.

Results and Discussion

Figure 14.1 shows the pattern of leaf-area increase with time at each node of the soybean plant. All points on a given day represent leaf area at each node from one sample plant. The sampling process was a destructive one, so we used different plants each day. The emergence of consecutive nodes is fairly regular, and the initial rates of area increase are similar; but the final leaf area increases significantly from one node to the next. This pattern greatly affects the penetration of radiation into the canopy and therefore the extent to which older nodes will contribute to the total carbon assimilated by each plant.

Figure 14.2 shows the decrease in incident flux density of photosynthetically active radiation at each of the first four nodes as a function of developmental

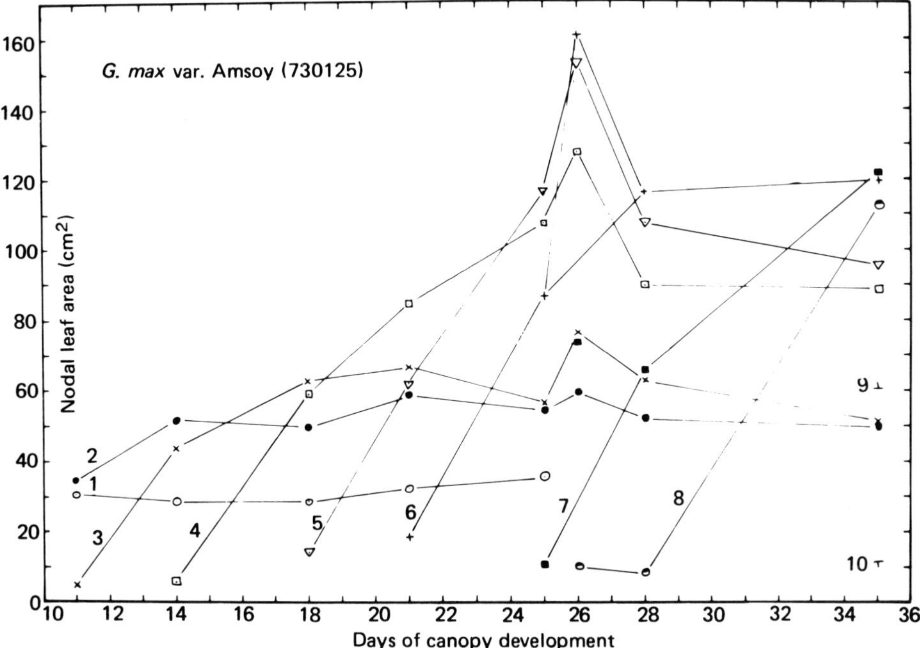

Fig. 14.1. Leaf area at each node as a function of time in developing soybean canopy.

time of the canopy. Each point represents the mean of ten measurements in the canopy at the respective node position; the vertical bars indicate one standard deviation around the sample mean. The variance of the average radiation available to any given node appears to be greatest in the time span during which the radiation is decreasing most rapidly. This can be at least partly accounted for by the fact that intermediate levels of the canopy are subjected to a high frequency of gaps in the canopy (see Anderson, 1966, for a full account of gap frequency as a function of height in the canopy). The combination of direct radiation flecks and shading by upper leaves contributes to a highly variable radiation flux density at intermediate canopy levels.

Figure 14.3 shows the time course of photosynthesis under standard conditions at each of the first five nodes of soybean. The conditions are saturating levels of radiation (produced by a 300-W incandescent lamp filtered through 3 cm of water and two sheets of Plexiglas 3.2 mm thick separated by a 5-cm air space), a leaf temperature of 28°C, 300 μl liter $^{-1}$ CO_2 in air, and a dew-point temperature in the air supplied to the sample chamber of 12.5°C. Each symbol in Fig. 14.3 represents a sample taken from replicate canopies. These replicates were raised under similar environmental conditions but at different times. The data shown in this figure indicate a relatively uniform maximum rate of photosynthesis per unit area. These data show a significant change in photosynthetic capacity with developmental time, and the pattern of change is similar from one node to the next. This pattern of

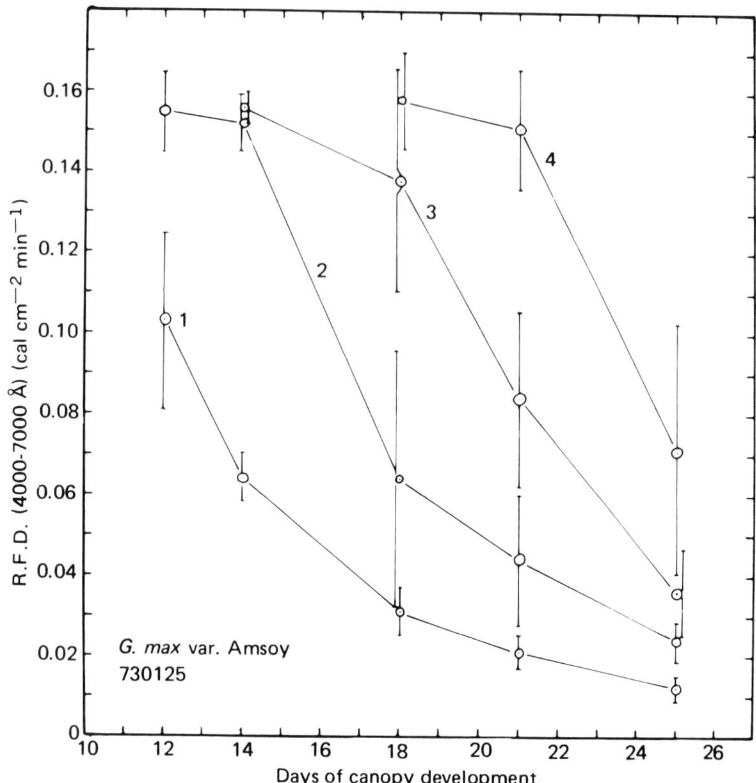

Fig. 14.2. Flux density of photosynthetically active radiation striking each node as a function of time in developing canopy. Each point represents the mean of 10 measurements in the canopy; the vertical bars indicate one standard deviation around the sample mean.

a gradual decrease in photosynthetic capacity with age was previously reported by Das and Leopold (1963) for the primary leaves of bean.

Figure 14.4 shows the relationship between relative photosynthetic capacity and relative magnitude of radiation available for photosynthesis. The most striking feature of this representation of the data is that the younger nodes do not reach their maximum photosynthetic capacity until the available radiation has been critically reduced by shading. This effect would be compensated for to some extent under natural conditions by the higher levels of radiation at the top of the canopy, although we predict that many leaves in natural plant canopies would reach their maximum photosynthetic capacities at a time when shading by other leaves and stems reduces the flux of photosynthetically active radiation well below the saturation level.

Figure 14.5 shows the relationship between net photosynthesis and stomatal plus boundary-layer conductance values (reciprocal of resistance). These values of conductance were calculated from measurements of transpiration made simul-

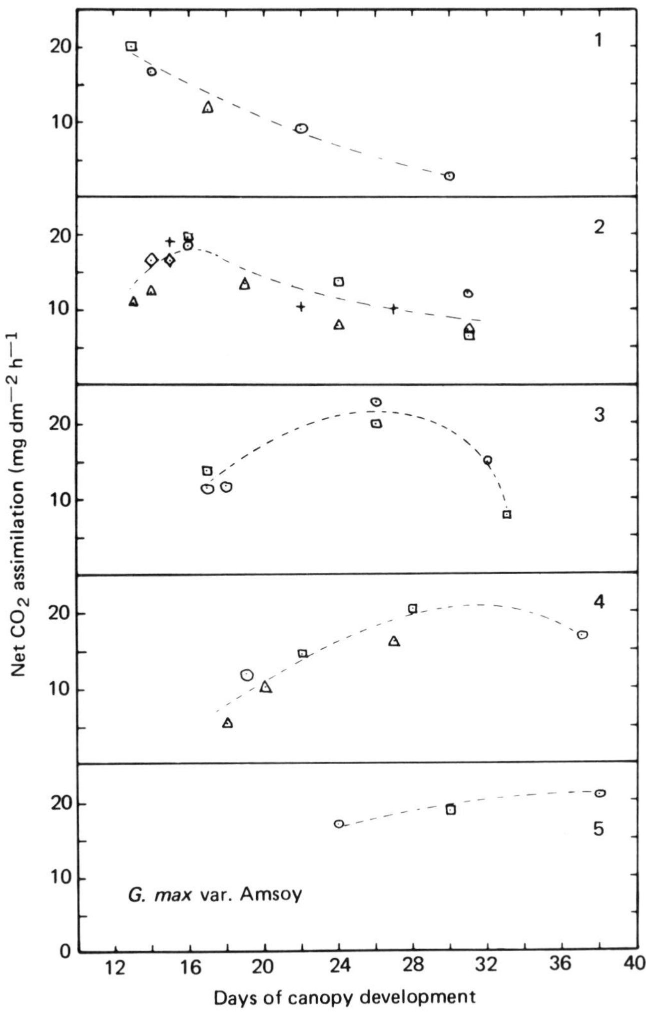

Fig. 14.3. Rate of net photosynthesis at each node under standard conditions (see the text) as a function of time in developing canopies. Various symbols represent samples from replicate canopies.

taneously with the measurements of net photosynthesis. In tissue younger than 24 days, there is a strong, linear correlation between net assimilation and stomatal plus boundary-layer conductance. Conductance from the free air to the substomatal cavity may exert a controlling influence on net assimilation during this time period. However, data from days 24 and 31 indicate that even though high conductance values persist, assimilation is being reduced by some other component or process. Such a component or process could include lowered conductance (increased resistance) to CO_2 diffusion across cell and chloroplast boundaries or reduced rate of carboxylation of the initial acceptor in the CO_2 assimilation scheme. If the enzyme

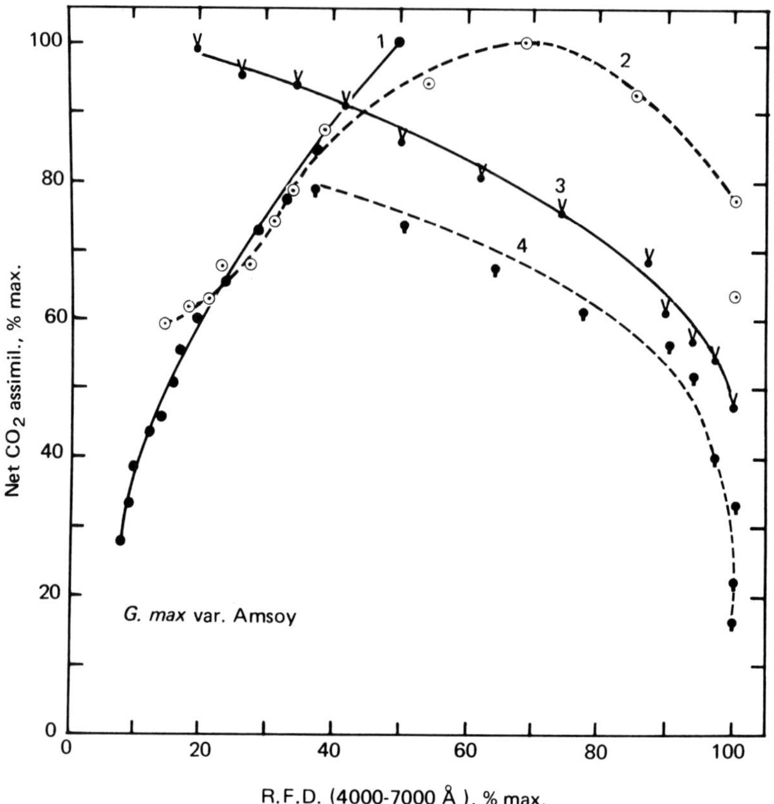

Fig. 14.4. Percent of maximum photosynthetic capacity as a function of percent maximum photosynthetically active radiation available at the first four nodes of a developing soybean canopy.

carbonic anhydrase is active in the transport of CO_2 across membranes, reduced conductance to CO_2 could be caused by the lowered rate of synthesis of this enzyme or a decrease in its activity or both. Likewise, the carboxylation step would be slowed by a decrease in the rate of the carboxylation enzyme synthesis or by a decrease in its activity or both. [Smillie (1962) showed that in peas the activity of ribulose-1,5-diphosphate carboxylase rises and falls in close parallel with the photosynthesis rate measured under standardized conditions. Dickmann (1971) found that the development of photosynthetic capacity in cottonwood closely paralleled the activity of the carboxylase enzyme.] Lowered activity of these enzymes could result from a generalized decrease in protein synthesis in aging leaves. In this case one would expect a corresponding decrease in a wide range of metabolic activities during the time photosynthetic capacity decreases. This is strongly suggested but not proved by data such as that presented by Smillie (1962) and Geronimo and Beevers (1964). Both show a decrease in the activity of all enzymes tested during the period in which the photosynthetic capacity is decreasing.

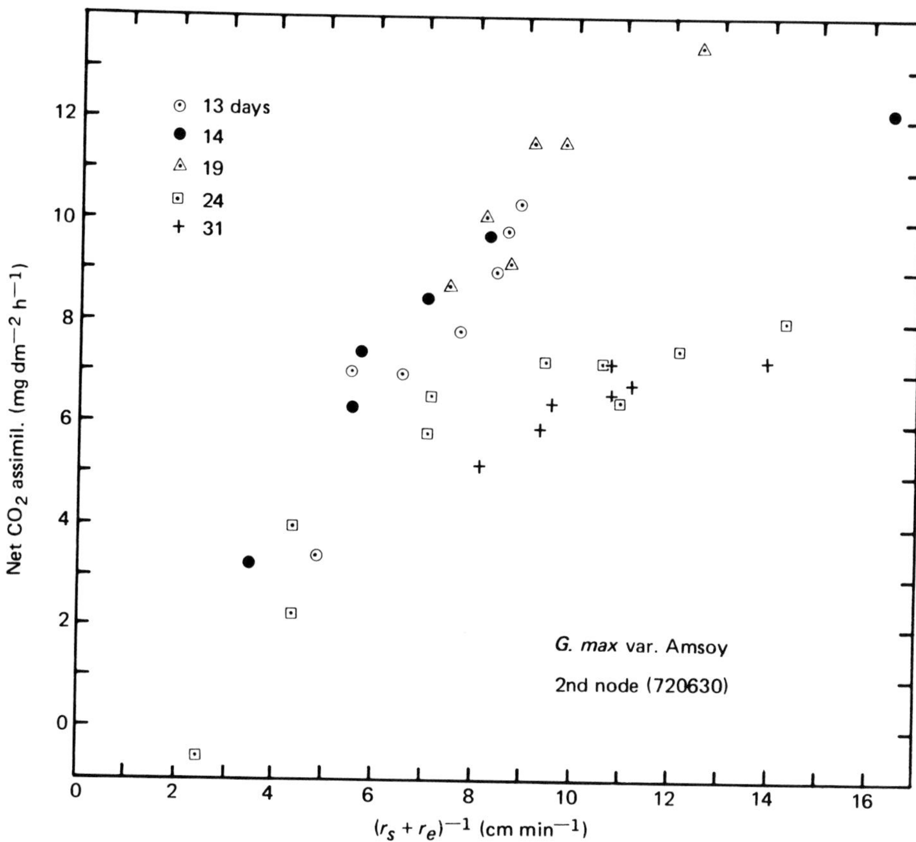

Fig. 14.5. Net photosynthesis as a function of stomatal and boundary-layer conductance at different stages of development of the second node of the soybean.

Figure 14.6 indicates that the normal age-related decrease in photosynthetic capacity of the soybean can be arrested by preventing development elsewhere in the plant. In this experiment leaflets above the second node were excised at the stem before any significant leaf expansion had taken place at each of these nodes. Other treatments included cutting the stem just above the second node and artificially irradiating the second node leaflets of plants, which were otherwise allowed to develop normally. These data suggest that the age-related decrease in photosynthetic capacity is more strongly influenced by the physiological state of the tissue than by the level of radiation reaching it. We predict that enzyme activities in leaves showing sustained high photosynthetic capacity remain high in comparison with those in control plants, which gradually decrease. Radiation levels may have a direct effect on the activity of certain enzymes, although we do not expect this to be the primary control mechanism. It is possible that younger leaflets release compounds during their development which inhibit physiological activity at nodes lower on the stem, along the lines of apical dominance in the control of branching

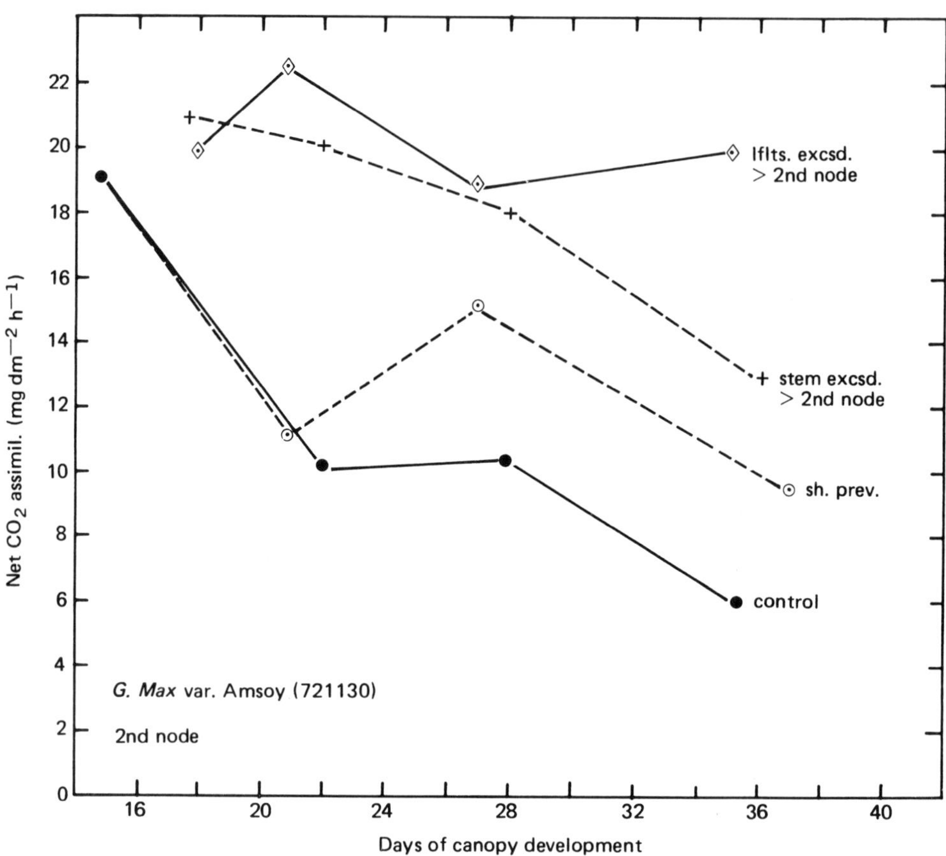

Fig. 14.6. Effect of partial defoliation on the photosynthetic capacity of the second node leaf of the soybean.

in plants (see Thimann, 1972). This would ensure that the youngest (and most highly irradiated) leaflets would also be the most active physiologically and presumably receive a large share of plant resources. Another mechanism which could account for the data in Fig. 6 is that the transport system for carrying materials from roots to leaves always favors younger leaflets—perhaps in response to a steep gradient between roots and young leaflets with respect to a specific metabolite. Thrower (1967) completely prevented the normal senescence of the first four leaves of soybean by removing the apical bud just above the fourth leaf. She also found that translocation from the fourth leaf of disbudded plants was significantly higher than from matched control fourth leaves as the control leaves approached senescence. In disbudded plants, 26 percent of the label assimilated by the fourth leaf via $^{14}CO_2$ was exported, whereas only 14 percent was exported by control fourth leaves. In recently expanded leaves at the fourth node, 28 percent of the assimilated label was exported. It is not clear whether the higher level of export by younger leaves and those whose senescence was delayed results from higher photosynthetic rates of the leaves themselves or from a more efficient translocation process from the leaves.

Figure 14.7 shows the difference in chlorophyll content between leaves from control and defoliated plants. Leaves from defoliated plants seem to show a continued increase in chlorophyll content throughout the lifetime of the leaflet. We interpret this result to be a secondary effect of the experimental treatment of these plants and not a likely cause of the sustained high level of photosynthetic capacity. We also noted that the texture of leaflets from partially defoliated plants was more coarse than that of control plant leaflets. Similar effects were found by Woolhouse (1967) with re-greening of senescent leaves of *Perilla frutescens*.

Table 14.1 indicates the extent to which each node contributes to the net photosynthesis of the entire plant at different stages of development. This table was compiled from data on the rate of net photosynthesis as a function of photosynthetically active radiation flux density. These data were taken at regular intervals throughout the lifetime of leaflets at each node. The data were combined with the

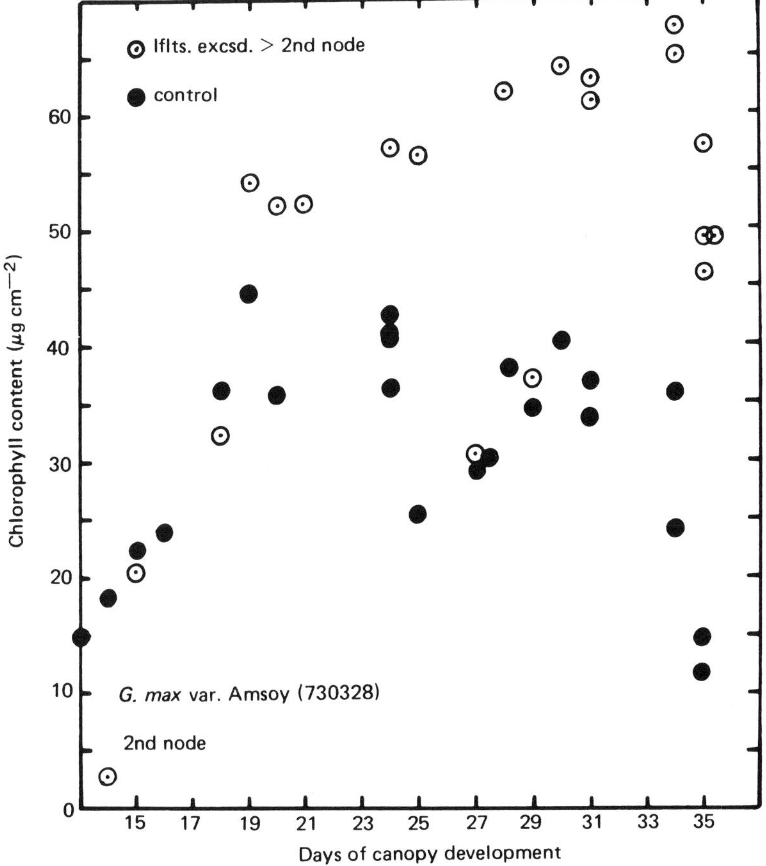

Fig. 14.7. Chlorophyll content of the second node of soybean as a function of time for partially defoliated and control plants.

Table 14.1 Net Photosynthesis Rate at Each Node at Various Stages of Canopy Development[a]

	Day Number							
Node	**11**	**13**	**15**	**17**	**19**	**21**	**23**	**25**
1	5.97(0.68)	4.50(0.33)	2.79(0.15)	1.55(0.07)	0.99(0.04)	0.70(0.03)	0.39(0.01)	
2	2.80(0.32)	7.97(0.58)	9.75(0.53)	7.25(0.34)	4.70(0.19)	3.38(0.12)	2.10(0.06)	1.50(0.03)
3		1.36(0.10)	5.67(0.31)	10.71(0.50)	12.19(0.50)	9.75(0.36)	6.31(0.18)	3.41(0.08)
4			0.15(0.01)	1.97(0.09)	6.05(0.25)	9.57(0.35)	14.11(0.41)	12.00(0.27)
5					0.68(0.03)	3.60(0.13)	9.23(0.27)	20.33(0.46)
6						0.24(0.01)	2.12(0.06)	7.31(0.16)
7								0.09(0.00)
	8.77	13.83	18.36	21.48	24.60	27.23	34.25	44.64

[a] Units are mg CO_2 h^{-1}.

time course of radiation striking each node, as shown in Fig. 14.2, to determine the actual rate of photosynthesis per unit leaf area at each node throughout a portion of the developmental time of the canopy. Finally, the rate per unit area was multiplied by total leaf area at the respective node to obtain the values shown in Table 14.1. Numbers given in parentheses are the decimal fraction of the total plant photosynthesis carried out by the node indicated.

Each node attains a higher maximum rate of photosynthesis per hour than the previous node. This is due almost entirely to the increase in leaf area from one node to the next in this system. The times at which the photosynthetic maxima occur for each node are separated by 4 days in this table. At the time each of the first five nodes reaches its photosynthetic maximum, that node accounts for approximately one-half of the total plant assimilation.

As for broad-leaved plant canopies, such as soybean, which have rapid attenuation of short-wavelength radiation as it passes through the canopy, it is not surprising that the period of time is quite short during which the physiological–biochemical capacity for photosynthesis is at or near maximum. Very soon after the leaflets of a given node are fully expanded, the flux density of photosynthetically active radiation reaching that node drops well below saturation. Maintaining a physiological state of high photosynthetic potential would be of no benefit. The system reported here is consistent with the suggestion (Woolhouse, 1967) that fraction I protein synthesis (60–70 percent of which is carboxydismutase) is turned off around the time of full leaf expansion. Thereafter the photosynthetic capacity falls according to the rate at which fraction I protein is degraded.

It would be of particular interest to discover if any direct connection exists between a component or components of the physical environment of a developing plant leaf and the rate at which specific proteins are synthesized in it. Such a connection would go far, I believe, in explaining the mechanism of physiological–biochemical adaptation in plants. If such a connection does exist, it would necessarily be complex to avoid overreaction to minor fluctuations in the environment.

Acknowledgments

The author is indebted to Kirk A. MacKay for assistance in the experiments and for many helpful discussions. Data on chlorophyll content of leaf tissue were provided by Paul Nemeth. Theodore Brown assisted in the experiments. This work was supported by the Charles L. and Francis K. Hutchinson Fund for Botany and by the Wallace C. and Clara A. Abbott Memorial Fund at The University of Chicago.

References

Anderson, M. C.: 1966. Stand structure and light penetration. II. A theoretical analysis. J. Appl. Ecol. *3*, 41–54.

Das, T. M., Leopold, A. C.: 1963. Photosynthesis and leaf senescence. Plant Physiol. *38* (Suppl.), xiii.

Dickmann, D. L.: 1971. Chlorophyll, ribulose-1,5-diphosphate carboxylase and hill reaction activity in developing leaves of *Populus deltoides*. Plant Physiol. *48* (2), 143–145.

Geronimo, J., Beevers, H.: 1964. Effects of aging and temperature on respiratory metabolism of green leaves. Plant Physiol. *39*, 786–793.

Smillie, R. M.: 1962. Photosynthetic and respiratory activities of growing pea leaves. Plant Physiol. *37* (6), 716.

Thimann, K. V.: 1972. The natural plant hormones. *In* Plant physiology, a treatise (ed. F. C. Steward), vol. VIB: The physiology of development: the hormones. New York: Academic Press.

Thrower, S. L.: 1967. The pattern of translocation during leaf ageing. Soc. Expt. Biol. Symp. *21*, 483–506.

Woolhouse, H. W.: 1967. The nature of senescence in plants. Soc. Expt. Biol. Symp. *21*, 179–214.

15

Energy Exchange and Plant Survival on Disturbed Lands

RICHARD LEE, WILLIAM G. HUTSON, AND STEPHEN C. HILL

The Problem

Man's activities, especially the extraction of mineral resources, frequently create barren lands difficult to revegetate. Newly exposed surfaces tend to be physically unstable, chemically inferior, and microclimatologically unfavorable for plant life. In particular, surface-temperature extremes severely restrict the establishment and growth of native tree seedlings and grasses.

The microclimates of disturbed lands vary greatly, depending on the mesoclimate, topographic aspect, and spoil hydrologic and thermal properties. Coal spoils frequently contain materials that tend to reduce the short-wavelength reflectivity of the surface, increasing the radiant energy load and maximum surface temperature. As a result, the darker spoils revegetate more slowly than others, and may remain barren for many years.

Surface temperature is a key variable in energy-exchange processes, and emerging plants are coupled closely to the energetic processes of their substrata. In the evolutionary scheme, plants have been restricted to rather well-defined thermal niches that vary among the species. This variability suggests, among other protective mechanisms, that some plant types are more loosely coupled to their immediate thermal environments.

These observations suggest that any practical modification aimed at increasing the reflectivity of spoil surfaces could create more favorable microclimates for plants, and that criteria used in the selection of plant species for spoil reclamation should include consideration of plant energy-coupling parameters. The objectives of the research reported here were (1) to observe the effects of coal-spoil reflectivity modifications on net radiant energy exchange and maximum surface temperatures, and survival rates of selected species; and (2) to evaluate the energy-exchange characteristics of the species.

Previous Work

Most earlier studies have centered on the description of spoil surface temperatures and moisture conditions, and are adequately reviewed by the authors cited here. It is generally understood that surface temperatures on coal spoils frequently exceed tolerable limits for plant life (Schramm, 1966). Deely (1970) observed that maximum surface temperatures consistently reached 50–55°C on dry, light-spoil material, and 65–70°C on dark spoils during the summer months in Pennsylvania. In humid regions, spoil moisture content apparently is not a limiting factor once plant root systems are well established, but new seedlings often die because the surfaces dry rapidly following rain (Thompson and Hutnik, 1971).

Apparently no comprehensive evaluation of the energy budgets of coal spoils has been attempted. Munn's (1973) annotated bibliography lists more than 600 references on strip mining alone, only three of which relate to studies of microclimate. The most basic and detailed study was that reported by Deely (1970), who found typical values of long-wavelength emissivity, ε; short-wavelength reflectivity, r; and thermal conductivity, h_b, for 22 spoil types. His data, summarized for major spoil groups, are given in Table 15.1.

Hutson (1972) and Hill (1973) measured the components of incoming and outgoing short- and long-wavelength radiation on coal spoils in West Virginia. The short-wavelength reflectivities varied between about 0.05 and 0.22 for dark and light spoils, respectively. They found that maximum surface temperatures could be predicted, based on energy-budget considerations, from a knowledge of the surface reflectivities and heat-transfer coefficients.

Methods

Study sites were selected near Morgantown, West Virginia, on typical light- and dark-colored coal mine spoils. The light spoils were primarily of sandstone or limestone origin, and the dark spoils were poor-quality coal wastes, or a mixture of coal waste and sandstone spoil. Square 10-m^2 plots were established on both limestone and coal wastes to measure the radiant energy exchange and surface temperatures. Rectangular 2.7×3.0 m (8-m^2) plots were delineated for plant survival studies.

Samples of the surface spoil materials were obtained for laboratory analysis of pH and other characteristics. Subsequently, on both light and dark spoils, alternate plots in each group were treated with common whitewash (i.e., a 20:10:1 mixture of water, hydrated lime, and white portland cement), applied as a spray using about 500 cm^3 m^{-2}.

We measured radiation and surface temperature during calm midday periods (i.e., within 2–3 h of solar noon) on clear days during the summers of 1970, 1971, and 1972. "Economical" radiometers (Tanner *et al.*, 1969) with four sensing heads were used to measure the components of radiation, and independent observations

Table 15.1 Long-Wavelength Emissivity, ε; Short-Wavelength Reflectivity, r; and Thermal Conductivity, h_b, for Major Spoil Groups

			h_b (ly-cm min^{-1} °C^{-1})	
Spoil Group	ε	r	**Oven Dry**	**6 Days After Rain**
Light sandstone	0.981	0.328	0.034	0.202
Light mineral shale	0.982	0.236	0.039	0.226
Dark mineral shale	0.984	0.124	0.057	0.189
Dark organic shale	0.977	0.105	0.047	0.086
Bituminous coal	0.988	0.059	0.011	0.063

Source: After Deely (1970).

of the incoming short-wavelength flux were obtained with Eppley and silicon-cell pyranometers. Maximum surface temperatures were observed with the Barnes PRT-10L radiation thermometer. Supplementary data were obtained from local weather stations, a portable wind-speed indicator, and indicator crayons (Tempilsticks).

The planting and seeding were done during May 1971 and 1972. Seedlings of red oak (*Quercus rubra*), black locust (*Robinia pseudoacacia*), Austrian pine (*Pinus nigra*), and Virginia pine (*Pinus virginiana*) were planted in separate plots, 20 seedlings per plot (60- by 60-cm spacing), replicated 10 times each on control and treated plots. Red oak and Virginia pine were seeded according to the same plan. Tree-of-heaven (*Ailanthus altissima*) and a grass mixture (*Festuca* and *Lolium perenne*) were broadcast-seeded, 20 g m^{-2}. The number of live tree seedlings in each group of 10 replications were counted at the end of the first growing season (late September). Grass cover (%) was estimated using a point-transect sampling technique.

Thermal Climate

Mean values of the radiation components for five midday periods during the summer of 1970 are given in Table 15.2. Wind speeds during observation periods were generally within 50–100 cm s^{-1}. The general validity of the data was confirmed by repetition of the observations during the summers of 1971 and 1972. The values given were obtained for the lightest and darkest spoils tested, and probably reflect extremes for spoil groups in the area.

Short-wavelength radiation intensities usually were within the range 0.9–1.6 ly min^{-1}. The values for light and dark spoils were obtained on different days and are not directly comparable. It is clear, however, that the radiation load (i.e., net radiation, R_n) for a treated dark spoil is significantly less than that for a natural light spoil. The treatment increased the albedo (i.e., $100r$) from about 5 to 33 percent for the dark spoil, and from about 22 to 40 percent for the light spoil.

Table 15.2 Mean Values of Radiation Components (ly min^{-1}) for Natural and Treated Light and Dark Spoils

Observation	Symbol	Light Spoil		Dark Spoil	
		Natural	Treated	Natural	Treated
Short-wavelength radiation					
Incoming	S_i	1.230	1.230	1.312	1.312
Reflected	S_0	0.269	0.487	0.060	0.428
Net	S_n	0.961	0.743	1.252	0.884
Long-wavelength radiation					
Incoming	L_i	0.524	0.524	0.515	0.515
Outgoing	L_0	0.676	0.653	0.834	0.727
Net	L_n	−0.151	−0.129	−0.319	−0.212
Net all-wave radiation	R_n	0.809	0.614	0.933	0.672
Short-wavelength reflectivity	r	0.219	0.396	0.046	0.326

Observed air and surface temperatures are given in Table 15.3. Air temperatures varied from about 18 to 36°C. Mean air temperatures at screen height were slightly lower, as might be expected, over the treated plots, even though they represented small oases in the general microclimate. The temperature of natural light-spoil surfaces did not deviate greatly, on the average, from air temperature, but individual plot and point maxima exceeded air temperature by from 6 to 13°C. It is noteworthy that the surface temperatures of treated dark spoils were significantly higher than those of natural light spoils, in spite of their lower net radiation. Surface temperatures of individual plots and points on dark spoils exceeded the air temperature by from 22 to 30°C.

The effect of surface reflectivity, r, changes on spoil surface temperature, T_s, was determined indirectly from the regression of T_s on absorbed short-wavelength radiation, $(1 - r)S_i = S_n$. The equations,

$$\text{light spoil:}\quad T_s = 23.42 + 4.798S_n \tag{1}$$

$$\text{dark spoil:}\quad T_s = 18.45 + 20.028S_n \tag{2}$$

Table 15.3 Air and Surface Temperatures (°C) for Natural and Treated Light and Dark Spoils

Temperature	Symbol	Light Spoil		Dark Spoil	
		Natural	Treated	Natural	Treated
Air (mean)	$\bar{T}_a$	26.9	26.6	29.3	28.2
Surface (mean)	$\bar{T}_s$	28.8	26.2	45.6	34.6
Difference	$\bar{T}_s - \bar{T}_a$	1.9	−0.4	16.3	6.4
Plot maximum	T_s	38.0	32.6	58.8	46.0
Point maximum	T_s	45.0	39.0	66.0	53.0

indicate that surface temperature differences between natural and treated spoils increase as a function of S_n, as shown in Fig. 15.1. The predicted differences are not quite as great as the observed differences (Table 15.3).

Coal mining spoils dry very rapidly at the surface following rain, and maximum surface temperatures generally increase by from 2 to 4°C d^{-1} (Deely, 1970). Once a dry layer forms at the surface, evaporation is negligible (about 2 mm $week^{-1}$) and can be ignored as a means of energy dissipation. This permits a simplified energy-budget equation:

$$H + B + L_0 = S_n + L_i \tag{3}$$

where H and B are sensible heat losses by convection and conduction, in that order, and each of the terms on the left increases with surface temperature.

The rate of change of surface temperature with $S_n + L_i$ for coal spoils was obtained from the regressions

$$\text{light spoil:}\quad T_s = 16.75 + 7.833(S_n + L_i) \tag{4}$$

$$\text{dark spoil:}\quad T_s = 3.49 + 22.949(S_n + L_i) \tag{5}$$

where the reciprocals of the coefficients are equivalent to the sum of heat-transfer coefficients for H, B, and L_0, i.e.,

$$\text{light spoil:}\quad \frac{d(S_n + L_i)}{dT_s} = 0.128 \text{ ly min}^{-1}\,°\text{C}^{-1} \tag{6}$$

$$\text{dark spoil:}\quad \frac{d(S_n + L_i)}{dT_s} = 0.044 \text{ ly min}^{-1}\,°\text{C}^{-1} \tag{7}$$

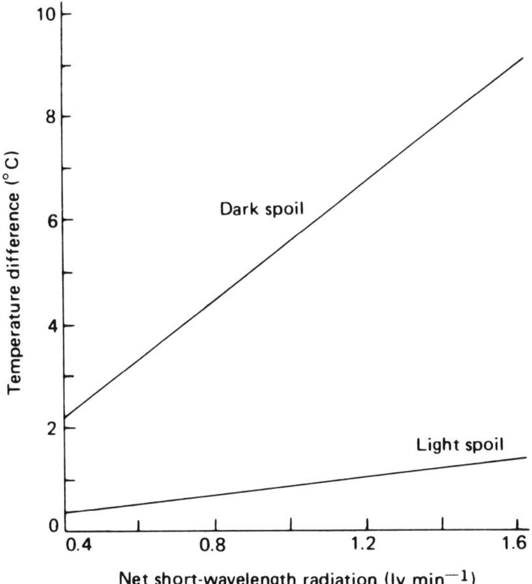

Fig. 15.1. Surface-temperature difference between natural and treated light and dark spoils as a function of net short-wavelength radiation.

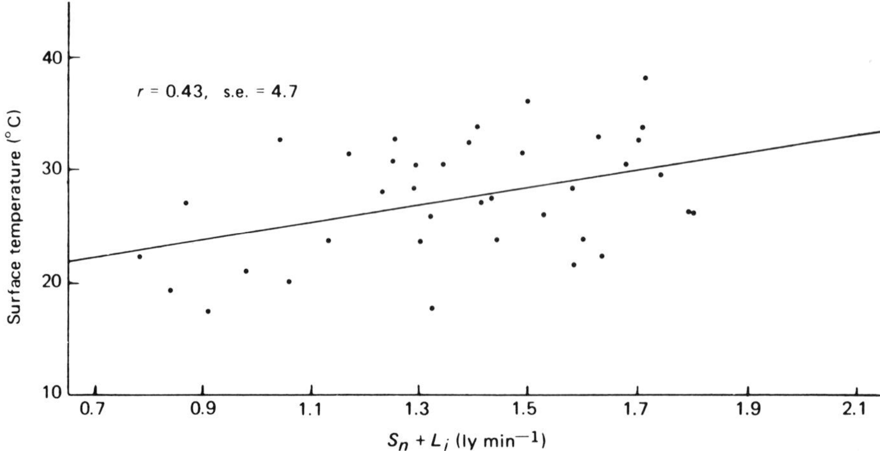

Fig. 15.2. Surface temperature as a function of net short- and long-wavelength incoming radiation (light spoil).

Since the rate of change of L_0 with temperature is virtually 0.01 ly min^{-1} °C^{-1} in the range 20 to 60°C, and the convection coefficient at wind speeds between 50 and 100 cm s^{-1} is probably of a similar magnitude (Geiger, 1965), the conduction coefficients, h_b, must have been about 0.02 ly min^{-1} °C^{-1} (dark spoil) and 0.10 ly min^{-1} °C^{-1} (light spoil). These values are within the ranges given in Table 15.1. The degree of uncertainty in the values can be inferred from Figs. 15.2 and 15.3, which show the scatter of points about the regressions [Eqs. (4) and (5)].

Plant Survival

The relative effects of increased surface reflectivity (reduced surface temperature) on the first-year survival rates of tree seedlings and grasses are listed in Table 15.4. Survival of the planted hardwood species, red oak and black locust, was improved by the treatment, especially on the dark spoil. The planted conifers, Austrian and Virginia pines, however, apparently were adversely affected; on the dark spoil, Austrian pine survival was reduced by 30 percent. This contrasts sharply with the improved survival of seeded Virginia pine. Moreover, the seeded hardwoods, red oak and tree-of-heaven, and the grasses survived more frequently on treated plots.

To explain the differential survival rates of the species, it is instructive to consider net radiation, R_f, for the foliage in individual instances as given by

$$R_f = (a_f)(S_i + S_0) + (L_i + L_0)(\varepsilon_f) - 2\varepsilon_f \sigma T_f^4 \tag{8}$$

where the subscript, f, indicates foliage surface. Assuming characteristic midday values, i.e., mean values from Tables 15.2 and 15.3, and foliage short-wavelength absorptivity, a_f (Kreith and Gates, 1966) and long-wavelength emissivity, $\varepsilon_f = 0.95$, solutions to Eq. (8) appear as in Table 15.5 for foliage at air temperature. The net

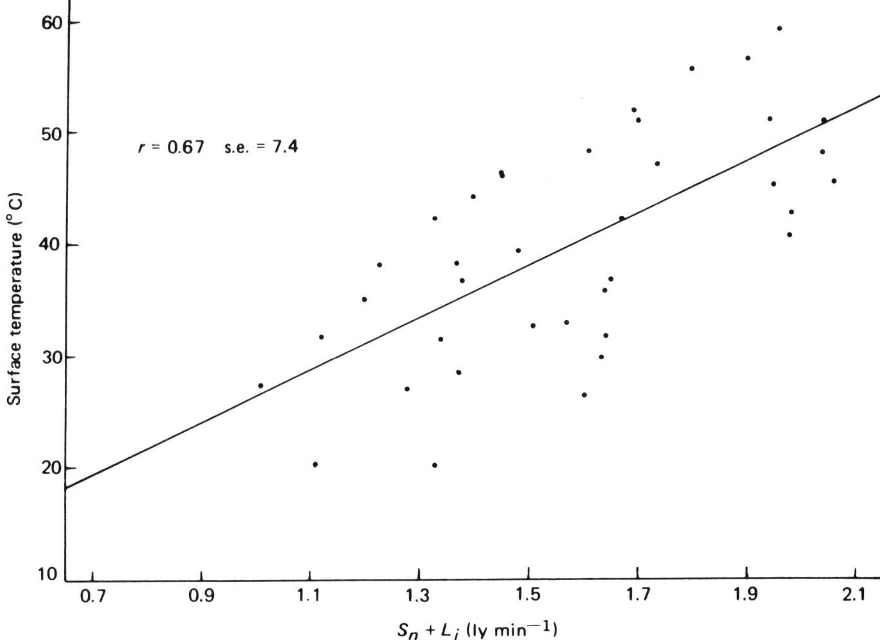

Fig. 15.3. Surface temperature as a function of net short- and long-wavelength incoming radiation (dark spoil).

Table 15.4 Relative Effects of Increased Surface Reflectivity on Plant Survival

	Light Spoil				Dark Spoil			
Species	**pH**[a]	*C*[b]	*T*[c]	*T/C*	**pH**	*C*	*T*	*T/C*
Planted								
Red oak	B	76	91	1.20	—	—	—	—
Black locust	A	45	46	1.02	A	47	81	1.72
Austrian pine	—	—	—	—	A	168	118	0.70
Virginia pine	A	64	58	0.91	A	92	86	0.93
Seeded								
Red oak	—	—	—	—	A	33	50	1.52
Tree-of-heaven	B	45	70	1.56	—	—	—	—
Grasses	B	25	36	1.44	A	2	8	4.00
Virginia pine	A	29	68	2.34	A	71	82	1.15

[a] pH: A, acid (3.1–3.6); B, basic (7.7).
[b] *C*: Number of surviving plants per 80 m^2, or percent grass cover, at end of first growing season on control plots.
[c] *T*: Number or surviving plants per 80 m^2, or percent grass cover, at end of first growing season on treated plots.

Table 15.5 Foliage Net Radiation of Tree Seedlings Growing on Natural and Treated Light and Dark Coal Spoils

		Net Radiation (ly min^{-1})			
		Light Spoil		Dark Spoil	
Species	a_f	C	T	C	T
Red oak	0.49	0.616	0.705	0.654	0.752
Black locust	0.39	0.465	0.534	0.517	0.578
Virginia pine	0.89	1.215	1.392	1.203	1.448

radiation for leaves at temperatures exceeding air temperature would be reduced by about 0.01 ly min^{-1} $°C^{-1}$ temperature difference.

It is apparent that the net radiation load of the pine foliage greatly exceeds that for the hardwoods during midday periods. Moreover, although the treatment significantly reduced spoil surface temperatures, it increased foliage net radiation. Apparently the additional stress was responsible for the poor survival of planted pine seedlings on the treated plots.

In spite of the additional stress, however, seeded Virginia pine survived more frequently on treated plots. To explain the apparent discrepancy, it is helpful to consider relative efficiencies of the foliage types in dissipating heat energy by convection. The juvenile leaves of newly emerged (seeded) pines are isolated needles. with characteristic dimensions (diameters) of about 0.05 cm. At wind speeds between 50 and 100 cm s^{-1}, the thickness of the leaf boundary layer must be about 0.015 cm (Gates, 1962). This means that the coefficient for forced convection will be approximately 0.24 ly min^{-1} $°C^{-1}$.

The clusters of needles on planted 2-year-old pines restrict convection losses. Gates *et al.* (1965) found that pine branch coefficients, h_c, for forced convection could be described by

$$h_c = (18.0 + 0.71V)10^{-3} \tag{9}$$

where V is the wind speed in cm s^{-1}. At $V = 100$, $h_c = 0.09$ ly min^{-1} $°C^{-1}$, or about one-third as great as that for single needles.

Interpretation

The thermal climates of disturbed lands contribute significantly to the difficulties of reclamation. High surface temperatures on the darker spoils can be reduced dramatically by any treatment that increases the surface reflectivity. Common whitewash is effective and inexpensive, but other coloring agents should be considered.

Some of the common hardwood species, especially black locust, and grasses, are able to survive on high-stress sites because of their low foliar absorptivities for short-wavelength radiation. Newly emerging seeded conifers are able to dissipate heat energy efficiently by convection, and their survival is not limited by the high reflectivity of treated spoils. Grass blades probably benefit from both low absorptivity and efficient convective exchange.

Energy-exchange characteristics of plants used to revegetate disturbed lands have not received attention proportional to their usefulness in assessing the relative merits of different species.

References

Deely, J. D.: 1970. High surface temperatures on sunlit strip-mine spoils in central Pennsylvania. College Park, Pa.: Pennsylvania State Univ. M.S. thesis.

Gates, D. M.: 1962. Energy exchange in the biosphere. New York: Harper & Row.

———, Tibbals, E. C., Kreith, F.: 1965. Radiation and convection for ponderosa pine. Am. J. Bot. *52*(1), 66–71.

Geiger, R.: 1965. The climate near the ground. Cambridge, Mass.: Harvard Univ. Press.

Hill, S. C.: 1973. Microclimatic modifications affecting first- and second-year survival of tree seedlings and grasses. Morgantown, W.Va.: W. Virginia Univ. M.S. thesis.

Hutson, W. G.: 1972. Microclimate modifications affecting the reforestation of surface-mined lands. Morgantown, W.Va.: W. Virginia Univ. M.S. thesis.

Kreith, F., Gates, D. M.: 1966. The micro-environment of broad-leaf plants—convection, radiation, and transpiration. *In* Inst. Environ. Sci. 1966 Ann. Techn. Meeting Proc., pp. 209–214.

Munn, R. F.: 1973. Strip mining. An annotated bibliography. Morgantown, W.Va.: W. Virginia Univ. Library.

Schramm, J. R.: 1966. Plant colonization studies on black wastes from anthracite mining in Pennsylvania. Trans. Am. Phil. Soc. *56*(1), 1–194.

Tanner, C. B., Federer, C. A., Black, T. A. Swan, J. B.: 1969. Economical radiometer theory performance, and construction. Madison, Wis.: Univ. Wis. College Agric. Life Sci. Res. Rept. *40*, 1–86.

Thompson, D. N., Hutnik, R. J.: 1971. Environmental characteristics affecting plant growth on deep-mine coal refuse banks. Dept. Environ. Resources Commonwealth of Pennsylvania Spec. Res. Rept. *SR-88*, 1–81.

PART IV

Theoretical Models of Animals

Introduction

David M. Gates

The ecology of animals is complex and more difficult to study than the ecology of plants because of their movements. As with plants, however, many environmental variables interact with the animal through the exchange of energy, solids, and fluids (including gases). Animal metabolism and water loss may vary a great deal with environmental conditions. Because of the complex interaction of so many environmental variables and animal properties, it is essential to have the appropriate theoretical framework to tie them all together. The energy relationship of an animal with its environment is an appropriate theoretical framework or model by which to understand some aspects of animal ecology. Many questions concerning animal distribution, competition, behavior, feeding strategies, and predator–prey relationships can be addressed through the concept of energy exchange.

Heat transfer between an animal and its environment is a straightforward problem in theoretical physics, but it is complicated in the case of an animal because of body shape, anatomical complexities, and physiological control mechanisms. Although the basic problem concerning the exchange of heat between an animal and its immediate environment can be formulated in terms of basic physical principles, the microclimate of the environment in the immediate vicinity of an animal is exceedingly complex, which creates considerable difficulty for the animal ecologist. The radiation field is highly variable with time and space and is spectrally and geometrically complex. The wind profile in the vicinity of an animal is extremely irregular, usually turbulent, sometimes laminar, and enormously variable. The animal ecologist is confronted with the difficult problem of what, exactly, he should measure and then what to do with the measurements.

The complexities of heat exchange between an animal and its environment have caused some animal ecologists to use oversimplified approaches to the subject (e.g., a Newton's law formulation). A theoretical model of heat transfer between an animal and its environment is not only absolutely essential in the interest of coherent insight into many animal ecology problems, but is also necessary to define the methods of measurement. Our first theoretical models were time-independent (e.g., steady-state models) and "lumped-parameter" rather than "distributed-parameter" models. They gave us considerable insight into problems of animal ecology. However, it is now critical to develop much more elaborate theoretical models which resemble the animal as much as possible, with all its complications of shape, color, size, structure, and physiology in the complex environment of wind, temperature, radiation, and humidity. George Bakken, a physicist, has been working with our biophysical ecology group on this problem. Through detailed theoretical modeling of heat transfer between the animal and its environment, he has gained great insight into a large set of ecological questions. But beyond that, he has formulated some extremely valuable methods for the assessment of an animal's microclimate and its impact on the animal. He uses casts of animals as temperature sensors in the environment and shows from the cast that the living animal has a physiological offset temperature. The cast responds to all the complex environmental factors in essentially the same way that the live animal does, except for physiological response. The theoretical model and analysis show us exactly the relationship between animate and inanimate objects, and which environmental factors must be measured.

A theoretical model, if it properly represents an animal, even as a first approximation, can be used to understand the effect of change of one or more variables or parameters on the animal. The steady-state "lumped-parameter" heat-transfer model of animals, which Warren Porter and I originated several years ago, has been very useful for this purpose. James Spotila, trained in animal ecology, worked with our group on the problem of the influence of body size and insulation on body temperature. He describes this work here and shows that too low a body temperature for homeotherms places an extremely heavy evaporative-water-loss demand on animals in warm climates and a reduction in physiological efficiency. Too high a body temperature places too much of a metabolic demand on animals in cool climates, and the instability of proteins, lipids, and cell membranes sets an upper temperature limit.

Sylvia S. Morhardt was a member of the biophysical ecology group at the Missouri Botanical Garden and Washington University, where she completed her Ph.D. thesis on the energetics of the Belding ground squirrel. Detailed observations were made of this squirrel in its native habitat in the Sierra Nevada mountains of California, and thorough measurements were made of its microclimate during the late spring and summer months. In addition, she obtained good physiological information from laboratory measurements with the Belding ground squirrel concerning body temperature, metabolic and water-loss rates, breathing rate, heart beat, and body characteristics. From all this information, Morhardt was able to demonstrate the position of the Belding ground squirrel within the theoretically predicted climate space diagram as a function of the time of day. This is one of the most detailed studies ever made of the actual position of an animal in its habitat to the theoretical climate space diagram established from basic physiological information.

Richard Tracy has been associated with Porter's group at the University of Wisconsin. He has advanced the theoretical modeling of animal heat budgets and in particular has published commentaries on the misuse by animal ecologists of

Newton's law. He demonstrates clearly how theoretical modeling of animal heat budgets based on fundamental physical laws gives one great insight into animal adaptation and behavior, exemplifying how good theoretical analysis must be intimately combined with laboratory experiments and field observations. Here he analyzes the energy and water balance between an amphibian and its environment. Warren Porter, who has done a great deal to advance the methodology of biophysical ecology and mechanistic models of animals, has joined forces with C. R. Tracy and engineers J. W. Mitchell and W. A. Beckman to advance the techniques of biophysical ecology, a subject that requires the methodology of physics, mathematics, and heat-transfer engineering. Their paper describes some of their work concerning predator–prey relationships as revealed by heat-transfer analysis.

An enormous amount of information concerning animal energetics can be inferred from paleontological studies of fossil remains. Robert T. Bakker gives an incisive review of this information in his paper. Many of the questions raised by Bakker show the need for very careful and thorough theoretical analysis of these difficult problems. Many of the examples concerning the energetics of animals at various stages of evolution can be subjected to theoretical analysis using mechanistic models. I am confident that, when this is done, many anomalies and some discrepancies that are now difficult to explain will be resolved. The use of the term "climate space" in this paper is not exactly the same concept as originated by Porter and myself. Here it refers to the temperature of the environment, whereas we intended it to be a three- or four-dimensional space, consisting of air temperature, radiation, wind, and perhaps humidity. In any event, this extremely interesting and challenging paper by Bakker gives us great insight into the evolution of animals and their bioenergetics.

16

Heat-Transfer Analysis of Animals: Some Implications for Field Ecology, Physiology, and Evolution

George S. Bakken and David M. Gates

Introduction

Mathematical modeling studies in science are not an end in themselves, but rather a tool. The fundamental purpose of mathematical analysis is to provide generalized, intellectually tractable insight into the operation and interaction of the complex factors involved in the physical and biological process under study. Two often contradictory requirements must be met. First, the model must include all relevant factors, with sufficient detail to give predictive precision for validation and practical applications of the analysis. Second, the analysis must be simple enough to be intellectually tractable and give a clearer subjective understanding than the raw data and knowledge of the individual processes involved. The second requirement implies that the model be constructed so that it is analytically soluble. For the complex nonlinear processes in biology, this is seldom possible without sacrificing precision. This is certainly true in thermal analysis of animals.

The best procedure in such a case is to proceed by a series solution that gives successive approximations to the real situation. The first-order approximation can be designed to carry the principal factors and have an analytic solution to fulfill the second requirement. Higher-order terms in the series and numerical solutions may then be used to give predictive precision when required.

In the past, mathematical analyses of heat transfer and temperature regulation in animals have fallen into two classes. First, most physiologists and physiological ecologists use "Newton's law of cooling." This gross oversimplification of the heat-transfer process states only that the heat transfer between an object and its environment is linearly proportional to a constant and a temperature difference. Neither the constant (Tracy, 1972) nor the temperature difference is well defined in the natural environment, and thus Newton's law provides no insight into the actual processes and interactions of heat transfer in animals. At the other extreme are the very detailed studies of heat transfer and temperature distribution in the human body, with emphasis on control processes and mechanisms. The symposium edited by Hardy *et al.* (1970) surveys recent work. Precise models of this sort are specific for the particular animal and usually require computer evaluation. Such a model, although precise, provides little more general insight into the principles and biological implications of thermoregulation than direct empirical experimentation on the animal.

We have followed an intermediate approach, pioneered in ecology by the work of Porter and Gates (1969), wherein the contributions to the total thermal budget of the animal are separated by mode (conduction, convection, radiation, metabolism, etc.) but averaged over the whole animal or "lumped." The use of lumped parameters assumes the mathematically false relation

$$\int f(x)\cdot g(x)\,dx = \left[\int f(x)\,dx\right]\cdot\left[\int g(x)\,dx\right] \qquad (1)$$

However, the relation is usually approximately correct and may be further corrected by introducing factors that convert the actual average of a parameter to an effective value which gives more nearly correct results. This procedure is well developed in the treatment of radiation (Kreith, 1965, pp. 217–234) with shape factors. During the remainder of this discussion of lumped parameters, we shall assume any such factor to be included so that all symbols represent effective rather than true average values. This is not to suggest that a distributed-parameter analysis may not be necessary for some studies. However, this lumped-parameter analysis is quite adequate for investigating large-scale trends in comparative physiological ecology and evolution.

Our general intent here is to outline the formalism that we have developed for the lumped-parameter model of the thermal energy budget of an animal, including the time-dependent effects of heat storage by the heat capacity of body tissues and body-temperature-dependent metabolic heat production. We shall discuss a few of the more interesting implications of the theory for simplified field instrumentation for studies of thermal ecology and time-dependent thermoregulatory behavior, and some general implications for the evolutionary history of thermoregulation. We shall not develop the theory in detail, but instead will present the results and qualitative arguments for the biological interpretations.

The Model: Description and Steady-State Solution

General Description

Heat transfer is a complex process, involving a number of mechanisms that operate simultaneously. The underlying physics of these mechanisms is well known except for the more complex cases of convective heat transfer. The reader is referred to a good general introduction to heat-transfer theory, such as Kreith (1965), for the essential principles of heat transfer. The application of heat-transfer theory to animals is complex because of the unusual properties of biological materials, especially fur (Davis and Birkebak, 1974). Good introductions to the application of heat-transfer theory to animals can be found in Birkebak (1966), Porter and Gates (1969), and Tracy (1972).

We have used the electrical analog representation of heat flow in the development of the model (Kreith, 1965, pp. 18–20). Essentially, this assumes that heat flow is linear, at least for a modest range of temperature differences. Then heat flow follows approximately the general relation: flow = (potential difference)/resistance. Thus heat flow may be regarded as equivalent to a current flowing in an electrical circuit, temperature difference as equivalent to voltage difference, and thermal resistance or conductance equivalent to an electrical resistor. Consequently, a thermal circuit can be represented by the same symbols as an electrical circuit, as shown in Figs. 16.1 and 16.2. The basic analytic device used in the development is Kirchhoff's law, which states that the energy flowing into a node (T_x) in a circuit diagram is equal to the energy flowing out of the node (i.e., conservation of energy). In heat-flow analysis, nodes correspond to isothermal (or nearly so) surfaces of the body, particularly the interfaces between the core and the skin, skin and pelage, and pelage and the environment. In a distributed-parameter model the surface elements are infinitesimal and strictly isothermal, whereas in the lumped-parameter model the surfaces are only approximately so. Figure 16.1 shows representative thermal circuits for various classes of animals. These models lump all heat capacity in an isothermal core. This is a reasonable assumption for animals with effective pelage and very large animals but results in errors in moderate-size, uninsulated animals, where thermal diffusion can be significant. (Colbert *et al.*, 1946, show an example of this in the delayed thermal rise and death of a small *Alligator mississippiensis*.)

The simplest nontrivial circuit, Fig. 16.1a, will serve to illustrate the development of the model. Kirchhoff's law applied to the surface node, (T_r), assumes the form (heat leaving surface) = (heat flowing to surface), or

$$H(T_r - T_a) + \sigma\varepsilon T_r^4 + E_r = K(T_b - T_r) + Q_a \tag{2}$$

where K = conductivity from the surface element at temperature T_r to the body core at temperature T_b

H = convection coefficient

T_a = air temperature

Thermal radiation from the surface to the environment is given by the $\sigma\varepsilon T_r^4$ term, where the Stefan–Boltzmann constant $\sigma = 8.12 \times 10^{-11}$ cal cm^{-2} °K^{-4} min^{-1}, and the emissivity ε ranges from 0 to 1, with biological materials typically 0.95–0.99. Evaporative cooling at the surface is denoted by E_r and total absorbed radiation by Q_a. The sign convention regards metabolic heat production and heat flow from the core to the environment as positive.

A second equation results from applying Kirchhoff's law to the T_b node which

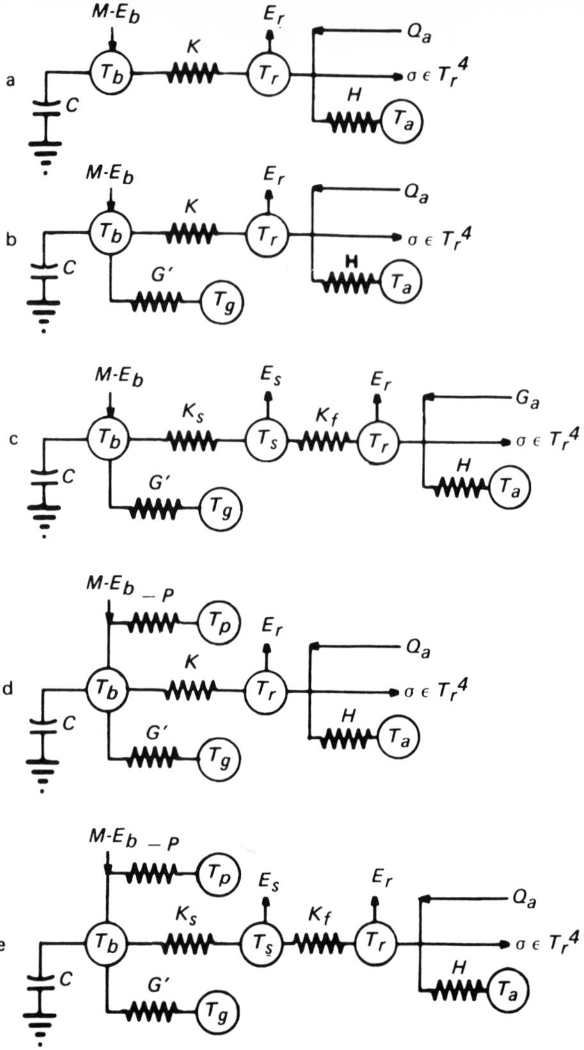

Fig. 16.1. Thermal circuit diagrams for the electrical analog representation of various classes of animals. The connection points or nodes represent parts of the animal or environment with a well-defined temperature such as the body core (T_b) or the ground temperature (T_g). Each node is labeled with the thermal potential or temperature of the corresponding part of the animal or the environment. The resistor symbols represent the thermal conductance between the parts of the animal or environment represented by the nodes connected by the resistance symbol. The capacitor symbol represents the heat capacity of the body of the animal. Simple arrows indicate simple thermal currents or heat flow which are not directly related to a temperature or thermal conductance. Diagram a represents a naked animal with negligible conduction to the ground. Diagram b represents a naked animal with significant conduction to the ground. Diagram c represents an animal with both pelage and significant conduction to the ground. In diagrams a–c, metabolism and evaporative water loss are assumed constant. In diagrams d and e, metabolism and respiratory evaporation are assumed to be linearly dependent on body temperature in the neighborhood of some body temperature $T_b = T_p$, so $(M - E_b) \rightarrow (M - E_b)_p + P(T_b - T_p)$. As explained in the text, this is equivalent to placing an imaginary conductance $-P$, connecting the body core with an environmental heat source/sink at a temperature T_p, as shown.

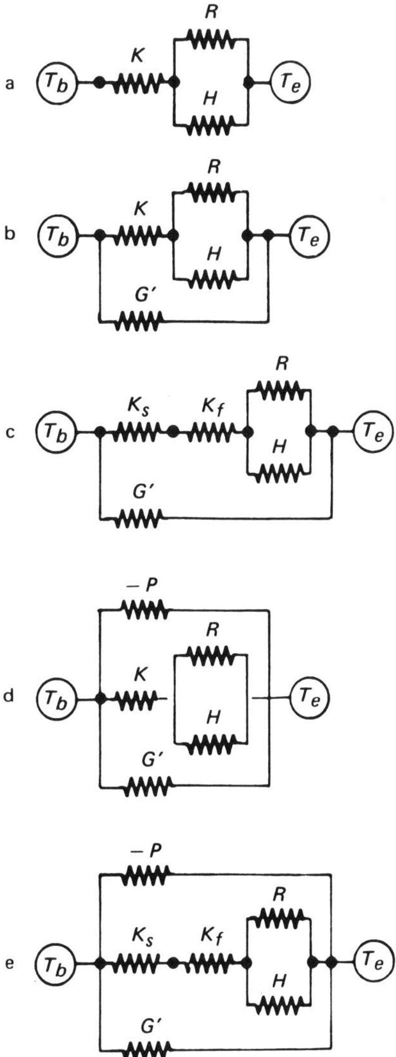

Fig. 16.2. Equivalent circuits for the same classes of animals as in Fig. 16.1. The air, ground, radiant, and surface temperatures have been formed into an equivalent environmental temperature T_e called the "operative environmental temperature." The thermal resistances between the parts of the animal and between the animal and the environment have been formed into an equivalent circuit of thermal resistance I, equivalent to the insulation parameter measured in metabolism chamber experiments. The net metabolic heat production $(M - E_b)$ is the thermal current flowing through this circuit. Surface and skin water loss E_r and E_s see somewhat different equivalent circuits and equivalent temperatures, and must be suitably combined with $(M - E_b)I$ to give the temperature difference between the body core and the operative environmental temperature, $T_\Delta = T_b - T_e$, as described in the text.

corresponds to the isothermal core of the animal. This has the form (heat stored) + (heat flowing out) = (net heat production) or

$$C\frac{dT_b}{dt} + K(T_b - T_r) = M - E_b \tag{3}$$

where C = heat capacity of the body
M = rate of metabolic heat production
E_b = rate of heat loss due to respiratory evaporation
dT_b/dt = rate of change of T_b with time

For the moment net metabolic heat production $(M - E_b)$ will be assumed constant and independent of T_b and T_a. All coefficients, including E_r, E_b, M, and Q_a, are referred to a unit surface area in the lumped-parameter analysis. A convenient choice of units is to express temperature in degrees Kelvin (°K = °C + 273.16); K, H, and other heat-transfer coefficients in cal cm^{-2} °K^{-1} min^{-1}; heat capacity C in cal cm^{-2} °K^{-1}; and physiological terms such as M and the various E's in cal cm^{-2} min^{-1}.

Although Eqs. 2 and 3 can be solved simultaneously as they stand, the T_r^4 radiation term introduces excessive complexity. To facilitate a solution, we have chosen to linearize the radiation term by expanding it about a mean surface temperature $\overline{T}_r$ and dropping higher-order terms. Then

$$\sigma\varepsilon T_r{}^4 = (4\sigma\varepsilon\overline{T}_r{}^3)T_r - 3\sigma\varepsilon\overline{T}_r{}^4 \tag{4}$$

This linearization differs from the usual procedure used in engineering and is dictated by the lack of a well-defined radiant temperature in the natural environment. It is convenient to denote the linearized radiation coefficient $4\sigma\varepsilon\overline{T}_r{}^3$ by R and form a term $Q_n = (Q_a + 3\sigma\varepsilon\overline{T}_r{}^4)$. The error term in the linearization of radiation is about $6\sigma\varepsilon\overline{T}_r{}^2(T_r - \overline{T}_r)$, which limits the range of validity to $(T_r - \overline{T}_r) < 10$ to 30°C, depending on the overall significance of the thermal radiation term. Then Eq. 2 has the form

$$H(T_r - T_a) + RT_r + E_r = K(T_b - T_r) + Q_n \tag{2a}$$

Steady-State Solution

The steady-state solution of Eqs. 2a and 3 corresponds to the model analyzed by Porter and Gates (1969) and is useful in developing the notation and some ideas for instrumentation. Solving Eqs. 2a and 3 with $dT_b/dt = 0$ for T_b by eliminating T_r gives

$$T_b = (M - E_b)\frac{K + H + R}{K(H + R)} - E_r\frac{1}{H + R} + \frac{HT_a + Q_n}{H + R} \tag{5}$$

To simplify notation, we may define a quantity I equivalent to the insulation parameter of Scholander *et al.* (1950):

$$I = \frac{K + H + R}{K(H + R)} \tag{6}$$

This is a thermal resistance (units cm^2 °K min cal^{-1}). Further, we may define an "operative environmental temperature" T_e:

$$T_e = \frac{HT_a + Q_n}{H + R} \tag{7}$$

T_e may be identified as the temperature of an inanimate object of zero heat capacity with the same size, shape, and radiative properties as the animal and exposed to the same microclimate. It is also equivalent to the temperature of a blackbody cavity producing the same thermal load on the animal as the actual nonblackbody microclimate, and therefore may be regarded as the true environmental temperature seen by that animal. Finally, we may define a "physiological offset temperature" T_Δ to represent the amount an animal may offset T_b from T_e by utilizing metabolism, evaporation, and insulation.

$$T_\Delta = (M - E_b)I - E_r\left(\frac{1}{H + R}\right) \tag{8}$$

With these definitions the heat flow from the animal can be represented by the equivalent circuits given in Fig. 16.2. I is the equivalent resistance of the network connecting T_b and T_e in Fig. 16.2. The expression for body temperature becomes

$$T_b = T_e + T_\Delta \tag{9}$$

The quantities I and T_e are equivalent to the operative heat-exchange coefficient, I_o, and the operative temperature, T_o, concepts used in the human physiology literature (Herrington *et al.*, 1937; Kerslake, 1972, p. 66). The parameters in I and I_o, T_e and T_o, are identical for Eqs. 6 and 7, but the insulation and the operative environmental temperature for the more general thermal circuits of Fig. 16.1 are not equivalent to I_o and T_o.

In general an animal has conduction to the ground and may have external pelage. To incorporate this in the model, K must be broken into two parts, K_s for the skin and K_f for the pelage. Similarly, evaporation takes place from both the skin and the pelage surface, denoted by E_s and E_r, respectively. Conduction to the ground acts independently from other heat-loss mechanisms and is denoted by a separate coefficient G' transmitting heat to the ground at a temperature T_g. The final circuit is shown in Fig. 16.1c. Applying Kirchhoff's law to the nodes at the core, skin–fur interface, and fur surface gives three simultaneous equations with a solution identical to that of Eq. 9. However, I, T_e, and T_Δ are now defined as follows:

$$K_{sf} = \frac{K_s K_f}{K_s + K_f} \tag{10}$$

$$I = \frac{K_{sf} + H + R}{(K_{sf} + G')(H + R) + K_{sf}G'} \tag{11}$$

$$T_e = \frac{K_{sf}(HT_a + Q_n) + G'T_g(K_{sf} + H + R)}{(K_{sf} + G')(H + R) + K_{sf}G'} \tag{12}$$

$$T_\Delta = (M - E_b)I - E_s\frac{K_{sf} + [K_s/(K_s + K_f)](H + R)}{(K_{sf} + G')(H + R) + K_{sf}G'} - E_r\frac{K_{sf}}{(K_{sf} + G')(H + R) + K_{sf}G'} \tag{13}$$

Net metabolic heat production $M - E_b$ is again assumed constant and independent of T_b. Equation 12 is a more general form of the operative environmental temperature than Eq. 7, and the difference from the operative-temperature concept of human physiology is clear.

Notice that if E_s and E_r are assumed to be zero, Eq. 13 is formally equivalent to Newton's law of cooling:

$$T_\Delta = T_b - T_e = I(M - E_b) \tag{13a}$$

The errors inherent in analyzing physiological data with Newton's law are quite clear from a brief consideration of Eqs. 11 and 12. The insulation of an animal is often considered a property of the animal. However, Eq. 11 shows that I is a function of the convection coefficient, which depends in part on the wind velocity of the environment, and the conduction coefficient to the substrate G', which will depend on the posture of the animal and the properties of the substrate, particularly the conductivity and the heat capacity. Conversely, the temperature of the environment is not the air temperature or the enclosure temperature, but depends on all the thermal properties of the environment and properties of the animal, such as K_{sf}, and size, shape, and posture, which strongly affect the values of G' and H. Thus we have the interesting situation in which the supposed property of the animal, I, is determined in part by the environment (as noted by Tracy, 1972), and the temperature of the environment, T_e, is determined by the properties of the animal. This should be a sufficient warning against the use of oversimplified analysis and should clearly indicate that heat transfer in animals shows the strong interactions between the animal and its environment so characteristic of all ecological problems.

Applications in Instrumentation

A frequent criticism of more detailed studies of heat transfer in animals is that the resulting model is too complex to be used in field studies of thermal ecology, where the primary interest is biology, not heat transfer. This criticism has some merit, and consequently we have tried to cast the result of our model in more general terms. The $T_b = T_e + T_\Delta$ formulation is a result. We have also tried to develop simple instrumentation that would be less expensive, complex, and heavy than the usual micrometeorological instruments but yet give results with general validity which are of use in more specialized studies. In particular, some method of making an integrated reading of the thermal stress on the animal, ideally the operative environmental temperature T_e, is needed to replace the largely meaningless "black-bulb thermometer" reading commonly used as an overall heat-stress index.

The identification of T_e as the "true" environmental temperature and its equivalence to the temperature of an inanimate model of the animal suggests that it would be useful to measure T_e directly with a cast of the animal as a "T_e thermometer." This measurement is similar to the use of the Vernon (1932) globe thermometer, which is designed (Kerslake, 1972, pp. 67–72) to approximate the response of a human by adjusting H and R to give the same value of T_e (as in Eq. 7) as a human. This instrument and its dubious progeny, the black-bulb thermometer, are of limited use since they only approximate the actual values of H and R and completely omit some factors (primarily G') in Eq. 12. Most of these limitations are eliminated if an object that exactly duplicates the animal except for metabolism and water loss is used as a thermometer and placed in the same position in the environment as the animal under study. The exact method of construction will depend on the study to be conducted. Most commonly it will be desirable to have a rapid response so that the operative environmental temperature may be monitored continuously. This requires a low heat capacity, which can be best achieved by using a thin metal cast, painted to duplicate the radiative properties of the animal or covered by the integument of the animal using taxidermy techniques. A technique for making thin shell metal casts of reptiles and amphibians is discussed in the Appendix to this chapter. Alternatively, it may be desirable to duplicate the time response of the animal so that the heat capacity of the animal must be duplicated. It must be remembered that the heat capacity of the *whole* animal must be duplicated, and this requires that the heat capacity per unit volume, not per unit weight, be duplicated. Tissue has a whole-animal average of approximately 0.7–0.8, and this range is duplicated by various brass and bronze alloys which could be used to make a solid cast. Also, the hollow cast could be filled with a suitable composite material, with the composition adjusted to match the heat capacity of the animal.

The primary utility for a cast as a T_e thermometer is in the study of small reptiles and amphibians. Lizards control their body temperature closely (Cowles, 1940; Cowles and Bogert, 1944; Heath, 1964, 1965, 1968) by behavioral means—that is, by adjusting their position and posture in the environment so that the time-average value of T_e falls inside the preferred range of body temperatures. Minor variation in shade, skin color, substrate, exposure to the sky, wind velocity, and temperature strongly influence T_e for a small animal near the ground. The distance scale is much too small to allow normal micrometeorological instrumentation. Under these conditions, several T_e thermometers placed at various locations in the environment would provide a "map" of operative environmental temperatures available to the lizard. This map would greatly facilitate behavioral studies.

To demonstrate the range of operative environmental temperatures present in an apparently homogeneous environment, a number of T_e thermometers were constructed in the form of thin-walled metal cylinders with a 4:1 length-to-diameter ratio. Diameters ranged from 0.64 to 15.5 cm. The site was a sparsely vegetated sand spit on the shore of Douglas Lake, Cheboygan County, Michigan, near The University of Michigan's Biological Station's Alfred M. Stockard Lakeside Laboratory. This site provided good exposure to solar radiation and wind during most of the day.

Typical results are shown in Table 16.1. One series of cylinders was placed on 0.64-cm square balsa wood blocks so that they were 2–3 mm above the sand substrate. Another series was hung 20 and 90 cm above the substrate by cotton thread. All were in an area roughly 1 by 2 m and most were within an area $\frac{1}{2}$ by 1 m. The shaded cylinder was placed on leaf and twig litter under a low (0.5-m-high) bush that provided broken shade. Within this area, light gray 1.6-cm-diameter cylinders

Table 16.1 Operative Environmental Temperatures of Cylinders of Various Diameters, Colors, and Orientations in a Sandy Area Exposed to Full Sun and Wind[a]

Cylinder	Diameter (cm)	Position	Exposure	Height (cm)	Color	T_e (°C)
1	1.6	ESE–WNW	Sun	0.3	White	37.5
2	1.6	ESE–WNW	Sun	0.3	Black	50.0
3	1.6	ESE–WNW	Sun	0.3	Gray	45.5
4	1.6	NNE–SSW	Sun	0.3	Gray	46.0
5	1.6	Vertical	Sun	0.3	Gray	41.0
6	1.6	ESE–WNW	Sun	20.0	Gray	39.5
7	1.6	ESE–WNW	Sun	90.0	Gray	36.0
8	1.6	ESE–WNW	Shade	1.0	Gray	28.3
9	0.64	ESE–WNW	Sun	0.3	Gray	41.5
10	1.6	ESE–WNW	Sun	0.3	Gray	45.5
11	2.86	ESE–WNW	Sun	0.3	Gray	45.5
12	3.81	ESE–WNW	Sun	0.3	Gray	47.5
13	15.50	ESE–WNW	Sun	1.0	Gray	47.0

[a] A small shrub 0.5 m high provided broken shade for cylinder 8. All cylinders were in an area of 1 × 2 m. Microclimate parameters for this area were: $T_{air} = 26°C$; $T_{ground} = 38.5°C$; wind at 20 cm, 67 cm s^{-1}; wind at 90 cm, 310 cm s^{-1}; solar radiation in an 0.3- to 3-μ band, 1.3 cal cm^{-2} min^{-1}. August 17, 1972, 14:15 to 14:16. Douglas Lake Cheboygan County, Michigan.

placed at the surface could experience an operative environmental temperature range from 28 to 46°C.

To validate the concept further, we have constructed thin shell casts of the common western fence lizard, *Sceloporus occidentalis*, by the method described in the Appendix. Since the net metabolic heat production is very small, the temperature of the lizard should approximate T_e under equilibrium conditions. Comparisons of the lizard with the cast are given in Fig. 16.3.

The use of this technique with small endotherms is less straightforward since the high internal heat production will increase $\bar{T}_r$ and alter R and Q_n. In this case a model must be constructed with a heater and covered by the integument of the animal; the model attached to a temperature controller set to T_b. If the heat capacity C is known, I may be determined from a cooling curve (see the next section). Then measuring the heater power ($M - E_b$) allows T_Δ to be measured and T_e thus to be determined from Eq. 9. Of course, heater power is equivalent to metabolic rate and thus is a direct index of environmental stress. However, determination of T_e gives a more intuitive result since temperature is an intensive parameter, whereas heater power or metabolism is an extensive parameter that is hard to use comparatively.

We wish to emphasize here that animals larger than the space occupied by a complete set of micrometeorological instruments are best studied directly; the diseconomy of substituting a brass moose for proper instruments is fairly clear.

The general concept of a thin metal cast as meteorological instrumentation and the development of the necessary technology suggest another important application. Convective heat transfer is one of the primary modes of heat loss in smaller

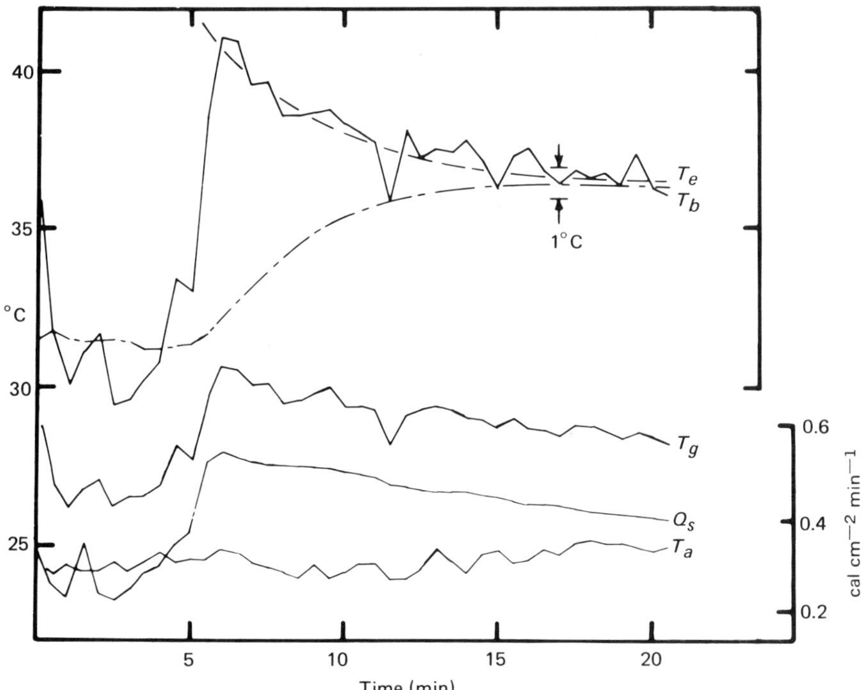

Fig. 16.3. Comparison of the body temperature T_b of a 9-g lizard (*Sceloporus occidentalis*) with the operative environmental temperature T_e measured by a fast-response thin metal cast of another lizard of the same size and species. To ensure uniform microclimate conditions, the animals were placed on a flat styrofoam substrate ($G' \cong 0$). Note the rapid response of the hollow cast, which has a time constant of less than 1 min, with 90 percent response in less than 2 min. Note also that when the cast and the lizard are in thermal equilibrium (arrow), T_e and T_b agree within $\frac{1}{4}$°C. This good agreement is due to a good match between the absorptivity of the cast and the lizard. This match is partly fortuitous, since the lizard can, like most lizards, actively vary its absorptivity. Other lines in the figure are the air temperature T_a, the styrofoam surface temperature T_g, and the incident solar radiation (Eppley pyranometer) Q_s.

animals, and knowledge of the convection coefficient of an animal in different air-flow conditions in the natural habitat is essential. However, the air flow in outdoor conditions is complex and ill defined, with large-scale velocity variations (gusts) and small-scale microturbulence from vegetation and other irregularities. Consequently, convection coefficients typically show 50 percent turbulent enhancement and poor (± 50 percent) correlation with air velocity when measured outdoors (Pearman *et al.*, 1972). A further problem for smaller animals is that the wind velocity and air temperature near the ground follow a logarithmic profile and thus vary over the animal.

This situation complicates the usual procedure for developing correlations between laboratory data and field conditions. The usual procedure for applying convection coefficient data is to express the data in terms of dimensionless units, called the Reynolds number (Re), Nusselt number (Nu), and Prandtl number (Pr). A full description of the use and interpretation of the dimensionless quantities is given in most heat-transfer texts (e.g., Kreith, 1965). Very briefly, and with great oversimplification, the Prandtl number is a constant of the fluid, the Reynolds number is a dimensionless velocity, and the Nusselt number is a dimensionless convection coefficient. They are related as follows:

$$\mathrm{Nu} = \mathrm{Re}^n \mathrm{Pr}^m$$

so a plot of Nu versus Re can be used to find the convection coefficient for objects of any size in any wind velocity as long as the objects are of similar shape and similarly oriented to air flow and the air flow has the same general character with respect to velocity profiles and turbulence.

Because of these complexities, the usual procedure of calculating the convection coefficient from Reynolds number–Nusselt number plots made using wind-tunnel data and field measurements of wind velocity becomes complex. Extensive measurements of the mean wind velocity, gusts, and small-scale microturbulence to within a few millimeters of the surface would be required to determine the convection coefficient accurately. In most cases this effort would not be justified by the improvement in the accuracy of the convection coefficient determination, particularly since the variability in all the relevant parameters under field conditions is great. However, the environment often shows consistencies in the level of gustiness, turbulence, wind direction, and velocity in relation to the surface topography, and the consequences for the convection coefficient could well be biologically significant. Also, the general problem of the details of convection in the outdoor environment is of considerable intrinsic interest.

For example, Porter *et al.* (1973) have shown how the lizard *Dipsosaurus dorsalis* exploits the air-velocity profile by climbing higher in bushes to obtain greater convective cooling as the day becomes progressively hotter toward noon. It would be interesting to know if variations in turbulence level may be exploited similarly.

A phenomenon that needs further investigation is the effect of size on the convection coefficient of animals near the ground. For free-stream conditions, it is well known that the convection coefficient decreases approximately as the square root of the diameter of the object. However, at or near the surface of the ground the velocity profile must also be considered. A large object near the surface of the ground will extend farther above the surface and thus into the higher-velocity region of the profile. Thus a large object experiences a higher average air velocity than a smaller object. Since the convection coefficient increases roughly as the square root of the air velocity, this will increase the convection coefficient of the larger object relative to the smaller, opposing the free-stream tendency of the convection coefficient to decrease with size. The net effect is for the convective heat loss of small and large animals to differ by less than might be expected. Table 16.2 shows the consequences for T_e.

This effect also suggests that essentially upright or long-legged animals would

Table 16.2 Equilibrium Temperatures of Gray Cylinders Outdoors over a Sandy Substrate at The University of Michigan's Biological Station at Douglas Lake, Cheboygan County, Michigan

Diameter (cm)	Surface Level[a] T_e (°C)	Above Surface[b] T_e (°C)
0.64	33.0	30.5
1.59	36.5	34.0
2.86	37.0	35.0
3.81	37.0	36.25
15.5	36.75	39.0

[a] Surface level (3 mm): solar radiation 1.0 ly, air temperature 22°C, wind 100–150 cm s^{-1}.
[b] 90 cm above surface: solar radiation 0.8 ly, air temperature 27.5°C, wind 40–80 cm s^{-1}.
Note that the temperatures of the cylinders, equal to the operative environmental temperatures T_e for the cylinders, are well separated 90 cm above the ground by the decrease in the convection coefficient with increasing diameters. However, at the surface the cylinders are in a velocity gradient and the larger cylinders experience a higher average wind velocity, counteracting the decrease in convection coefficient with size. The result is that the operative environmental temperature is essentially the same for all the cylinders except for the smallest (0.64 cm), even though the meteorological conditions (higher radiation, higher wind speed, and larger difference between cylinder and air temperature) would be expected to give a larger differential in T_e. This effect is also partly due to the lower wind velocity and hence reduced convection coefficients at the surface, so that the significance of thermal radiation, which is the same for all size cylinders, is relatively greater.

have a substantial convective advantage over animals of similar size that have horizontal posture or short legs.

The effort required for a fundamental study of the convection coefficient in a turbulent flow showing a logarithmic profile cannot be justified for general field studies where the fundamental interest is biology rather than heat transfer. A simpler means of determining the convection coefficient with good accuracy is required. We suggest that the best procedure for most thermal ecology studies of small animals near the surface is not to develop Reynolds number–Nusselt number correlations in the wind tunnel and then determine the field convection coefficient by measuring wind velocity and using the Re–Nu correlation. Instead, the convection coefficient would be determined directly by using a pair of gold-plated, polished, metal casts, one heated and the other unheated, as a field convection coefficient meter. The procedure is identical to the use of the heated-bulb-thermometer anemometer.

The procedure is to place the casts on a substrate of low thermal conductivity, such as styrofoam, so that G is essentially zero. Then the insulation I is given by Eq. 6. Since the emissivity of polished gold is low (ε = 0.05–0.02), the radiative-heat-transfer coefficient $R = 4\sigma\varepsilon\overline{T}_r^3$ will be around 4×10^{-4} cal cm^{-2} s^{-1} °C^{-1} compared with $H = 5$ to 10×10^{-3} cal cm^{-2} s^{-1} °C^{-1} for cylinders 2–10 cm in diameter in laminar flow of 10 cm s^{-1}. Thus, even for low air velocities, R may be assumed zero as a first approximation, in which case I becomes

$$\lim_{R \ll H} I = \lim_{R \ll H} \frac{(K + H + R)}{K(H + R)} = \frac{K + H}{KH} \tag{14}$$

Since the cast is made of metal, K is very large and $K + H \cong K$. Thus

$$\lim_{K \gg H} I = \lim_{K \gg H} \frac{K + H}{KH} = \frac{K}{KH} = \frac{1}{H} \tag{15}$$

The heated cast is equivalent to an animal with metabolism so that, since E_b and E_r are zero for the polished gold cast, Eq. 13a gives for the cast

$$T_b^{\text{cast}} = T_e^{\text{cast}} + \frac{M^{\text{cast}}}{H} \tag{16}$$

where M^{cast} is just the power dissipated in the heater. Since T_e^{cast} is just the temperature of the unheated gold-plated cast (as discussed for the "T_e" thermometer

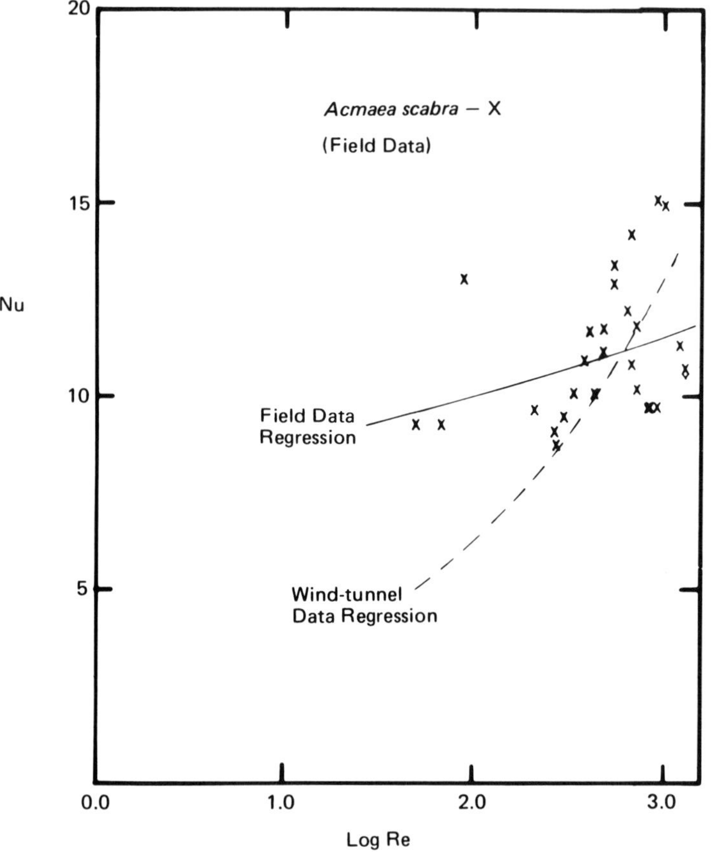

Fig. 16.4. Plot of the Reynolds number Re (a dimensionless wind velocity) versus the Nusselt number Nu (a dimensionless convection coefficient) for a gold-plated cast of the intertidal limpet *Acmaea scabra* (Gastropoda). Re is determined for the height of the shell and the wind velocity measured 2 cm above the substrate and extrapolated to the apex of the shell by using a logarithmic velocity profile determined by auxiliary measurements at 5 and 10 cm above the substrate. Note that, owing to turbulent, unstable air flow, the actual Nusselt number differs significantly from the value that would be calculated using wind-tunnel data and the Re value measured in the field.

application of the painted hollow-metal cast), all the parameters in Eq. 16 are known, and, to a first approximation,

$$H = \frac{T_{\text{hot cast}} - T_{\text{cold cast}}}{M^{\text{cast}}} \tag{17}$$

Accurate determinations will require corrections for the small errors (around 10 percent or less) that result from residual conduction and radiation. Because the corrections are small, they need not be known with great precision.

The heater in the heated gold-plated cast may be operated in either the constant-current (M_{heater} = constant) or constant-temperature ($T_b^{\text{cast}} = T_{\text{hot cast}}$ = constant) mode. The constant-current mode requires less equipment but gives a slow response, since an appreciable amount of time is required to dissipate the heat stored in the cast. Response time can be reduced by using a thin metal cast, as described in the Appendix. More rapid response is attainable with the constant-temperature mode, but a temperature-control unit must be used and the heater power suitably monitored.

To test the utility of this approach, we constructed gold-plated silver casts of the intertidal limpets *Acmaea scabra* and *Acmaea digitalis* (Gastropoda) with heaters and thermocouples for S. E. Johnson's study (Chapter 31, this volume). Figure 16.4 shows data that Johnson obtained in the field. The site was a protected cove on the Hopkins Marine Station grounds which had very turbulent and unpredictable wind flow with 90–180° fluctuations in wind direction and comparable velocity fluctuations. There is little correlation between the wind velocity at the apex of the shell and the convection coefficient (Reynolds number and Nusselt number do not correlate), although the data are not inconsistent with wind-tunnel results (dashed line). This is a feature of the environment and not the technique, since the Reynolds–Nusselt number plot in Fig. 16.5, using data that Johnson took in a wind tunnel with a cast replica of the rock surface for the wind-tunnel floor, shows good results. The instruments and methods used were otherwise identical for both sets of data.

Figures 16.4 and 16.5 clearly show that, had the convection coefficient been determined from wind-velocity measurements and the wind-tunnel Re–Nu plot (Fig. 16.5), the convection coefficient would frequently have been 40 or 50 percent in error from the actual field convection coefficient (Fig. 16.4). Although some additional effort is required to construct the casts, the equipment is as easy to use as a conventional heated thermometer anemometer.

The Model: Time-Dependent Solutions

Abrupt Change in Microclimate

We defined the parameters T_e, T_Δ, and I for both furred and naked animals with and without conduction to the ground in Eqs. 5–7 and 10–13. The application of Kirchhoff's law to any of the thermal circuits in Fig. 16.1 results in two or three

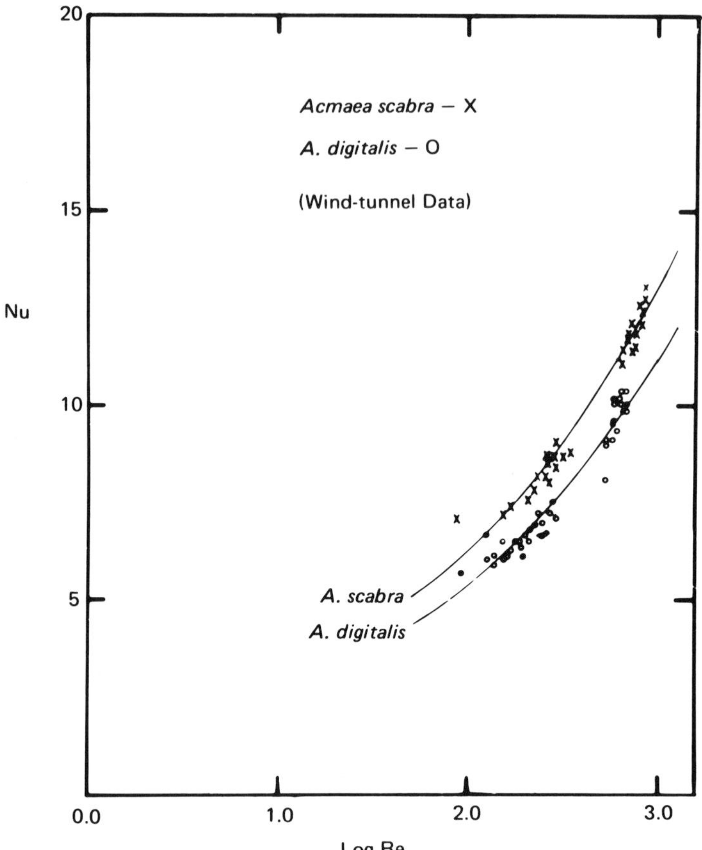

Fig. 16.5. Plot of Reynolds number–Nusselt number data taken from gold-plated casts of the shell of two species of intertidal limpets (Gastropoda). Note that Nu and Re correlate well under stable wind-tunnel conditions. A plaster replica of the natural rock substrate was used as the floor of the wind tunnel to duplicate the natural level of turbulence in the air flow at the surface. The Reynolds number was determined as in Fig. 16.4.

simultaneous equations equivalent to Eqs. 2a and 3 for circuit 1A. All but one of the equations are strictly algebraic, and thus the set may be reduced to a single ordinary differential equation. The proper definition of T_e, T_Δ, and I for each case represented by the various circuits in Fig. 16.1 allows the resulting differential equation to be written in the same form for all cases:

$$\frac{dT_b(t)}{dt} + \frac{T_b(t)}{IC} = \frac{T_e + T_\Delta}{IC} \tag{18}$$

Equation 18 may be solved by expressing T_e and T_Δ as time-dependent functions that describe the environment and the animal's physiological response.

The most useful solution is obtained for the case where an animal of body temperature $T_b = T_e + T_\Delta$ moves into a new microclimate ($T_e \rightarrow T_e'$) and changes its physiological state ($T_\Delta \to T'_\Delta$) and allows its body temperature to vary with time $[T_b \to T_b(t)]$ until a new equilibrium temperature $T'_b = T'_e + T'_\Delta$ is reached as $t \to \infty$. This corresponds to shuttling behavior in an ectotherm or a heating–cooling-curve experiment on an ectotherm. The situation is equivalent to a resistor–capacitor discharge in electronics and would be expected to show a simple exponential decay. This is indeed the case, as applying the boundary conditions $T_b(t = 0) = T_e + T_\Delta$ and $T_b(t \to \infty) = T'_e + T'_\Delta$ gives the solution

$$T_b(t) = (T'_e + T'_\Delta) + [(T_e + T_\Delta) - (T'_e + T'_\Delta)]e^{-t/\tau} \tag{19}$$

where the time constant $\tau = IC$. The time constant is the simplest (in a mathematical sense) description of the response of a system showing exponential behavior, and is numerically the time required for $(1 - 1/e) \simeq 63$ percent of the total change to occur.

Shuttling Between Two Microclimates

This time-dependent solution is applicable to shuttling behavior, where an animal has access to two microclimates with corresponding activity states. In the hotter microclimate, the equilibrium body temperature is denoted $T'^h_b = T'^h_e + T'^h_\Delta$, and in the colder microclimate, the equilibrium body temperature is denoted $T'^c_b = T'^c_e + T'^c_\Delta$. The animal has a range of preferred body temperatures with upper and lower set points for thermal responses. We may denote the upper set point by T^h_b and the lower set point by T^c_b, and assume that both values lie between T'^h_b and T'^c_b. As a first approximation to real behavior, we may assume that the behavior is purely thermoregulatory with minimum activity. Then the animal will remain in the colder microclimate until $T_b(t) = T^c_b$ and then move into the warmer microclimate. When $T_b(t) = T^h_b$, the animal will return to the cooler microclimate, completing the cycle. This behavior is diagrammed in Fig. 16.6.

This type of behavior may be approached analytically by using Eq. 19 to compute the body temperature as a function of time in each microclimate or the total time that can be spent in each microclimate as a function of the set points and the microclimate parameters. Equation 19 may be rewritten in the form

$$T_b(t) = T'^h_b + (T^c_b - T'^h_b)e^{-t/\tau_h} \tag{19a}$$

for the animal in the warmer microclimate. The elapsed time when $T_b(t) = T^h_b$ and the animal returns to the cooler microclimate is

$$t_h = -\tau_h \ln \frac{T^h_b - T'^h_b}{T^c_b - T'^h_b} \tag{20}$$

A similar calculation may be done to find the time spent in the cool microclimate, with the result

$$t_c = -\tau_c \ln \left(\frac{T^c_b - T'^c_b}{T^h_b - T'^c_b}\right) \tag{20a}$$

In practice the animal will not be concerned solely with thermoregulation, and other considerations will result in behavior that is less regular. Cabanac and

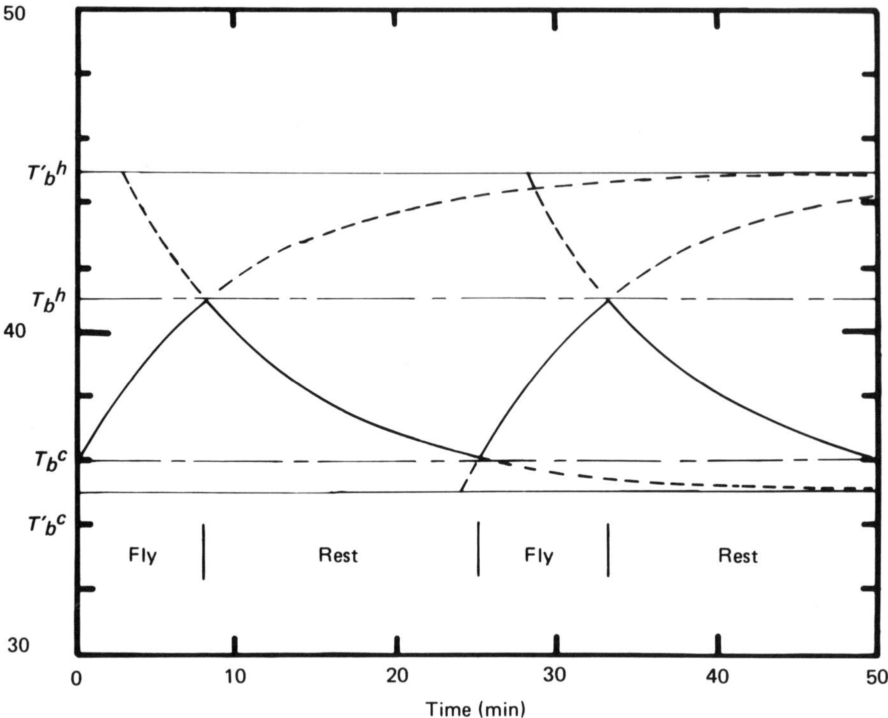

Fig. 16.6. Plot of body temperature T_b against time for idealized shuttling behavior. The upper and lower set points for body temperature are T_b^h and T_b^c. The equilibrium body temperature in the hot environment is $T_b'^h$ and, in the cold environment, $T_b'^c$. The legend indicates the activity state for a hummingbird shuttling in response to hot conditions as described in the text.

Hammel (1971, Fig. 1) and Hammel *et al.* (1967, Fig. 1) show laboratory shuttling behavior in the lizard *Tiliqua scincoides*.

The most obvious condition besides thermoregulation that would influence the behavior of a shuttling animal is the presence of a food resource in either a hot or cold microclimate. The animal might then be concerned with maximizing the time available in whichever microclimate contained the food resource. To study this case, a duty cycle $\mathscr{D}$ may be defined which gives the percentage of total active time that may be spent in the food-containing microclimate as a function of the microclimate and physiological parameters. For the case with the resource in the warm environment,

$$\mathscr{D}_h = \frac{\Delta t_h}{\Delta t_h + \Delta t_c} \tag{21}$$

and for the food resource in the cold environment,

$$\mathscr{D}_c = \frac{\Delta t_c}{\Delta t_h + \Delta t_c} \tag{21a}$$

Since the time spent in each part of the environment and the duty cycle depends on the choice of T_b^h and T_b^c, a wide range of strategies is possible. The duty cycle for the time spent in the hotter environment is a maximum for T_b^c just slightly less than T_b^h, but the duration of each excursion approaches zero. Since food-gathering efficiency often increases with the length of each excursion, optimizing strategies exist for maximum foraging per unit time under any given set of T_b^h, T_b^c, $T_b'^h$, and $T_b'^c$. These strategies may well be employed by lizards, turtles, and other animals, so this sort of analysis is of considerable utility in understanding the details of shuttling behavior and its effect on foraging.

The study of shuttling behavior by thermal modeling is not restricted to the study of ectotherms. Many endotherms show a labile body temperature and use it to facilitate excursions into conditions that cannot be tolerated for an extended period of time. For a large animal such as a camel, the "shuttling excursion" can even be all day (Schmidt-Nielsen, 1964).

In some cases a detailed study of the thermal energy budget, including behavioral aspects such as shuttling, is required to answer such ecological questions as that which relates to the nature of the limiting factor on an organism. One possible study is suggested by a study of Stiles (1971) on the time budget of Anna's hummingbird, *Calypte anna*. Stiles noted that the percentage of total time spent flying decreased as both the air temperature and the black-bulb temperature (which roughly indicates the amount of solar radiation present) increased. The percentage of total time spent flying, in fact, was almost the same on different days with essentially the same air temperature and radiation (black-bulb temperature) conditions, although the partitioning of activities during flight was very different. On days during the breeding season a large fraction of the flight time was spent in territorial defense; during the nonbreeding season most of the flight time was devoted to feeding (Stiles, 1971, Fig. 1).

The agreement of the total percentage of the time spent flying despite rather different biological behavioral requirements suggests that the hummingbird experienced excessive heat stress due to high air temperature and radiation combined with high metabolic heat production while flying. To remain active, the hummingbird apparently used a shuttling strategy, as diagrammed in Fig. 16.6.

It is possible to use data from the literature for several species to evaluate Eqs. 19–21, and the results (see Fig. 16.6) are also consistent with this interpretation. However, attempting to interpret data from several different species of hummingbird and the microclimate significance of the temperature of a test tube of unspecified size, "painted with flat black enamel, placed in full sun and out of the wind" (Stiles, 1971), is dubious at best.

Alternatively it may be argued that nighttime temperatures, and consequently thermoregulatory demands on metabolic reserves, are covariant with daytime temperatures, and that less flight time was needed to obtain food on hot days. The difference between feeding time during the breeding and nonbreeding season could be ascribed to the more abundant food resources available during the breeding season, and the very close agreement in total flying time could be fortuitous.

Simple regression-type analysis would probably confirm both hypotheses, and thus this study illustrates a case in which a complete thermal- and food-energy-budget analysis is required to understand the actual ecological factors influencing the behavior of an animal.

General field studies of the thermal aspects of shuttling, foraging, time budgets, and behavior are complicated by the wide range of T_e values in the environment noted earlier. It will be necessary to have a fairly complete map of the T_e values in the study area to understand this class of problems fully, and apply Eqs. 19–21 to separate thermoregulatory from other behavior. The use of casts as "T_e thermometers" is probably the only economical method of making such a map.

Metabolism and Cooling-Curve Experiments

A common experiment in the study of the physiological ecology of ectotherms is the heating–cooling-curve experiment. This experiment normally consists of equilibrating the animal at one temperature and then placing it in a hotter or colder environment and monitoring the body temperature until it reaches a new equilibrium. The curve is then evaluated using Newton's law of cooling, usually in the form

$$\frac{dT_b(t)}{dt} = -\frac{1}{IC}[T_b(t) - T_A] \tag{22}$$

where T_A is the "ambient" temperature, usually taken to be the air or wall temperature of the enclosure where the heating or cooling takes place. The usual parameter determined is the "cooling constant" or conductance $1/I$, expressed in a variety of units.

Several inaccuracies are associated with this procedure, as can be seen by writing Eq. 18 in the same form:

$$\frac{dT_b(t)}{dt} = -\frac{1}{IC}[T_b(t) - (T'_e + T'_\Delta)] \tag{22a}$$

First, as indicated by Eqs. 6 and 11, the insulation I or thermal conductance $1/I$ is not a constant of the animal but depends greatly on the variable environmental factors that determine H, R, and G'. The resulting errors are significant, as Tracy (1972) has discussed. In addition to his comments on convection (H) and radiation (R), I wish to point out that conduction to the substrate may often be the dominant mode of heat loss, so that any variation in the nature of the substrate may create serious errors. Spray and May (1972) found that the rate of heating and cooling of a turtle (*Chrysemys picta*) differed by 32 percent, depending on the clay content of the substrate. This large difference indicates that conduction to and through the substrate was a major avenue of heat transfer. However, the type, thickness, temperature, and thermal conductivity of the substrate is rarely specified in cooling-curve experiments.

Second, note that the endpoint temperature for the cooling or heating curve is not the air or surface temperature or even the operative environmental temperature, but the sum of the operative environmental temperature T'_e and the physiological offset temperature T'_Δ. If the data are plotted using T_a or T'_e as the endpoint T_A, any T'_Δ resulting from metabolic heat production or evaporative water loss will produce further error in the determination of I. This correction was noted for metabolic heat production by Bartholomew and Tucker (1963), but the effect of evaporative water loss was not noted. Their method of correcting I was to take the slope of the semilog plot from the usual method of plotting $T_b - T_a$ and apply the correction to the resulting value of I. This involves the extrapolation of a curved line and is subject to inaccuracy. The proper procedure is to determine the true endpoint $T_e + T_\Delta$ by letting the animal come to thermal equilibrium and taking the resulting equilibrium body temperature as the endpoint, rather than T_A or T'_e. The semilog plot will then always give a straight line unless T'_e or T'_Δ varies during the experiment.

Thus far in the analysis we have assumed that M, E_b, and E_r are constant, with the consequence that they affect the endpoint temperature and the rate of change of body temperature dT_b/dt but, with proper data analysis, do not affect I or τ. However, the metabolic and evaporative water-loss rates normally depend on body temperature, typically in an exponential fashion. Spotila *et al.* (1973) have noted that a body-temperature-dependent metabolic rate or evaporative water-loss rates will affect I and τ. Strunk (1971) reached the same conclusion, but confused the time constant τ with the rate of change dT_b/dt and made numerous serious calculational and interpretive errors which largely invalidate his work (Bakken and Gates, 1973).

The effect is qualitatively as follows. For M, E_b, E_r, and E_s constant, T'_Δ will be constant at a fixed value determined by the heat-transfer characteristics of the animal and the environmental conditions. However, when any of the M and E parameters depend on T_b, the endpoint body temperature, $T'_b = T'_e + T'_\Delta$, will change during the heating or cooling process, and approach $(dT'_\Delta/dT_b < 0)$ or recede from $(dT'_\Delta/dT_b > 0)$ the body temperature as it attempts to reach equilibrium.

For example, in a heating-curve experiment, the body temperature will increase at a rate appropriate to a simple constant endpoint, $T'_b = T'_e + T'_\Delta$, with M and E's having the values appropriate to the initial body temperature. However, for $0 < dT'_\Delta/dT_b < 1$, the endpoint temperature T'_b will steadily increase to a final equilibrium value T'^{eq}_b since T'_Δ increases with the increase of net metabolic heat production. Since the rate of change of body temperature initially will correspond to a low temperature difference, $T_b(t) - T'_b < T_b(t) - T'^{\text{eq}}_b$, it will take longer for the body temperature to come to equilibrium than if the large value of T_Δ corresponding to T'^{eq}_b were present initially. The result is that the animal will appear to have a time constant larger than IC when $\log\,[T_b(t) - T'^{\text{eq}}_b]$ is plotted against time.

The exact details of the behavior of T_b will depend on the rate of increase of T'_Δ with T_b, dT'_Δ/dT_b. The case just described assumes that $0 < dT'_\Delta/dT_b < 1$. If $dT'_\Delta/dT_b = 1$, the body temperature will increase at a constant rate until heat death results. If $dT'_\Delta/dT_b > 1$, T'_Δ increases faster than T_b, the body temperature will increase at an increasing rate, corresponding to explosive heat death. If $dT'_\Delta/dT_b < 0$, the body temperature will equilibrate more rapidly than for a positive or constant value of dT'_Δ/dT_b, and the time constant will appear smaller than IC.

The situation where the animal is cooling ($dT_b/dt < 0$) may be similarly analyzed, with the same result: $dT'_\Delta/dT_b > 0$ increases τ and $dT'_\Delta/dT_b < 0$ decreases τ.

In general, the dependence of T_Δ on T_b will be exponential. An analytic solution is possible for this case, but it is much simpler and more instructive to consider the case where the net metabolic heat production ($M - E_b$) is linearly dependent on T_b while E_s and E_r remain constants. For this case we take some body temperature $T_b = T_p$ as a linearization point and form the T_b dependence of ($M - E_b$) as follows:

$$M - E_b = (M - E_b)_p + P(T_b - T_p) \tag{23}$$

where $(M - E_b)_p$ is the value of $(M - E_b)$ at $T_b = T_p$ and $P = d(M - E_b)/dT_b$.

The term $P(T_b - T_p)$ has the same mathematical form as the conductive and convective heat-transfer terms in the equations that result from applying Kirchhoff's law to the circuits in Fig. 16.1a–c. Thus the variable metabolism term $P(T_b - T_p)$ may be regarded as placing a fictitious negative conductance $-P$ (a negative conductance is equivalent to an amplifier), connecting T_b with a heat source of temperature T_p, as diagrammed in Fig. 16.1d and e. A real conductance of this sort would be expected to change the insulation and thus the time constant, so the mathematically equivalent metabolic heat production dependent on body temperature would be expected to do the same.

For the most general case where the animal has pelage, Kirchhoff's law may be applied to the diagram in Fig. 16.1e to give three equations, which may be solved simultaneously as before. The solution may be arranged as in Eqs. 9 and 19, but the parameters, T_e, T_Δ, and I assume apparent values that differ from the body-temperature-independent-metabolism cases in Eqs. 11–13. These apparent values will be denoted by a superscripted a. The apparent insulation is

$$I^a = \frac{K_{sf} + H + R}{(K_{sf} + G' - P)(H + R) + K_{sf}(G' - P)} = \frac{I}{1 - IP} \tag{24}$$

where I is as given by Eq. 11. Further,

$$T_e^a = \frac{K_{sf}(HT_a + Q_n) + G'T_g(K_{sf} + H + R)}{(K_{sf} + G' - P)(H + R) + K_{sf}(G' - P)} = \frac{T_e}{1 - IP} \tag{25}$$

$$\begin{aligned} T_\Delta^a = {} & [(M - E_b)_p - PT_p]I^a - E_s \frac{K_{sf} + [K_s/(K_s + K_f)](H + R)}{(K_{sf} + G' - P)(H + R) + K_{sf}(G' - P)} \\ & - E_r \frac{K_{sf}}{(K_{sf} + G' - P)(H + R) + K_{sf}(G' - P)} \\ = {} & \frac{(T_\Delta)_p}{1 - IP} - PT_pI^a \end{aligned} \tag{26}$$

Here $(T_\Delta)_p$ is T_Δ evaluated at $T_b = T_p$.

From Eq. 24 we can clearly see that when ($M - E_b$) increases as T_b increases (i.e., $P > 0$), I^a will be increased, increasing $\tau^a = I^aC$, as was argued qualitatively. For $M - E_b$ decreasing as T_b increases, $P < 0$ and I^a and τ^a are reduced. Explosive body-temperature rise and heat death occur when I^a, and thus τ^a, become negative, so the exponent in Eq. 19 becomes positive. This occurs when

$$P \geqq \frac{(K_{sf} + G')(H + R) + K_{sf}G'}{K_{sf} + H + R} = \frac{1}{I} \tag{27}$$

For a nonfurred animal, K_{sf} in Eq. 10 reduces to the skin conductivity K_s. Since K_s is much greater than K_f or K_{sf}, a much larger positive P is required to produce explosive heat death than for a furred animal. Thus an animal insulated with pelage is much more sensitive to the effects of a metabolic rate dependent on body temperature than a naked animal, a fact that is of general importance to the evolution of endothermy as well as its simple application to explosive heat death.

The values of the parameters were earlier indicated to be apparent rather than actual values. The situation can be clarified by considering an animal with body-temperature-dependent net metabolic heat production in thermal equilibrium with $T_b = T_b^{\text{eq}}$. Then the net metabolic heat production is indistinguishable from that of an animal with a constant net metabolic heat production $(M - E_b)^{\text{eq}} = (M - E_b)_p + P(T_b^{\text{eq}} - T_p)$. Since the metabolic rate is constant, Eqs. 11–13 apply, with Eq. 13 having the form

$$T_\Delta^{\text{eq}} = [(M - E_b)_p + P(T_b^{\text{eq}} - T_p)]I - E_s \frac{K_{sf} + [K_s/(K_s + K_f)](H + R)}{(K_{sf} + G')(H + R) + K_{sf}G'}$$

$$- E_r \frac{K_{sf}}{(K_{sf} + G')(H + R) + K_{sf}G'} \tag{13b}$$

Then the equilibrium body temperature is

$$T_b^{\text{eq}} = T_e + T_\Delta^{\text{eq}} \tag{28a}$$

$$= T_b^a + T_\Delta^a \tag{28b}$$

where Eq. 28a uses the terms in Eqs. 11, 12, and 13, and Eq. 28b uses the terms in Eqs. 24–26. The two expressions for T_b^{eq} may be shown to be formally identical by solving Eq. 28a, which contains T_b^{eq} on both the right and left sides of the equal sign, for T_b^{eq}. The resulting expression is identical with Eq. 28b. Thus there is no conflict between the solutions in the steady-state case. However, the measured values of the parameters τ and I will differ depending on whether the measurement is made by a steady-state method (Eq. 13a) or cooling-curve analysis.

The problem presented by the apparent values of T_Δ^a, T_e^a, and I^a may be simplified somewhat and the effect of body-temperature-dependent net metabolic heat production clarified with the aid of the discussion above, which shows the identity of the body-temperature-dependent and constant $M - E_b$. Recall that the time-dependent solution for a sudden change of environment and activity level was equivalent to Eq. 19 and, in the present case,

$$T_b(t) = (T_e'^a + T_\Delta'^a) + [(T_e^a + T_\Delta^a) - (T_e'^a + T_\Delta'^a)]e^{-t/\tau^a} \tag{19b}$$

where $\tau^a = I^aC$. Since both the initial and the final conditions represent thermal equilibrium, net metabolic heat production is constant in both T_Δ and T'_Δ. Then the equivalent expressions of Eqs. 28a and 28b may be used to rewrite Eq. 19a as

$$T_b(t) = (T_e' + T_\Delta'^{\text{eq}}) + [(T_e + T_\Delta^{\text{eq}}) - (T_e' + T_\Delta'^{\text{eq}})]e^{-t/\tau^a}$$

or

$$T_b(t) = T_b'^{eq} + (T_b^{eq} - T_b'^{eq})e^{-t/\tau^a} \tag{19c}$$

The cooling curve of an animal with net metabolic heat production dependent on body temperature, when plotted with respect to the final equilibrium body temperature $T_b'^{eq}$ as suggested for the constant-metabolism case, gives just a straight line in the usual semilog plot. However, the slope of the line corresponds to the apparent time constant τ^a. The apparent insulation $I^a = I/1 - IP$ thus will differ from the values measured by a cooling-curve experiment on a dead, non-evaporating animal or determined by a steady-state measurement of metabolic rate versus metabolism chamber temperature on a live animal. The effect of body-temperature-dependent net metabolic heat production may be summarized as follows:

1. The equilibrium body temperature T_b^{eq} for a given set of conditions will be a function of the relation of the net metabolic heat-production temperature coefficient P to the various thermal conductivities for the various heat loss pathways, as given by Eq. 28b.
2. The metabolism coefficient P acts as a negative conductance or amplifier of body-temperature variations for $P > 0$. $P > 0$ implies that net metabolic heat production increases with a T_b increase.
3. The net effect of $P > 0$ is to decrease the denominator of Eqs. 24–26 and thus to increase the apparent insulation I^a, time constant τ^a, and the total amount of change in T'_Δ resulting from a given change in environmental conditions $T_e \rightarrow T'_e$. The combined increase in the total temperature change and in the time constant will increase the time required for the animal to come to a new equilibrium body temperature in response to a change in environmental conditions.
4. The conditions for explosive heat death, where T_b increases at an increasing rate, corresponds to the time constant becoming negative. This occurs when $P > 1/I$.
5. The effects of body-temperature-dependent metabolic heat production are augmented by small values of K_{sf} and G', that is, in a well-insulated animal. This requires a well-insulated animal such as a bird or mammal to maintain $M - E_b$ essentially constant by adjusting evaporative water loss to offset M. (In a more complete analysis it can be shown that cutaneous water loss E_s and E_r can be used as well as E_b, but the analysis is greatly complicated.)

Metabolism, Cooling Curves, and the Evolution of Thermoregulation

The question of the evolution of thermoregulation has been one of considerable interest, and has resulted in a great deal of speculation and argument. However, most of the thermoregulatory organs involve soft tissues that are rarely preserved in the fossil record. Consequently, the majority of hypotheses for the mode of evolution of thermoregulation by endothermic means have rested on analogy with highly evolved living forms. This procedure is fraught with considerable hazard, since, unlike the laws of physics, chemistry, and most processes of geology, uniformitarianism cannot always be assumed with respect to the physiology of animals. This is a consequence of the applicability of uniformitarianism to some of the fundamental aspects of biology, specifically genetics, selection, and the continued process of evolution.

Generally we wish to suggest that a more profitable approach might be based on principles where uniformitarianism may be assumed more safely. Thus we would base speculation about the evolution of endothermy on the physics of heat transfer, biochemical processes, and general physiological processes that need not be considered specific to one phyletic line. This necessarily involves model building and analysis, and the thermal-energy-budget theory such as we have outlined here is immediately applicable. The particular power of the model-building approach is that it need not be restricted to consideration of organisms with living prototypes, but allows consideration of hypothetical animals that possess morphological structures and combinations of properties not present in existing forms.

The proper execution of a study of this nature requires the careful analysis of the various processes of heat transfer and the physiology of thermoregulation as seen in living forms that possess features to be included in the model animal. It requires careful synthesis of the hypothetical animals to be investigated. Here we can sketch only the simplest sort of analysis of this type, but we think it gives some interesting insights into some pathways of evolution of endothermy, differing somewhat from present speculation.

The dynamics of cooling and the results of cooling-curve studies have often been related to the evolution of endothermy. Generally the concept has been that the initial steps toward homeothermy occurred to prolong the time that an animal could spend at its preferred body temperature for activity or digestion (Bartholomew and Tucker, 1963). The ability to control the rate of heat gain and loss would allow an animal with, for example, a high preferred body temperature to warm quickly and cool slowly, thereby increasing the time spent at the preferred body temperature. This is shown in Eqs. 19–21, where a large τ_c and a small τ_h would give a longer duty cycle in the cold environment. As noted in the preceding section, body-temperature-dependent net metabolic heat production results in effective time constants τ^a that differ significantly from the constant metabolism value $\tau = IC$. This effect should be included in any discussion of this path for the evolution of endothermy.

The usual hypothesis (Bartholomew and Tucker, 1963; Dawson and Hudson, 1970) is that a slow elevation of the basal metabolic rate resulted in an increased capacity to control the rate of change of body temperature. This was amplified by vasoconstrictive control of I and hence τ. Further evidence cited for this mode of evolution of endothermy is provided by the ontogeny of endothermy in altricial animals (principally rodents and birds), where a thermoregulatory capability is shown in metabolism chamber experiments before effective insulation develops in the form of pelage (Dawson and Hudson, 1970; Dawson and Evans, 1960). Finally, it has been suggested (Gunn, 1942; Cowles and Bogert, 1944; Cowles, 1958; Bartholomew and Tucker, 1963) that insulation in the absence of significant metabolic heat production would be maladaptive since it would slow the rate of warming of animals with a high preferred body temperature and, especially in a cold seasonal climate, prevent the attainment of adequately high body temperature.

Although widely accepted, these arguments are by no means unassailable, even on biological grounds. First, it is not clear that the basal metabolic rate is the primary factor in homeothermy. The response of birds and mammals to cold stress includes a large component of muscular thermogenesis by shivering. The most obvious difference between birds and mammals and lower vertebrates is the presence of nonshivering thermogenesis; nevertheless, shivering thermogenesis certainly is the logical candidate as the primitive mechanism since it is available more widely and has been observed in reptiles, notably breeding pythons (Vinegar *et al.*, 1970). By far the most spectacular example of endothermy in the absence of a high basal metabolic rate is the intermittent endothermy of the flight muscles in several orders of insects, notably the sphingid moths (Heinrich, 1971).

The ontogenetic arguments based on the development of altricial birds and animals are also misleading since these animals, at the time of the onset of thermoregulatory behavior, are normally well insulated by the nesting material provided by the parents, and typically several young form a thermal unit in the nest with a higher aggregate heat production than a single individual. Endothermy is a much more economical proposition under these conditions than when the young animals are fully exposed outside the nest in a metabolism chamber. The thermoregulatory behavior seen under metabolism conditions is probably a laboratory artifact representing an emergency reaction to unusual conditions that would very likely be ruinously expensive in metabolic cost as a routine strategy.

The argument for the maladaptive value of fur for reptiles is generally based on heliothermic desert lizards, who typically maintain a high active body temperature by behavioral means (Cowles and Bogert, 1944; Heath, 1964, 1965; Brattstrom, 1965). Operative temperatures in the preferred range of body temperatures for basking lizards, 35–42°C, exist only during the daylight hours when Q_a is large enough to produce an operative environmental temperature substantially greater than air temperature (see Eqs. 7 and 12). A short time constant is necessary to allow T_b to reach the preferred value while such operative temperatures are available. However, an animal with a lower preferred body temperature, in the 20–30°C range, would find exposure to the operative temperatures present in full sun highly stressful.

Many groups of animals with primitive characteristics show such a low preferred body temperature. Brattstrom (1965) lists the mean active body temperature of various reptile groups. Turtles have T_b around 28.4°C, but this depends on the group and can be considerably lower. Snakes have a mean T_b of 25.6°C and lizards as a whole, including nondesert groups, have a mean T_b of only 29.1°C, with the Gekkonidae, Anguidae, and Xantusidae all below 25°C. The rhynchocephalian *Sphenodon punctatum* has a very low active body temperature, somewhat below 20°C. Finally, the monotremes, which are the most primitive mammalian group, have a very low upper temperature tolerance, with thermoregulation failure at an air temperature of approximately 30°C. In particular, the platypus, *Ornithorhynchus anatinus*, suffers heat stress for a metabolism chamber temperature (roughly equal to T_e) above 25.0°C (Smyth, 1973). Most of these groups have ecological specializations that may disqualify them from being obviously better models of therapsids than heliothermic lizards. However, when considered with the low preferred body temperatures of most amphibians, it seems as reasonable to postulate a low (20–30°C) body temperature for therapsids as the high (35–40°C) value derived from heliothermic lizards.

With this assumption, it is more reasonable to suppose, as did Cowles (1940, 1946), that the evolution of insulating pelage preceded endothermic thermoregulation as a defense against heat stress induced by solar radiation. As a result of its large time constant, a well-insulated animal with limited net metabolic heat production would assume a mean body temperature near the daily average operative environmental temperature (Spotila *et al.*, 1973), thus escaping the stress of large daily body-temperature fluctuations. This is undoubtedly an adaptive benefit, since the selection of a constant body temperature is far more widespread than the selection of a high body temperature. This adaptation would be of particular value to a large, slow-moving animal living in an exposed habitat with little shade. In this case good body insulation giving a very long time constant would protect the animal from overheating while foraging in the direct sunlight all day.

A contemporary model would be the land tortoise *Gopherus* and the box turtle *Terrapene*. These animals often forage exposed to the sun but are rather slow moving and cannot escape to shade quickly by flight or rapid digging. Spray and May (1972) have shown that *Gopherus polyphemus* and *Terrapene carolina* use the reverse strategy in controlling their heating and cooling rates from that described earlier for lizards. Because of the problem associated with heat stress, they adjust their circulation to minimize the heating rate and maximize the cooling rate, so that they may maximize the time available for foraging in the sun. The capacity for altering the rate of heat transfer is quite as remarkable as that reported for the Galapagos marine iguana, *Amblyrhynchus cristatus* (Bartholomew and Lasiewski, 1965). *T. carolina* cooled twice as fast as it heated, whereas *A. cristatus* heated twice as fast as it cooled. *G. polyphemus* cooled about $1\frac{1}{2}$ times as fast as it heated. The insulation provided by the shell of the turtles is a significant factor in rejecting solar radiation, since it may be heated above the temperature that would injure living tissue. Not only does it add insulation to that possible by vasoconstriction, but it also allows vasoconstriction to reduce K to values that would result in surface temperatures exceeding the injury value if the tissue did not have the extra protection.

Summing up the preceding argument, we are suggesting that a proto-mammal might have evolved some form of hair as a protection against solar radiation, as suggested by Cowles (1940, 1946). The insulation evolved as a solar radiation shield would then be a preadaptation which allows a relatively minor increase in metabolic heat production to result in an adaptively significant T_{Δ}. When the animal was subjected to cold stress as well as heat stress, which would occur in a seasonal midlatitude or "temperate" climate, selection for increased metabolic heat production could act to produce a recognizably mammalian form. The increased metabolic heat production would most probably have been muscular activity since an elevated basal rate would result in heat stress under warm conditions.

One possible objection is that, under warm conditions, the metabolic heat generated by normal locomotor activity of an insulated animal would also elevate T_b, producing heat stress. If an initially reptilian metabolism with a low overall rate and dependence on anaerobic metabolism for bursts of activity is assumed, the average rate of heat production will be low enough that evaporative cooling by panting or saliva spreading could reduce the net heat production to an essentially zero or even negative (net evaporative cooling) value. These mechanisms are well developed in many reptiles, including crocodilians, lizards, and turtles, for dealing with an external heat load. Evaporative cooling mechanisms would presumably be a necessary preadaptation to allow the animal to enter hot, high-radiation environments that would provide subsequent selection for insulation. As an example, the wool insulation in sheep, which use panting as the primary mechanism for evaporative cooling, decreases the heat load significantly in the presence of strong solar radiation (Yeats, 1967).

As we suggested earlier, the effect of body temperature on metabolic rate must be considered in connection with the evolution of endothermy. This can be illustrated by the following hypothetical animal and thermoregulatory behavior.

The animal is assumed to have evolved some form of insulation in response to strong solar radiation, as suggested by Cowles (1940, 1946). The animal is also assumed to have an essentially reptilian metabolism with a low capacity for aerobic metabolism and a consequent dependence on anaerobic metabolism for activity. The endogenous heat production that results from activity would be small and could be easily dissipated by panting and saliva spreading. During cooler conditions, the animal would not be subject to heat stress, and less evaporative cooling would be required. This would be the case during cool seasonal conditions for a large animal or cool daily conditions for a small animal.

An animal of this sort would be able to gain a selective advantage by the following strategy, which would represent an early form of endothermy. If this animal, adapted to strong solar radiation, encountered cooler conditions, the insulation would retard the rate of warming of the animal so that it would not become active until late in the day. Foraging and prey pursuit would result in the production of metabolic heat. Since the animal is in an assumed condition where only the lower part of the active body temperature range will have been obtained, it will probably not use increased evaporative cooling to offset metabolic heat production, and the body temperature will tend to remain constant or rise. The animal will be assumed to cease foraging after the end of the daylight hours as T_b falls. It will then be faced with the problem of maintaining adequate body temperature for digestion. The contemporary gecko *Gehyra variegata* models aspects of this behavior (Bustard, 1967).

This problem has been described by Cowles and Bogert (1944) for a reptile attempting to adopt nocturnal habits to escape seasonal peaks in heat stress during the day. The body temperature of a small, uninsulated animal would fall rapidly at night, and digestion would probably cease as in modern reptiles. The result could be the salmonella-type food poisoning commonly seen in reptiles maintained at excessively cool temperatures, resulting from the continued activity of microorganisms in the digestive tract. The situation would be somewhat better for the hypothetical insulated reptile. First, the larger value of the insulation I would give a longer time constant $\tau = IC$, which would increase the time that the body temperature would be adequate for digestion. Further, since the animal would not be using evaporative cooling $(M - E_b)$ and P would have positive values. The metabolic rate and heat production would be well above the standard rate, owing to digestion and assimilation. It is thus quite possible for the temperature coefficient of metabolism, P, to be significant relative to the thermal conductance of the animal, $1/I$, if the animal is somewhat insulated, as postulated. This would result in a further increase of the time constant $\tau = IC$ to $\tau^a = IC/(1 - IP)$ as discussed earlier, particularly in Eqs. 19b and 24. This further increases the time available for digestion. Because of the amplifying effect on P and I, the value of fur-like insulation on the assumed proto-endotherm is thus greater than is evident when only the effect of insulation on the simple time constant IC and the temperature difference T_Δ that can be maintained in steady state are considered. Physical activity during the digestive period would also tend to maintain body temperature at a level adequate for digestion.

Other scenarios of this sort can also be used to illustrate the amplifying effect of the temperature coefficient on insulation, which could lead to selection for increased insulation, increased aerobic metabolism, and ultimately endothermy as we know it. The scenario described is particularly attractive since it illustrates how selective forces resulting from falling seasonal temperatures would tend toward the nocturnality that has frequently been postulated for therapsid proto-mammals and early mammals (Jerison, 1971). The capacity for nocturnal foraging would open new niches closed to animals with a relatively high body-temperature requirement and without a usable heat-production and storage capacity.

A key feature of this scenario for the evolution of endothermy is the implied selection pressure for increased aerobic metabolism. If the hypothetical animal is assumed to be a predator, it is probable that the pursuit and capture of the prey would utilize anaerobic metabolism and result in an oxygen debt. The repayment of the debt would be complicated by the aerobic metabolism requirements of the elevated body temperature and digestion. This would result in selection for increased aerobic capacity.

Selection for increased aerobic capacity would be even greater in a sequel to the preceding scenario. The animal can be assumed to have improved his insulation and aerobic capacity somewhat in response to the selective pressures outlined above. The hypothetical animal could then forage nocturnally by using its insulation to retain the heat generated by locomotor activity during foraging, or by strictly heat-producing muscular contractions similar to the proto-shivering of breeding pythons. This would closely parallel the strategy used by sphinx moths to maintain muscular coordination during nocturnal foraging.

In addition to its thermoregulatory significance, an increased aerobic capacity would allow more activity which would be useful in foraging and escape from predators. This could be independently subject to positive selection and contribute to the evolution of increased aerobic capacity, or act as a preadaptation for the evolutionary processes outlined above.

An increase in basal metabolic rate to the level associated with contemporary mammals would result if it is assumed that the scope of aerobic metabolism is limited so that an increase in active metabolism results in an increase in basal aerobic metabolism. The increase in active aerobic metabolic capacity would thus be the key to endothermy and the elevated basal metabolic rate seen in mammals and birds. W. R. Dawson (private communication) has also suggested that an increase in active aerobic metabolism led to endothermy, and this mechanism is implied in the work of Heath (1968), Bakker (1971), and McMahon (1973).

This work has been supported in part by The Ford Foundation and the Atomic Energy Commission Grant AEC–AT(11–1)–2164.

Appendix: Construction of Hollow Metal Casts of Small Reptiles and Amphibians

The process for constructing hollow metal casts of small reptiles and amphibians is one of the variants of a process known as electroforming, a process most commonly used in preparing master molds for phonograph records, typefaces, and masters for coins and medallions. Basically, the process consists of preparing a master, usually negative, mold of conductive material or a material that may be treated to make the surface conductive. The surface is then heavily plated with copper or other suitable metal to the desired thickness and detached from the mold. Owing to the complex shape of a small animal and the difficulties inherent in plating strongly concave surfaces and assembling the resulting highly fragile separate halves of the cast, an alternative procedure commonly used in art work is preferable. The procedure is to make a positive cast of the object to be electroformed in a low-melting-point substance such as wax or a low-melting-point alloy such as Wood's alloy (melting point 70–80°C) and plating the surface to the desired thickness. This process sacrifices the fine detail possible in the negative mold process. The loss is not great because the coating may be much thinner than the desired thickness of the final cast if Wood's alloy is used, since a film of the alloy adheres to the outer electroplated layer and reinforces it after the bulk of the core has been melted out and the cast cooled. For a small (10–50 g) lizard, an electroplated copper layer of 0.04 mm is adequately rigid to withstand the process of melting out the core. The final cast is about 0.5 mm thick and thus easily handled as long as it is cooler than the melting point of the Wood's alloy. An additional advantage of Wood's alloy is that it may be cast in algin-base dental plaster, and thus live animals may be used as models. We shall now summarize the procedure.

Step 1

Prepare a molding box slightly larger than the animal. The sides should slope 10–20° from bottom to top to facilitate removal of the mold. The top and bottom should have three or four holes about 1–1.5 cm in diameter and be removable, with hooks on the sides of the box for heavy rubber bands to secure them in place when required.

The molding may be done with any of the quick-setting alginate impression materials used in dentistry. We have obtained good results with Jeltrate type I fast-set material (L. D. Caulk Co., Division of Dentsply International, Inc., Milford, Delaware 19963). The usual formulation used in dentistry does not flow easily enough; we find that a ratio of 100 ml of water to 60 ml by volume of powder, mixed for 5 s in a blender, gives better properties in this application. A blender of average size can prepare batches containing 150–300 ml of water. Larger or smaller batches do not mix well and require the simultaneous use of several blenders. Rinsing immediately after mixing a batch with cold water by blending for 20 s with the container half full greatly simplifies the following steps, since three mixes are required in quick succession.

To facilitate handling, chill or otherwise anesthetize the animal to be used as a model. While the animal is cooling, prepare a substrate layer in the molding box about 2 cm thick using the molding material. After the animal is ready and the substrate layer has set, prepare enough molding material to pour a second layer that will cover the legs, tail, and part of the back of the animal to a point just anterior to the hind legs. Proper immersion may require that the mold be tilted as shown in Fig. 16.7. Pour the molding material before immersing the animal to ensure that the ventral region will be reproduced. Immerse the animal in the molding material and force the feet and tail to the substrate layer, and allow the molding compound to set. Placing four fingertips in the material to a depth of 5–10 mm will provide "registration pins," which greatly facilitate reassembly of the mold for casting. Then the third layer is mixed and poured and the lid forced in place, with excess molding material escaping through the holes provided. The layers and position of the animal are diagrammed in Fig. 16.7. After layer 3 has

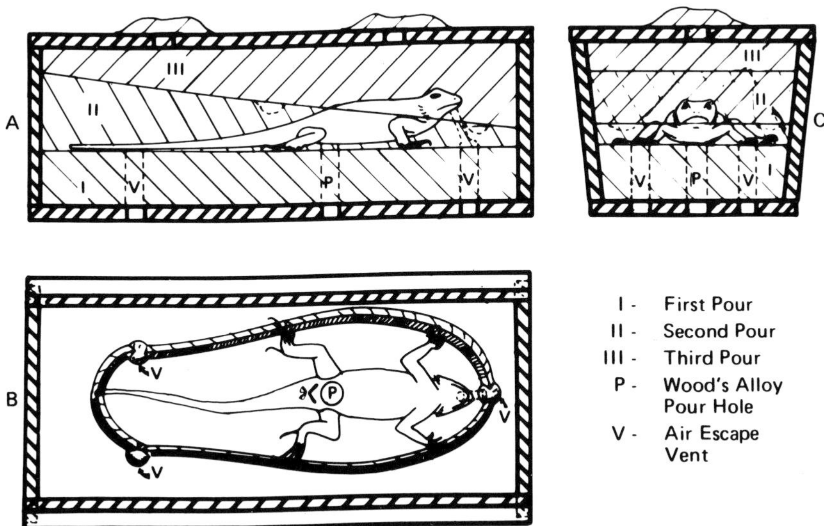

Fig. 16.7. Diagram of the molding box and arrangement of the successive layers of alginate dental investment compound for preparing a mold from a live lizard.

set, remove the lids and mold, and lift layer 3 from the posterior end until the back is exposed, and then press forward to slip the mold off the head and neck. The animal may now resume breathing. The elapsed time since immersion will vary with the speed of the investigator and the time required for the molding compound to set. The setting time is strongly dependent on the concentration of the mix and the water temperature, with quicker setting at higher concentrations of powder and higher water temperatures. Typical times will run from 2 to 6–7 min.

Separate layers 1 and 2 and locate the toes and tail on the ventral side of layer 2. Remove enough molding material from the tips of the toes to allow air to escape when the Wood's alloy is poured. Slit the mold along the ventral side of the tail and free the tail to prevent autotomy. Flex the mold and use forceps to remove the legs. The animal will usually have rewarmed by this time and will be most helpful in assisting its removal.

Step 2

Using a cork borer, make a hole through the first two layers into the ventral region of the mold to pour the alloy. Make at least two additional holes through layer 1 for vents and cut grooves in layer 2 to connect them with the toes, tip of the tail, and nose to allow air to escape during the pour as shown in Fig. 16.7. Heat the mold to 50–60°C. The best way is to boil the mold in water. Pat dry with towels and reassemble the mold in the box. A layer or two of wet paper towels under the cover will help compress the joints in the mold and minimize the amount of flash on the final cast. Melt the Wood's alloy, heat to 130–150°C, and pour slowly into the ventral opening. Immerse the mold in cold water to cool the cast to below 40°C (this usually takes about $\frac{1}{2}$ h) and remove. Trim the flash (extra metal forced into the joints in the mold and in the pour and vent holes). If the legs and tail are incomplete, the vents must be enlarged, new vents added, and/or the mold heated more before the pour. Excessive spattering during the pour or bubbles and an irregular crystalline surface on parts of the cast indicate that the alloy was too hot when poured or that it cooled too slowly. Reconstruct missing toes (it is very difficult to reproduce toes in small animals using this technique) by inserting copper wire heated with a soldering iron and coated with nonacid flux into suitably placed holes drilled in the feet and shaping to the proper length and curvature. *Caution!* Wood's alloy contains toxic heavy metals, including bismuth and cadmium, and should be melted under a hood. Gloves should be worn during trimming.

Step 3

Prepare a copper plating solution in advance. The following formulation gives good results:

Distilled water	1 liter
Copper sulfate	
($CuSo_4 \cdot 5H_2O$)	250 g
Sulfuric acid (conc.)	75 g
Thiourea	0.005 g

Filter the prepared solution through activated carbon after preparation and before each use to remove organic impurities that cause brittle crumbly deposits. The electroplating apparatus consists of a 0 to 8-V 15-A dc power supply and a magnetic-bar stirring apparatus. Concentrations and materials are not critical. See an electroplating text such as W. Blum and G. Hogaboom (*Principles of Electroplating and Electroforming*, 3rd ed. New York: McGraw-Hill Book Company, 1949) for more details.

All traces of grease, oxide, and other nonconductive coatings must be removed from the cast by cleaning with a strong degreasing solvent such as trichloroethylene and pickling in a dilute sulfuric or hydrochloric acid solution. Just before plating, the cast should be electrocleaned in a weak acid electrolyte such as dilute sulfuric acid by connecting the cast to the negative terminal of the power supply and a stainless-steel electrode to the positive terminal. Placing the electrode and cast in the solution with about 5 to 6 V applied results in the liberation of hydrogen at the cast, which reduces any remaining oxide coating. After all trace of oxide has been removed, rinse the cast under running water and distilled water. Store in distilled water while preparing the copper plating apparatus.

Select a container large enough to immerse the cast horizontally, and cut a piece of copper sheeting with an area of twice the area of the cast to serve as the anode. The copper anode should be carefully degreased to prevent organic contamination of the plating solution and enclosed in filter paper or thoroughly washed muslin to retain the assorted bits of crud that evolve during the plating operation. Fill the beaker with plating solution, immerse the cast in the center of the container by hanging by a 20–24 AWG copper wire, and connect the cast to the negative terminal and the anode to the positive terminal of the power supply. Apply 4 to 7 V with continuous stirring until the desired thickness of plating is achieved. The exact time for a given thickness is a function of voltage and temperature. For a small lizard, 4–5 V at 30°C for $2\frac{1}{2}$ h gives good results. Lumpy deposits on the nose and feet indicate that the anode is too large or the voltage too high.

Step 4

Rinse the plated cast thoroughly and neutralize any remaining acid with soap or sodium bicarbonate. Drill or grind small openings in the tip of the tail, under the nose, and in the tip of each leg. Select a beaker large enough to allow the cast to be immersed vertically, fill with water, and bring to a boil. Suspend the cast by a piece of wire and immerse vertically in the water. As the cast heats, the Wood's alloy will melt and run out the lowest opening. Gentle vertical shaking will assist in removing the alloy, but the cast is very fragile at this point and care should be used. After the cast has cooled, the film of Wood's alloy adhering inside the copper plate will reinforce the hollow cast so that it may be handled with relative impunity. If a pure copper cast is desired, the Wood's alloy may be rubbed with powdered graphite before plating.

Step 5

The cast may now have thermocouples attached internally, the holes for draining the Wood's alloy closed, and any missing toes reconstructed by drilling small holes in the feet and inserting copper wire and fixing with Wood's alloy solder or epoxy glue. The cast may now be painted to match the solar absorptivity of the model animal and is ready for use.

In many cases it will be desirable to know the specific heat capacity of the cast to allow the measurement of the overall heat-transfer coefficient of the cast by a cooling-curve determination. The specific heat of the components differ significantly, with Cu = 0.093, Wood's alloy = 0.0352, and paint = 0.2 to 0.4 (data from C. D. Hodgman, 1962. (ed.), *Handbook of Chemistry and Physics*, 44th ed., pp. 2352, 2378, 2379, 1962, Cleaveland: Chemical Rubber Publishing Co.) so the weight of each component must be known. In this case the cast must be weighed before and after electroforming, after melting out the core, after installing the thermocouples, and after painting.

References

Bakken, G. S., Gates, D. M.: 1974. Notes on "Heat loss from a Newtonian animal." J. theoret. Biol. *45*, 283–292.

Bakker, R. T.: 1971. Dinosaur physiology and the origin of mammals. Evolution *25*, 636–658.

Bartholomew, G. A., Lasiewski, R. C.: 1965. Heating and cooling rates, heart rate, and simulated diving in the Galapagos marine iguana. Comp. Biochem. Physiol. *16*, 573–582.

———, Tucker, V. A.: 1963. Control of changes in body temperature, metabolism, and circulation by the Agamid lizard *Amphibolurus barbatus*. Physiol. Zoöl. *36*, 199–218.

Birkebak, R. C.: 1966. Heat transfer in biological systems. Intern. Rev. Gen. and Expt. Zool. *2*, 269–344.

Brattstrom, B. H.: 1965. Body temperatures of reptiles. Am. Midland Naturalist *73*, 376–422.

Bustard, H. R.: 1967. Activity cycle and thermoregulation in the Australian gecko *Gehyra variegata*. Copeia *1967*, 753–758.

Cabanac, H. P., Hammel, H. T.: 1971. Peripheral sensitivity and temperature regulation in *Tiliqua scincoides*. Intern. J. Bioclimatol. Biometeor. *15*, 239–243.

Colbert, E. H., Cowles, R. B., Bogert, C. M.: 1946. Temperature tolerances in the American alligator and their bearing on the habits, evolution and extinction of the dinosaurs. Bull. Am. Museum Nat. Hist. *86*(7), 329–373.

Cowles, R. B.: 1940. Additional implications of reptilian sensitivity to high temperature. Am. Naturalist *74*, 542–561.
———: 1946. Fur and feathers: a response to falling temperature? Science *103*, 74–75.
———: 1958. Possible origin of dermal temperature regulation. Evolution *12*, 347–357.
———, Bogert, C. M.: 1944. A preliminary study of the thermal requirements of desert reptiles. Bull. Am. Museum Nat. Hist *83*(5), 261–296.
Davis, L. B., Jr., Birkebak, R. C.: 1973. On the transfer of energy in layers of fur. Biophys. J. *14*, 249–268.
Dawson, W. R., Evans, F. C.: 1960. Relation of growth and development to temperature regulation in nestling vesper sparrows. Condor *62*, 329–340.
———, Hudson, J. W.: 1970. Birds. *In* Comparative physiology of thermoregulation (ed. G. C. Whittow), vol. 1, pp. 223–310. New York: Academic Press.
Gunn, D. L.: 1942. Body temperature in poikilothermic animals. Biol. Rev. *17*, 293–314.
Hammel, T. T., Caldwell, F. T., Abrams, R. M.: 1967. Regulation of body temperature in the blue-tongued lizard. Science *156*, 1260–1262.
Hardy, J. D., Gagge, A. P., Stolwijk, J. A. J. (eds.): 1970. Physiological and behavioral temperature regulation. Springfield, Ill.: Charles C Thomas.
Heath, J. E.: 1964. Head-body temperature differences in horned lizards. Physiol. Zoöl. *37*, 273–279.
———: 1965. Temperature regulation and diurnal activity in horned lizards. Univ. Calif. Berkeley Publ. Zool. *64*(3), 97–136.
———: 1968. The origins of thermoregulation. *In* Evolution and environment (ed. E. T. Drake), pp. 259–278. New Haven: Yale Univ. Press.
Heinrich, B.: 1971. Temperature regulation of the sphinx moth, *Manduca sexta*, Part I. J. Expt. Biol. *54*, 141–152.
Herrington, L. P., Winslow, C. E. A., Gagge, A. P.: 1937. The relative influence of radiation and convection upon vasomotor temperature regulation. Am. J. Physiol. *120*, 133–143.
Jerison, H. J.: 1971. More on why birds and mammals have big brains. Am. Naturalist *105*, 185–189.
Kerslake, D. M.: 1972. The stress of hot environments. New York: Cambridge Univ. Press.
Kreith, F.: 1965. Principles of heat transfer. Scranton, Pa.: International Textbook.
McMahon, T.: 1973. Size and shape in biology. Science *179*, 1201–1204.
Pearman, G. I., Weaver, H. L., Tanner, C. B.: 1972. Boundary layer heat transfer coefficients under field conditions. Agric. Meteor. *10*, 83–92.
Porter, W. P., Gates, D. M.: 1969. Thermodynamic equilibria of animals with environment. Ecol. Monogr. *39*, 227–244.
———, Mitchell, J. W., Beckman, W. A., DeWitt, C. B.: 1973. Behavioral implications of mechanistic ecology. Oecologia *13*, 1–54.
Schmidt-Nielsen, K.: 1964. Desert animals: physiological problems of heat and water. London: Oxford.
Scholander, P. F., Walters, V., Hock, R., Irving, L.: 1950. Body insulation of some arctic and tropical mammals and birds. Biol. Bull. *99*, 225–236.
Smyth, D. M.: 1973. Temperature regulation in the platypus, *Ornithorhynchus anatinus* (Shaw). Comp. Biochem. Physiol. *45A*, 705–715.
Spotila, J. R., Lommen, P. W., Bakken, G. S., Gates, D. M.: 1973. A mathematical model for body temperatures of large reptiles: implications for dinosaur ecology. Am. Naturalist *107*, 391–404.
Spray, D. C., May, M. L.: 1972. Heating and cooling rates in four species of turtles Comp. Biochem. Physiol. *41A*, 507–522.
Stiles, F. G.: 1971. Time, energy, and territoriality of the Anna hummingbird (*Calypte anna*). Science *173*, 818–821.
Strunk, T. H.: 1971. Heat loss from a Newtonian animal. J. theoret. Biol. *33*, 35–61.

Tracy, C. R.: 1972. Newton's law: its application for expressing heat losses from homeotherms. BioScience *22*, 656–659.

Vernon, H. M.: 1932. The measurement of radiant heat in relation to human comfort. J. Ind. Hyg. Toxicol. *14*, 95–111.

Vinegar, A., Hutchison, V. H., Dowling, H. G.: 1970. Metabolism, energetics, and thermoregulation during breeding of snakes of the genus *Python* (Reptilia, Boidae). Zoologica *55*, 19–48.

Yeats, N. T. M.: 1967. The heat tolerance of sheep and cattle in relation to fleece or coat character. *In* Biometeorology (ed. S. W. Tromp, W. H. Weihe), Proc. Third Intern. Biometeor. Congr., pp. 464–470. Oxford: Pergamon Press.

17

Body Size, Insulation, and Optimum Body Temperatures of Homeotherms

JAMES R. SPOTILA AND DAVID M. GATES

Introduction

The purpose of this study is to determine quantitatively the role of physical and physiological properties of homeotherms in the adaptation of these animals to their physical environment. A theoretical model is developed to describe the interaction of an animal's body size and insulation with the physical characteristics of its environment and the role of all these factors in determining the metabolic rate, body temperature, and evaporative water-loss rate of the animal.

Exchange of heat energy between an animal and its environment determines the energy level at which the animal can function. A poikilotherm's internal heat production is low and the animal lacks effective insulation, so its deep body temperature follows closely the effective temperature of its microclimate. A homeotherm generates relatively large quantities of heat and has high-quality insulation, so its body temperature remains relatively independent of environmental heat load for a wide range of conditions. Maintenance of a constant, relatively high body temperature is advantageous in integrating physiological and biochemical systems with different thermal coefficients (Carey and Lawson, 1973) and may give homeotherms numerous competitive and survival advantages, which enable them to exploit niches unavailable to poikilotherms (Spotila *et al.*, 1972).

The body temperature, metabolic rate, rate of evaporative water loss, body size, and insulative properties of homeotherms appear to be so interrelated as to suggest a causal relationship. This is evident in geographic size variation in birds (James, 1970) and woodrats (Brown and Lee, 1969) and in the effects of body size and shape on the energetic efficiency of weasels (Brown and Lasiewski, 1972). McNab (1971) questions the significance of animal–energy–environment interactions and the role of body size in the heat-energy transfer and climatic adaptation of homeotherms. Ricklefs (1973) comments on surface–volume ratios and size variation in cold-blooded vertebrates, concluding that predator–prey strategies have a greater effect on geographic variation in body size than do temperature relationships. These conflicting conclusions emphasize an important point. Although significant progress has been made (Porter and Gates, 1969; Heller and Gates, 1971), there is still no theoretical analysis that can serve as a framework for available physiological and physical data related to metabolism, size, and climatic requirements of animals.

Many authors have attempted to explain the details of mammalian and avian adaptations by generating analyses of the energetics of homeothermism, based almost entirely on empirical data. Tabulation of detailed physiological data has led to the development of many allometric equations (McNab, 1970; Brody and Proctor, 1932; Calder, 1968; Kleiber, 1932, 1961; Herreid and Kessel, 1967; Schmidt-Nielsen, 1972; Yarbrough, 1971; Kendeigh, 1969; Lasiewski and Dawson, 1967), but has not provided a clear insight into the interaction of animal physical properties, the physical environment, and the thermoregulatory processes of homeotherms. Empirical formulas to describe metabolic rate as a function of size usually take the form of $M = KW^b$, where M is the metabolic rate, K is a constant (~ 10), W is weight, and b is a constant (~ 0.7). Although the development of such a formula often suggests some interesting ideas, it tells us little about the mechanisms involved in thermoregulation and the transfer of heat energy between the animal and its environment.

Most discussions of the energetics of homeothermy have been plagued by reliance on Newton's law of cooling to describe heat transfer. McNab (1970) attempted a theoretical analysis to explain the effect of body weight on the energetics of temperature regulation. He employed allometric equations and ratios of allometric equations to determine the applicability of Newton's law to this problem. Yarbrough (1971) employed similar methods to examine thermoregulation of small birds. Calder and King (1972) demonstrated defects in McNab's theoretical treatment. Problems with Yarbrough's interpretations have been pointed out by Calder (1972). Kleiber (1972a, 1972b) has discussed problems associated with the misapplication of Newton's law to studies of homeotherms and its confusion with Fourier's law of heat flow by many zoologists and physiologists. Birkebak (1966) and Strunk (1971) suggested that the heat exchange of animals be described by the use of models based on the physical principles of heat transfer and the thermal properties of animals. Tracy (1972) demonstrated the difficulties and pitfalls inherent in the use of Newton's law as a basis for discussions on the evolutionary adaptations of homeotherms and predictions of heat loss from homeotherms under natural conditions. His calculations make clear the reasons why reliance on the use of Newton's law in the past has led to questionable conclusions about the role of insulation and body size in the overall heat loss of animals.

By carefully analyzing the exchange of heat energy between homeotherms and their physical environment, we can quantify the importance of animal properties such as size, shape, quality of insulation, and metabolic rate in determining the ability of animals to adapt to their physical environment. This approach also provides insight into some of the factors that determine the optimum body temperature of homeotherms. By basing our model on the physics of heat-energy transfer, we gain a much better understanding of the mechanisms involved in animal–energy–environment interactions than could be obtained by reliance on empirical formulas alone.

Methods

The basic mathematical model of environment–animal energy exchange is based on the physics of heat transfer and known physical and physiological properties of animals. For a full discussion of animal-heat-energy-budget equations, see the papers by Birkebak (1966), Bartlett and Gates (1967), Porter and Gates (1969), Heller and Gates (1971), and Spotila *et al.* (1972, 1973).

According to the first law of thermodynamics, for any animal under steady-state conditions, energy into and energy generated by the animal must equal the energy flowing out of the animal. We describe the animal's energy budget by the equation

$$Q_{\text{abs}} + M = \varepsilon\sigma T_r^4\,(^\circ\text{K}) + k\,\frac{V^{0.6}}{\text{OD}^{0.4}}\,(T_r - T_a) + E \tag{1}$$

where Q_{abs} = energy absorbed from the environment (cal cm^{-2} min^{-1})
M = heat generated by metabolism (cal cm^{-2} min^{-1})
ε = emissivity (0.97)
σ = Stefan–Boltzmann constant (8.13×10^{-11} cal cm^{-2} min^{-1} °K^{-4})
T_r = surface temperature (°K or °C)
k = constant [1.95×10^{-3} cal cm^{-2} min^{-1} °C^{-1} (s cm^{-1})]
V = wind speed (cm s^{-1})
OD = outside body diameter, including fur or feathers (cm)
T_a = air temperature (°C)
E = evaporative water loss (cal cm^{-2} min^{-1})

As a general case, we assume that we are dealing with a very simple homeotherm that exchanges energy with its environment primarily by radiation, convection, and evaporation. We lump all water loss as one term (E). In this first approximation model, to reduce the complexity of calculations, we assume that the animal is cylindrical in shape, with no appendages. The ratio of length to diameter is 4. Insulation consists of a layer of fur or feathers, the thickness of the layer defined as $C \cdot D$, where D is core diameter with no fur or feathers and C is a constant proportion of D for all sizes of D in a given set. Insulation (I) can be defined by the expression

$$I = \frac{C \cdot D}{K} \tag{2}$$

where K is conductivity of insulation (cal cm^{-1} min^{-1} °C^{-1}). Since

$$T_r = T_b - I(M - E)$$

we can substitute for T_r in Eq. 1 and obtain

$$Q_{\text{abs}} + M = \varepsilon\sigma\left[T_b - \frac{CD}{K}(M - E)\right]^4 + k\frac{V^{0.6}}{\text{OD}^{0.4}}\left[T_b - \frac{CD}{K}(M - E) - T_a\right] + E \quad (3)$$

In this equation we solve for the term $M - E$. This is the heat generated by metabolism minus the heat dissipated by evaporative water loss and is the net heat production within the animal. It is a measure of the animal's physiological thermoregulatory activity. Using this equation we can predict the effect of the independent environmental variables, such as radiation, air temperature, and wind speed, on the dependent variables of the animal: body temperature, metabolic rate, and water loss. The interaction among these variables will be influenced by the size of the animal, the thickness of its insulation, and the conductivity of its insulation. Changes in mass flow rates of blood to external tissues and local differences in insulation can greatly affect heat-flow patterns within the animal.

Although this model does not address these problems, it does allow us to examine in detail specific facets of the animal–energy–environment interaction. It accounts for the roles of body size and insulation in the thermoregulation of animals and outlines the general parameters that limit the ability of an animal to function as a homeotherm.

Results

Animals with very thin insulation ($C = 0.01$) cannot generate enough heat to survive under cold conditions with a clear sky at night (Fig. 17.1A). Small animals are at a particular disadvantage, and even a shrew could not maintain a high body temperature under these conditions. Increased insulation ($C = 0.1$) brings the $M - E$ requirement down within a reasonable range (Fig. 17.1B). Differences between calculated curves and points based on experimental data are due to the different C values. Large size and insulation both act to decouple an animal from its physical environment and decrease demands on its physiological capabilities.

A similar pattern is seen for animals exposed to the full sun at high air temperatures, although $M - E$ values are now negative (Fig. 17.1C and D). Demands on evaporative water loss are particularly high for small animals. Large animals ($D = 100$ cm) depend less on evaporative water loss for cooling. Even without thick insulation, their large size serves to isolate them from the environment. Since homeotherms normally do not have a negative $M - E$ under steady-state conditions, these graphs indicate that the animals for which we present actual data cannot survive the high-energy conditions used in Fig. 17.1C and D. Examination of their climate spaces (Porter and Gates, 1969) confirms this prediction.

The interaction of insulation and body size is shown in Fig. 17.2A and B. Under low-energy conditions an increase in D from 2 to 14 cm has an effect on $M - E$ equivalent to increasing C from 0.01 to 0.1. For high-energy conditions, D must be increased to 35 cm to get the same effect. Thick insulation is particularly important to small homeotherms but is relatively less effective in large animals.

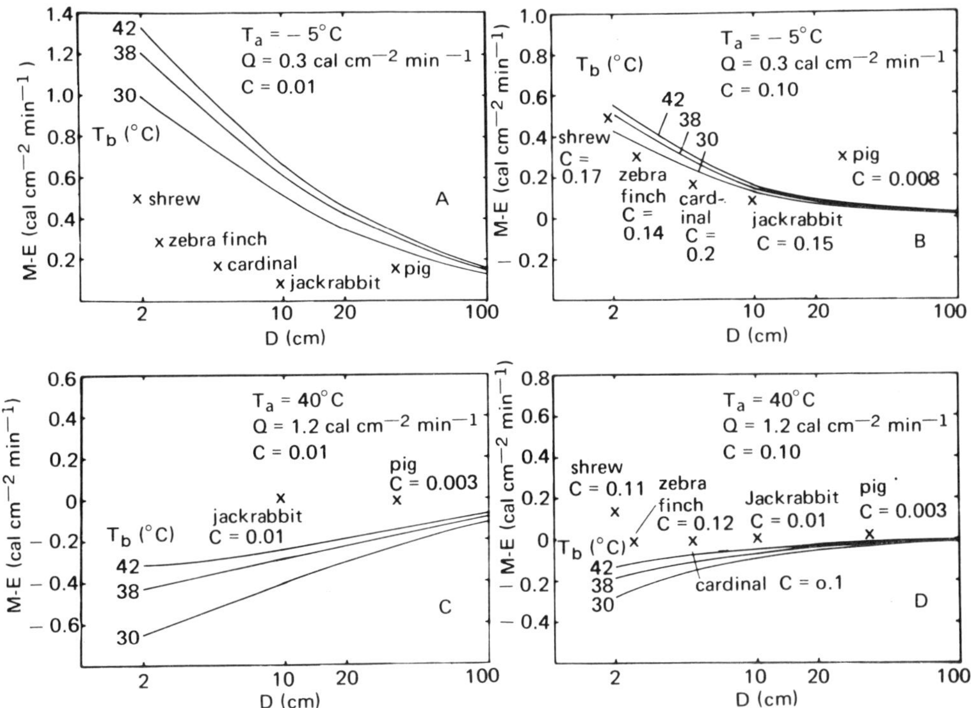

Fig. 17.1. Effect of body size and insulation on the $M - E$ requirement of homeotherms. A. Cold conditions, clear sky at night, low insulation. B. Cold conditions, clear sky at night, increased insulation. C. Full sun, high air temperature, low insulation. D. Full sun, high air temperature, increased insulation. Insulation consists of a layer of fur or feathers, the thickness of the layer defined as $C \cdot D$, where D is the inside diameter with no insulation and C is a constant proportion of D for all sizes of D in a given set. Data for real animals are taken from Porter and Gates (1969).

Discussion

Environmental heat load, the size of an animal, and its insulative properties place specific demands on an animal's metabolism and evaporative water-loss rates. Small homeotherms must compensate for their lack of bulk and insulation by large adjustments in metabolism and evaporative water loss under high- and low-energy conditions. Large size allows a homeotherm to maintain a relatively constant rate of metabolism and water loss when exposed to widely varying heat loads.

Contrary to the conclusions of Scholander (1955) and McNab (1971), body size is an important factor in the climatic adaptation of homeotherms. Large size reduces the effect of convection, thus reducing heat transfer between the animal and its environment and lowering demands on an animal's physiological mechanisms of heat generation and dissipation (Fig. 17.1). As demonstrated by Scholander *et al.* (1950), insulation is also very important in thermoregulation. It is most effective on

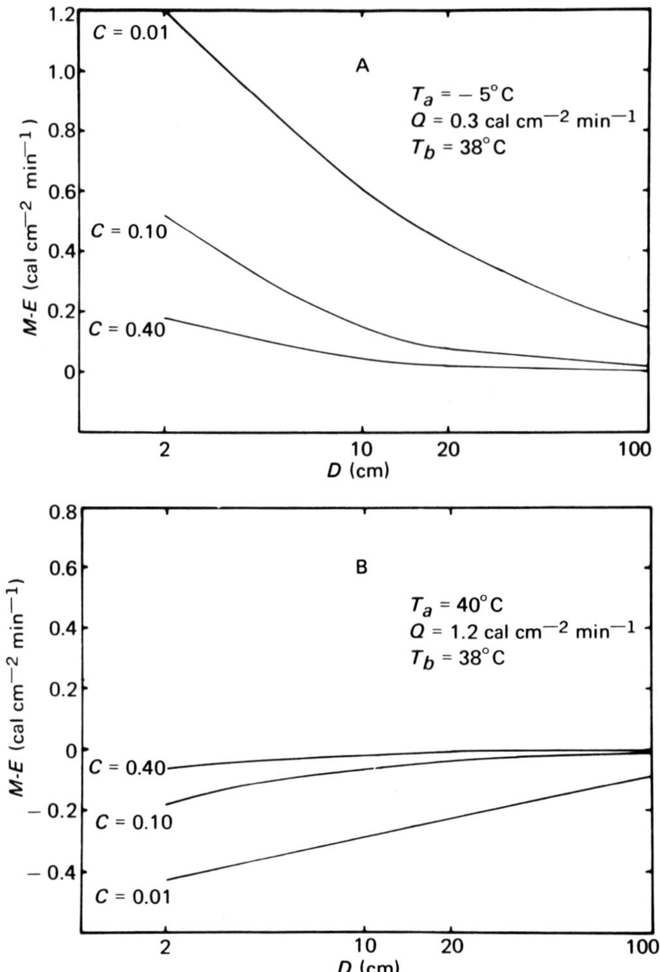

Fig. 17.2. Interaction of body size and insulation and their effect on the $M - E$ requirement of homeotherms. A. Cold conditions, clear sky at night. B. Full sun, high air temperature. For an explanation of values of C and D, see Fig. 17.1.

small homeotherms and relatively less effective on large animals since they already have reduced physiological demands due to their size (Fig. 17.2).

Both insulation and body size act to isolate the animal from its surroundings, thus allowing its internal processes to proceed at a pace independent of fluctuating ambient conditions. Therefore, among homeotherms, we would expect to find the largest and/or best insulated members of a species or genus in fluctuating and/or extreme environments and smaller representatives in stable and/or moderate environments. This has been shown to be the case in some birds (James, 1970) and mammals (McNab, 1971, Fig. 4, *Felix concolor*; Fig. 8, *Sciurrus carolinesis* and

Glaucomys volans and *G. sabrinus*). In poikilotherms, we would expect to find the largest members of a species or genus in stable environments and the smallest members in fluctuating environments. That is because a stable environment would provide a steady year-round growth rate, while a fluctuating environment might prevent growth during part of the year. Lindsey (1966) stated that, among some poikilothermic vertebrates, the proportion of species with large adult size tends to increase from the equator toward the poles. But he dealt with entire faunas. No one has yet formally analyzed the geographic variation in body size in a species or genus of poikilothermic vertebrates.

Maintenance of a high, constant body temperature under conditions of both high- and low-energy input is evidently advantageous for homeotherms but it is very costly in terms of energy production and dissipation (Fig. 17.1). Most mammals regulate their body temperature between 36 and 40°C, and most birds regulate between 38 and 42°C. These animals are prevented from regulating at a higher temperature by the possibility of heat death. The upper limit for animal life is 45 to 51°C (Brock, 1967). When body temperatures approach this range, enzymes pass their optima (Fig. 17.3A), lipids change in state, and cell membranes become increasingly permeable (Prosser and Brown, 1961, p. 239). Animals regulate at temperatures where the thermal decay of protein is small (Fig. 17.3B), so that the protein replacement rate can be satisfied without burdening the energy-production system (Morowitz, 1968, pp. 114–116). Heat death in psychrophilic bacteria is due to damage to the cell membrane and subsequent lysis. The susceptibility of animals to high temperature may be due in large part to the thermal sensitivity of their more complicated membrane systems (Brock, 1967). Even if heat death were not a problem, regulation of body temperature above 42°C would require extensive insulation and/or a very high metabolic rate when the animal was exposed to cold conditions (Fig. 17.1A and B). This, in turn, would cause increased demands on the animal's food supply. If the food supply was inadequate for maintenance under these conditions, the animal would be forced to enter torpor. McNab (1969) has shown that this is the cause for torpor and hibernation in temperate bats.

Some bats regulate at body temperatures below 35°C, and *Tachyglossus* and *Cyclopes* regulate near 30°C (McNab, 1969), but most homeotherms regulate at higher temperatures. Regulation at a low body temperature under high-energy input conditions would require very high rates of evaporative water loss and low rates of metabolism (Fig. 17.1C and D). With surface temperature below ambient temperature, the animal could not lose heat by conduction or convection, and the effect of reradiation would be reduced. It would have to rely on evaporative water loss as the main avenue of heat dissipation and would run the risk of desiccation if it did not remain near a stable supply of water. This would restrict its ability to exploit many terrestrial habitats. In addition, it would either have to rely on extensive panting or else maintain sparse insulation to increase respiratory and cutaneous water loss. Both alternatives would make it difficult for the animal to remain internally isolated from its environment and thus would increase selective pressures for relaxation of homeothermy. Finally, thermoregulation at a low body temperature would reduce the efficiency of the animal's physiological systems and

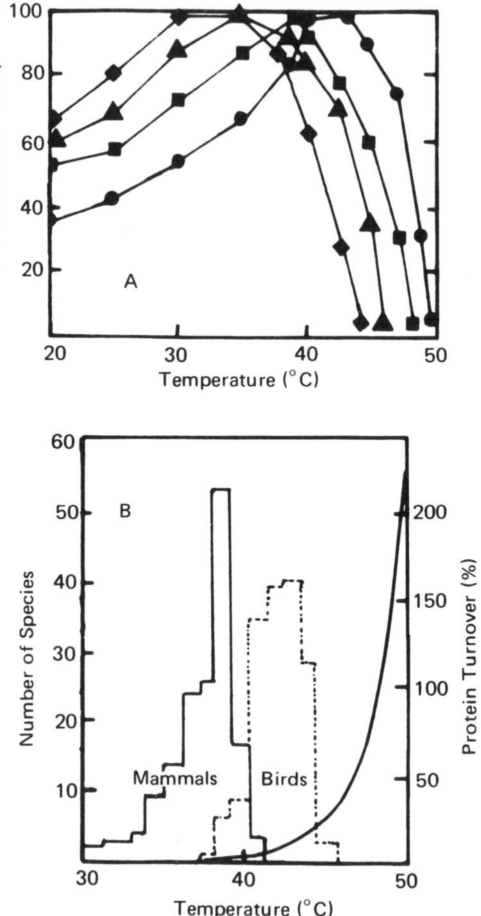

Fig. 17.3. A. Effects of temperature on the relative activity of adenosine triphosphatase from four species of lizards (redrawn from Licht, 1967, p. 138). B. Temperatures at which birds and mammals thermoregulate, and the effect of temperature on the thermal stability of proteins. (Redrawn from Morowitz, 1968, p. 115.)

place it at a competitive disadvantage with homeotherms that regulate at higher temperatures.

It appears that selection has acted to balance the advantages of a stable body temperature, as near the optimum for successful functioning of an animal's total integrated system as possible, against the danger of heat death and the energetic cost of changes in metabolism, water loss, body size, and insulation. This has resulted in the maintenance of high body temperatures by most homeotherms and in a trend toward increased body size and/or quality and quantity of insulation in most birds and mammals.

Comparative physiologists have spent much time and effort arguing the merits of the "surface law" and whether metabolic rate is best described as a function of the two-thirds or three-fourths power of body weight. The basic difficulty inherent in these discussions has been insistence on the use of oversimplified expressions, such as Newton's law of cooling, to describe the heat-energy transfer between animals and their environment. Reliance on Newton's law has led to questionable conclusions concerning the role of physical and physiological properties of animals in the heat-energy balance between animals and their environment. One example is Scholander's (1955) conclusion that insulation is more important than body size in overall heat loss from an animal (see Birkebak, 1966; Cremers and LeFebvre, 1965). In thinking in terms of surface area and the ratio of surface area to body weight, Scholander missed the significance of size and its importance in reducing the effect of convection in the energy exchange of animals. Although the energy-budget analysis presented in this study is not sufficiently developed to account for all the variation found in the thermoregulatory strategies of homeotherms, it does provide a useful tool for studying the mechanisms involved in animal energetics. Whereas in the past, physiologists and ecologists discussed surface/volume ratios and power-law formulas, we can now map out the specific contribution of radiation, convection, evaporation, insulation, etc., to the heat exchange between an animal and its environment and to the ability of that animal to function as a homeotherm.

Note added In Press

Recent information leads us to make the following additional comments.

First, Hochacka and Somero (*Strategies of Biochemical Adaptation*, W. B. Saunders, 1973) have presented strong arguments that other aspects of kinetics, especially those related to substrate affinity, may be more informative in describing enzyme function under normal cellular conditions than those described in Fig. 17.3A. Second, in discussing the very real effect of body size on the adaptation of homeotherms to climate, we do not mean to suggest that factors such as insulation and vasomotor changes are not important. A change in fur thickness, probably at a small energy cost, has the same effect as a very large change in body size, which probably occurs at a much higher energy cost. So it is not surprising that selection has resulted in many homeotherms with thick fur while large size has been selected in only a few homeotherms (see Fig. 17.2). However, the effect of body size is real and should not be lightly dismissed.

Acknowledgments

This study was supported in part by a grant from the Ford Foundation. During completion of the manuscript, J. R. Spotila was supported by a Faculty Fellowship and Grant in Aid from the Research Foundation of the State University of New York. We thank Paul Licht for critically reviewing this chapter.

References

Bartlett, P. N., Gates, D. M.: 1967. The energy budget of a lizard on a tree trunk. Ecology *48*, 315–322.

Birkebak, R. C.: 1966. Heat transfer in biological systems. Intern. Rev. Gen. and Expt. Zool. *2*, 269–344.

Brock, T. D.: 1967. Life at high temperatures. Science *158*, 1012–1019.

Brody, S., Procter, R. C.: 1932. Growth and development with special reference to domestic animals. Further investigations of surface area in energy metabolism. Mo. Res. Bull. *116*.

Brown, J. H., Lasiewski, R. C.: 1972. Metabolism of weasels: the cost of being long and thin. Ecology *53*, 939–943.

———, Lee, A. K.: 1969. Bergmann's rule and climate adaptation in woodrats (*Neotoma*). Evolution *23*, 329–338.

Calder, W. A.: 1968. Respiratory and heart rates of birds at rest. Condor *70*, 358–365.

———: 1972. Heat loss from small birds: analogy with Ohm's law and a re-examination of the "Newtonian model." Comp. Biochem. Physiol. *43A*, 13–20.

———, King, J. R.: 1972. Body weight and the energetics of temperature regulation: a re-examination. J. Expt. Biol. *56*, 775–780.

Carey, F. G., Lawson, K. D.: 1973. Temperature regulation in free-swimming blue-fin tuna. Comp. Biochem. Physiol. *44A*, 375–392.

Cremers, C. J., Le Febvre, E. A.: 1965. Thermal modeling applied to animal systems. Trans. Am. Soc. Mech. Engrs., Ser. C, J. Heat Transfer *88*, 125–130.

Heller, H. C., Gates, D. M.: 1971. Altitudinal zonation of chipmunks (*Eutamias*): energy budgets. Ecology *52*, 424–433.

Herreid, C. F., Kessel, B.: 1967. Thermal conductance in birds and mammals. Comp. Biochem. Physiol. *21*, 405–414.

James, F. C.: 1970. Geographic size variation in birds and its relationship to climate. Ecology *51*, 365–390.

Kendeigh, S. C.: 1969. Tolerance of cold and Bergmann's rule. Auk *86*, 13–25.

Kleiber, M.: 1932. Body size and metabolism. Hilgardia *6*, 315–353.

———: 1961. The fire of life. New York: Wiley.

———: 1972a. Body size, conductance for animal heat flow and Newton's law of cooling. J. Theoret. Biol. *37*, 139–150.

———: 1972b. A new Newton's law of cooling? Science *178*, 1283–1285.

Lasiewski, R. C., Dawson, W. R.: 1967. A re-examination of the relation between standard metabolic rate and body weight in birds. Condor *69*, 13–23.

Licht, P.: 1967. Thermal adaptation in the enzymes of lizards in relation to preferred body temperatures. *In* Molecular mechanisms of temperature adaptation (ed. C. L. Prosser), pp. 131–145. Am. Assoc. Advan. Sci. Publ. *84*.

Lindsey, C. C.: 1966. Body sizes of pokilothermic vertebrates at different latitudes. Evolution *20*, 456–465.

McNab, B. K.: 1969. The economics of temperature regulation in neotropical bats. Comp. Biochem. Physiol. *31*, 227–268.

———: 1970. Body weight and the energetics of temperature regulation. J. Expt. Biol. *53*, 329–348.

———: 1971. On the ecological significance of Bergmann's rule. Ecology *52*, 845–854.

Morowitz, H. J.: 1968. Energy flow in biology. New York: Academic Press.

Porter, W. P., Gates, D. M.: 1969. Thermodynamic equilibria of animals with environment. Ecol. Monog. *39*, 245–270.

Prosser, C. L., Brown, F. A., Jr.: 1961. Comparative animal physiology. Philadelphia: Saunders.

Ricklefs, R. E.: 1973. Ecology. Newton, Mass.: Chiron Press.

Schmidt-Nielsen, K.: 1972. How animals work. New York: Cambridge Univ. Press.

Scholander, P. F.: 1955. Evolution of climatic adaptations in homeotherms. Evolution *9*, 15–26.

———, Hock, R., Walters, V., Irving, L.: 1950. Adaptation to cold in arctic and tropical mammals and birds in relation to body temperature, insulation, and basal metabolic rate. Biol. Bull. *99*, 259–271.

Spotila, J. R., Lommen, P. W., Bakken, G. S., Gates, D. M.: 1973. A mathematical model for body temperatures of large reptiles: implications for dinosaur ecology. Am. Naturalist *107*, 391–404.

———, Soule, O. H., Gates, D. M.: 1972. Biophysical ecology of the alligator—heat energy budgets and climate spaces. Ecology *53*, 1094–1102.

Strunk, T. H.: 1971. Heat loss from a Newtonian animal. J. Theoret. Biol. *33*, 35–61.

Tracy, C. R.: 1972. Newton's law: its application for expressing heat losses from homeotherms. BioScience *22*, 656–659.

Yarbrough, C. G.: 1971. The influence of distribution and ecology on the thermoregulation of small birds. Comp. Biochem. Physiol. *39A*, 235–266.

18

Use of Climate Diagrams to Describe Microhabitats Occupied by Belding Ground Squirrels and to Predict Rates of Change of Body Temperature

SYLVIA S. MORHARDT

Introduction

Climate diagrams were described by Porter and Gates (1969) as a means of defining important environmental extremes outside of which an animal with given physical and physiological properties cannot maintain thermal equilibrium.

The present work, using the Belding ground squirrel, expands on the climate-diagram concept to show how it may be used to describe the thermal environment of any microhabitat. To do this, the shape of the animal, the orientation of this shape to direct solar radiation, and the absorptivity of the animal must be known or approximated. Dependence of the average absorbed radiation on the actual size of the animal is avoided by using proportions that describe the shape of the organism. The model described in this work is a cylinder with hemispherical ends. Orientations of the model described include the long axis of the shape remaining (1) perpendicular to the direction of the sun, (2) parallel to the direction of the sun, and (3) vertical with respect to gravitational forces. Climate diagrams using each of these orientations were calculated for each of several microhabitats throughout a day. Therefore, each climate diagram displays for each microhabitat a daily record of combinations of air temperature and average absorbed radiation encountered by any organism approximating the shape of the model with the specified orientations to direct solar radiation.

Climate diagrams that describe the thermal environment of an animal in a particular microhabitat throughout the day may be overlaid on climate diagrams that describe environmental restrictions on the animal's capability to maintain thermal equilibrium. For the squirrels studied, the thermal load in warm microhabitats during the early afternoon in some cases exceeded the squirrel's capacity to lose heat, particularly at low wind speeds. The difference in average absorbed radiation between the point on the climate diagram that describes the microhabitat occupied by the animal, and the maximum limit of the animal for maintaining thermal equilibrium at the air temperature of the microhabitat and an appropriate wind speed, represents excess absorbed radiation, which will result in a rise in body temperature. An equation is presented that can be used to approximate the rate of change in body temperature which the animal in the nonequilibrium situation will experience. If the animal's body temperature and acceptable range of body temperatures are known, it is possible to estimate the length of time the animal may remain in the microhabitat where he cannot maintain thermal equilibrium. Similar calculations could be used to determine the rate of cooling of an overheated animal that has moved into the shade or other less extreme microhabitat.

Since many animals, including homeotherms, have highly variable body temperatures, combined use of climate diagrams for the animal and for the microhabitat can help to relate behavioral and physiological aspects of temperature regulation to the precise microclimates in which species normally live. Also, abilities to remain in thermal equilibrium within a given microhabitat may be compared for differently sized animals of a given species, or for different species having similar shapes.

Study Area and Habits of the Squirrels

The Belding ground squirrel is known as either *Citellus beldingi beldingi* or *Spermophilus beldingi.* Belding squirrels are diurnal herbivores, and are inactive during a hibernation period of approximately 8 months. The animals do not hoard food for the winter season (Grinnell and Dixon, 1918; Heller and Poulson, 1970); winter survival depends on fat deposits accumulated during the previous summer season of feeding. The squirrels, especially the young, must have favorable microhabitats and feeding opportunities to fatten sufficiently for survival of the hibernation period.

Belding squirrels are somewhat colonial, especially when populations inhabit large open areas. Individuals appear to interact little, but respond to warning whistles and frequently enter common burrows.

The population studied occupied an alpine meadow at 8950-ft elevation near the South Fork of Bishop Creek on the eastern slope of the Sierra. The meadow included approximately 12,500 m^2 of grasses and was bordered by higher rocky, shrubby areas vegetated primarily with sage. A small lake and aspen groves were nearby. Figure 18.1 is a map of the study area. In rocky places there were numerous large granite boulders, and the soil was thin and dry. Many ground squirrel burrows occurred near boulders on the border of the meadow. Numerous pathways defined the most frequently traveled routes of squirrels between burrows. These are shown in Fig. 18.2.

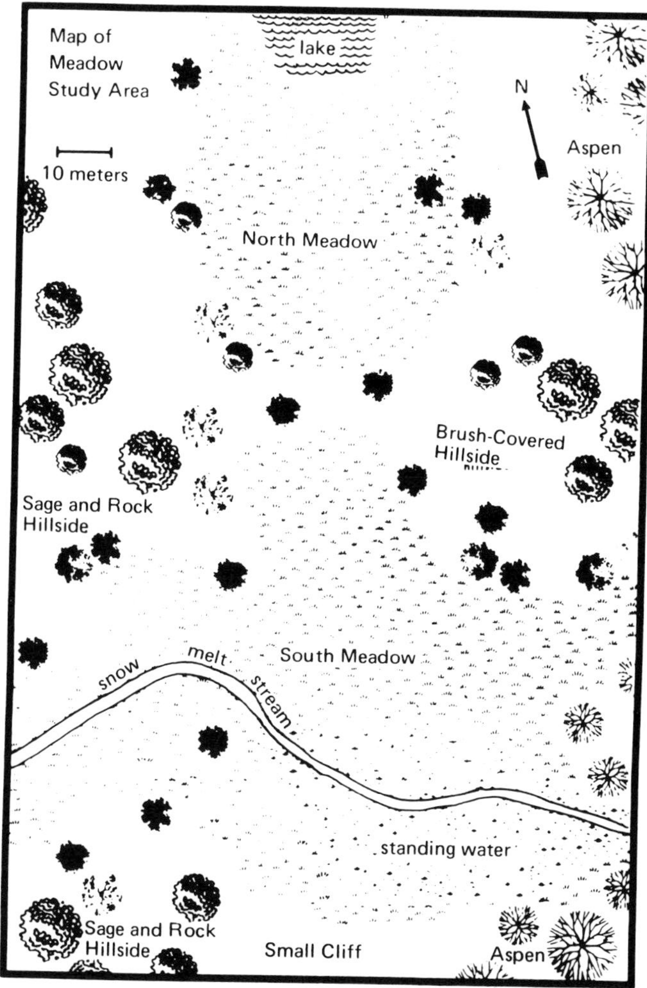

Fig. 18.1. Meadow study area. This map shows the majority of the meadow study area, indicating the vegetation, substrate, and the moisture characteristics as they existed in mid-May 1970.

Preparation of Climate Diagrams for Microhabitats

Purpose

A natural population of squirrels was studied so that microhabitats meaningful to free-ranging animals could be characterized. The microclimate of each microhabitat utilized by the squirrels was examined during each of three typical daily weather patterns. Climate diagrams were derived that describe the hourly combinations of air temperature and absorbed radiation that would be experienced by

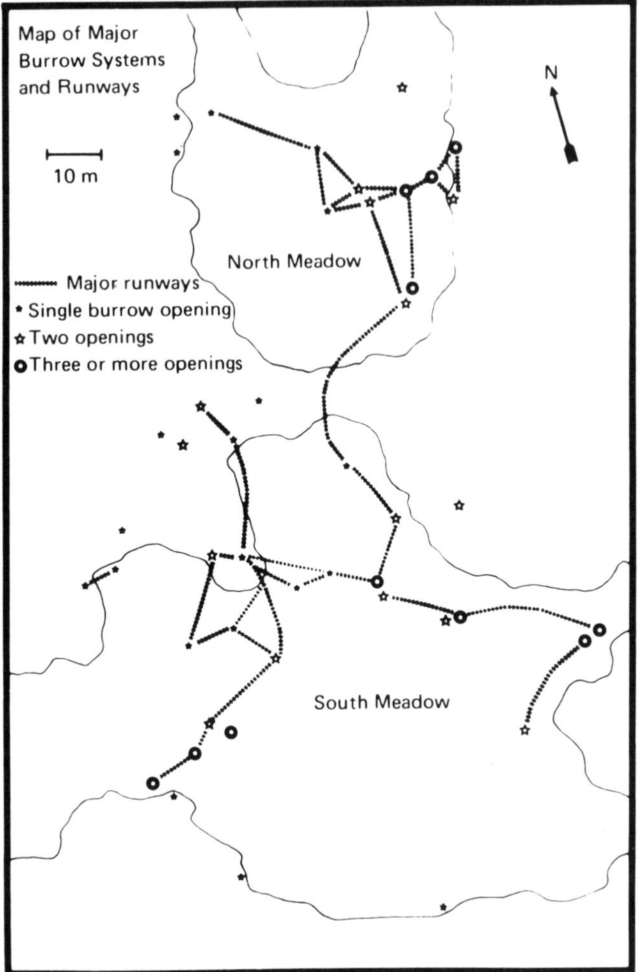

Fig. 18.2. Major burrow systems and runways. This map encompasses the same area and is drawn to the same scale as Fig. 18.1. Where burrow openings existed, there were often several in the immediate vicinity. The number of openings is indicated by the type of star. The dotted lines between burrow openings represent major runways visible in the grass in mid-May 1970.

a squirrel occupying the particular site. Since absorbed radiation depends on properties of the squirrels and their orientation to direct solar radiation, methods are given for including these considerations in the compilation of climate diagrams for the microhabitat.

The advantage of using the climate diagram to describe microhabitat is that the diagram can be compared directly with the type of climate diagram introduced by Porter and Gates (1969), which describes environmental limitations to an animal's ability to maintain thermal equilibrium. The next section of this paper discusses the relationship between the two types of climate diagrams and thermoregulatory responses of the animals.

Methods

The study site was established on May 13, 1970. Squirrels were trapped using Havahart live traps set near burrow complexes and feeding areas. Trapped animals were removed from the traps quickly, sexed, and weighed to the nearest 0.1 g on a triple-beam balance. Before being released, squirrels were marked with distinguishing symbols painted on their fur using the Nyanzol A dye mixture reported by Melchior and Iwen (1965). Observations of the animals were made and recorded on 16 days throughout the season, both on trapping days and on days when the meadow was left undisturbed.

All instrumentation used at the field study site was battery-operated and portable since there was no external power source available, and no equipment could be left in the field when not in use. For these reasons, no continuous recordings of microclimate were possible, but measurements were made on days that were judged to be typical, as well as under more extreme weather conditions.

Air and burrow temperatures were measured with a Yellow Springs Instrument Co. thermister thermometer, Model 43 TD. Wind speeds were determined with a Hastings Air-Meter. Relative humidity was measured with a portable Honeywell relative humidity indicator, Model W611A. Solar radiation was measured with an Eppley pyranometer, Model 8-48, with a wavelength sensitivity between about 0.28 and 2.5 μ, connected to a sensitive portable potentiometer (Pyrotest Potentiometer, West Instrument Corp.). Thermal radiation from the atmosphere and the ground was measured with a Barnes infrared radiometer, Model PRT 10L. The temperature readings from the radiometer were converted to blackbody caloric equivalents, using the tables published by the Smithsonian Institution (List, 1968).

The reflectivity of both the darker dorsal fur and the lighter lateral fur on each of three live squirrels was measured with a recording scanning spectrophotometer (Beckman Instruments, Model DK-11A). The reflectivity, r, of the fur was recorded between 0.3 and 2.5 μ. To obtain the average reflectivity of the fur to sunlight, 90 points, at wavelength intervals of 20 mμ and greater, were read from the spectrophotometer recording and used in a computer program which multiplied the reflectivity of the fur at each point times the intensity of solar radiation at the same wavelength, and then calculated the mean reflectance. The spectral distribution of sunlight used was that corresponding to a clear, dry, summer day at noon.

Results and Discussion

Observations indicated that Belding ground squirrels leave their burrows only between the hours of 0700 and 1800 solar time. In the morning there was plenty of light before 0700, but squirrels did not emerge from their burrows until the sun rose above the mountains and shone directly on their burrows. There was a gradual disappearance of individuals from the meadow in the late afternoon, so that by dusk all squirrels were (presumably) in their burrows.

During the hours of above-ground activity, essentially six different types of microhabitats were frequently utilized by ground squirrels in the study area.

During different periods of the summer, ground squirrels spent different proportions of their time in each of these microhabitats. The microhabitats and the changes in utilization by the squirrels are shown in Fig. 18.3. In the early part of the summer, there was no green grass, but later, squirrels spent most of their time feeding in the green meadows. Very little of the squirrels' time was spent in the relatively protected habitats of the shade or the burrow.

The radiation and air temperature of each microhabitat were measured during three different types of daily weather patterns. A perfectly clear day near the summer solstice was chosen to include the hottest environments. A stormy day near the end of the summer was chosen to include the coldest conditions. A typical day is represented by the combination of a clear morning and an intermittently cloudy afternoon. In Fig. 18.4A–G, each line on a graph of a given general weather pattern was derived from a series of measurements taken over a period of 3–5 days of appropriate weather. Data points from these days were combined so that the curves represent conditions during an idealized composite day.

Each graph in Fig. 18.4A–G represents the radiant environment that exists under different combinations of general weather conditions and the typical microhabitats frequented by the animals (described in Fig. 18.3). Since the microhabitat fully within the burrow is relatively independent of the immediate general weather conditions, the combinations of the burrow microhabitat and general weather conditions are not given.

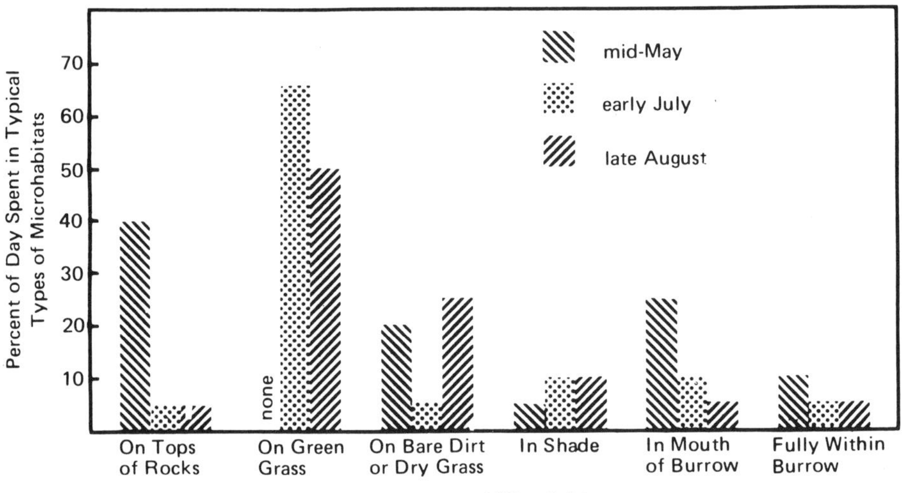

Fig. 18.3. Seasonal changes in the utilization of different types of microhabitats by Belding ground squirrels. This histogram shows the percentage of time between emerging from the burrow in the morning and retiring to the burrow in the evening which was spent by Belding ground squirrels in each of six different types of microhabitat. Percentages are shown for three different periods of the squirrels' active seasons: mid-May, early July, and late August. (From Morhardt and Gates, 1974. Copyright, the Ecological Society of America.)

On clear days, because of the high values for direct solar radiation, it might be expected that an animal could not maintain thermal equilibrium with its environment. Also, the microhabitats available to the squirrels are most different from one another when the direct solar radiation is most intense. For these reasons, the thermal microclimate of each of the microhabitats is described for the clear-day situation. Figure 18.4A–C share the same solar radiation and atmospheric thermal radiation because animals in each situation are fully exposed to the upper hemisphere. The air temperature over grass (Fig. 18.4B) is slightly lower than that over rocks or dirt (Fig. 18.4A and C) because of evaporation from the transpiring vegetation. Similarly, the ground-surface temperature, T_g, is somewhat lower for the grassy microhabitat. The temperature increase of the granite rock surface (Fig. 18.4A) lags behind the temperature increase of the dirt surface (Fig. 18.4C) and does not reach as high a maximum during the day or cool as rapidly in the late afternoon as does the exposed soil surface at the burrow entrances.

An animal in the shade on a clear day (Fig. 18.4D) may experience the same air temperature as an animal in the open but will not receive energy from direct solar radiation. Since the substrate in the shade does not receive direct solar radiation, its surface temperature will not rise above air temperature. In the mouth of the burrow, the animal is in partial shade and, for convenience of calculation, one-third of the animal is considered to be subjected to the radiant environment described by Fig. 18.4C, and two-thirds subjected to blackbody radiation at the temperature of the air 5–10 cm within the burrow. This air temperature is shown as the sole variable of Fig. 18.4E. Temperatures between 6 and 18°C were measured at 20 cm and farther into the burrows, but these temperatures are not related to the immediate general climatic conditions.

For a typical day with afternoon clouds, only the microclimate on green grass is represented (Fig. 18.4F) because the grass represents the microhabitat where animals spend most of their time and because the grass microclimate differs very little from that of the other microhabitats which are fully exposed to solar radiation. On a stormy day when there is no direct solar radiation and there may be intermittent rain, there is very little difference between the different microhabitats above ground. Again, the only example illustrated for a stormy day is that in the grass, since the animals spend most of their time feeding, even in the rain.

Ultimately, the effect of the radiant microclimate on an animal depends on certain physical properties of the animal itself—specifically its geometry and the absorptivity of its surface. The absorptivity of animal surfaces to infrared thermal radiation is nearly 1.0 (Birkebak, 1966), but absorptivities to solar radiation vary considerably and must be measured for each species. On the three *C. beldingi* used in the determination of absorptivity, the darker fur on the middle of the back had a higher average absorptivity (79.1 percent) than the lighter fur on the sides (68.9 percent). An absorptivity to solar radiation of 74 percent (the average of back and side values) is used in all further calculations, even though slight variations probably occur with sun angle, with orientation of the fur to the sun, and with precise spectral quality of the solar radiation.

The shape of an animal determines the proportion of its body that will receive upward and downward directed streams of radiation, and the orientation of this

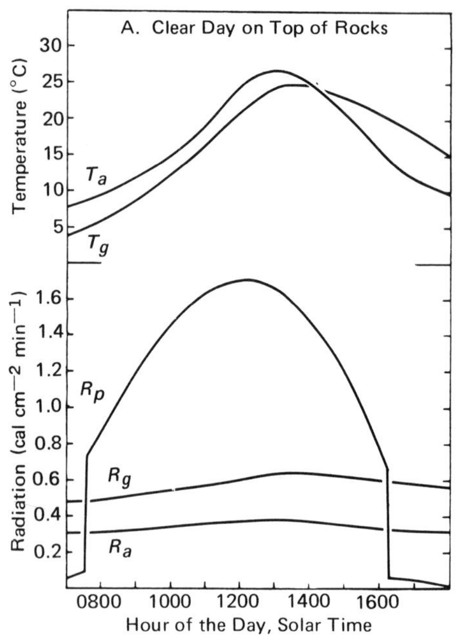
A. Clear Day on Top of Rocks
Temperature (°C)
Radiation (cal cm−2 min−1)
T_a
T_g
R_p
R_g
R_a
0800 1000 1200 1400 1600
Hour of the Day, Solar Time

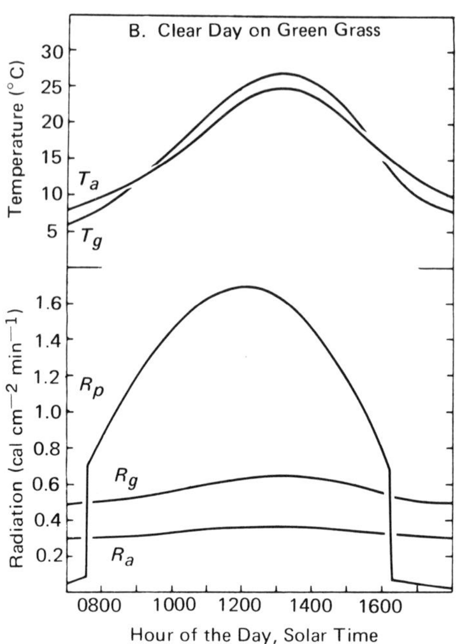
B. Clear Day on Green Grass
Temperature (°C)
Radiation (cal cm−2 min−1)
T_a
T_g
R_p
R_g
R_a
0800 1000 1200 1400 1600
Hour of the Day, Solar Time

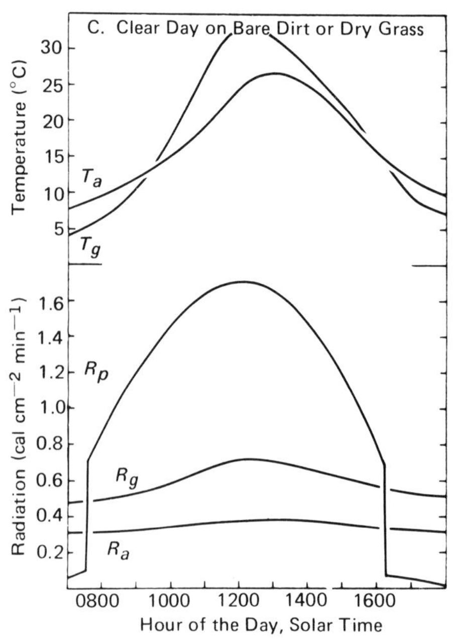
C. Clear Day on Bare Dirt or Dry Grass
Temperature (°C)
Radiation (cal cm−2 min−1)
T_a
T_g
R_p
R_g
R_a
0800 1000 1200 1400 1600
Hour of the Day, Solar Time

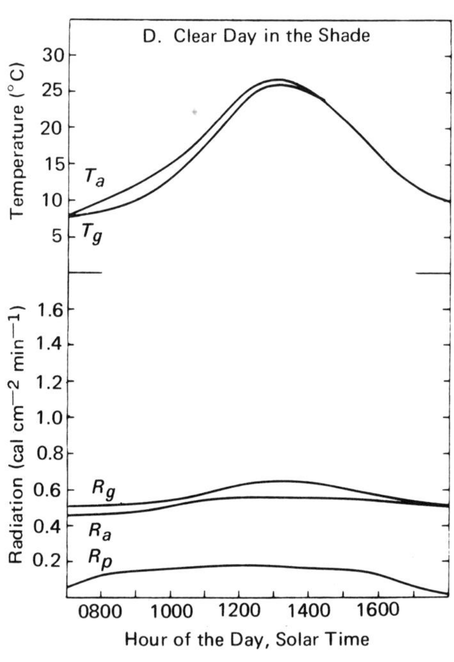
D. Clear Day in the Shade
Temperature (°C)
Radiation (cal cm−2 min−1)
T_a
T_g
R_g
R_a
R_p
0800 1000 1200 1400 1600
Hour of the Day, Solar Time

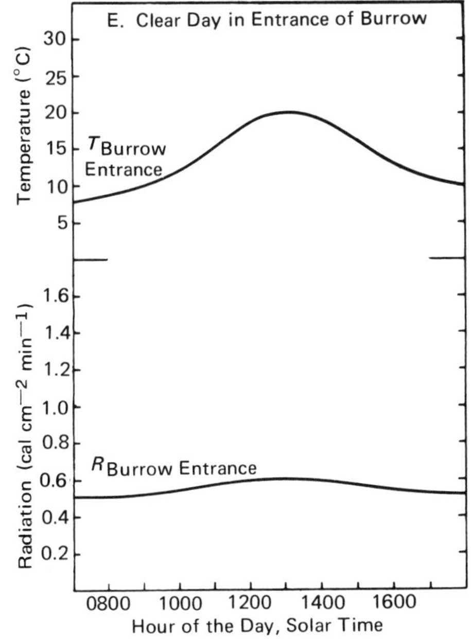

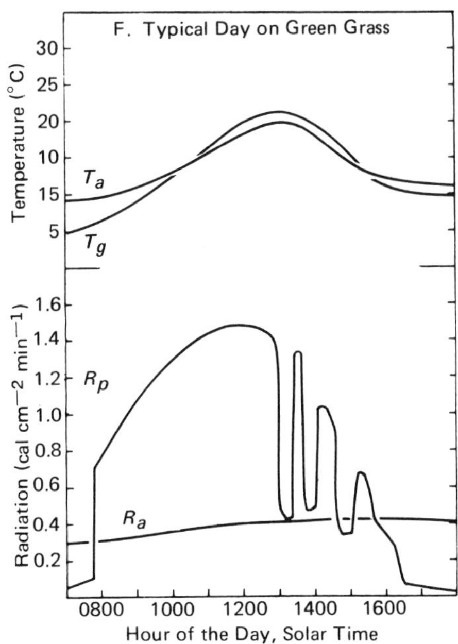

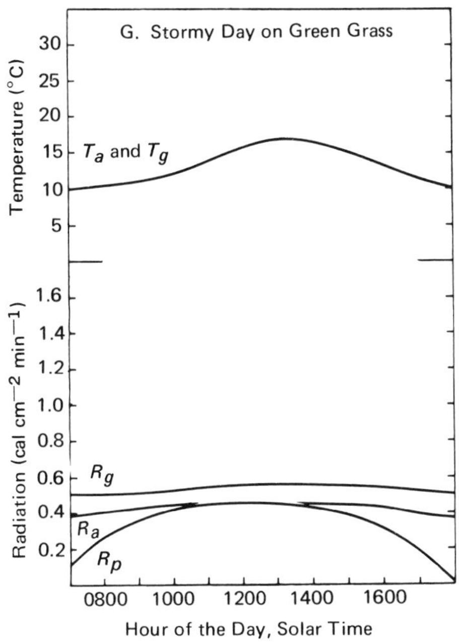

Fig. 18.4. Daily variation in the radiant environment in different microhabitats under different general weather patterns. Each figure represents the daily variation in the radiant environment for a different combination of microhabitat and weather pattern. Upper portions of these figures describe daily fluctuations of air temperature (T_a) and temperature of the surface of the substrate (T_g). In Fig. 18.4E only the temperature of the air 5–10 cm inside the burrow entrance ($T_{\text{burrow entrance}}$) is plotted. Lower portions of Fig. 18.4A–G describe daily fluctuations of solar radiation as a combination of energy from direct sunlight and energy from scattered skylight (R_p). Fluctuations of downward thermal radiation from an average atmospheric temperature (R_a) and upward thermal radiation from an average substrate temperature (R_g) are also given. $R_{\text{burrow entrance}}$ is the blackbody radiation at $T_{\text{burrow entrance}}$. (From Morhardt and Gates, 1974. Copyright, the Ecological Society of America.)

shape to the direction of the sun strongly affects the amount of direct solar radiation absorbed by the animal. The climate diagram is presented by Porter and Gates (1969) as a means of describing the range of climatic variables in which a given animal can survive. A climate diagram shows the range of possible values of average radiation absorbed by an animal (Q_{abs}, on the abscissa) of a given shape and absorptivity for any given air temperature (on the ordinate). The lowest Q_{abs} for an animal at any air temperature would occur under a clear, cold night sky when the animal was receiving only infrared radiation from the ground (which would be close to air temperature) and the atmosphere (which might be as cold as -25 or -30°C under clear summer alpine conditions). The highest Q_{abs} would exist for an animal at any given air temperature during the most intense direct solar radiation when the animal's body was oriented in such a way that a maximum of surface area was exposed to direct sunlight. The absorptivity of the animal to solar radiation and the reflectivity of the substrate also affect the position of the line representing maximum absorbed radiation of the animal at each ambient temperature. In a climate diagram, all the possible radiant environments that an animal of a specific shape and absorptivity might encounter are included between the lines of minimum and maximum absorbed radiation. The relationship between temperature and radiation for the blackbody condition is absolute because the absorptivity of animals to thermal radiation is 1.0, and because the nondirectional quality of the radiation in the blackbody situation eliminates the dependence of Q_{abs} on the shape and orientation of the animal. Even though emissivities of objects surrounding an animal in nature would not be perfect, variables in a complex radiant environment can combine to result in an effective blackbody situation; i.e., a situation where the radiation received by the animal is equivalent to that which would be emitted by uniform surroundings acting as perfect blackbody radiators at the same temperature as the air.

To construct a climate diagram, it is necessary to convert environmental radiation, as measured by field instrumentation, to the average absorbed radiation (Q_{abs}) of an animal of a given shape and absorptivity at specific orientations to the sun. The effect of the actual size of the animal can be eliminated by using proportions of surface area receiving radiation to total surface area.

The general equation used to estimate the maximum or minimum radiation absorbed by a squirrel in any given radiant environment is as follows:

$$Q_{abs} = azS + 0.5[as + ar(S + s) + R_g + R_s] \qquad (1)$$

where Q_{abs} = average radiation absorbed by the geometrical solid chosen to represent the animal
a = average absorptivity of fur to solar radiation
S = direct solar radiation
s = indirect solar radiation
r = reflectivity of the substrate
R_g = average thermal radiation from the lower hemisphere (ground)
R_a = average thermal radiation from the upper hemisphere (atmosphere)
z = defined in Eq. 2

In this study a cylinder of length L and diameter D, with hemispherical ends was used to approximate the shape of a squirrel. The total surface area of such a cylinder is $\pi LD + \pi D^2$, but for the purposes of calculating Q_{abs}, the proportion of projected area to total surface area is used. (The projected area of an object is effectively the area of the shadow cast by the object on a surface normal to the direction of a point source which is illuminating the object from a great distance.) In calculating maximum and minimum Q_{abs}, the cylinder was considered oriented either with the long axis perpendicular to the direction of the sun ($Q_{abs\text{-}max}$) or parallel to the direction of the sun ($Q_{abs\text{-}min}$). The projected area of this shape when oriented with its long axis perpendicular to the direction of the sun is equivalent to $LD + \pi D^2/4$. Therefore, the ratio of the projected area to the total surface area is

$$\frac{LD + \pi D^2/4}{\pi LD + \pi D^2} = z \qquad \text{in Eq. 1 for } Q_{abs\text{-}max} \tag{2}$$

When the length of the animal is fixed at 2.5 times the diameter, the proportion equals 0.299 for $Q_{abs\text{-}max}$. Similarly, the projected area of the shape oriented with its long axis parallel to the direction of the sun is equivalent to $\pi D^2/4$, and the proportion of the projected area to total area is as follows:

$$\frac{\pi D^2/4}{\pi LD + \pi D^2} = z \qquad \text{in Eq. 1 for } Q_{abs\text{-}min} \tag{3}$$

Using the length to diameter ratio of 2.5, the proportion z reduces to 0.072 for $Q_{abs\text{-}min}$.

The forms of radiation, other than direct solar radiation, represented in Eq. 1 are considered to be of uniform intensity from all directions, either above or below the animal. Therefore, the energy absorbed from these relatively uniform sources of radiation may be adequately described by considering that only the top half of the animal absorbs indirect solar and atmospheric thermal radiation and only the bottom half of the animal receives reflected solar and ground thermal radiation. In each case, one-half the surface area of the animal is absorbing the radiation, and, therefore, the sum of all the terms is multiplied by 0.5.

The value used for r is determined by the reflectivity of the substrate of the microhabitat in which the radiation values are being considered. The value $r = 0.20$ is used for rock and green grass, $r = 0.10$ for dirt. The absorptivity of fur to sunlight, a, is considered to be constant at 0.74 for *C. beldingi*. R_g and R_a may be read directly from Fig. 18.4A–G, but the values of solar radiation measured by the pyranometer, R_p, are not the same as the $S + s$ values used in Eq. 1. When Eq. 1 is used, the shape representing the animal is oriented in a constant relationship to the direction of the sun regardless of the altitude of the sun in the sky. A pyranometer, however, measures the radiant energy incident on a fixed area in a horizontal plane regardless of the altitude of the sun, so the measured intensity of the sun decreases as the sine of the angle between the sun and horizon.

Pyranometer readings may be converted to the values desired for the calculation of radiation absorbed by the animal by using the following equation:

$$S + s = \frac{R_p - s}{\sin \theta} + s \tag{4}$$

where θ is the angle between the sun and the horizontal. These angles may be determined for any latitude at any time of the year and day by use of the Smithsonian Meteorological Tables (List, 1968). Since direct and indirect solar radiation were not measured separately, a standard relationship between the two was used. When direct solar radiation was present, it was considered to be 90 percent of the total. When direct solar radiation was not present, the total measured solar radiation is indirect and Eq. 4 does not apply.

An animal is certainly not always oriented in a constant relationship to the direction of the sun, but the positions used in calculating the solutions to Eq. 1 give the maximum and minimum amounts of absorbed radiation that an animal of that shape can experience. Calculations may be derived for other orientations of an animal or for more complex animal shapes by using the principles explained above.

The somewhat complex case of the picketing ground squirrel is used as an example of an intermediate orientation, because this is a position frequently assumed by the Belding ground squirrels, especially on the tops of rocks or the bare dirt around the burrow entrances.

The equation used to calculate the average radiation absorbed by a vertically positioned ground squirrel, $Q_{\text{abs-v}}$, was the following:

$$Q_{\text{abs-v}} = ayS + 0.4285[ar(S + s) + as + R_g + R_a] + 0.143(as + R_a) \tag{5}$$

In this case, y is the term that represents the proportion of the projected area of the shape when exposed to direct sunlight to the total surface area of the shape. This proportion includes consideration of the angle of the sun from the horizontal, θ, and is described by the following equation:

$$y = \frac{3\cos\theta + (\pi/8)(1 + \sin\theta)}{3.5\pi} \tag{6}$$

The derivation of Eq. 5 is similar in principle to that of Eq. 1, but a somewhat different shape is used to describe the vertical animal since the posterior of the picketing ground squirrel is opposed to the substrate. The shape chosen for the vertical animal was a cylinder of diameter D and length $L + (D/2)$, with a hemispherical top. The total surface area of this shape is $\pi D(L + D/2) + \pi D^2/2$ and is equivalent to the total surface area of the shape used for maximum and minimum exposures. The projected area of the cylindrical portion of the shape is $D\cos\theta \times (L + D/2)$. The projected area of the hemispherical top plus the projected basal portion of the cylinder (which is half an ellipse) varies between the area of a circle of diameter, D, when $\theta = 90°$, and the area of a semicircle, when $\theta = 0°$. This variation is described by writing the projected area of the hemispherical top plus basal area of the cylinder as $(1 + \sin\theta)\pi D^2/8$. The entire proportion of projected area to total area is written

$$\frac{D(L + D/2)\cos\theta + (1 + \sin\theta)\pi D^2/8}{\pi DL + \pi D^2} \tag{7}$$

When the relationship $L = 2.5D$ is used, this proportion reduces to that of Eq. 6.

The radiation other than direct solar radiation is also absorbed differently by the shape used for the picketing squirrel than by the other shape. Since only the

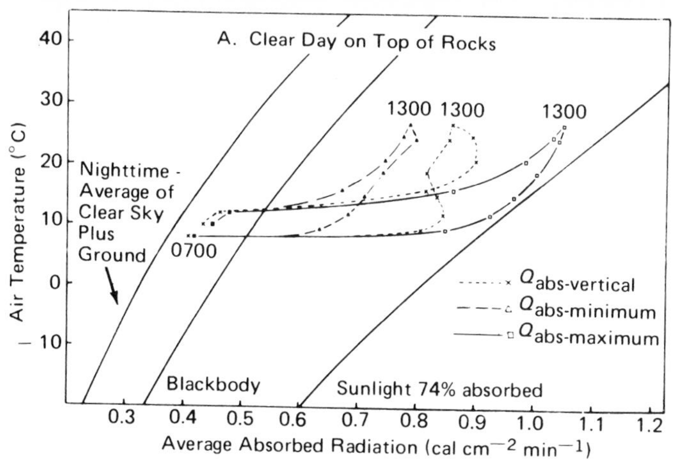
A. Clear Day on Top of Rocks
Air Temperature (°C)
Average Absorbed Radiation (cal cm−2 min−1)
1300 1300
1300
Nighttime - Average of Clear Sky Plus Ground
0700
Qabs-vertical
Qabs-minimum
Qabs-maximum
Blackbody
Sunlight 74% absorbed

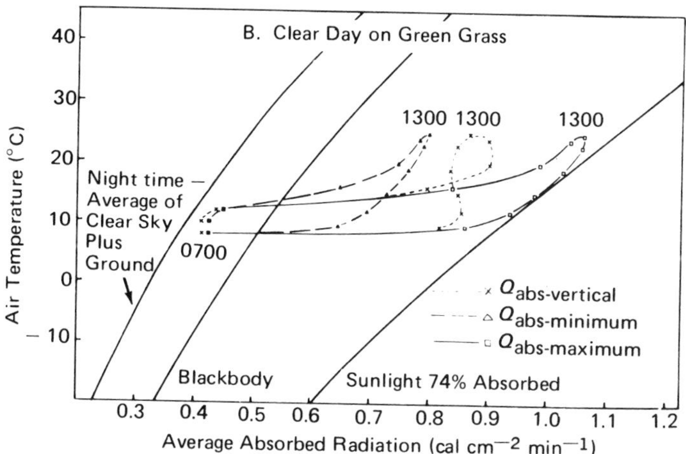
B. Clear Day on Green Grass
Air Temperature (°C)
Average Absorbed Radiation (cal cm−2 min−1)
1300 1300
1300
Night time — Average of Clear Sky Plus Ground
0700
Qabs-vertical
Qabs-minimum
Qabs-maximum
Blackbody
Sunlight 74% Absorbed

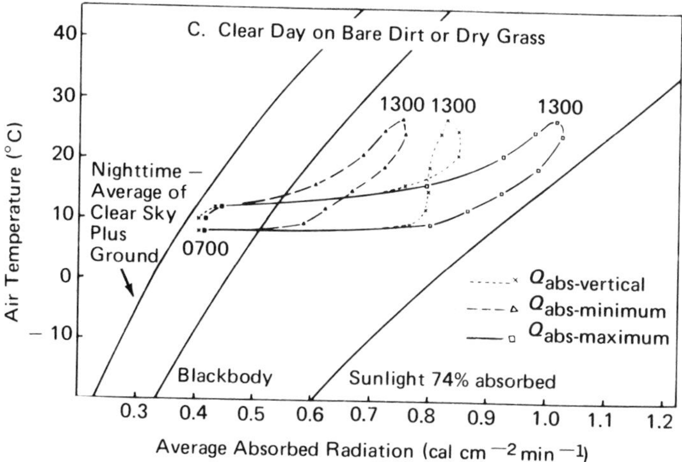
C. Clear Day on Bare Dirt or Dry Grass
Air Temperature (°C)
Average Absorbed Radiation (cal cm−2 min−1)
1300 1300
1300
Nighttime — Average of Clear Sky Plus Ground
0700
Qabs-vertical
Qabs-minimum
Qabs-maximum
Blackbody
Sunlight 74% absorbed

Fig. 18.5

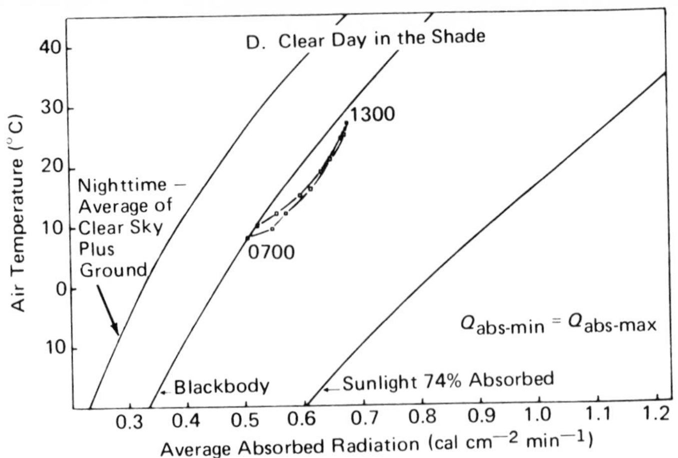
D. Clear Day in the Shade
Air Temperature (°C)
40
30
20
10
0
10
1300
0700
Nighttime — Average of Clear Sky Plus Ground
Blackbody
Sunlight 74% Absorbed
$Q_{abs\text{-}min} = Q_{abs\text{-}max}$
0.3 0.4 0.5 0.6 0.7 0.8 0.9 1.0 1.1 1.2
Average Absorbed Radiation (cal cm^{-2} min^{-1})

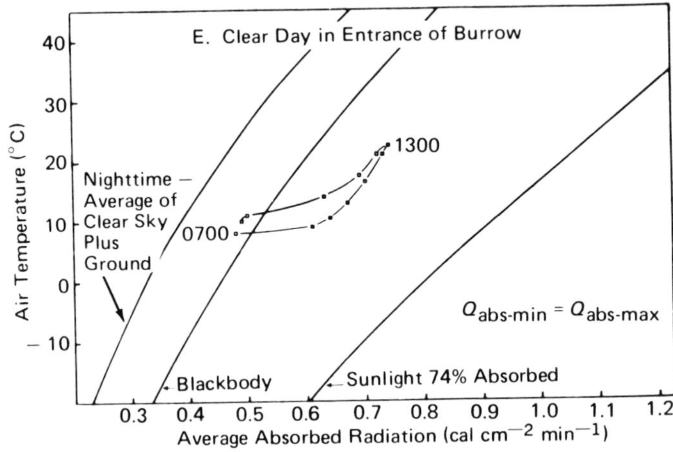
E. Clear Day in Entrance of Burrow
Air Temperature (°C)
40
30
20
10
0
– 10
1300
0700
Nighttime — Average of Clear Sky Plus Ground
Blackbody
Sunlight 74% Absorbed
$Q_{abs\text{-}min} = Q_{abs\text{-}max}$
0.3 0.4 0.5 0.6 0.7 0.8 0.9 1.0 1.1 1.2
Average Absorbed Radiation (cal cm^{-2} min^{-1})

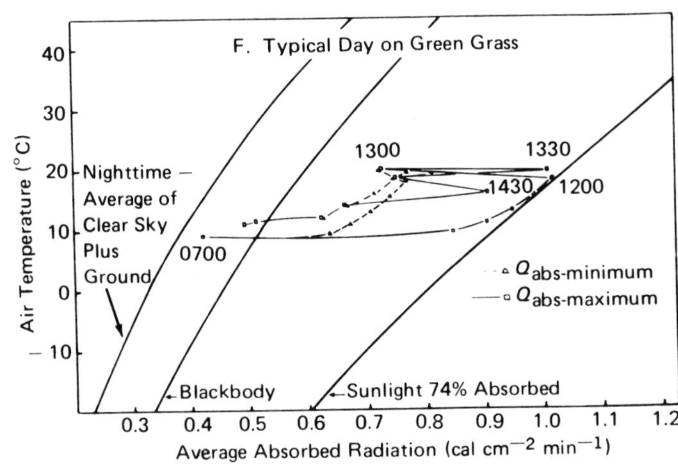
F. Typical Day on Green Grass
Air Temperature (°C)
40
30
20
10
0
– 10
1300
1330
1430
1200
0700
Nighttime — Average of Clear Sky Plus Ground
$Q_{abs\text{-}minimum}$
$Q_{abs\text{-}maximum}$
Blackbody
Sunlight 74% Absorbed
0.3 0.4 0.5 0.6 0.7 0.8 0.9 1.0 1.1 1.2
Average Absorbed Radiation (cal cm^{-2} min^{-1})

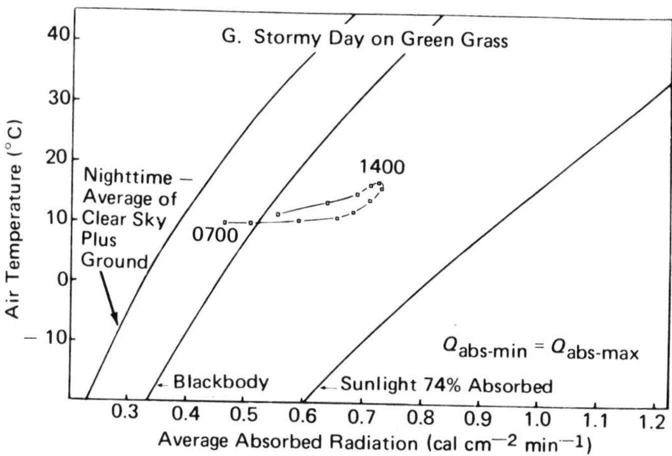

Fig. 18.5. Climate diagrams for the habitat. The data indicate the total amount of radiation absorbed by a geometrical model of a ground squirrel under the combinations of air temperature and radiation shown in Fig. 18.4A–G. The model is oriented in three different ways with respect to direct solar radiation to show how absorbed radiation differs at the same air temperature and the same time of day with differences in orientation toward the sun. The amount of radiation absorbed is greatest ($Q_{abs\text{-}maximum}$) when the long axis of the model is normal to the direction of the sun and least ($Q_{abs\text{-}minimum}$) when the hemispherical end of the model is toward the sun. The amount of radiation absorbed when the squirrel remains vertical ($Q_{abs\text{-}vertical}$) is intermediate between the extremes. Data points are taken directly from Fig. 4A–G at hourly or half-hourly intervals, and are identified at selected points by showing the hour of the day in solar time adjacent to the points. The line for sunlight 74 percent absorbed represents the maximum amount of radiation that could be absorbed by the model in direct sunlight at any air temperature. The blackbody curve indicates the intensity of radiation from a blackbody at any air temperature, and the curve labeled "nighttime average of clear sky plus ground" indicates the minimum of energy likely to be absorbed by the model when exposed to a night sky radiating at a temperature that is cooler than the air.

hemispherical end of the shape receives only indirect solar and atmospheric thermal radiation, the sum of these two quantities is multiplied by the proportion of the surface area of the hemisphere to the surface area of the total shape. The cylindrical portion of the shape receives an average of indirect solar (downward) plus reflected solar (upward), in addition to an average of atmospheric thermal (downward) plus ground thermal (upward) over its entire surface. The sum of these averages is multiplied by the proportion of the surface area of the cylindrical part of the shape to the surface area of the entire shape.

The general format of the climate diagram is used in Fig. 18.5A–G to describe the combinations of air temperature and Q_{abs} encountered by ground squirrels in each of their typical microhabitats at each hour throughout the day. Examples of each orientation of squirrels to the sun are included. There is a climate diagram corresponding to each set of thermal environments described in Fig. 18.4A–G, since the climate diagrams were derived from the hourly data shown in those figures. Since there is no direct solar radiation in Fig. 18.5D or G, maximum and minimum absorbed radiations are identical. For Fig. 18.5E, the calculations for the animal partially in the burrow were done only from the $Q_{abs\text{-}max}$ values for an animal one-third in the sun.

Comparisons among Fig. 18.5A–G will show which of the microhabitats have the most extreme thermal environments, and the importance of the direct solar radiation term. Figure 18.5A–C represent the most sustained of the high-Q_{abs} environments and the greatest difference between the maximum and minimum absorption of direct solar radiation. Figure 18.5F shows a high Q_{abs} until noon, but then the trend is broken by intermittent heavy cloud cover, during which the Q_{abs} falls off to a level equivalent to only indirect solar radiation. On all the climate diagrams in which direct solar radiation is a component, the large increase in Q_{abs} between 0700 and 0800 h is due to the fact that the sun is behind the mountains at 0700 but in full view by 0800. During the same hour the air warms very little, so that the Q_{abs} is almost solely responsible for the large change in the thermal environment between 0700 and 0800. A similar situation occurs during the hour that the sun goes behind the mountains in the evening.

The lowest values of Q_{abs} occur when there is no direct solar radiation at any time during the day (Fig. 18.5D and G). On a clear day (Fig. 18.5D), the air temperature is the same in both shade and full sun, but in the shade there is no direct solar component of the radiation, and the value used for R_a is actually nearer the blackbody value for the temperature of the object (usually a shrub) that is creating the shade than to the blackbody value for the temperature of the clear sky. On the stormy day (Fig. 18.5G), the air temperature does not reach as high a value as on the clear day, but the amount of indirect solar radiation is greater since the cloud cover scatters the direct sunlight rather than cutting it out completely.

When direct solar radiation contributes to the Q_{abs}, the orientation of the animal to the direction of the sun is very important in determining the radiation load. In some cases orientation is almost as important in determining an animal's Q_{abs} as is his choice of microhabitat. However, the animal's biological needs, such as feeding, require that much of its time be spent in certain microhabitats and, also, limit the possibilities for different orientations to the sun. The suitability of an animal to a given habitat will depend partially on its ability to thermoregulate while fulfilling its needs to eat, reproduce, escape enemies, maintain territory, or perform other functions.

Climate diagrams describing thermal microclimates do not include the environmental variables of wind speed and water-vapor density. However, these factors are used in solving the energy-budget equation of animals, and their effects are plotted on climate diagrams of the animals.

From the present work, only a few broad generalizations can be made about the wind speeds occurring at the study site. The air is often very still (less than 25 cm s^{-1}) just before and during sunrise on the meadow. Shortly after the sun starts shining on the meadow, a slight breeze (about 40–50 cm s^{-1} at a height of 10 cm above the ground) begins, and continues to increase until it is about 200 cm s^{-1} near the ground in the middle of the morning. Air speeds of at least about 100 cm s^{-1} are likely to continue throughout the day, with frequent gusts up to 300 or even 500 cm s^{-1}. Animals on the tops of rocks would usually be exposed to wind speeds 50–100 percent greater than those occurring near the surface of the dirt or grass. Wind speeds of 10, 100, and 1000 cm s^{-1} are used in the solutions of the energy-budget equation.

Relative humidities were measured at the study site on only a few occasions, but it is probably safe to assume that water-vapor densities were usually low on clear days, with higher values occurring only during and following afternoon showers or on stormy days. The full range of water-vapor densities was probably between about 4 and 20 mg of water vapor per liter of air, with the higher values occurring in tall green grass, in burrows, or during stormy weather. The rate of evaporative water loss from the animal may be determined experimentally at any desired value of ambient water-vapor density, and the appropriate rate then used in the solution of the energy-budget equation.

Combined Use of Climate Diagrams for the Animal and for the Microhabitat

Purpose

Climate diagrams for Belding ground squirrels were calculated using data from laboratory experiments designed to yield simultaneous values for metabolic rate (M), rate of evaporative water loss (E), body temperature (T_b), and insulation (I), for squirrels subjected to a broad range of environmental temperatures (see Morhardt, 1971; Morhardt and Gates, 1974). The climate diagrams for the squirrels are compared to climate diagrams for microhabitats to determine the sets of natural climatological variables under which Belding squirrels can maintain thermal equilibrium. Metabolic rates, body temperatures, and some other physiological variables of an animal in the field can be estimated by using the combined climate diagrams. Rates of change in body temperature of an animal in the field under nonequilibrium conditions are predictable using the combined climate diagrams and the equation presented.

Methods

A range of measured values was used to calculate climate diagrams for squirrels using the equation and methods presented by Porter and Gates (1969). Solutions to the energy-budget equation were plotted on climate diagrams for the standard wind speeds of 10, 100, and 1000 cm s^{-1}, representing still air, a gentle breeze, and a good wind, respectively. Only two examples of the sets of physiological and physical variables used to solve the energy-budget equation are presented here and are indicated on the diagrams (see Morhardt, 1971, for the full range of values examined).

Results and Discussion

Figure 18.6 shows the climate diagram for a large Belding squirrel combined with the range of values of T_a and Q_{abs} that could occur on a clear day when the animal was exposed to direct solar radiation. The stippled area corresponds to the possible range of combinations of T_a and Q_{abs} that could occur on a clear day on tops of rocks, on green grass, or on bare dirt, as shown in Figs. 18.5A–C. These microhabitats were chosen because they represent the conditions in which *C. beldingi* might experience thermal stress. Any of the climate diagrams for microhabitats (Fig. 18.5A–G) can be superimposed on climate diagrams for the squirrels to indicate in which microhabitats and at which times of day squirrels might be subject to thermal stress.

Figure 18.7 helps to illustrate how climate diagrams may be used to estimate the length of time which an animal may remain in a microhabitat unsuitable for establishing thermal equilibrium. If a ground squirrel of D = 5.3, with the climate diagram illustrated in Fig. 18.7, were grazing in the grass on a clear day at noon

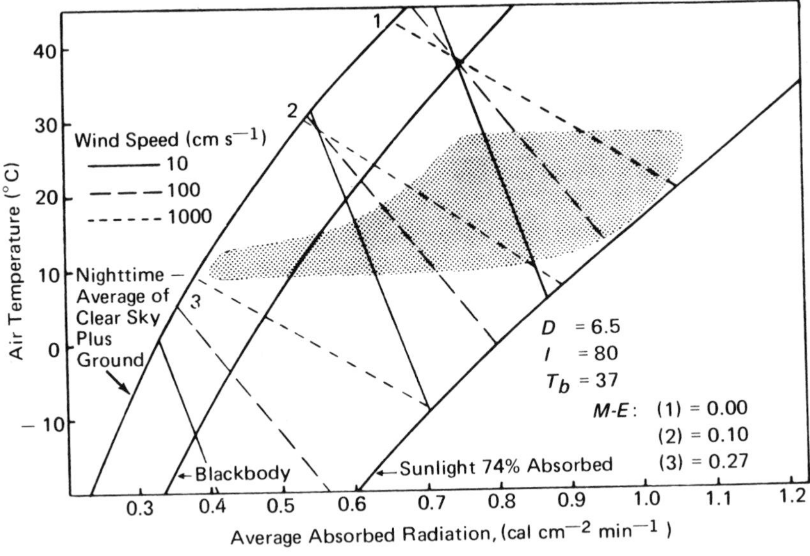

Fig. 18.6. Climate diagram for a ground squirrel superimposed on the range of combinations of T_a and Q_{abs} occurring on a clear day. Each of the three numbered groups of lines on this figure was calculated using the different values for M minus E indicated in the legend. Constant values for T_b, I, and D, as indicated, were used for the calculations. The three lines in each group represent the combinations of T_a and Q_{abs} that balance the energy-budget equation at the three corresponding wind speeds shown in the legend when the values for D, I, T_b, and $M - E$ are those stated for the appropriate group of lines. The stippled area includes the full range of combinations of T_a and Q_{abs} that an animal could encounter on a clear day on the green grass, on the tops of rocks, or on the bare dirt. Values for M and E are in cal cm^{-2} min^{-1}, for I in cm^2 min cal^{-1}, for D in cm, and for T_b in °C. (From Morhardt and Gates, 1974. Copyright, The Ecological Society of America.)

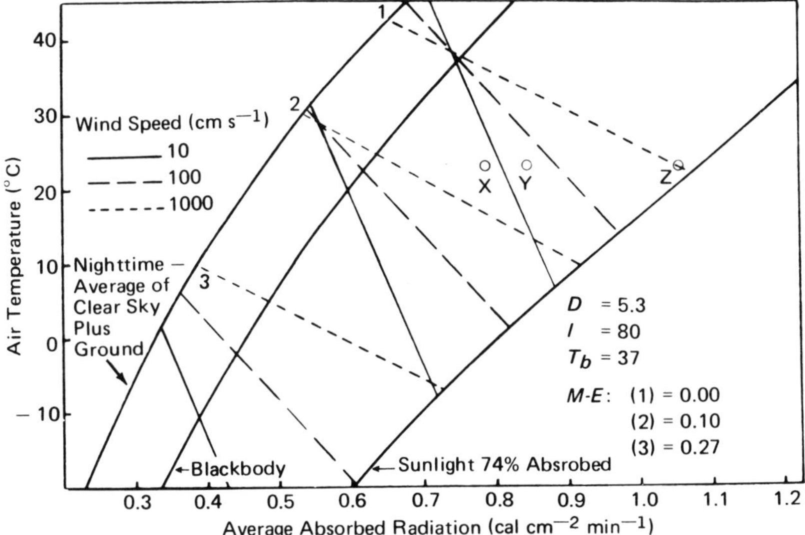

Fig. 18.7. Climate diagram for a ground squirrel and for a warm microhabitat at noon. Each of the three numbered groups of lines on this figure was calculated using the different values for M minus E indicated in the legend. Constant values of T_b, I, and D as indicated were used for all the calculations. The three different lines in each group represent the combinations of T_a and Q_{abs} that balance the energy-budget equation at the three corresponding wind speeds shown in the legend when the values for D, I, T_b, and $M - E$ are those stated for the appropriate group of lines. Points X, Y, and Z indicate three of the possible combinations of T_a and Q_{abs} that a squirrel would experience in the grass on a clear day at noon. Point Z is at $Q_{abs\text{-}maximum}$, point X is at $Q_{abs\text{-}minimum}$, and point Y is $Q_{abs\text{-}vertical}$. The text discusses transient states of animals at these points.

with its body oriented perpendicular to the direction of the sun ($Q_{abs\text{-}max}$), its thermal microclimate would be represented by point Z, where it could not maintain thermal equilibrium unless $V = 1000$. Points X and Y represent thermal environments for the animal oriented parallel to the direction of the sun ($Q_{abs\text{-}min}$), and the vertical animal ($Q_{abs\text{-}v}$), respectively. Only point X is within the range of thermal microclimates where the animal can maintain equilibrium in still air ($V = 10$). If the animal were at Y when $V = 10$, he would be absorbing about 0.035 cal cm^{-2} min^{-1} more energy than he could lose at that temperature and wind speed. The net input of energy to the animal in any environment where equilibrium cannot be established, can be determined by the difference between the Q_{abs} at the point in the thermal microhabitat that is being occupied, and the maximum Q_{abs} at the same T_a and appropriate V under which thermal equilibrium can be established. In Fig. 18.7 points X, Y, and Z are all at $T_a = 23°C$. The difference in Q_{abs} between point Y and the point where the line of $V = 10$ in group 1 crosses $T_a =$ 23°C is 0.035 cal cm^{-2} min^{-1}. The rate of change in T_b that would result from the net gain or loss of heat can be estimated by the equation

$$\frac{\Delta T_b[°\text{C}]}{\Delta t[\text{min}]} = \frac{\Delta Q[\text{cal cm}^{-2}\text{ min}^{-1}]\cdot\text{surface area}[\text{cm}^2]}{\text{body weight}[\text{g}]\cdot c[\text{cal g}^{-1}\ °\text{C}^{-1}]} \tag{8}$$

In Eq. 8, ΔQ is the difference between total energy gained and total energy lost, and c is the specific heat of animal tissue. The value $c = 0.85$ (derived from information in Kleiber, 1961) is used in this study. The body weight of the animal can be easily measured, and for this work, a functional relationship between body weight, surface area, and diameter was established. If Eq. 8 is used to determine the rate of change in T_b that the animal in Fig. 18.7 with $D = 5.3$ will experience while there is a net gain in his energy balance of 0.035 cal cm^{-2} min^{-1}, it can be shown that this animal's T_b will change initially at the rate of 0.06°C min^{-1}. (A Belding ground squirrel of $D = 5.3$ weighs about 220 g and has a surface area of about 325 cm^2). If the squirrel could tolerate an increase in T_b of 3°C, he could remain in thermal microhabitat Y for more than 50 min while $V = 10$. The same animal absorbing maximum radiation (point Z) at $V = 10$ would have a net gain in energy of about 0.20 cal cm^{-2} min^{-1} and could remain only about 9 min before his T_b increased 3°C. The rate at which an animal can lose excess heat in a cool environment can be estimated in a similar manner.

Heat gain is a more critical problem than heat loss for a mammal, particularly a hibernator, since the active T_b is closer to the upper lethal temperature than to the lower lethal temperature. Also, for *C. beldingi* during the summer, the thermal environment during sunny days is likely to cause heat stress. If 41°C is considered the upper permissible T_b, an animal starting with a T_b of 37°C can gain heat until its T_b has risen 4°C. A 250-g animal, with a rate of heat gain of 0.05 cal cm^{-2} min^{-1}, could wait approximately 48 min before it needed to avoid any further net gain in energy. A smaller animal would have slightly less time, and a larger animal, more time.

Conclusions

By combining the information shown on the climate diagram for a microhabitat with that given on the climate diagram for an animal, one may determine whether the animal can establish thermal equilibrium in that particular microhabitat at any time of the day. By comparing the climate diagrams of two or more species with the same microhabitat climate diagram, one can predict whether the species are energetically capable of occupying the same microhabitat during the same hours of the day. This information could be helpful in determining the probability of predator–prey encounters, interspecies competition for particular resources, and other multispecies interactions.

An analysis of energy exchange under particular environmental conditions may be used to determine whether habitat preferences and particular behavioral patterns of a species are necessary for the energy balance of the animals. If the animal can establish thermal equilibrium in the environment being considered without selecting specific microhabitats or exhibiting certain types of behavior, then his activities may be attributed to other causes.

This study has shown that Belding ground squirrels are well adapted energetically to the microclimatic conditions that prevail diurnally during the summer in the high mountain meadows of the Sierra Nevada. A comparison of climate diagrams for the microhabitats with those for the squirrels shows that *C. beldingi* could easily withstand much colder environments but not necessarily much hotter ones. However, any environment that is energetically below the thermal neutral zone of an animal requires that metabolic heat production be increased if the body temperature is to remain constant. An increase in metabolic rate utilizes calories that might otherwise be stored as body fat, and to add enough fat to last through an unpredictably long winter, squirrels must spend their summers eating and gaining weight as rapidly as possible.

A medium-sized squirrel maintaining a constant body temperature of 37°C can spend about 6 h of the hottest day feeding in the grass even at moderate wind speeds and with a body orientation such that direct solar radiation is maximally absorbed. By allowing his body temperature to increase to 40°C, the squirrel could increase the length of time he could stay in the grass at $Q_{\text{abs-max}}$ by about 1 h. By utilizing microhabitats where he could lose heat quickly (such as the burrow or the shade) or by changing his orientation to the sun so that Q_{abs} was less than maximum, the squirrel could spend almost the whole day in the grass. Microhabitats can be used selectively to maintain environmental conditions that will require the least expenditure of metabolic energy, but the observations from this study show that most of the activity of Belding ground squirrels need not be directly related to behavioral thermoregulation.

Acknowledgments

This research was supported, in part, by a Ford Foundation Fellowship administered through David M. Gates and the Missouri Botanical Garden. Computer time was paid for by the National Science Foundation, grant G-22296 to the Washington University computer facilities. David M. Gates supplied certain advice and equipment that contributed to this work.

References

Birkebak, R. C.: 1966. Heat transfer in biological systems. Intern. Rev. Gen. and Expt. Zool. *2*, 269–344.

Grinnell, J., Dixon, J.: 1918. Natural history of the ground squirrels of California. Calif. State Comm. Hort. Bull. *7*, 587–595.

Heller, H. D., Poulson, T. L.: 1970. Circannian rhythms. II. Endogenous and exogenous factors controlling reproduction and hibernation in chipmunks (*Eutamias*) and ground squirrels (*Spermophilus*). Comp. Biochem. Physiol. *33*, 357–383.

Kleiber, M.: 1961. The fire of life. New York: Wiley.

List, R. J.: 1968. Smithsonian meteorological tables. Smithsonian Inst. Misc. Collections *114*. Washington: Am. Inst. Press.

Melchior, T. R., Iwen, A.: 1965. Trapping, restraining, and marking arctic ground squirrels for behavioral observations. J. Wildlife Management *29*, 671–678.

Morhardt, S. S.: 1971. Energy exchange properties, physiology and general ecology of the Belding ground squirrel. St. Louis, Mo.: Washington Univ. Ph.D. diss.

———, Gates, D. M.: 1974. Energy exchange analysis of the Belding ground squirrel at its habitat. Ecol. Monogr. *44*, 17–44.

Porter, W. P., Gates, D. M.: 1969. Thermodynamic equilibria of animals with the environment. Ecol. Monogr. *39*, 245–270.

19

Water and Energy Relations of Terrestrial Amphibians: Insights from Mechanistic Modeling

C. RICHARD TRACY

Introduction

Amphibians are, in general, the least studied of the vertebrates. However, because amphibians are the most primitive terrestrial vertebrates, and therefore most amphibian species are coupled to aquatic environments during some or all of their life cycles, the relationship of these animals to their environmental water has been studied extensively. Yet, in a sampling of 68 papers dealing with water relations of amphibians (Table 19.1), 67 papers reported new experimental and/or descriptive data, only two papers appear to have incorporated experimental data into mechanistic models of water transfer, and none viewed the water relations of amphibians in terms of a total water-budget model. Furthermore, almost no research has been reported that illuminates the mechanistic interrelatedness of the total water- and energy-balance systems.

In this paper I shall review a recent set of mechanistic models of the water and energy budgets of terrestrial anurans (Tracy, 1972). I shall also discuss model simulations of water and energy exchanges between anurans and their natural environments in an attempt to revisit several ecological and physiological concepts of the thermal and hydric relations of amphibians.

The Models

Mathematical models of the water and energy budgets of a generalized terrestrial amphibian have been discussed in detail elsewhere (Tracy, 1972). However, I shall abstract the salient points of these models used in the discussion that follows.

Table 19.1 Sampling of Papers Dealing with Various Aspects of Amphibian Water Budgets

Type of Study	No. Studies	Descriptive Data Taken	Synthesis into Mechanistic Model of Water Exchange
(1) Evaporation	12	12	2 (?)
(2) Water uptake from free water	6	6	0
(3) Water uptake from soil	4	4	0
(4) Dehydration tolerance, partitioning of body water, electrolyte physiology, other hydric adaptations	30	29	0
(5) Water economics	14	14	0
(6) Interrelatedness of temperature and water relations	2	2	0
	68	67	2 (?)

(1) Adolph, 1927, 1932, 1933; Campbell and Davis, 1971; Chew and Dammann, 1961; Heatwole *et al.*, 1969, 1971; Jameson, 1965; Lasiewski and Bartholomew, 1969; Machin, 1969; Spight, 1967, 1968a.

(2) Hevesy *et al.*, 1935; Johnson, 1969; Jorgensen, 1949; Parsons, 1970; Stille, 1958; Stirling, 1877.

(3) Dole, 1967a; Heatwole and Lim, 1961; Walker and Whitford, 1970; Spight, 1968b.

(4) Balinsky *et al.*, 1961; Deyrup, 1964; Elkan, 1968; Farrell and McMahon, 1969; Gordon *et al.*, 1961; Gray, 1928; Greenwald, 1972; Heatwole *et al.*, 1971; Krogh, 1939; Lee and Mercer, 1967; Littleford *et al.*, 1947; Mayhew, 1965; McClanahan, 1964, 1967, 1972; McClanahan and Baldwin, 1969; Ruibal, 1962a, 1962b; Ruibal *et al.*, 1969; Sawyer, 1951; Schmid and Barden, 1965; Schmidt-Nielsen and Forster, 1954; Shoemaker, 1964, 1965; Shoemaker *et al.*, 1969, 1972; Shoemaker and Waring, 1968; Thorson, 1956, 1964; Thorson and Svihla, 1943.

(5) Bentley, 1966; Bentley *et al.*, 1958; Cohen, 1952; Lee, 1968; MacMahon, 1965; Main and Bentley, 1964; Packer, 1963; Parsons, 1970; Ray, 1958; Rey, 1937; Schmid, 1965; Spight, 1968; Thorson, 1955; Warburg, 1965.

(6) Lillywhite, 1971b; Spotila, 1972.

Water-Balance Model

The water-balance model is based on the physical principle of the conservation of mass. From this principle we know that for any arbitrarily defined system (e.g., a frog), the water transferred into the system must equal the water transferred out plus any water stored within the system's boundaries. Thus a very general equation of the water balance of an amphibian can be written

$$\dot{m}_s = \dot{m}_a + \dot{m}_u + \dot{m}_0 + \dot{m}_{st} \tag{1}$$

(see Table 19.2 for definitions of the terms). This equation indicates that the only source of water for the amphibian is from water conducted through soil and entering the animal through its integument. In fact, since most terrestrial amphibians do not drink, their usual water source is the moisture in soil (Bentley, 1966; Dole,

1967a). The mechanisms by which water is transferred through soil and into the amphibian are usually somewhat complex; however, an equation based on an analytic solution of traditional soil-water-flow models (Gardner, 1965; Tracy, 1974), using water transport into the animal as a boundary condition, gives an approximation of the non-steady-state water transport between an amphibian and its soil environment:

$$\dot{m}_s = \dot{m}_{eq}\left[1.0 + \frac{K_{frog}(A_v/\pi)^{0.5}e^{s^2}\,\mathrm{erfc}(s)}{k_{soil}}\right] \tag{2}$$

where

$$\dot{m}_{eq} = \frac{\psi_{soil} - \psi_{frog}}{(2A_v K_{frog})^{-1} + (2A_v^{0.5}k_{soil})^{-1}}$$

$$s = \frac{1.0 + K_{frog}(A_v/\pi)^{0.5}}{k_{soil}}\sqrt{\frac{Dt}{A_v/\pi}}$$

Equation 2 will describe the water uptake by an amphibian from a soil substrate that undergoes a transient change in soil moisture as a result of the amphibian's water exchange with the soil. A detailed discussion of the restrictions associated with Eq. 2 is given elsewhere (Tracy, 1972); the most important restriction is that the equation is only exactly correct if the water potential of frog remains constant during the period of time in which the animal's substrate changes water content in response to the animal's presence. (Water potential, ψ, is the potential energy per unit quantity of water. If is arbitrarily defined as mechanical work required to transfer a unit mass of water from a standard reference state, where ψ is taken as zero, to a situation where there is a potential with a defined value. A pool of pure water is defined as the standard reference state where $\psi = 0$. See Rose, 1966, for details of the water-potential concept.)

Most terrestrial amphibians exude a mucous covering over their entire bodies which is usually able to exchange water with the atmosphere as rapidly as a pure water surface. An animal that possesses such a mucous coating can lose water by evaporation extremely rapidly. The mechanism by which evaporative exchange of water between a "wet"-skinned animal and its environments occurs is relatively simple and can be described by the equation

$$\dot{m}_a = A_s h_D(\rho_s - \rho_a) \tag{3}$$

The other major source of water loss for amphibians is urination, and while fully hydrated frogs can apparently eliminate water by urination as rapidly as they can absorb it (Schmidt-Nielsen and Forster, 1954), dehydrated frogs do not generally urinate while rehydrating (Krogh, 1939). As a modeling approximation, it is perhaps safe to assume that most amphibians avoid urinating until fully hydrated and until their water uptakes exceed their water losses.

Whenever an amphibian loses or gains weight by desiccation or rehydration, this weight change must be accounted for as a rate of change of water storage. A

Table 19.2 Symbols

A_c = surface area of an imaginary plane in the amphibian's core (cm^2)
$\bar{a}_i$ = mean animal-surface absorptivities to the radiant fluxes
A_s = animal's surface area from which water vapor is exchanged (cm^2)
A_v = area through which water and energy is conducted from soil to the animal (cm^2)
c_p = specific heat of the animal (cal g^{-1} $°K^{-1}$)
D = soil-water diffusivity (cm^2 min^{-1})
e = base of the natural logarithm
$\dot{E}_{ex}$ = latent heat loss from the respiratory surfaces of the animal (cal min^{-1})
erfc x = complementary error function of x (see Flugge, 1954, for definitions and tabular values of the complementary error function)
F = radiational view factor of the animal's surface to the environment (assumed to be 0.9 for frogs sitting in the stereotyped sitting position)
h_c = convective heat-transfer coefficient (cal cm^{-2} min^{-1} $°K^{-1}$)
h_D = mass-transfer coefficient (cm min^{-1})
J = surface area/volume ratio correction factor of the shape factor for conduction of heat through a frog
k_a = thermal conductivity of air (cal cm^{-1} min^{-1} $°K^{-1}$)
K_{frog} = hydraulic conductance of the frog (g cm^{-2} min^{-1} mb^{-1})
k_{soil} = capillary conductivity of the soil (g cm^{-2} min^{-1} mb^{-1})
k_t = thermal conductivity of animal tissue (cal cm^{-1} min^{-1} $°K^{-1}$)
L = latent heat of vaporization of water (cal g^{-1} $°K^{-1}$)
$\dot{m}$ = metabolic heat production of the animal (cal min^{-1})
$\dot{m}_a$ = net water-vapor flux to the atmosphere (g min^{-1})
$\dot{m}_{eq}$ = water flux at soil-moisture profile equilibrium (g min^{-1})
$\dot{m}_O$ = net of all "other" water fluxes (i.e., fluxes other than $\dot{m}_s$, $\dot{m}_a$, and $\dot{m}_u$) to or from the frog (g min^{-1})
$\dot{m}_s$ = net liquid-water storage (g min^{-1})
$\dot{m}_{st}$ = rate of water storage (g min^{-1})
$\dot{m}_u$ = water flux by urination (g min^{-1})
$\dot{Q}_{abs}$ = total absorbed radiation from all sources (cal min^{-1})
$\dot{q}_{cond1}$ = rate of heat flow by conduction from the "exposed" surface of the animal to its core (cal min^{-1})
$\dot{q}_{cond2}$ = rate of heat flow by conduction from the animal's core to its surface in contact with the substrate (cal min^{-1})
$\dot{q}_{conv}$ = rate of heat flow from the animal by convection (cal min^{-1})
$\dot{q}_{evap}$ = rate of latent heat flow from the animal by evaporation (cal min^{-1})
$\dot{q}_{rad}$ = rate of thermal (long-wave infrared) radiation heat flow from the animal (cal min^{-1})
$\bar{r}$ = reflectivity of the substrate to direct and scattered sunlight
R_g = long-wave infrared radiation from the substrate (cal cm^{-2} min^{-1})
R_s = long-wave infrared radiation from the sky (cal cm^{-2} min^{-1})
S = direct solar radiation (cal cm^{-2} min^{-1})
s = scattered solar radiation (cal cm^{-2} min^{-1})
T_a = air temperature (°K)
t = time
$\bar{t}$ = transmissivity of the vegetational canopy to direct and scattered sunlight
T_c = animal's core temperature (°K)
T_s = animal's surface temperature (°K)
$\dot{V}$ = volume flux of air expired from respiratory surfaces (cm^3 min^{-1})
W = standard weight of the animal (g)
α = animal's absorptivity to long-wave infrared radiation
ε = animal's emissivity to long-wave infrared radiation
θ_{frog} = water content of the frog (g of H_2O per g of standard weight)
π = 3.1416
ρ_a = vapor density of the air (g cm^{-3})
ρ_s = vapor density at the animal surface (g cm^{-3})
ρ_l = vapor density of expired air (g cm^{-3})
σ = Stephan–Boltzmann constant (cal cm^{-2} min^{-1} $°K^{-4}$)
ψ_{frog} = water potential of the animal (mb)
ψ_{soil} = water potential of the soil (mb)

water-storage-rate equation that relates the water content of the frog to its water potential is

$$\dot{m}_{st} = W \frac{d\theta_{\text{frog}}}{d\psi_{\text{frog}}} \frac{\partial\psi_{\text{frog}}}{\partial t} \tag{4}$$

Energy-Balance Model

The energy-balance model, designed to predict an animal's surface and core temperatures as functions of environmental variables, consists of energy balances on two systems: (1) that part of the animal's skin surface that comes in contact with the atmosphere, and (2) the animal's core. The energy balance of the "exposed" surface is given by

$$\dot{Q}_{\text{abs}} = \dot{q}_{\text{rad}} + \dot{q}_{\text{conv}} + \dot{q}_{\text{evap}} + \dot{q}_{\text{cond 1}} \tag{5}$$

where $\dot{Q}_{\text{abs}}$ is the total rate of radiation absorbed from all sources given by (Porter and Gates, 1969)

$$\dot{Q}_{\text{abs}} = \bar{a}_1 A_1 S + \bar{a}_2 A_2 s + \bar{a}_3 \bar{r} A_3 (S + s) + \bar{a}_4 \bar{t}_4 A_4 (S + s) + \alpha\sigma(A_5 R_g + A_6 R_s) \tag{6}$$

(See Table 19.2 for a definition of terms.) The energy loss by thermal radiation to the environment is

$$\dot{q}_{\text{rad}} = \varepsilon\sigma A_s F T_s^4 \tag{7}$$

Convective heat loss is

$$\dot{q}_{\text{conv}} = A_s h_c (T_s - T_a) \tag{8}$$

Evaporative heat loss is

$$\dot{q}_{\text{evap}} = L A_s h_D (\rho_s - \rho_a) \tag{9}$$

And conductive heat loss from the animal's "exposed" surface to its core is

$$\dot{q}_{\text{cond1}} = \left[J \left(\frac{2\pi A_c A_s}{A_s - A_c} \right)^{0.5} \right] k_t (T_s - T_c) \tag{10}$$

(where the bracketed term is a shape factor that accounts for the fact that heat is conducted from a relatively large "exposed" surface area to a much smaller area that contacts the substrate).

The energy balance of the animal's core is

$$\dot{q}_{\text{cond1}} + \dot{M} = \dot{q}_{\text{cond2}} + \dot{E}_{\text{ex}} + \dot{q}_{\text{stor}} \tag{11}$$

where the heat conducted to the animal's core from its "exposed" surface, $\dot{q}_{\text{cond1}}$, plus the metabolic heat generated, $\dot{M}$, are balanced by the heat conducted from the core to the substrate, shown by the equation

$$\dot{q}_{\text{cond2}} = 0.5 \left[J \left(\frac{2\pi A_v A_c}{A_c - A_v} \right)^{0.5} \right] k_t (T_c - T_v) \tag{12}$$

(where the coefficient, 0.5, accounts for the fact that the animal's surface which contacts the substrate reaches a temperature roughly the average between the animal's core temperature and the substrate temperature. The substrate temperature is defined as the temperature of the substrate uninfluenced by the presence of the animal) plus respiratory water loss:

$$\dot{E}_{\text{ex}} = L\dot{V}(\rho_l - \rho_a) \tag{13}$$

plus thermal energy storage in the core,

$$\dot{q}_{\text{stor}} = Wc_p \frac{dT_c}{dt} \tag{14}$$

Model Components

The water- and energy-budget models outlined in the previous sections are simply sets of equations based on the physical principles of the conservation of energy and mass (Fig. 19.1). The terms of these equations consist of independent or environmental variables, dependent or organismal variables, and parameters or attributes of the organism. When these modeling components are separated, as in Table 19.3, it becomes immediately obvious what components of the environment are involved in the determination of the organismal variables. Furthermore, the physical and physiological attributes of the organism directly involved in integrating the thermal and hydric environment of the organism are specifically identified.

Table 19.3 also lists some of the areas of ecological significance associated with selected organismal variables and organism–environment interactions. In this paper, amphibians' body temperatures, hydration levels, and rate of change of hydration level were selected as important dependent variables of the overall water and energy dynamic system because of their significance in establishing (1) environmental constraints on organisms' dispersion, (2) organisms' evolutionary adaptations to thermal and hydric environments, and (3) the role of the environment in bioenergetic processes of organisms and populations.

Insights from Systems Simulation

Elsewhere (Tracy, 1972), all the parameters involved in the energy- and water-budget equations (identified in Table 19.3) have been evaluated for the leopard frog, *Rana pipiens*. Several of these parameters are not specific to leopard frogs; rather, certain of these parameters are applicable to most frogs and toads not greatly different in shape from the leopard frog. The nonspecific parameters include the relationship between area and size of the anuran (size is measured by weight or length), the surface/volume relationship, and the convection coefficients for heat and mass. The rest of the parameters are likely to be specific to each species. In this

Table 19.3 Variables, Parameters, and Areas of Significance for Water- and Energy-budget Modeling

Independent Variables	Parameters	Dependent Variables	Significance
(1) Direct solar radiation	(1) Areas (a) Receiving direct solar radiation (b) Receiving scattered solar and atmospheric thermal radiation (c) Receiving thermal radiation from the substrate (d) Involved in convection and evaporation (e) Involved in conduction	(1) Core body temperature	(1) Environmental constraints on local and geographic distributions
(2) Scattered solar radiation	(2) Spectral absorptivity	(2) Hydration level	(2) Ecological, behavioral, and physiological adaptations to local microclimates
(3) Incident thermal radiation	(3) Surface area/volume ratio	(3) Rate of change of hydration level	(3) Local environmental effects on bioenergetics
(4) Air speed	(4) Convection coefficients (a) For convection of mass (b) For convection of heat		
(5) Air temperature	(5) Water potential of the animal		
(6) Relative humidity	(6) Hydraulic conductance of the animal		
(7) Substrate temperature			
(8) Substrate water potential			
(9) Substrate hydraulic conductivity and diffusivity			

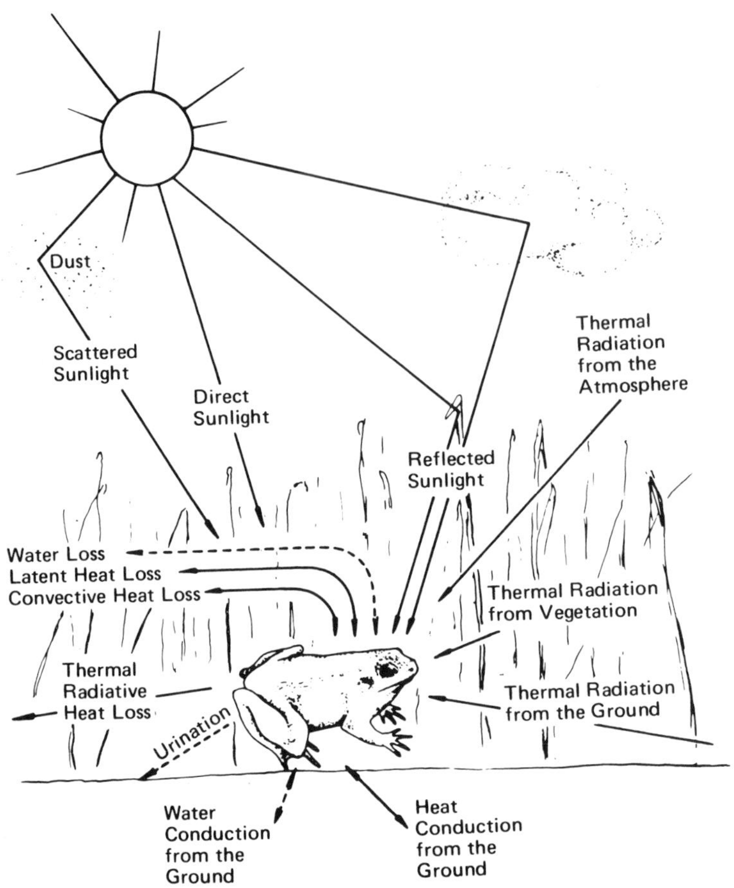

Fig. 19.1. Schematic summary of the exchanges of energy and water between a frog and its environment.

section of the paper I shall discuss computer simulations of a leopard frog's response to idealized microclimates at Pellston, Michigan, and attempt to discuss the ecological implications that stem from this sort of analysis.

The idealized microclimates were generated from the means of 10 years of air-temperature data (1950–1960) taken at the Pellston Airport; solar radiation estimates calculated from the computer program SOLRAD (McCullough and Porter, 1971) with the assumptions that direct and scattered radiation are depleted by 80 percent by the vegetation in "typical" leopard frog habitat (Dole, 1965a, 1965b); wind velocity, assumed to remain constant at 20 cm s^{-1} (approximately 0.5 mph); relative humidity, assumed to reach a high of 100 percent and at sunrise assumed to drop in proportion to the rise in air temperature to a low of 75 percent at 0200 h (C.S.T.) when the air temperature peaked; and substrate temperature, assumed to be equal to air temperature.

Temperature Relations

A simulation of the body temperatures of a 60-g leopard frog in the "idealized" Pellston, Michigan, microclimates (Fig. 19.2) illustrates that northern Michigan frogs, on the average, can probably avoid freezing body temperatures only as early in the spring as late April or early May when these frogs typically emerge from hibernation (Dole, 1967b). However, the environmental cue for emergence from hibernation is unknown and to my knowledge no experiments have been reported that shed light on this subject.

At this emergence time in the spring, the frogs appear to be able to achieve body temperatures that approach 20°C, which is only about 10°C cooler than achievable

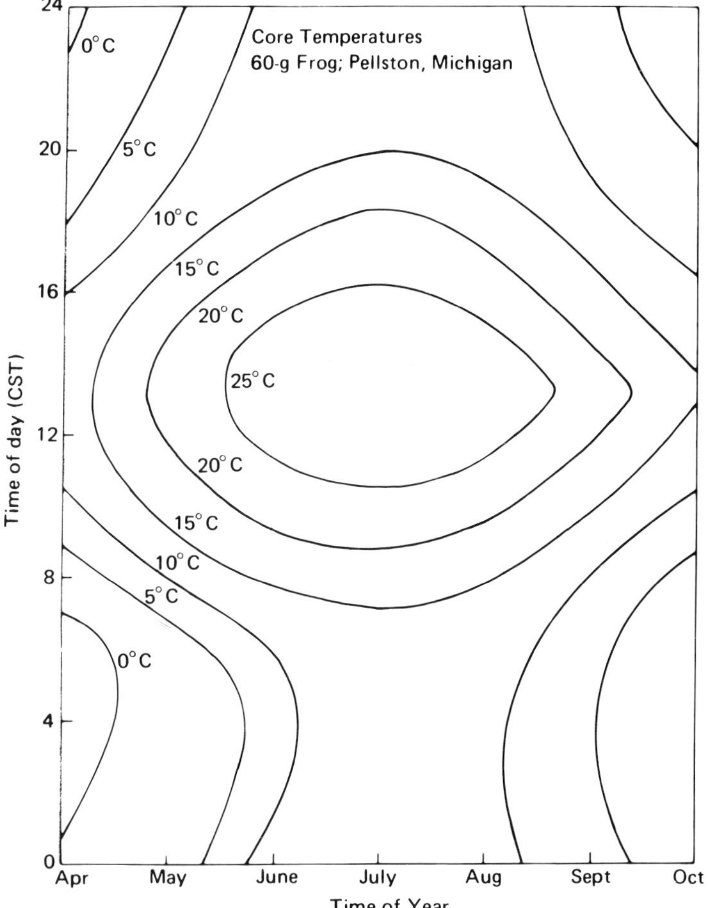

Fig. 19.2. Model-simulated core body temperatures of a 60-g leopard frog in "idealized" microclimates at Pellston, Michigan.

in the warmest summer months. Furthermore, the highest body temperature achieved in the "idealized" microclimate simulations is only about 27°C, which is well within the upper lethal limit for leopard frogs. These observations illustrate the relatively narrow range of body temperatures available to moist-skinned amphibians as compared with the body-temperature ranges of many reptiles. This is due to the damping influence of the high evaporation rates concomitant with "wet" skins (Tracy, 1972). This comparison between reptiles and amphibians is interesting when one considers the second item in the significance column of Table 19.3, concerning evolutionary adaptations to microclimates. Many reptiles, which can potentially experience wide body temperature ranges, have developed elaborate physiological (Norris, 1967) and behavioral (DeWitt, 1967) adaptations, to deemphasize this large temperature range. However, experimental evidence for any such adaptations among amphibians is very scarce (Lillywhite, 1971a). Lillywhite (1970) reported behavioral observations of free-ranging bullfrogs that he interpreted as thermoregulatory. His evidence was primarily the fact that his animals did not undergo a very great body temperature variation throughout the daylight hours. However, the simulations presented in Fig. 19.2 and elsewhere (Tracy, 1974) suggest that one should not expect a great body-temperature variation for moist-skinned amphibians even if they had no special thermoregularity adaptations.

One test of Lillywhite's contention that the low body-temperature variance in free-ranging bullfrogs indicates a behavioral thermoregulatory adaptation would be to force bullfrogs to remain in a homogeneous natural environment in which they would have little opportunity, behaviorally, to exploit thermal environmental differences to see if their body-temperature variance would be similar to that of frogs fully free to thermoregulate behaviorally. This is not a definitive test for the existence or precision of behavioral thermoregulation, but it would roughly indicate the extent to which daily body-temperature variance is damped by purely passive processes such as uncontrolled evaporation from "wet" skin. I recently performed just such an experiment in which I enclosed adult bullfrogs in hardware cloth ($\frac{1}{4}$-in. mesh) pens approximately 50 cm in diameter. One pen, containing a 650-g bullfrog, was totally unshaded from sunlight throughout the day, and another, also containing a 650-g bullfrog, was set up in a dense, shading canopy of quack grass approximately 1 m deep. Both pens were set up on black organic soil of a disturbed grassy border of the Nielson Marsh on the University of Wisconsin campus. The frogs were forced to remain in the pens from 0800 to 1900 h (C.D.S.T.) on a hot cloudless day (June 13, 1973). The body temperatures of the frogs were measured hourly with a Schulthesis rapid-recording thermometer inserted 2–4 cm cloacally. Total hemispherical solar radiation in the sunlit pen was approximated by measuring the radiation in a duplicate sunlit pen, not containing a frog, with a Eppley pyranometer. The direct component of this hemispherical solar radiation was calculated from program SOLRAD (McCullough and Porter, 1971). The solar radiation in the shaded pen was approximated by Eppley measurements in the grass canopy outside the pen. Air and wet-bulb temperatures in the sunlit pen were approximated by aspirating a Bacharack sling psychrometer approximately 5 cm above ground just outside the pen. Air temperatures in the shaded pen were measured with a copper-constantan thermocouple. Wet-bulb temperature in the canopy was measured with a fine-wire (36-gauge copper-constantan) psychrometer. Air speed in each pen was measured with a Hastings omnidirectional hot-wire anemometer set at approximately 5 cm above the ground. Substrate temperatures were measured with the Schultheis thermometer inserted into the soil 2–4 mm beneath the surface. After the hourly body temperature and microenvironment measurements were made, water was poured on the soil in the pens to ensure a constantly saturated soil substrate (substrate soil-water potential, measured with a null balancing tensiometer, never fell below −100 mb).

The results of this experiment (Fig. 19.3) show that the frog, which was forced to sit in the sunlit pen, had a body temperature range of only 4°C during the 9-h period between 1000 and 1900 h (C.D.S.T.). Furthermore, the predicted body temperatures, based on the environmental measurements made during the experiments and modeled with the assumption that the animal remained motionless in the sterotyped sitting position (Tracy, 1972), mimicked those measured on the living animal quite well. This suggests that the narrow body-temperature range of the sunlit bullfrog was not actively regulated by the frog. This compares with

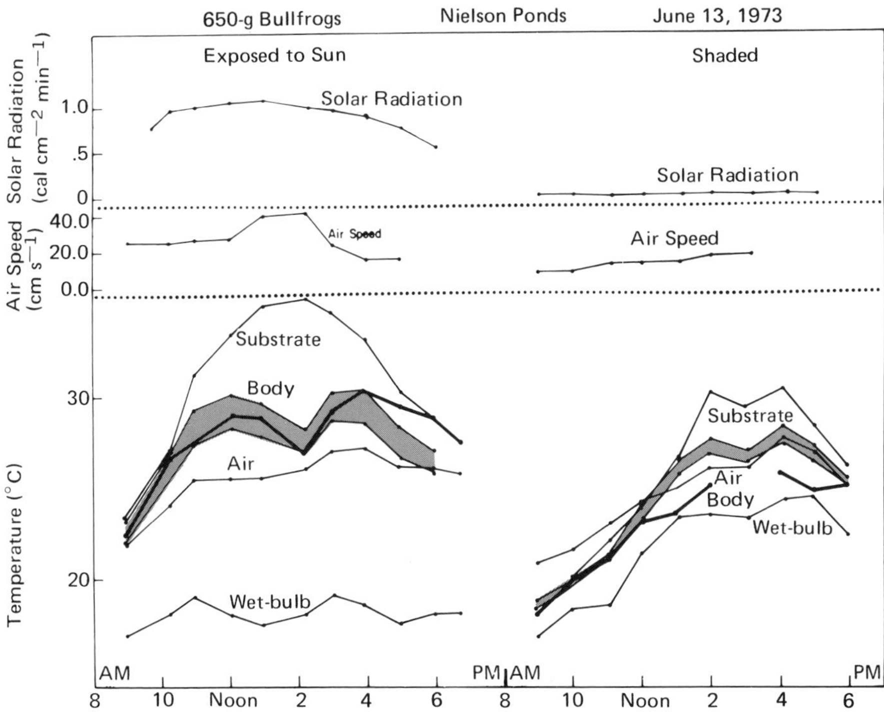

Fig. 19.3. Results of body-temperature experiments with two 650-g bullfrogs confined in shaded and sunlit hardware-cloth pens in the Nielson Marsh on the University of Wisconsin campus. The shaded band represents model predictions of body temperature based on the environmental variables measured during the experiment. The lower bounds of the shaded band assumes that there is no turbulence in the ambient air and the upper bounds of the band assumes a 50 percent enhancement in convective heat transfer as a result of turbulence in the air.

Lillywhite's observations of a free-ranging bullfrog in California that "for most of the day (ca. 1000–1800 hr) the [observed bullfrog's] body temperature was maintained within a relatively narrow range (4.6°C) by alternately emerging from and re-entering the water."

In my experiment the frog that was forced to remain in the shade (Fig. 19.3) had a larger body-temperature range than the sunlit frog. (The explanation for this has to do with the fact that the principal energy sources for the shaded frog were from convection and thermal radiation, whereas the main energy source for the sunlit frog was solar radiation. How amphibians' body temperatures are related to components of the thermal environment and environmental heat flows is analyzed elsewhere; Tracy, 1972.) Therefore, if it were physiologically advantageous for amphibians to have body temperatures restricted to a relatively narrow range (Bartholomew, 1970), it would appear that they could achieve this restricted body-temperature range by remaining in view of the sun throughout the day. However, the sunlit frog in my experiment became visibly stressed by dehydration by midday. Just before noon, the sunlit frog stopped producing urine and also began to place more of its ventral surface in contact with the saturated substrate in postures that have been called the "water absorption response of an anuran" (Stille, 1958). Therefore, it seems important to measure the advantages of "behavioral thermoregulation" attained by passively remaining in the sun, against the disadvantages of physiological damage due to dehydration.

Two conclusions appear appropriate from my analysis of the temperature relations of amphibians. First, the evidence that adult amphibians employ behavioral thermoregulation by shuttling between sun and shade or between sun and water does not appear to be substantial in view of the evidence that it is apparently unnecessary for amphibians to do more than sit passively in view of the sun to maintain relatively narrow body-temperature ranges throughout daylight hours. Second, it appears that the temperature relations of amphibians will have to be carefully evaluated in light of the interrelated water relations of these animals.

Water Relations

A simulation of the evaporative water loss rates of a 60-g leopard frog in the "idealized" Pellston, Michigan, microclimates (Fig. 19.4) illustrates the wide range of evaporating environments to which amphibians can be subjected. For example, in early July, a 60-g Pellston frog could lose water at rates ranging from the inconsequential rate of less than 0.1 g h^{-1} throughout most of the night, to the highly desiccating rate of 4.0-g or 6 percent per hour near midday (this daytime rate would be even higher if the frog were fully exposed to the sun; the simulation assumes 80 percent shading from the sun's radiation). Figure 19.5 further illustrates that these evaporation rates are so great that there appears to be no time during the year that leopard frogs can go without a source of water for 24 h and not dehydrate to at least 85 percent of their fully hydrated weight. The analysis presented in Fig. 19.5 also shows that since the evaporation rate for the modeled frog is so low at

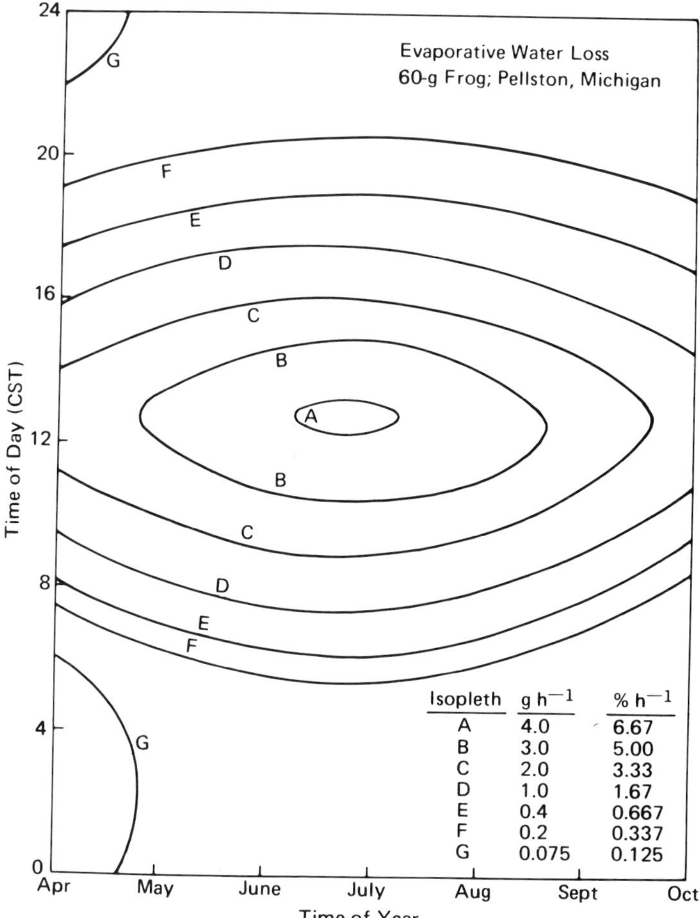

Fig. 19.4. Model-simulated evaporative water-loss rates of a 60-g leopard frog in "idealized" microclimates at Pellston, Michigan.

nighttime, a well-hydrated frog can go without a water source throughout the night and well into the daylight hours before being threatened by severe desiccation. However, it appears to be impossible for frogs to abandon their water source before noon and not become severely dehydrated before evening.

Since the usual source of water for leopard frogs is moist soil, and leopard frogs appear to be strongly tied to their soil-water source, environments whose soils cannot supply water fast enough for frogs to balance their water budgets should be regarded as unsuitable leopard frog habitats, and hence an environmental constraint on the local and geographic distribution of these frogs (see "significance" column of Table 19.3). Figure 19.6 illustrates an analysis of the soil-water potentials necessary to balance exactly the evaporation rates of Fig. 19.4 for 60-g frogs dehydrated to 97 and 85 percent of their fully hydrated weights (the soil is assumed to be a sandy loam with the water-transport characteristics reported in

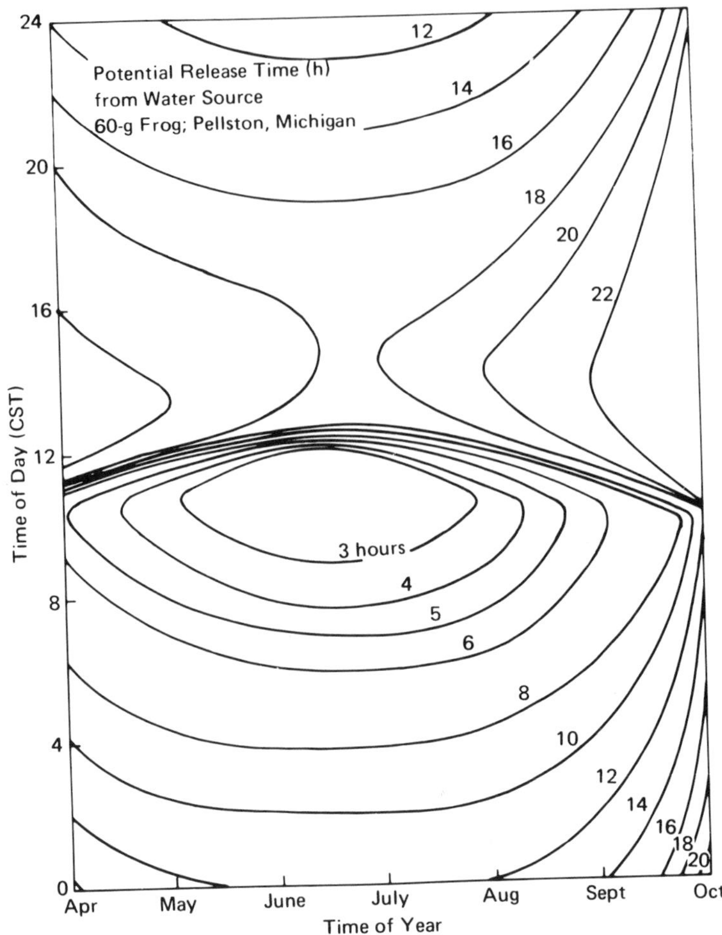

Fig. 19.5. Model-simulated "release" times necessary for a fully hydrated 60-g leopard frog to lose 15 percent of its body weight by evaporation at the rates given in Fig. 19.4 for "idealized" microclimates at Pellston, Michigan.

Gardner, 1965). This analysis shows that the range of soil moistures providing suitable habitats for leopard frogs is apparently very limited. In fact, there appears to be no time during the year in which a leopard frog can balance its water budget on soils as dry as most soils at field capacity (most rain-wetted soils are considered to have water potentials of -330 mb after gravitationally induced draining is complete; Spight, 1967). Therefore, one should not expect to find leopard frogs at great distances from habitats, such as marshes, where soils are poorly drained.

In Dole's 1965a, 1965b, 1967a, 1967b, 1971) exhaustive studies of the natural ecology of leopard frogs, he described the usual dispersion of leopard frogs in Budzinski's pasture near Pellston, Michigan, as being confined to "permanently wet" soil. Dole (1967a) reported soil-water contents for this permanently wet area ranging from 37 to 375 g of H_2O/g of dry soil. A recent calibration of the soil in

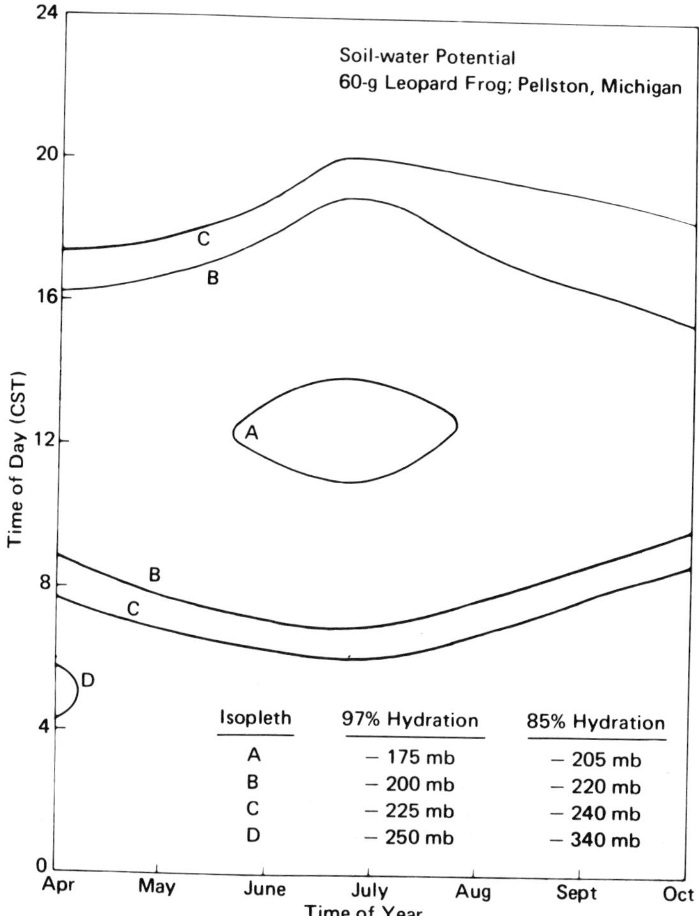

Fig. 19.6. Model-simulated soil-water potentials necessary to balance the evaporation rates given in Fig. 19.4 for a 60-g leopard frog in "idealized" Pellston, Michigan, microclimates at equilibrium body hydration levels of 97 and 85 percent.

Budzinski's pasture (Tracy, 1974) shows that even the driest soil of the permanently wet area within which leopard frogs are usually confined is saturated (i.e., $\psi = 0$ mb). Therefore, it indeed appears that leopard frogs are often severely restricted by their water relations, as pointed out from simulations with the water-balance model.

Bioenergetics

If one combines energy-budget models, which describe the body temperatures of an amphibian in terms of environmental variables (Tracy, 1974), with physiological models, which describe the metabolic rates of an amphibian in terms of its body temperature (Hutchinson *et al.*, 1968), the resultant model will describe the metabolism of the organism in terms of environmental variables. With such a model I have integrated the "standard" metabolic rate (defined as the metabolic rate of the animal in the absence of specific dynamic action and exercise) of a leopard frog living its third season in the idealized Pellston, Michigan, microclimates (Fig. 19.7). From Fig. 19.7 we can see that the seasonal metabolic requirement for this idealized 2-year-old frog is only 58 kcal (another 1–2 kcal is utilized during the hibernation period in winter; Daniel Rittschof, University of Michigan, unpublished data). If we assume that the efficiency with which leopard frogs digest their ingested energy is 73 percent (Gary Smith, Furman Univ., personal communication), then the frog's seasonal ingested energy requirement is approximately 80 kcal. Furthermore, if we assume that the frog's food items contain about 5 kcal (a caloric equivalent of the dry weight of prey), then the frog's seasonal food requirement is roughly 16 g of prey dry weight. This energy requirement should be

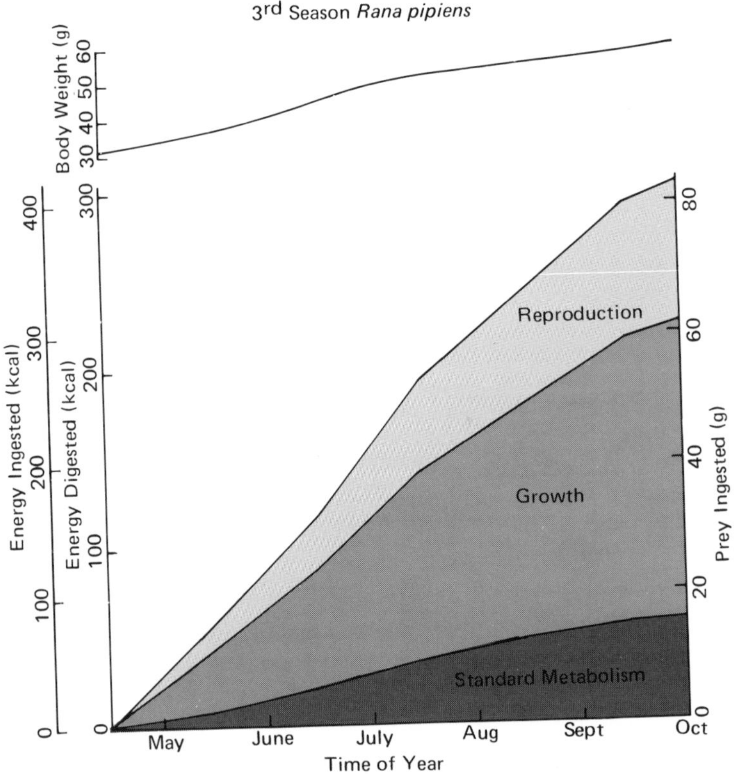

Fig. 19.7. Model-simulated seasonal energetic requirements of a 60-g leopard frog in "idealized" Pellston, Michigan, microclimates. Seasonal costs for growth and reproduction were approximated from body and ovary growth rates for Pellston leopard frogs living in natural environments (Daniel Rittschoff, Dept. Zool., Univ. Michigan, unpublished data). The units of the prey-ingested axis are grams of dry weight.

regarded as minimal since it does not account for exercise, specific dynamic action, growth, or reproduction.

Rittschoff (unpublished data) has described the natural body growth of leopard frogs, from a Pellston population, through their first two seasons. If you project this two-season growth trend through the third season, it appears that a leopard frog 72 mm long (snout–vent length) weighing 32 g at the beginning of its third season would grow to 86 mm and almost double in weight, to 60 g, by the end of that season (Fig. 19.7). The food requirement for this amount of growth would be 45 g (dry weight) of prey (this assumes that all the digested energy can be used for growth at the rate of 1 g for each 5 kcal). This boosts the total prey requirement to 61 g (dry weight), which means that, if one assumes that the energy requirement for exercise in this very sedentary frog (Dole, 1965b) is low, the frog would require just over its own body weight in dry prey to survive and grow naturally through its third season.

There is an additional energy requirement placed on female frogs which produce their first egg masses during their third season. Rittschoff (unpublished data) quantified the relationship between body and ovary weights for leopard frogs, and from his relationship it appears that third-season female leopard frogs usually accumulate about 14 g of eggs that they lay in the spring of their fourth season. The food requirement for this egg production is about 21 g of prey dry weight (using the same assumptions as for growth), which boosts the food requirement for a third-season female to about 82 g (dry weight).

From the analyses presented in this section, certain educated speculations (hypotheses) concerning the potential impact of leopard frogs as predators and the environmental effects on the total energetic requirements of leopard frogs can be made. For example, it can be seen that the efficiency for the conversion of biomass from arthropod prey to leopard frogs might be as great as 51 percent for reproductively active female frogs and 45 percent for adult males (this certainly must represent an upper limit since exercise metabolism and specific dynamic action were considered to be negligible in the calculation of these figures). This implies that leopard frogs might not impart a great impact on their prey populations, because, as a lower limit, frog populations may require only about six times their own biomass in prey, live weight (this assumes that the prey contain 80 percent water). Furthermore, the component of leopard frogs' total seasonal metabolic requirement most directly related to the environment is the standard metabolism, which is apparently only about 20–26 percent of the total seasonal requirement. Therefore, slight differences in the thermal environments of leopard frogs may result in very little difference in these frogs' total metabolic requirements and hence very little difference in their impact on prey populations. This would imply that ecological studies concerned with carbon flows through amphibian populations might require much greater precision in the measurements of size and composition of the populations than in measurements of the thermal environments in which the animals live.

These educated speculations will certainly require further theoretical, experimental, and descriptive research, both in the laboratory and in the field.

Summary and Conclusions

A set of mathematical models, based on the physical principles of the conservation of energy and mass, has been used in this paper to model the water and energy exchanges between an amphibian and its environments. Simulations with these models were used to provide a new vantage point from which to view the temperature and water relations of amphibians. It was observed from modeling simulations and a new experiment, with bullfrogs, that behavioral thermoregulation, by shuttling between sunlight and shade or sunlight and water, probably offers moist-skinned amphibians little advantage in terms of added precision in regulating body temperature over simply remaining exposed to sun. However, model simulations illustrate that even shaded microenvironments provide stressfully desiccating environments for mucous-covered amphibians unless they can at least sit on substrates as moist as poorly drained soils that never dry to field capacity. Model simulations also illustrate that amphibians might be highly efficient in converting their prey into amphibian biomass and that an amphibian's temperature relations may not greatly affect this efficiency.

Considering amphibians' temperature, water, and bioenergetic relations as they appear from the analyses presented in this paper, and also considering these relations in terms of a single working system, one would conclude that amphibian behavioral temperature regulation in nature should be either crude or nonexistent. The highly plastic thermal physiology of amphibians (Hutchison, 1961; Hutchison and Ferrance, 1970; Brattstrom, 1963, 1968, 1970; Brattstrom and Lawrence, 1962; Dunlap, 1968; Farrell, 1971; Spotila, 1972) makes such a conclusion at least plausible. Another possible conclusion here is that the local dispersions and activity patterns (when activity patterns can be defined) of most terrestrial amphibians probably depend to a much greater extent on their water relations than on their temperature relations. But the pervading conclusion offered here is that modeling techniques provide a very powerful investigatory tool for obtaining new insights into ecological processes.

Acknowledgments

I would like to thank W. P. Porter for his tireless support of my biophysical ecological research. I also thank C. B. DeWitt, who helped free me from other professional obligations so that I could participate in the symposium of The University of Michigan's Biological Station. My thanks also go to Dan Rittschof for allowing me to use unpublished data from his Ph.D. dissertation. I would also like to thank W. R. Welch and my spouse, Bobbi, who have helped in the research. I gratefully acknowledge the assistance of Dorothy Ingle, Claire LeVine, and Joanne Robinson in the technical preparation of this paper.

This research was supported by an Atomic Energy Commission grant to W. P. Porter of the Department of Zoology, and a National Science Foundation grant to R. Bryson of the Institute for Environmental Studies, both at the University of Wisconsin.

References

Adolph, E. F.: 1927. The skin and the kidney as regulators of the body volume of frogs. J. Expt. Zool. *47*, 1–30.

———: 1932. The vapor tension relations of frogs. Biol. Bull. *62*, 112–125.

———: 1933. Exchanges of water in the frog. Biol. Rev. *8*, 224–240.

Balinsky, J. B., Cragg, M., Baldwin, E.: 1961. The adaptation of amphibian waste nitrogen excretion to dehydration. Comp. Biochem. Physiol. *3*, 236–244.

Bartholomew, G. A.: 1970. Energy metabolism. *In* Animal function: principles and adaptations (ed. M. S. Gordon). New York: Macmillan.

Bentley, P. J.: 1966. Adaptations of amphibia to arid environments. Science *152*(3722), 619–623.

———, Lee, A. K., Main, A. R.: 1958. Comparison of dehydration and hydration of two genera of frogs (*Heleioporus* and *Neobatrachus*) that live in areas of varying aridity. J. Expt. Biol. *35*, 677–684.

Brattstrom, B. H.: 1963. A preliminary review of the thermal requirements of amphibians. Ecology *44*(2), 238–255.

———: 1968. Thermal acclimation in anuran amphibians as a function of latitude and altitude. Comp. Biochem. Physiol. *24*, 93–111.

———: 1970. Amphibia. *In* Comparative physiology of thermoregulation (ed. G. C. Whittow), pp. 135–166. New York: Academic Press.

———, Lawrence, P.: 1962. The rate of thermal acclimation in anuran amphibians. Phisiol. Zool. *35*, 148.

Campbell, P. M., Davis, W. K.: 1971. The effects of various combinations of temperature and relative humidity on the evaporative water loss of *Bufo valliceps*. Texas J. Sci. *22*(4), 389–402.

Chew, R. M., Dammann, A. E.: 1961. Evaporative water loss of small vertebrates, as measured with an infrared analyzer. Science *133*, 384–385.

Cohen, N. W.: 1952. Comparative rate of dehydration and hydration in some California salamanders. Ecology *33*(4), 462–479.

DeWitt, C. B.: 1967. Precision of thermoregulation and its relation to environmental factors in the desert iguana, *Dipsosaurus dorsalis*. Physiol. Zool. *40*, 49–66.

Deyrup, I. J.: 1964. Water balance and kidney. *In* Physiology of the amphibia (ed. J. A. Moore). New York: Academic Press.

Dole, J. W.: 1965a. Spatial relations in natural populations of the leopard frog, *Rana pipiens* (Schreber) in northern Michigan. Am. Midland Naturalist *74*(2), 464–478.

———: 1965b. Summer movements of adult leopard frogs, *Rana pipiens* (Schreber) in northern Michigan. Ecology *46*(3), 236–255.

———: 1967a. The role of substrate moisture and dew in the water economy of leopard frogs, *Rana pipiens*. Copeia *1967*(1), 141–149.

———: 1967b. Spring movements of leopard frogs, *Rana pipiens* (Schreber) in northern Michigan. Am. Midland Naturalist *78*(1), 167–181.

———: 1971. Dispersal of recently metamorphosed leopard frogs *Rana pipiens*. Copeia *1971*(2), 221–228.

Dunlap, D. G.: 1968. Critical thermal maximum as a function of temperature of acclimation in two species of hylid frogs. Physiol. Zool. *41*, 432–439.

Elkan, E.: 1968. Mucopolysaccharides in the anuran defense against desiccation. J. Zool. *155*, 19–53.

Farrell, M. P.: 1971. Effect of temperature and photoperiod acclimations on the water economy of *Hyla crucifer*. Herpetologica *27*(1), 41–48.

———, MacMahon, J. A.: 1969. An eco-physiological study of water economy in eight species of tree frogs (*Hylide*). Herpetologica *4*, 279–294.

Flugge, W.: 1954. Four-place tables of transcendental functions. Oxford: Pergamon Press.

Gardner, W. R.: 1965. Rainfall, runoff, and return. Meteor. Monogr. *6*(28), 138–148.

Gordon, M. S., Schmidt-Nielsen, K., Kelly, H. M.: 1961. Osmotic regulation in the crab-eating frog (*Rana cancrivora*). J. Expt. Biol. *38*, 659–678.

Gray, J.: 1928. The role of water in the evolution of the terrestrial vertebrates. Brit. J. Expt. Biol. *6*, 26.

Greenwald, L.: 1972. Sodium balance in amphibians from different habitats. Physiol. Zool. *45*(3), 229–237.

Heatwole, H., Lim, K.: 1961. Relation of substrate moisture to absorption and loss of water of the salamander, *Plethodon cinereus*. Ecology *42*(4), 814–819.

———, Torries, F., Blasini de Austin, S., Heatwole, A.: 1969. Studies of anuran water balance. I. Dynamics of evaporative water loss by the coqui, *Eleutherodactylus portoricensis*. Comp. Biochem. Physiol. *28*, 245–269.

———, Cameron, E., Webb, G. J. W.: 1971. Studies on anuran water balance. II. Desiccation in the Australian frog, *Notaden bennetti*. Herpetologica *27*(4), 365–378.

Hevesy, G. V., Hofer, E., Krogh, A.: 1935. The permeability of the skin of frogs to water as determined by D_2O and H_2O. Skandinav. Archiv. *73*, 199–214.

Hutchison, V. H.: 1961. Critical thermal maxima in salamanders. Physiol. Zool. *34*, 92–125.

———, Ferrance, M. R.: 1970. Thermal tolerances of *Rana pipiens* acclimated to daily temperature cycles. Herpetologica *26*(1), 1–8.

———, Whitford, W. G., Kohl, M.: 1968. Relation of body size and surface area to gas exchange in anurans. Physiol. Zool. *41*(1), 65–85.

Krogh, A.: 1939. Osmotic regulation in aquatic animals. New York: Cambridge Univ. Press.

Jameson, D. L.: 1965. Rate of weight loss of tree frogs at various temperatures and humidities. Ecology *47*(4), 605–613.

Johnson, C. R.: 1969. Water absorption response of some Australian anurans. Herpetologica *25*(3), 171–172.

Jorgensen, C. B.: 1949. Permeability of the amphibian skin. II. Effect of moulting of the amphibian skin on the permeability to water and electrolytes. Acta Physiol. Scand. *18*, 171–180.

Lasiewski, R. C., Bartholomew, G. A.: 1969. Condensation as a mechanism for water gain in nocturnal desert poililotherms. Copeia *1969*(2), 405–407.

Lee, A. K.: 1968. Water economy of the burrowing frog, *Heleiporus eyrei* (Gray). Copeia *1968*(4), 741–745.

———, Mercer, E. H.: 1967. Cocoon surrounding desert-dwelling frogs. Science *157*, 87–88.

Lillywhite, H. B.: 1970. Behavioral temperature regulation in the bullfrog, *Rana catesbeiana*. Copeia *1970*(1), 158–168.

———: 1971a. Temperature selection by the bullfrog, *Rana catesbeiana*. Comp. Biochem. Physiol. *40A*, 213–227.

———: 1971b. Thermal modulation of cutaneous mucous discharge as a determinant of evaporative water loss in the frog, *Rana catesbeinana*. Z. Vergleich. Physiol. *73*, 84–104.

Littleford, R. A., Keller, W. R., Philips, N. E.: 1947. Studies on the vital limits of water loss in plethodontid salamanders. Ecology *28*, 440–447.

Machin, J.: 1969. Passive water movements through skin of the toad *Bufo marinus* in air and in water. Am. J. Physiol. *216*(6), 1562–1568.

MacMahon, J. A.: 1965. An eco-physiological study of the water relations of three species of the salamander genus, *Plethodon*. South Bend, Ind.: Univ. Notre Dame. Ph.D. diss.

Main, A. R., Bentley, P. J.: 1964. Water relations of Australian burrowing frogs and tree frogs. Ecology *45*(2), 379–382.

Mayhew, W. W.: 1965. Adaptations of the amphibian, *Scaphiopus couchi* to desert conditions. Am. Midland Naturalist *75*(1), 95–109.

McClanahan, L.: 1964. Osmotic tolerance of the muscles of two desert-inhabiting toads, *Bufo cognatus* and *Scaphiopus couchi*. Comp. Biochem. Physiol. *12*, 501–508.

———: 1967. Adaptations of the spadefoot toad, *Scaphiopus couchi* to desert environments. Comp. Biochem. Physiol. *20*, 73–99.

———: 1972. Changes in body fluids of burrowed spadefoot toads as a function of soil water potential. Copeia *1972*(2), 209–216.

———, Baldwin, R.: 1969. Rate of water uptake through the integument of the desert toad, *Bufo punctatus*. Comp. Biochem. Physiol. *28*, 381–389.

McCullough, E. M., Porter, W. P.: 1971. Computing clear day solar radiation spectra for the terrestrial ecological environment. Ecology *52*(6), 1008–1015.

Norris, K. S.: 1967. Color adaptation in desert reptiles and its thermal relationships. *In* Lizard ecology: a symposium (ed. W. W. Milstead), pp. 162–339. Columbia, Mo.: Univ. Missouri Press.

Packer, W. C.: 1963. Dehydration, hydration, and burrowing behavior in *Heleioporus eyrei* (Gray) (Leptodactyledae). Ecology *44*(4), 643–651.

Parsons, R. H.: 1970. Mechanisms of water movement into amphibians with special reference to the effect of temperature. Corvallis, Ore.: Oregon State Univ. Ph.D. diss.

Porter, W. P., Gates, D. M.: 1969. Thermodynamic equilibria of animals with environment. Ecol. Monogr. *39*, 245–270.

Ray, C.: 1958. Vital limits and rates of desiccation in salamanders. Ecology *39*(1), 75–83.

Rey, P.: 1937. Recherches expérimentales sur l'économie de l'eau chez les Batraciens. Ann. Physiol. *13*, 1081–1144.

Rose, C. W.: 1966. Agricultural physics. Oxford: Pergamon Press.

Ruibal, R.: 1962a. The adaptive value of bladder water in the toad, *Bufo cognatus*. Physiol. Zool. *35*(3), 218–223.

———: 1962b. Osmoregulation in amphibians from heterosaline habitats. Physiol. Zool. *35*, 133–147.

———, Tevis, L., Jr., Roig, V.: 1969. The terrestrial ecology of the spadefoot toad *Scaphiopus hammondii*. Copeia *1969*(3), 571–584.

Sawyer, W. H.: 1951. Effect of posterior pituitary extract on permeability of frog skin to water. Am. J. Physiol. *164*, 44–48.

Schmid, W. D.: 1965. Some aspects of the water economics of nine species of amphibians. Ecology *46*(3), 261–269.

———, Barden, R. E.: 1965. Water permeability and lipid content of amphibian skin. Comp. Biochem. Physiol. *15*, 423–427.

Schmidt-Nielsen, B., Forster, R. P.: 1954. The effect of dehydration and low temperature on renal function in the bullfrog. J. Cellular Comp. Physiol. *44*, 233–246.

Shoemaker, V. H.: 1964. The effects of dehydration on electrolyte concentrations in a toad, *Bufo marinus*. Comp. Biochem. Physiol. *13*, 261–271.

———: 1965. The stimulus for the water balance response to dehydration in toads. Comp. Biochem. Physiol. *15*, 81–88.

———, McClanahan, L., Jr., Ruibal, R.: 1969. Seasonal changes in body fluids in a field population of spadefoot toads. Copeia *1969*(3), 585–591.

———, Waring, H.: 1968. Effect of hypothalamic lesions on the water balance responses of a toad (*Bufo marinus*). Comp. Biochem. Physiol. *24*, 47–54.

———, Balding, D., Ruibal, R., McClanahan, L. L., Jr. : 1972. Urincotelism and low evaporative water loss in a South American frog. Science *175*, 1018–1020.

Spight, T. M.: 1967. Evaporation from toads and water surfaces. Nature *214* (5090), 835–836.

———: 1968a. The water economy of salamanders: evaporative water loss. Physiol. Zool. *41*(2), 195–203.

———: 1968b. The water economy of salamanders: exchange of water with the soil. Biol. Bull. *132*, 126–132.

Spotila, J. R.: 1972. Role of temperature and water in the ecology of lungless salamanders: Ecol. Monogr. *41*(1), 195–225.

Stille, W. T.: 1958. The water absorption response of an anuran. Copeia *1958*(3), 217–218.

Stirling, W.: 1877. On the extent to which absorption can take place through the skin of the frog. J. Anat. Physiol. *11*, 529.

Thorson, T. B.: 1955. The relationship of water economy to terrestrialism in amphibians. Ecology *36*(1), 100–116.

———: 1956. Adjustment of water loss in response to desiccation in amphibians. Copeia *1956*(4), 23–237.

———: 1964. The partitioning of body water in amphibia. Physiol. Zool. *37*(4), 395–399.

———, Svihla, A.: 1943. Correlation of the habitats of amphibians with their ability to survive the loss of body water. Ecology *24*(3), 374–381.

Tracy, C. R.: 1972. A model of the water and energy dynamic interrelationships between an amphibian and its environment. Madison, Wisc.: Univ. Wisconsin. Ph.D. diss.

Walker, R. F., Whitford, W. G.: 1970. Soil water absorption capabilities in selected species of anurans. Herpetologica *26*, 411–418.

Warburg, M. R.: 1965. Studies on the water economy of some Australian frogs. Austral. J. Zool. *13*, 317–330.

20

Environmental Constraints on Some Predator–Prey Interactions

WARREN P. PORTER, JOHN W. MITCHELL, WILLIAM A. BECKMAN, AND C. RICHARD TRACY

Introduction

The study of environmental relationships of living organisms has a long history. Environmental effects on animals were reported by Davenport and Castle (1896). Early work on reptiles was done by Atsatt (1939) and by Cowles and Bogert (1944). Quantitative engineering approaches to the energy balance of humans first began appearing with the work of Hardy and DuBois (1938). The extension of energy-balance modeling to outdoor situations and nonhuman animals was firmly established with the publication of papers by Bartlett and Gates (1967) and Norris (1967). Porter and Gates (1969) made some initial efforts to generalize energy-balance calculations for a variety of animals. Their development of a climate-space niche was useful in expressing physiological limits of animals as they related to various combinations of environmental parameters. But the concept is limited because, first, it is a steady-state model; second, animals rarely live at their physiological limits; and third, microhabitats available in time and space are not specified. Although a climate-space diagram represents a useful concept, ecological and behavioral limits needed to be defined. Transient energy-balance models were needed and had to include more biology along with physics and engineering. Microhabitat specification was needed. Tests needed to be done on live animals in both controlled environmental facilities and out of doors. Recently, these things have been done for a relatively simple desert environment and a desert lizard, *Dipsosaurus dorsalis* (Porter *et al.*, 1973). This paper outlines the usefulness of energy-balance concepts to examine predator–prey relationships. It is the basis for the extension of our investigations to other predator–prey systems, one of which we describe here. We shall attempt only a semiquantitative description here since the equations and experiments testing them have been published elsewhere.

Fig. 20.1. Mojave desert test site 2 mi southwest of Kelso, California.

The Mojave Desert was chosen for the development of integrated energy-balance models for both microclimate and the animals in those microclimates. Figure 20.1 illustrates the openness of this habitat and the sparsely leaved vegetation. The wind blows through these plants freely. Figure 20.2 is a qualitative description of the energy flows to and from the substrate and a desert lizard. These

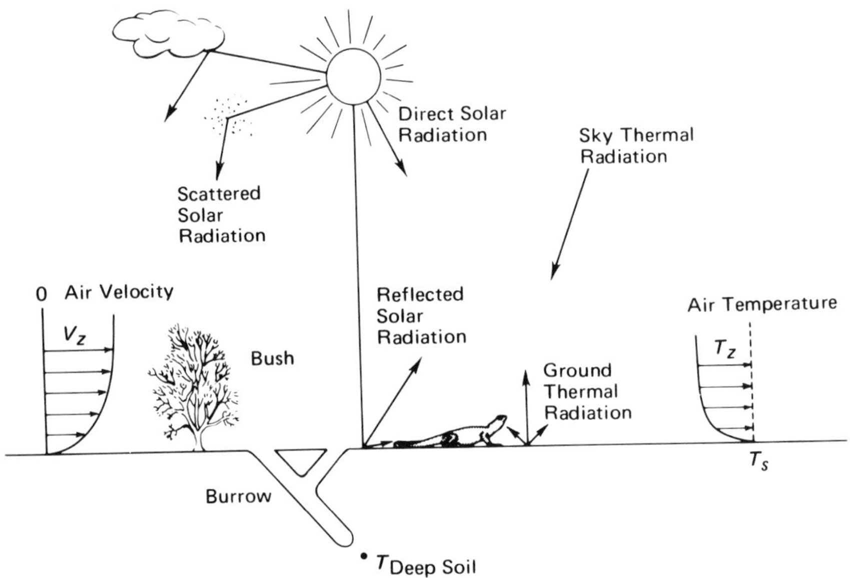

Fig. 20.2. Energy-flow diagram of the desert environment.

flows are described in detail elsewhere in this volume. The energy-balance equation is simply the first law of thermodynamics, which states that

$$E_{\text{in}} + E_{\text{gen}} = E_{\text{out}} + E_{\text{st}}$$

where E_{in} = solar absorbed
= direct solar + scattered solar + reflected solar
E_{gen} = metabolic heat production
E_{out} = net thermal radiation exchange with sky + net thermal exchange with ground + convection + conduction + evaporation
E_{st} = stored thermal energy

The mathematical definitions of these equations may be found in Porter *et al.* (1973). Each of the parameters in the above equations can be evaluated by methods described by Bartlett and Gates (1967) and Porter *et al.* (1973). Tests of model predictions on live desert iguanas in the wind tunnel, in the Biotron, and outdoors are described in Porter *et al.* (1973). These tests demonstrated that such equations are valid for predictions of field temperatures and activity patterns. Tracy (this volume) demonstrates that amphibian field temperatures can also be accurately predicted. Using our micrometeorology model to specify microclimates available throughout the day and throughout the year and knowing temperature preferences of an animal, one can use the energy-balance equation for an animal to predict gross daily and seasonal activity patterns. Figure 20.3 is a simulation for a typical July 15 in Palm Springs, California. One type of input data for the micrometeorological model was air temperature at 200 cm obtained from Weather Bureau climatological data. Program SOLRAD (McCullough and Porter, 1971) calculated clear-day solar radiation spectra for Palm Springs. Soil reflectivity, conductivity, soil roughness, and deep soil temperature were obtained either from sand taken from Palm Springs or from the literature. Deep soil temperature was inferred from the mean monthly average. Maximum and minimum air temperatures are the 10-year average (1950–1960). Air temperatures are assumed to fluctuate sinusoidally with a peak in early afternoon and a minimum approximately 1 h before sunrise. From these data, air temperatures above the soil surface, soil-surface temperature, and subsurface temperatures at various depths are computed using a finite-difference numerical solution technique described by Beckman (Porter *et al.*, 1973). The desert iguana in this simulation was assumed not to influence the microclimate and to be standing on the surface the entire day. Under these conditions, a solid line shows what the desert iguana's temperature should be over a 24-h period. Before approximately 7:30 A.M., we would expect the animal's temperature to be below its preferred minimum activity temperature of 38°C (DeWitt, 1967); thus we would predict that the animal would remain in its burrow where it is warmer before 0730 h. Once having emerged, the animal should warm rapidly along the line predicting its temperature between about 7:30 and 8:30 in the morning. Once the animal reaches 43°C, we could expect it to begin moving into and out of the shade in an effort to stay below the upper activity temperature level, since the animal begins to lose excessive amounts of water above 43°C. From approximately 8:30 until about 9:30, we could expect the animal to be shuttling in

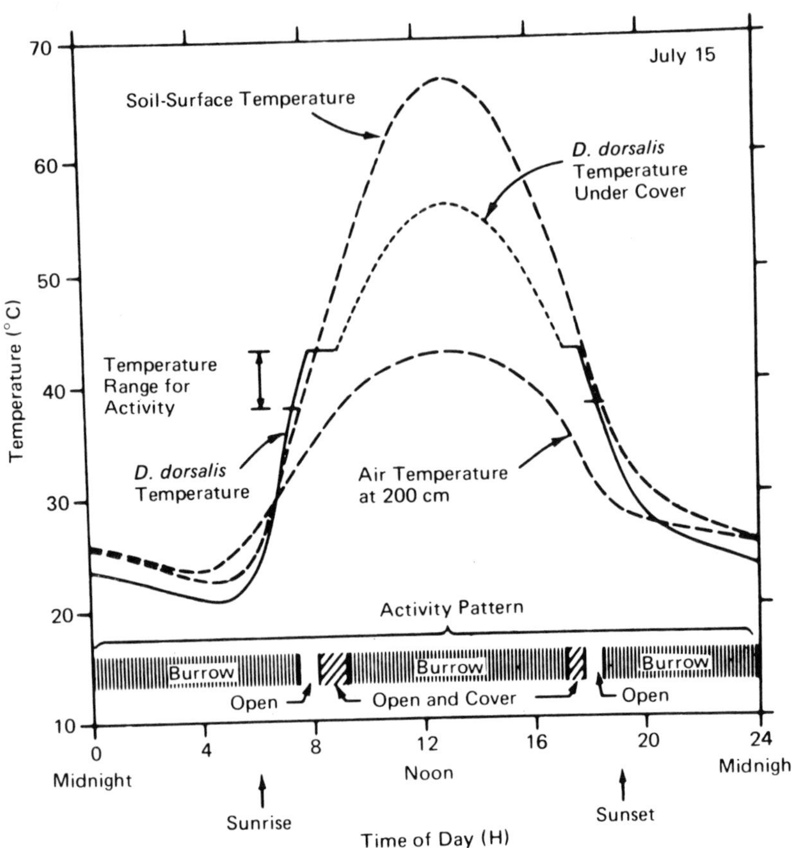

Fig. 20.3. Predicted environmental and desert iguana temperatures for Palm Springs, California (July 15).

and out of the shade, spending more and more time in the shade, until, at about 9:30, if the animal were constrained to the surface of the ground, it would overheat and die. We would then expect the animal to retreat to its burrow, where temperatures would be cooler and it could survive during the heat of the day. In this simplistic model, we would expect the animal to emerge again in the afternoon when temperatures had diminished, and we see a reversal of the sequence of behavioral events from being in deep shade to being out in the open with a final retreat into the burrow at about 1820 h.

Simulations for each month of the year yield predictions for activity patterns as shown in Fig 20.4. The elliptically shaped lines with temperature designations are isotherms of desert iguana temperatures at different times of the day and year. It is easiest to understand how these curves are constructed by looking at July 15 and comparing it with Fig. 20.3 for the same date. Starting at zero hours in Fig. 20.4 on July 15 and moving upward, one reaches the 38°C isotherm at 7:30 A.M. The animal could remain in the open about an hour as in Fig. 20.3. Specific days, when connected together, form isotherm patterns whose dimensions depend upon the

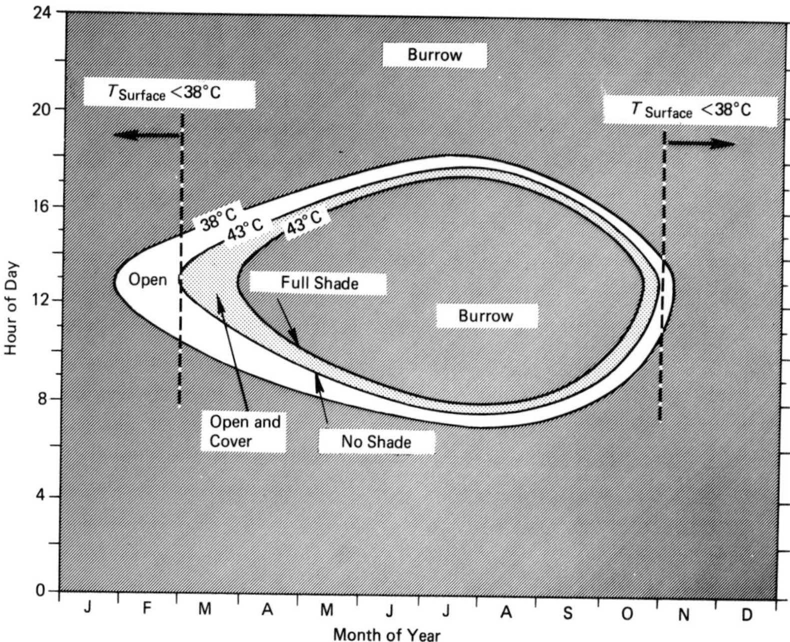

Fig. 20.4. Predicted seasonal behavioral pattern for the desert iguana at Palm Springs.

animal's temperature preferences and the physical climate for that area. For Palm Springs these predictions suggest that desert iguanas will not emerge until March 1. Surface temperatures do not exceed 38°C on the average before this date by our calculations. Activity could last until about November 1, when again surface temperatures should not exceed 38°C. This analysis also predicts a unimodal pattern of activity in the spring, followed by a bimodal pattern of activity in the middle of summer and a unimodal pattern again in fall. Figure 20.5 compares Mayhew's observations on activity in the Palm Springs area with our predictions. Particularly evident is the lack of afternoon activity in the middle of summer, which the animals could theoretically utilize if only temperature were considered. As Porter *et al.* (1937) show, this lack of afternoon activity is most probably due to a water constraint imposed on the animals by their own physiology and by the water content of the plants in their environment. The additional lizard temperature lines in Fig. 20.5 allow estimation of the effect of different minimal activity temperatures and the effect of climbing in bushes above the surface on activity time. In the same paper, food and water requirements are predicted, as well as a single hypothetical predator–prey activity pattern overlap.

Figure 20.6 illustrates a hypothetical predator–prey interaction where a low-temperature-preference ant is superimposed on the activity pattern of the desert iguana. The ant is assumed to be so close to the surface as to be at ground-surface temperature. One point of this hypothetical calculation is to demonstrate the usual

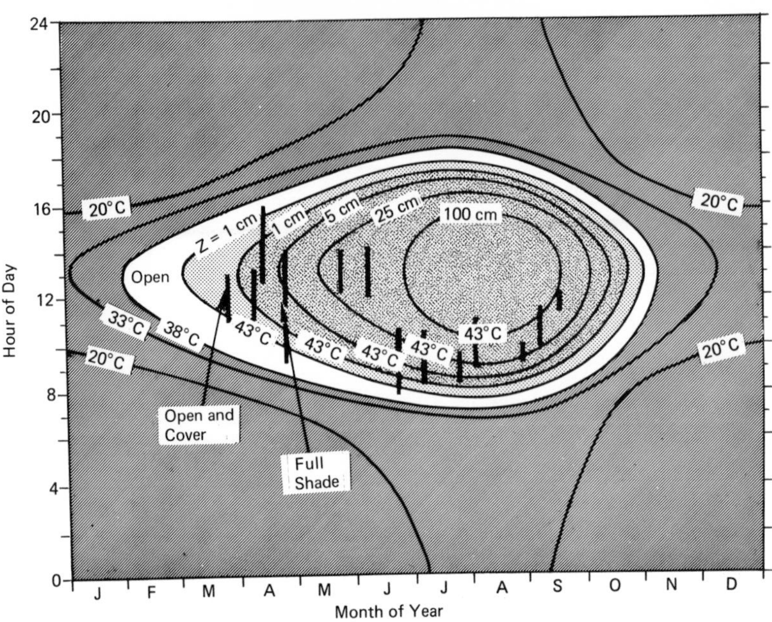

Fig. 20.5. Predicted seasonal behavioral pattern for the desert iguana for a wider variety of possible body temperatures and heights (z) above the ground.

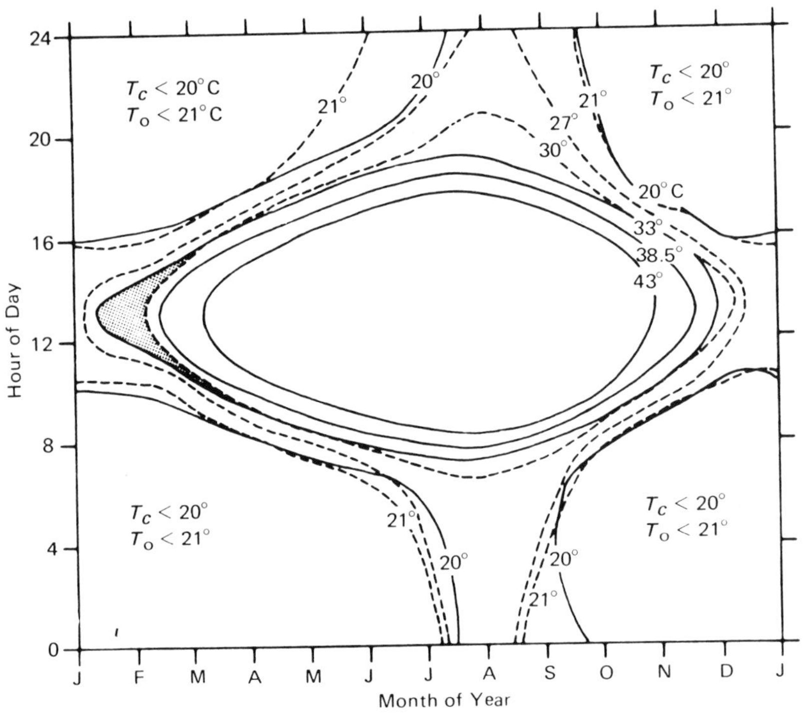

Fig. 20.6. Predicted amount of activity pattern overlap between a desert iguana and a low-temperature-preferring ant.

lack of interaction or encounter probability between a low-temperature-preference species and a high-temperature-preference organism. It also suggests that, if there are several potential prey species each with different preferred temperatures, there may well be seasonal changes in the prey taken by a predator which might be predictable. Since the desert iguana is herbivorous and prefers high temperatures, it rarely encounters potential predators with low temperature preference, except in unusual circumstances where it may accidentally wander close to a cool microclimate.

Garter Snake–Frog Predator–Prey Interactions

Holling (1966) has done a number of fascinating predator–prey studies. His papers prompted us to investigate potential ramifications of predator–prey analyses using activity pattern overlap. We examined garter snakes (*Thamnophis sirtalis*) and the leopard frog (*Rana pipiens*). These animals occur over a wide range of latitudes and, since they are near water or in water, they are potentially subject to thermal pollution effects which may alter their activity patterns and affect survival of frog eggs and tadpoles.

The northern location we chose was Pellston, Michigan. Climatological data are available for this area and extensive experiments have been done principally by Dole (1965) on *Rana pipiens*. Garter snake ecology was also done for this region by Carpenter (1952). As a southern location, we chose Aiken, South Carolina. Again climatological data are available, and extensive observations of both garter snakes and frogs have been made there. The capability for applying energy- and mass-balance equations to determine activity patterns of frogs was developed by Tracy (1972). It was clear, based on these frogs' wide temperature tolerance, that they would be active at virtually any time of the year except when the air temperature dips below freezing. The garter snake has much less time for activity, because of its minimum activity temperature of 15°C. An analysis of an energy-balance equation on this animal showed us that it was very tightly coupled to soil temperature, although we measured a solar absorptivity of 88 percent on our DK-2A spectroreflectometer. Since these snakes typically inhabit moist, tall grassy shady areas, we examined micrometeorological studies of dense vegetative cover by Rose *et al.* (1971), Byrne and Rose (1972), and Goncz and Rose (1972). Their studies indicate that air-temperature profiles down into deep vegetation are essentially equal to air temperatures at 200 cm above the ground. Since soils are moist where frogs and these snakes occur, we could expect that soil-surface temperatures should stay within a few degrees of air temperature. Furthermore, since sunlight would be greatly attenuated passing through dense vegetation cover, there should be little solar intensity to cause soil temperatures to rise. Thus we assumed as a first approximation that ground-surface temperatures were the same as air temperature at 200 cm. As we will see below, errors in predicted temperatures of this order of magnitude make little difference in the conclusions.

Figures 20.7–20.9 show calculations for the activity times of garter snakes in Michigan and South Carolina and an extreme case in South Carolina, where an arbitrary 10°C increase in air temperature has been added at all times to simulate an extreme situation of thermal pollution. The minimum of 15°C for activity seems to be typical for a variety of species of garter snakes. Carpenter (1952) found that the mean preferred temperature for *Thamnophis sirtalis sirtalis*, *T. s. sauritus*, and *T. butleri* is approximately 25.6°C. He found activity temperatures to lie between 20 and 30°C. Heckrotte (1961) demonstrated that *T. radix* is not active below 15°C. Stewart (1965) determined that *T. s. concinnus* and *T. ordinoides* prefer temperatures between 25 and 30°C, have a minimum preferred temperature of 17°C, and tolerate a maximum of 35°C. Kitchell (1969) demonstrated that *T. sirtalis* has a minimum preferred temperature of 16°C when shedding and has higher minimums for other activities. We have not yet adjusted the results for acclimation in this first simulation, although Jacobson and Whitford (1970) have shown acclimation effects on the physiology of *T. proximus*.

Figures 20.7–20.9 were computed using the 10-year average temperatures between 1960 and 1969 for each region. The predictions suggest that garter snakes should be active about 20 weeks in Michigan, about 40 weeks in Aiken, South Carolina, and should be active year round if Aiken should have 10°C thermal pollution. Since the climate fluctuates from day to day, particularly in northern

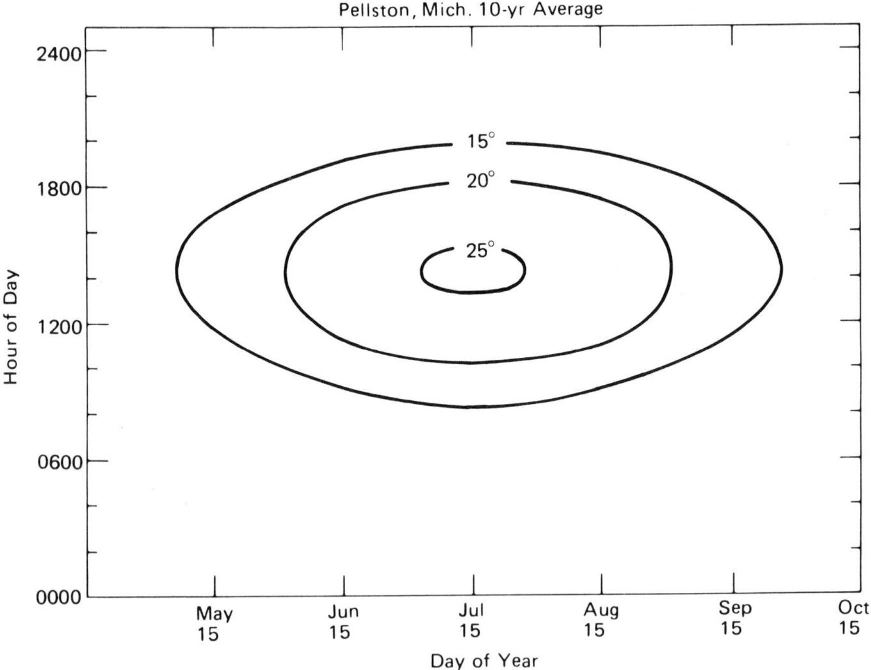

Fig. 20.7. Predicted seasonal behavioral pattern for garter snakes near Pellston, Michigan.

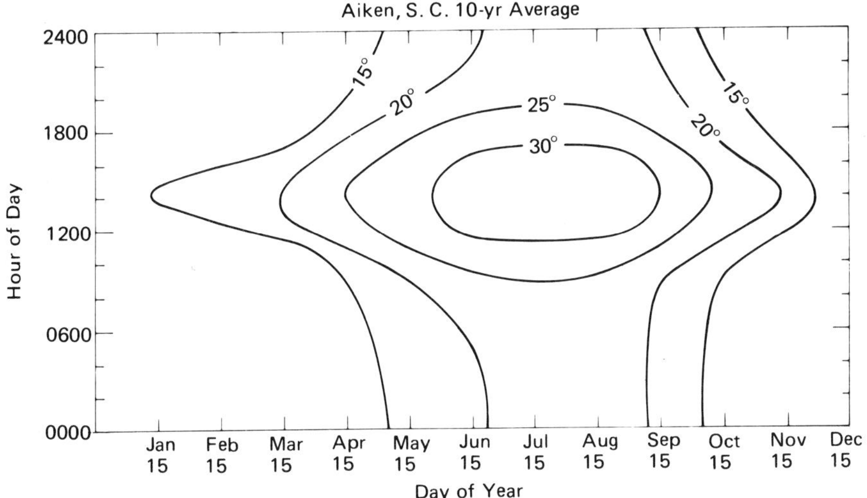

Fig. 20.8. Predicted seasonal behavioral pattern for garter snakes near Aiken, South Carolina.

regions, it would seem essential to examine the day-to-day sequence of temperatures, particularly in the spring and fall, when some days would not reach 15°C and therefore prevent garter snakes from actively hunting or digesting (Skoczylas, 1970). Figure 20.10 is Weather Bureau data taken for April and May in Pellston, Michigan, in 1960. This particular year had monthly averages that were nearly identical to the 10-year average for that location. The first day that exceeded 15°C maximum temperature was April 15. As late as May 23, there was a day whose

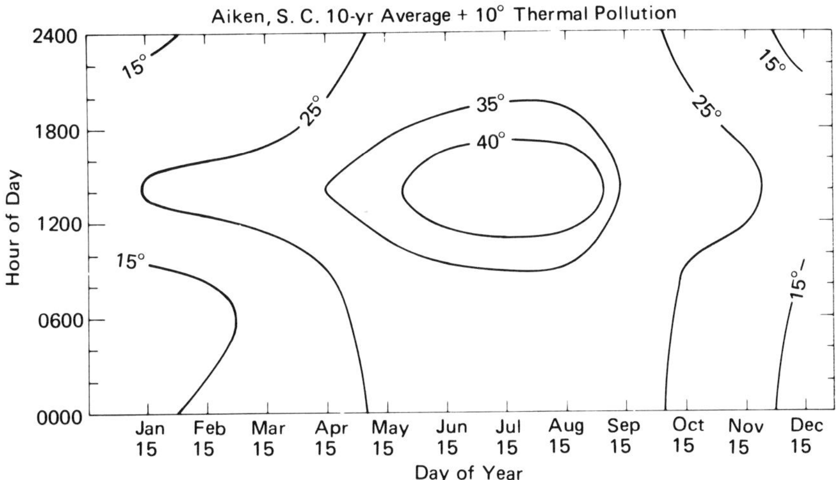

Fig. 20.9. Predicted seasonal behavioral pattern for garter snakes near Aiken if all air temperatures were 10°C higher due to thermal pollution.

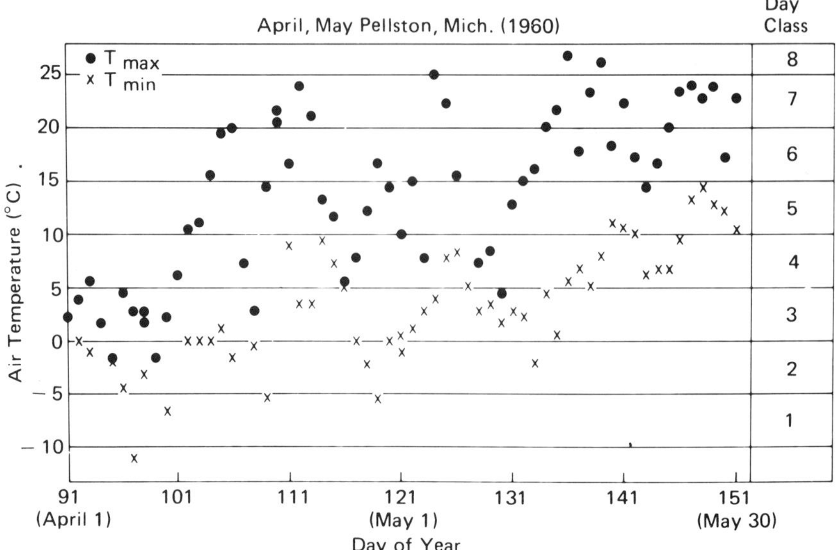

Fig. 20.10. Daily air-temperature data from Pellston, Michigan, in 1960.

maximum temperature did not exceed 15°C. Since garter snakes can apparently hunt only at body temperatures above 15°C, it seemed that the frequency and spacing of days exceeding 15°C might be of vital importance to an animal freshly out of hibernation that is trying to obtain enough food to survive. This concept is illustrated in Fig. 20.11, where the first part of May 1960 is examined in detail. The animal could not exceed 15°C at any time during the first 2 days and it was not until approximately 1000 h that the animal could begin to hunt. If prey is captured at 1600 h on May 4, only about 3 h would be available for digestion since the snake body temperature drops below 15°C at about that time. Since the follow-

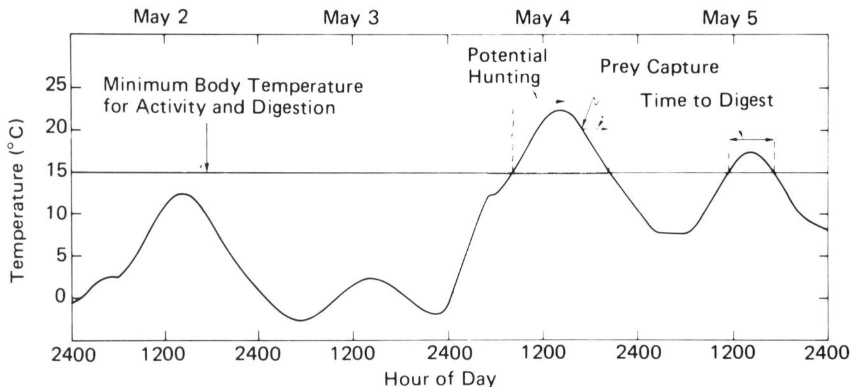

Fig. 20.11. Limits to hunting and digesting by garter snakes for a short sequence of days in 1960 near Pellston, Michigan.

ing day is cooler, only about 3 h is available during that day to carry on digestion. It is apparent that, for the animal to survive, the digestion rate must be rapid enough to permit the snake to obtain sufficient energy to make up for times when it is too cold to hunt or digest. Figures 20.10 and 20.11 show the potential calories available from digestion if we assume the 50-g snake has a full stomach and is exposed to average climatic conditions for Michigan and South Carolina. Also shown near the bottom of the graph is the metabolic requirement for the animal for each day of the year. Skozcylas' careful work, using x-rays and dissections, demonstrated that digestion was completely arrested at 5°C, was very slow or intermittent at 15°C, and was most rapid at about 25°C. Since there are no data in the literature on digestion as a function of temperature for the garter snake, we have used Skozcylas's 1970 data on *Natrix natrix* for our calculations of digestion for the garter snake. Unfortunately, he made no determinations of digestion at 30°C, and the digestion rates he observed at 35°C were lower than rates at 25°C. We therefore

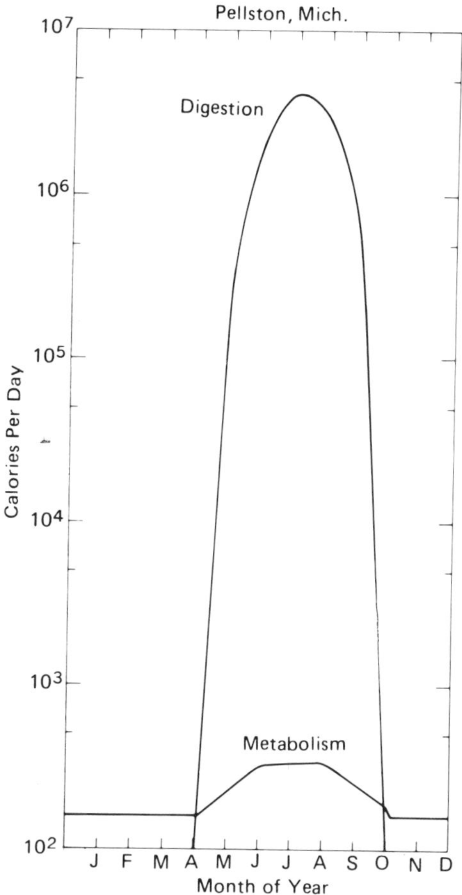

Fig. 20.12. Daily resting metabolism and potential daily maximum digestion using 10-yr-average days for Pellston, Michigan.

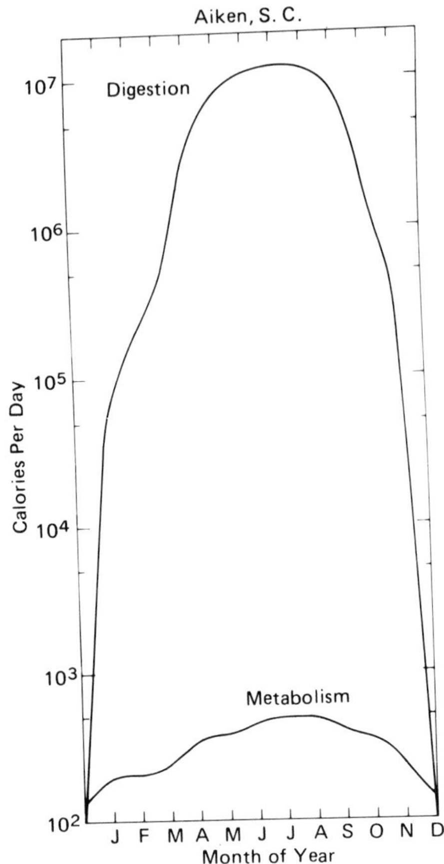

Fig. 20.13. Daily resting metabolism and potential daily maximum digestion using 10-yr-average days for Aiken, South Carolina.

assumed (1) a Q_{10} increase of 2 for digestion rates between 25 and 30°C; (2) a logarithmic decrease from a projected 30°C value to the measured 35°C value; and (3) that no digestion occurs above 40°C. Thus, if garter snake digestion rates are similar to those of *Natrix*, the rate of digestion is orders of magnitude higher than the rate of metabolism for any given day. In modeling the garter snake, therefore, the daily sequence of temperatures is not significant and averages are sufficient.

Integrated metabolism was computed (Porter and Tracy, 1974) using metabolic rates typical of snakes based on work by Jacobson and Whitford (1970), Dmi'el (1967), Dmi'el and Borut (1972), and Dmi'el (1972).

Integration of the area under the metabolism curves for Pellston, Michigan, and Aiken, South Carolina, yields the total energy requirement for a 50-g garter snake for an entire year, as shown in Table 20.1. A 50-g garter snake would need approximately 79,500 cal to maintain itself for an entire year. The total caloric requirement needed allows one to estimate the biomass required for the snake to maintain itself for a year. We assumed a low digestive efficiency of 36 percent, inferring this from data of Brown (1958). The reader can easily revise our calculations to reflect other digestive efficiencies. The results of Myer and Kowell (1971) on king snakes suggest that their digestive efficiencies are typically higher than those we are assuming. Furthermore, work by Vinegar *et al.* (1970) on pythons shows digestive efficiencies above 50 percent. Thus our biomass estimates for garter snake maintenance are probably high.

Table 20.1 shows that, if the biomass required for reproduction is considered, an additional 136 g of biomass is needed. This assumes a maximum production of young of 34 individuals annually (Hoffman, 1970). Assumptions about assimilation efficiency and percent water content of snake biomass are indicated in the table. The number of frogs needed to meet these biomass requirements varies from slightly more than three third-year frogs to 136 newly emerged frogs. It seems clear that the simplest feeding strategy for snakes would be to attempt to obtain the largest frogs possible and thereby reduce the amount of time needed for hunting. The consequences for the frog population might be very different depending on the feeding strategies of snakes.

A final consideration in this estimation of total biomass requirements for garter snakes would be an increase in size of the animal. We assume a doubling of weight due to growth and a 25 percent assimilation efficiency (Brown, 1958). We find an additional 200 g of biomass needed. This is nearly the same amount as required for annual maintenance in Michigan.

Total biomass requirements for maintenance, reproduction, and growth for 50-g garter snakes are about 550 g of wet weight of biomass for the Pellston, Michigan, area and approximately 700 g for Aiken, South Carolina. If maintenance only is considered, only half of the above biomass requirement is needed to support an animal for an entire year. Thus an animal in the Pellston, Michigan, area, eating only third-year frogs, could survive on six frogs per year and might consume as many as 14 third-year frogs in 1 year. With 20 weeks of activity time available, that would be an average of one frog a month to less than one frog a week. Similar feeding intervals would be expected for Aiken, South Carolina, where there are an average of 40 weeks of available activity time. Our calculations also suggest that a 10° increase in year-round temperatures in Aiken would double the maintenance requirements of garter snakes, but increase the overall annual requirement by only about 50 percent. This assumes no significant temperature effect on the biomass requirements of developing embryos.

Since an individual snake seems to consume relatively few frogs, we examined an active endotherm predator to compare its biomass requirements with those of the ectothermic garter snake. We chose mink to make some preliminary estimates of their biomass requirements. As Table 20.1 indicates, a mink's daily requirement for maintenance is approximately the same as the annual maintenance requirement for a 50-g garter snake. On an equivalent weight basis, 50 g of mink requires approximately 26 times more prey biomass for maintenance than a 50-g garter snake. All data used in the calculations were taken from Farrell and Wood (1968). We have not made any estimates of biomass requirements for reproduction and growth in mink, but the mass required must be significant. Since reproduction in both snake and mink takes place over a relatively narrow time period of about 2

Table 20.1 Biomass Requirements for 50-g Garter Snake, Northern and Southern United States (Including 10°C Thermal Pollution)

Maintenance	Cumulative Metabolism (cal)	Annual Constant Temperature That Would Give Same Cumulative Metabolism (°C)	Required g Dry Weight to Supply Needed Annual Calories
Pellston, Mich.	79,500	9.1	15.9
Aiken, S.C.	130,800	17.3	26.2
Aiken + 10°C	271,400	32.7	54.3

	Grams Needed if 36% Digestive Efficiency (D.E.) $\left(\text{D.E.} = \frac{\text{cal ingested} - \text{cal excreted}}{\text{cal ingested}}\right)$	80% Frog Water Content (Thorson, 1964)	Frogs Needed Annually		
			Mid 3rd yr: 40 g	Mid 2nd yr: 20 g	Newly emerged: 1 g
Pellston, Mich.	44 g dry wt.	221 g wet wt.	5.5	11	221
Aiken, S.C.	73 g dry wt.	363 g wet wt.	9.1	18	363
Aiken + 10°C	151 g dry wt.	754 g wet wt.	18.8	37	754

Reproduction

N_{max} = 34 young at 1 g each = 34 g wet weight = 6.8 g dry weight (80% snake water content; Prosser *et al.*, 1952).
25% assimilation efficiency (A.E.) of food → young = 136 g wet weight = 27.2 g dry weight.

$\left(\text{A.E.} = \frac{\text{cal into biomass}}{\text{cal ingested}}\right)$

	No. 40-g frogs	No. 20-g frogs	No. 1-g frogs
Pellston, Mich. Aiken, S.C. Aiken + 10°C	3.4	6.8	136

Growth

Snake mass: 50 g → 100 g wet weight; 50 g wet weight to be added; 10 g dry weight to be added for 25% assimilation efficiency.

200 g wet weight needed	40 g dry weight needed	
No. 40-g frogs	No. 20-g frogs	No. 1-g frogs
5	10	200

	Average Annual Available Activity Time (weeks)	Frogs Needed Annually		
		Mid 3rd yr: 40 g	Mid 2nd yr: 20 g	Newly emerged: 1 g
Totals = Pellston, Mich.	20	14	28	557
Aiken, S.C.	40	18	35	669
Aiken + 10°C	52	27	54	1090

Preliminary Estimates of Biomass Requirements for an 800-g Adult Mink (data from Farrell and Wood)

Maintenance	Required g Dry Weight to Supply Needed Daily Calories	Grams Needed if 75% Digestive Efficiency (Farrell and Wood)	Grams Wet Weight Needed Daily (80% Water–Frogs)	No. Frogs Needed Daily		
				3rd yr: 40 g	2nd yr: 20 g	Newly Emerged: 1 g
190×10^3 cal day^{-1}	38	50.7	253	6.3	12.7	253

Note: This animal must consume the equivalent of 31% of its body weight day^{-1}.

months, it must be then that hunting efficiency must be greatest to obtain several times the biomass needed for maintenance alone. Failure to obtain sufficient biomass for production of young will result in fewer offspring. Thus "hunting efficiency" would not only be a function of hunger but would also depend on the reproductive or endocrine state of the female. The short time interval for reproduction each year may well be the crucial period of time, determining population size the succeeding year.

Conclusions

These simple predator–prey analyses have several implications. First, neglecting reproduction and growth, a limiting case of biomass relationships between ectothermic predators and their prey could be a ratio of 1:4 rather than the 1:10 ratio usually assumed. Obviously such a low ratio could not long persist since it precludes predator reproduction. However, it is likely that a population of ectothermic predators could survive for at least 1 year without reproducing. A second implication is that endotherm biomass densities must be very much less than ectotherm predator biomass densities because of the more rapid metabolic rate of endotherms. Even though on a per-gram basis endotherm biomass requirements are 26 times those of an ectotherm on a per-gram basis, endotherms frequently outweigh ectothermic predators. Thus on an individual whole-organism basis, vertebrate ectotherm–endotherm predator ratios may approach 400 to 1.

Finally, it would seem that, if thermal pollution were to increase average temperatures by 10°C, we could expect a doubling of biomass maintenance requirements by ectothermic predators.

In summary, energy-balance equations for microclimates and animals can be used to predict activity times, food requirements, and potential predator–prey interactions. Preliminary analyses of a simple predator–prey system have raised many more questions than they have answered. The model has pointed to a lack of vital information in the literature on digestive efficiencies, growth rates in the field, biomass requirements for reproduction, and biomass requirements for early growth to maturity. Physiological measurements on a variety of active endotherm predators are also absent from the literature. Information on number of young produced as a function of food available to the mother of virtually any wild mammal is lacking. Such data are needed if we are to develop mechanistic models for multispecies predator–prey interactions in biomes.

References

Atsatt, S. R.: 1939. Color changes as controlled by temperature and light in the lizards of the desert regions of southern California. Univ. Calif. Los Angeles Publ. Biol. Sci. *1*, 237–276.

Bartlett, P. N., Gates, D. M.: 1967. The energy budget of a lizard on a tree trunk. Ecology *48*, 315–322.

Brown, E. E.: 1958. Feeding habits of the northern water snake, *Natrix sipedon sipedon* Linneaus. Zoologica *43*(2), 55–71.

Byrne, G. F., Rose, C. W.: 1972. On the determination of vertical fluxes in field crop studies. Agric. Meteor. *10*, 13–17.

Carpenter, C. C.: 1952. Growth and maturity of the three species of *Thamnophis* in Michigan. Copeia *1952*(4), 237–243.

Cowles, R. B., Bogert, C. M.: 1944. A preliminary study of the thermal requirements of desert reptiles. Bull. Am. Museum Nat. Hist. *83*(5), 266–296.

Davenport, C. B., Castle, W. E.: 1896. Studies in morphogenesis. III. On the acclimatization of organisms to high temperatures. Arch. Entwicklungsmech. Organ. *2*, 227–249.

DeWitt, C. B.: 1967. Precision of thermoregulation and its relation to environmental factors in the desert iguana, *Dipsosaurus dorsalis*. Physiol. Zool. *40*, 49–66.

Dmi'el, R.: 1967. Studies on reproduction, growth, and feeding in the snake *Spalerosophis cliffordi* (Colubridae). Copeia *1967*(2), 332–346.

———: 1972. Relation of metabolism to body weight in snakes. Copeia *1972*(1), 179–181.

———, Borut, A.: 1972. Thermal behavior, heat exchange, and metabolism in the desert snake *Spalerosophis cliffordi*. Physiol. Zool. *45*(1), 78–94.

Dole, J. W.: 1965. Summer movements of adult leopard frogs, *Rana pipiens* (Schreber) in northern Michigan. Ecology *46*, 236–255.

Farrell, D. J., Wood, A. J.: 1968. The nutrition of the female mink *Mustela vision*. II. The energy requirement for maintenance. Can. J. Zool. *46*, 47–51.

Goncz, J. H., Rose, C. W.: 1972. Energy exchange within the crop of townsville stylo, *Stylosanthes humilis* H.B.K. Agric. Meteor. *9*, 405–419.

Hardy, J. D., DuBois, E. F.: 1938. Basal metabolism, radiation, convection and vaporization at temperatures of 22 to 35°C. J. Nutr. *15*, 477–497.

Heckrotte, C.: 1961. The effect of the environmental factors in the locomotory activity of the plains garter snake *Themnophis radix radix*. Animal Behaviour *10*(3–4), 193–207.

Hoffman, L. H.: 1970. Observations on gestation in the garter snake, *Thamnophis sirtalis sirtalis*. Copeia *1970*(4), 779–780.

Holling, C. S.: 1966. The strategy of building models of complex ecological systems. *In* Systems analysis in ecology (ed. K. Watt), pp. 195–214. New York: Academic Press.

Jacobson, E. R., Whitford, W. G.: 1970. The effect of acclimation on physiological responses to temperature in the snakes, *Thamnophis proximus* and *Natrix rhombifera*. Comp. Biochem. Physiol. *35*, 439–449.

Kitchell, J. F.: 1969. Thermophilic and thermophobic responses of snakes in a thermal gradient. Copeia *69*(1), 189–191.

McCullough, E. M., Porter, W. P.: 1971. Computing clear day solar radiation spectra for the terrestrial ecological environment. Ecology *52*(6), 1008–1015.

Myer, J. S., Kowell, A. P.: 1971. Eating patterns and body weight change of snakes when fasting and when food deprived. Physiol. and Behaviour *6*, 71–74.

Norris, K. S.: 1967. Color adaptation in desert reptiles and its thermal relationships. *In* Symposium on lizard ecology (ed. W. Milstead), pp. 162–229. Columbia, Mo.: Univ. Missouri Press.

Porter, W. P., Gates, D. M.: 1969, Thermodynamic equilibria of animals with environment. Ecol. Monogr. *39*, 245–270.

———, Tracy, C. R.: 1974. Modeling the effects of temperature changes on the ecology of *Rana pipiens* and *Thamnophis sirtalis*. *In* Thermal ecology (eds. J. W. Gibbons, R. R. Sharitz), Oak Ridge: AEC Symp. Ser. CONF-730505, pp. 382–386.

———, Mitchell, J. W., Beckman, W. A., DeWitt, C. B.: 1973. Behavioral implications of mechanistic ecology (thermal and behavioural modeling of desert ectotherms and their microenvironment). Oecologia *13*, 1–54.

Prosser, C. L., Bishop, D. W., Brown, F. A., Jr., John, T. L., Wulf, V. J.: 1952. Comparative animal physiology. Philadelphia: W. B. Saunders.

Rose, C. W., Begg, J. E., Byrne, G. F., Goncz, J. H., Torsell, B. W. R.: 1971. Energy exchanges between a pasture and the atmosphere under steady and non-steady-state conditions. Agric. Meteor. *9*, 385–403.

Skoczylas, R.: 1970. Influence of temperature on gastric digestion in the grass snake, *Natrix natrix* L. Comp. Biochem. Physiol. *33*, 793–804.

Stewart, G. R.: 1965. Thermal ecology of the garter snakes *Thamnophis sirtalis concinnus* (Hallowell) and *Thamnophis ordinoides* (Baird and Girard). Herpetologica *21*(2), 81–103.

Thorson, T. B.: 1964. The partitioning of body water in amphibia. Physiol. Zool. *37*(4), 395–399.

Tracy, C. R.: 1972. A model of the water and energy dynamic interrelationships between an amphibian and its environment. Madison, Wisc.: Univ. Wisconsin. Ph.D. diss.

Vinegar, A., Hutchison, V. H., Dowling, H. G.: 1970. Metabolism, energetics, and thermoregulation during brooding of snakes of the genus *Python* (Reptilia, Boidae). Zoologica *55*, 19–48.

21

Experimental and Fossil Evidence for the Evolution of Tetrapod Bioenergetics

Robert T. Bakker

Introduction

Tetrapod evolution is the story of the interaction of individuals and populations with their primary producers, competitors, prey, and predators, all armed with various levels of metabolism and thermoregulatory mechanisms, within the environmental context of continents constantly fragmenting and coalescing as lithospheric plates rift, collide, or glide past one another, throwing up mountain ranges and continually changing the pattern of continental climates. The fossil record gives direct evidence of bioenergetic strategies in extinct organisms: functional morphology can identify adaptations for respiration, locomotor activity, and food processing; community paleoecology can interpret predator–prey ratios as gauges of the rate of energy consumption by the predators. Climate also leaves its mark in the rocks: cycles of wind, rainfall, and temperature govern the texture and geometry of sediments, the oxidation state of iron compounds, and the character of clay minerals.

Ectotherm versus Endotherm

Among living tetrapods the most striking contrast in bioenergetics is that between ectotherms, represented by lizards, and endotherms, represented by birds and advanced placental mammals. Since both birds and mammals evolved from ectothermic reptiles, consideration of this contrast is a useful starting point for reviewing the fossil record (Table 21.1).

Table 21.1 Bioenergetic Parameters: Lizards versus Birds and Mammals

Parameter	Lizards	Birds and Mammals	Units
Standard metabolism 38°C body temperature	$3.5(W)^{-0.25}$	$17(W)^{-0.25}$	cal $(g\cdot h)^{-1}$
Same	$47(W)^{-0.25}$	$146(W)^{-0.25}$	kcal $(g\cdot yr)^{-1}$
Yearly energy budget for population	$24–70(W)^{-0.25}$	$300–1400(W)^{-0.25}$	kcal $(g\cdot yr)^{-1}$; W = adult weight
Secondary productivity ÷ yearly energy budget	10–30%	0.3–3%	
Digestive efficiency (assimilation ÷ ingestion) for predators	70–90%	70–90%	
Maximum predator–prey standing crop ratio, if predator adult weight ≃ prey adult weight	10–80%	1–3%	
Cost of locomotion			
quadrupedal	$41(W)^{-0.4}(S) + 1.7(M_{std})$		cal $(g\cdot h)^{-1}$
Bipedal	$11(W)^{-0.2}(S) + 1.7(M_{std})$		S $(km\cdot h^{-1})$
Maximum energy available through anerobic glycolysis	0.15–0.3		cal g^{-1} of body weight
Maximum energy available through aerobic exercise metabolism	$10–25(W)^{-0.25}$	Rodents: $70–100(W)^{-0.25}$ Shrew: $170(W)^{-0.25}$ Birds and large mammals $260–450(W)^{-0.25}$	cal $(g\cdot h)^{-1}$

Standard Heat Production

Under standard conditions—resting and postabsorptive in the thermoneutral zone—heat production in mammals and birds is about 3.5 times higher than in reptiles of the same weight and body temperature (Dawson, 1972, 1973). The high standard metabolism (M_{std}) in endotherms may well be the result of thyroxin increasing the heat produced by the sodium pump at the cell membrane (Edelman and Ismail-Beigi, 1971). Whatever the biochemical mechanism, the M_{std} of mammals and additional heat produced without shivering in many species during cold stress does not require any behavioral activity, any static or dynamic work from the muscles, or any muscular contradictions at all (Jansky, 1973). Jansky has labeled mammalian M_{std} as "obligatory non-shivering thermogenesis (NSTG)"—an apt term—and the extra heat production during cold stress as "regulatory non-shivering thermogenesis" (Jansky, 1973).

Thermal Conductance and Homeothermy

Except during heat stress, thermal conductance of birds and mammals is usually much less than in reptiles. Although some tropical mammal species may be poor thermoregulators outside their preferred, equable climate space (Dawson, 1973), even among the most primitive mammals (monotremes) some species can maintain an air temperature–body temperature gradient of 10°C or more (Schmidt-Nielsen *et al.*, 1966); with NSTG and shivering, the gradient can be increased to 30°C. Some lizards are cited frequently as "incipient endotherms" since the oxygen consumption during intense exercise may exceed the M_{std} of a mammal of the same weight and temperature and may raise the lizard core temperature a degree or two (Bartholomew and Tucker, 1963; Asplund, 1970; Fig. 21.1). But since thermal conductance does not decrease significantly toward mammal values during such exercise, and the 1° or 2° increase in body temperature makes only a negligible difference in physiology, it seems unlikely that natural selection has been acting to incorporate the heat produced during exercise as a useful part of the lizard thermoregulatory mechanisms, at least not in the size range studied (body weight 100 g to several kilograms). Aerobic exercise metabolism does maintain a body core–air or water temperature gradient of up to 10°C or more in some swimmers and fliers—tuna and sharks (Carey and Teal, 1966, 1969), leatherback sea turtles (Mrosovsky and Pritchard, 1971), and large moths (Bartholomew and Epting, Chapter 22, this volume). Heat from muscle contractions seems to give exercise endothermy to brooding pythons (Hutchison *et al.*, 1966). These cases of exercise–metabolism endothermy require muscular contractions; endothermy lasts only as long as the periods of high activity. Bird–mammal endothermy is fundamentally different; except during hibernation-like torpor, high M_{std} is continuous, automatic, and requires no contractions.

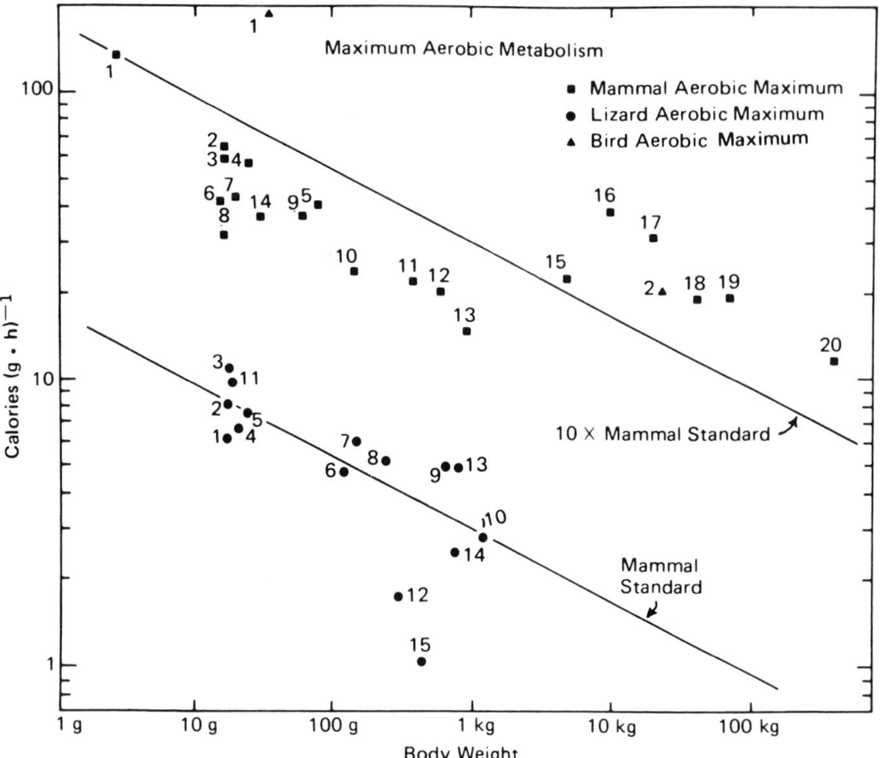

Fig. 21.1. Maximum aerobic (sustained) exercise metabolism as a function of body size in lizards (dots), mammals (squares), and birds (triangles). Lizards (body temperature in parentheses): 1, 2, *Gerrhonotus multicarinatus* (30); 3, 10, *Tupinambis nigropunctata* (30); 4, 5, 8, *Varanus niloticus* (30); 6, *Ctenosaura pectinata* (30); 7, 9, *Varanus exanthematicus* (30); 11, *Cnemidophorus sexlineatus* (35–40); 12, *Amphibolurus barbatus* (30); 13, *Varanus gouldi* (35–40); 14, *Iguana iguana* (30); 15, *Tiliqua scincoides* (40). References: 1–10, this paper; 11, Asplund (1970); 12, Bartholomew and Tucker (1963); 13, Bartholomew and Tucker (1964); 14, Moberly (1968); 15, Bartholomew *et al.* (1965). Mammals: 1, pigmy shrew; 2, common vole; 3, bank vole; 4, white mouse; 5, hamster; 6, meadow vole; 7, white-footed mouse; 8, red-backed mouse; 9, chipmunk; 10, hamster; 11, rat; 12, 13, guinea pig; 14, white mouse; 15–17, dog; 18, cheetah; 19, man; 20, horse. References: 1–5, Jansky (1965); 6–9, Wunder (1970); 10–17, 19, 20, Pasquis *et al.* (1970); 18, Taylor and Rowntree (1973b). Birds: 1, parakeet; 2, rhea. References: 1, Tucker (1966); 2, Taylor *et al.* (1971).

Yearly Energy Budgets

Most birds and advanced mammals regulate at a relatively high core temperature (37–42°C). Since M_{std} is higher for any given core temperature, and core temperatures are kept high much more continuously, the total yearly M_{std} of most birds and mammals is at least an order of magnitude greater than that of an ectotherm of the same weight. In mammals and in organisms as a whole, M_{std} seems to decrease with increasing body weight as the −0.25 power of body weight (Hemmingsen, 1960). For some groups for which data are available over a smaller size range than for mammals, such as lizards (Bartholomew and Tucker, 1964), other exponents have been calculated, but these may prove not to be significantly different from the well-documented −0.25. Total yearly energy budget per unit weight for individual organisms should also decrease as $(W)^{-0.25}$. Rapidly growing juveniles have higher rates of energy metabolism than adults of the same weight (Brody, 1945; Kleiber *et al.*, 1956). Thus the total yearly energy budget per unit standing crop (SC) of a breeding population theoretically should decrease with increasing adult size even more rapidly than $(W)^{-0.25}$, since reproductive rates decrease with increasing size and the percentage of the total SC represented by juveniles also decreases. Figure 21.2 is a plot of total yearly energy budget (assimilation)/caloric value of the standing crop as a function of adult body size; data are from various field and theoretical studies. Assimilation/SC is 20–30 times higher in endotherms than in ectotherms of the same weight.

Secondary Productivity

Secondary productivity—the assimilated energy fixed as new tissue in growth and reproduction—is a key parameter for using predator–prey ratios as bioenergetic indicators in the fossil record. In general, for animals with similar bioenergetics but differing in size, the ratio of secondary productivy (SP) to assimilation is a constant "Kleiber's Law"—Kleiber, 1961, p. 320), and SP/SC should decrease with increasing adult size parallel to assimilation/SC (Fig. 21.2, filled circles).

Vulnerability to predation strongly affects SP/SC ratios for any given body size. Predators (spiders, wolves, lions) tend to have lower SP/SC than their potential prey (Van Hook, 1971; Moulder and Reichle, 1972; Jordan and Botkin, 1971; Schaller, 1972; Fig. 21.2). Species that use highly cryptic, fossorial, or nocturnal habits or unpalatability to escape predation can have SP/SC ratios lower than their more exposed but similarly sized relatives. In diurnal lizards and most mammals and birds SP/SC for body size 10–100 g is 100–300 percent (Fig. 21.2; 9–23); but for night lizards (*Xantusia*) and some plethodont salamanders in this size range, cryptic habits reduce the SP/SC ratio necessary to maintain a steady-state population by an order of magnitude (Zweifel and Lowe, 1966; Organ, 1961). Species can escape predation by evolving tremendous size. Predation on ungulates of moose–deer–wildebeeste–zebra size by hyaenas, big cats, and canids can be very heavy, and SP/SC ratios of 20–30 percent are reached by the prey populations (Figs. 21.2: 29, 31, 32). But ungulates of rhino and elephant size (adult weight 2–5 tons) are too powerful to be taken often, even by lions (Schaller, 1972), and SP/SC ratios in these pachyderms are only 5–7 percent or lower, much less than would be predicted on the basis of $(W)^{-0.25}$ alone (Figs. 21.2: 33, 34). Rigorous climates with low primary productivity can suppress SP/SC. As mentioned, SP/SC may reach 20–30 percent in deer–moose-size ungulates under heavy predation. In muskoxen, whose adult weight is less than that of moose, reproductive potential and SP/SC seem

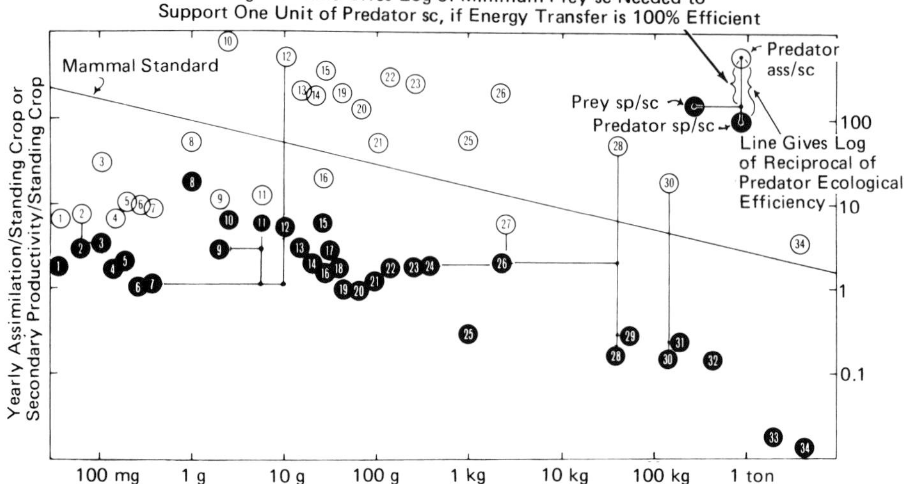

Fig. 21.2. Total yearly population energy budget per unit standing crop and secondary productivity per unit standing crop as functions of adult body weight. Filled circles: energy value of yearly secondary productivity ÷ energy value of standing crop. Open circles: yearly energy budget ÷ energy value of standing crop. The length of the vertical line from the center of the open circle of any potential predator to the horizontal line from the center of the solid circle of any potential prey gives the log of the minimum ratio of prey standing crop to predator standing crop necessary to support the predator population, assuming 100 percent efficiency in transfer of prey secondary productivity to predator assimilation. 1, 2, spiders; 3, spider prey; 4, isopod; 5, parasitic fly; 6, moth; 7, 8, grasshopper; 9, lizard *Uta*; 10, pigmy shrew; 11, lizard *Anolis*; 12, shrew; 13, deer mouse; 14, wood mouse; 15, old field mouse; 16, lizard *Sceloporus;* 17, bank vole; 18, vole; 19, least weasel; 20, sparrow; 21, swamp sparrow; 22, blue tit; 23, great tit; 24, gray squirrel; 25, owl; 26, jackrabbit; 27, lizard *Lacerta*; 28, wolf; 29, white-tailed deer; 30, lion; 31, wildebeeste–zebra (lion prey); 32, moose; 33, rhino; 34, elephant. References: 1, 2, 6, 10, 13, 15, 17, 20, 22, 23, Golley (1968); 7, 8, 12, 14, 16, 19, 24, 25, Varley (1970); 4, Moulder and Reichle (1972); 3, 5, Van Hook (1971); 9, Tinkle (1967), Alexander and Whitford (1968), Turner quoted in Golley (1968); 11, Kitchell and Windell (1972); 18, McNab (1963), Blair (1960); 21, modified from Golley (1968); 27, Avery (1971); 26, French *et al.* (1965); 29, Wagener (1970); 30, 31, Schaller (1972); 28, 32, Jordan and Botkin (1971); 33, Goddard (1967); 34, Hanks and MacIntosh (1972).

much lower than in moose (Freeman, 1971). Barren-ground tundra environments may limit the maximum assimilation per individual to values far below those of savannah or temperate-forest ungulates. Muskoxen communal defense against wolves seems unusually effective (Walker, 1968), a necessity for these animals since the maximum replacement rate is much less than in ungulates in more productive habitats.

Despite much higher energy budgets, SP/SC in endotherms is not significantly higher than in ectotherms of the same weight, if both are exposed to heavy predation. In other words, birds and mammals have paid a high price for the increased M_{std} which is the key to their endothermy: the "ecological efficiency," the ratio of

SP to total yearly assimilation, reaches 10 or 20 percent in many ectotherms, invertebrate and vertebrate (Golley, 1968; Varley, 1970) but a maximum of only a couple of percent in birds and mammals. High SP/assimilation rates, comparable to those of ectotherms, are possible in birds and mammals only in domesticated species, where growth rates are hyperaccelerated by selective breeding, high-protein foods and hormones, time from birth to maturity is sharply reduced, and heat loss from exercise metabolism is minimized by confinement.

Predator–Prey Ratios

A prey population with a given SP/SC ratio can support a much higher SC of ectothermic predators than of endothermic predators, because of the high energy demands of the latter. Because of ectothermy and their relatively low SP/SC ratios, spiders can reach a theoretical ratio of predator SC to prey SC of 60 percent or more. Field studies consistently have reported predator–prey SC ratios for spiders of 10–40 percent (Table 21.2). In an all-lizard predator–prey system with predator and prey of about the same size, a similar theoretical ratio is possible. In fact, since most lizards prey upon arthropods several orders of magnitude smaller than themselves, lizard SC could easily exceed their prey SC. Because of high assimilation/SC, a mammal SC could exceed its prey SC only if the adult size difference was enormous, as in blue whales preying on euphausiacean crustacea. If predator and prey are about the same size, maximum predator SC would be only a couple of percent of prey SC in birds or mammals. In India and Africa, hyaenas, canids, and big cats reach a maximum predator–prey SC ratio of only 1 or 2 percent (Bakker, 1972a; Schaller, 1972). The biophysical model of Porter *et al.* (Chapter 20, this volume) also predicts that maximum predator–prey SC ratio would be 20–30 times higher in an ectothermic predator (garter snake) than in an endotherm (mink).

Anerobic Exercise Metabolism

The total amount of energy available for a short burst of intense activity from anerobic glycolysis in lizards seems to be about the same as in man and the dog, about 0.15–0.3 $cal \cdot g^{-1}$ of body weight (Margaria, 1972; Cerretelli *et al.*, 1964; Bennett and Licht, 1972). There is no reason to suppose that the maximum amount of creatine phosphate (the anerobic fuel used for the most rapid activity bursts) which can be packed into the muscles of ectotherms would be any less than that of endotherms of the same weight. Also, for a given tissue temperature, the maximum rate of energy consumption per gram of skeletal muscle is probably not less in ectotherms than in endotherms. Thus one could predict that the maximum rate of anerobic exercise metabolism should not be lower in ectothermic reptiles as a whole than in birds or mammals. Since the energy consumption during locomotion can be calculated from the body weight and running speed (as described below),

Table 21.2 Predator–Prey Standing Crop Ratios

Predator	Ratio (%)	Prey	Environment	Reference*
Ectotherms				
Spiders	35	Potential prey in size range available to spiders	Forest floor, Oak Ridge, Tennessee	a
All cryptozoan predators, spiders, beetles, etc.	13	All cryptozoan herbivores and detritivores	Forest floor, Oak Ridge, Tennessee	a
Spiders	18	All prey species, including instars too large to be caught by spiders	Forest floor, Oak Ridge, Tennessee	b
Spiders	40	Potential prey in size range available to spiders	Forest floor, Oak Ridge, Tennessee	b
Spiders	16	All arthropods except spiders	Canadian pasture	b
Spiders	43	All arthropods except spiders	Woodland, England	b
Parasitic flies, beetles	13	Winter moth larvae	Clover stand	c
Dimetrodon (very primitive mammal-like reptile)	54	All possible prey species, mostly semiaquatic amphibians	Semiarid tropical delta, Early Permian, Texas; palaeolatitude 5°S	d
Transitional from ectotherm to endotherm				
Early therapsids (advanced mammal-like reptiles); anteosaurs, pristerognathids	10–17	Early therapsids: dicynodonts, tapinocephalians, titanosuchians	Intracontinental basin, Late Permian, South Africa; temperate, postglacial climate; palaeolatitude 70°S	e
Gorgonopsians (early therapsids)	8–10	Pareiasaurs	Semiarid alluvial plain, Late Permian, North Uralian region; warm climate; palaeolatitude 20°N	f
Gorgonopsians	20	Dicynodonts	Intracontinental basin, Late Permian, Lake Nyasa; warm temperate postglacial climate; palaeolatitude 65°S	g
Endotherms				
Tyrannosaurs	3	Duckbill, horned, and armored dinosaurs	Subtropical, mesic delta, Late Cretaceous Alberta; palaeolatitude 50°N	e
Cats, dogs, hyaenodons	4.5	Ungulate mammals	Oligocene South Dakota, warm temperate climate	e
Cats, dogs	3	Ungulate mammals	Pliocene, Nebraska, temperate climate	e
Lions, hyaenas	1–2	Ungulate mammals	Ngorongoro Crater, Tanzania	h

* a, Moulder and Reichle (1972); b, Van Hook (1971); c, Varley (1970); d, this paper; e, Bakker (1972a); f, Amalitzky (1922) and Pravoslavev (1927a, 1927b); g, Huene (1942, 1950) and Parrington (personal communication); h, Schaller (1972). Palaeolatitudes from Robinson (1973) and Russell (1973).

anerobic exercise metabolic rates can be computed from data on locomotor activity. In fact, the top sprinting speed in small lizards is as high or higher than in small mammals (about 30 km·h^{-1}; Oliver, 1955; Snyder, 1956; Layne and Benton, 1954; Belkin, 1961).

Aerobic Exercise Metabolism

The maximum sustained, aerobic exercise metabolism is an order of magnitude greater in birds and advanced mammals than in lizards and other ectotherms (Fig. 21.1). The function of high aerobic exercise metabolism is twofold: (1) to permit high levels of sustained activity; (2) to permit rapid payment of oxygen debt so that bursts of anerobic metabolism can be spaced more closely in time. Since birds and mammals must acquire an order of magnitude more food per unit time than ectothermic reptiles, it is not surprising that most endotherms are geared for more nearly continuous exercise.

For small, active lizards a 10-s burst of highest running speed would incur an oxygen debt that could be paid back in a few tens of minutes. Several such bursts in quick succession would require a recovery period of up to 1 h (Bennett and Licht, 1972), and if most of the creatine phosphate and glycogen stores had been spent the animal would be incapacitated and very vulnerable during this time. However, small animals usually have microhabitats—burrows, bushes, rocks—to hide in, on, or under during physiological incapacitation. Increasing body size presents a problem: aerobic exercise metabolism and the speed of oxygen-debt payment decrease with increasing weight (Fig. 21.1); although the energy consumption per gram during running at an even speed decreases for all speeds with increasing weight, the energy consumed during accelerating at the beginning of a sprint probably remains a constant per gram of body weight km^{-1}·h^{-2}, and for a large ($>$100-kg) animal, acceleration consumes most of the energy in a short run; the availability of hiding places decreases with increasing body size. It would be possible to "build" a 300-kg lizard with the long, powerful limbs of a lion, pack it with enough anerobic energy stores to sprint at 50–70 km·h^{-1}, and us it to chase zebra and wildebeeste. But this lion–lizard would not be able to enjoy its prey after an exhaustive run, even if the hunt were successful: during the long oxygen-debt-recovery period, the lion–lizard probably would lose its prey to a competitor or fall victim to predation itself. Although no quantitative field data are available on locomotor activity, the largest living lizard, the Komodo dragon (*Varanus komodoensis*), has much shorter limbs with much less appendicular musculature than do cats or canids of the same weight and seems geared for much slower top speeds.

Bioenergetic Gauges for Fossils

When is a fossil sample a faithful index of life predator–prey SC ratios? Both biological and geological criteria must be satisfied:

1. High predator–prey SC rates will be present in ectotherms only if the predator population is not regulated at a size much smaller than that possible if all the potential prey SP could be utilized; in other words, the predator population should approach the food-limiting condition.

2. Since large animals can maintain steady-state populations with greater consistency despite short-term fluctuations of primary productivity or water, because lower energy and water turnover in larger animals permits longer fasts, a predator–prey system of very large body size would approach the food-limiting condition more consistently than one of small body size.

3. All the prey species in the sample must be readily available to predators. Generally, solitary carnivores—cats, Komodo dragons—kill readily prey up to twice their own body size (Auffenberg, 1970). Evolution of saber teeth (smilodonty) or group hunting increases the ratio of prey–predator body size.

4. Sedimentological processes during carcass transport and burial will sort fragments according to size and density; therefore, predator and prey should not differ greatly in size and robustness of body and bones.

5. Predator and prey should be preserved in the same facies (depositional environment).

6. Since carcass disintegration is a function of transport distance from death site to final burial site, predator and prey should be in the same state of disarticulation and wear.

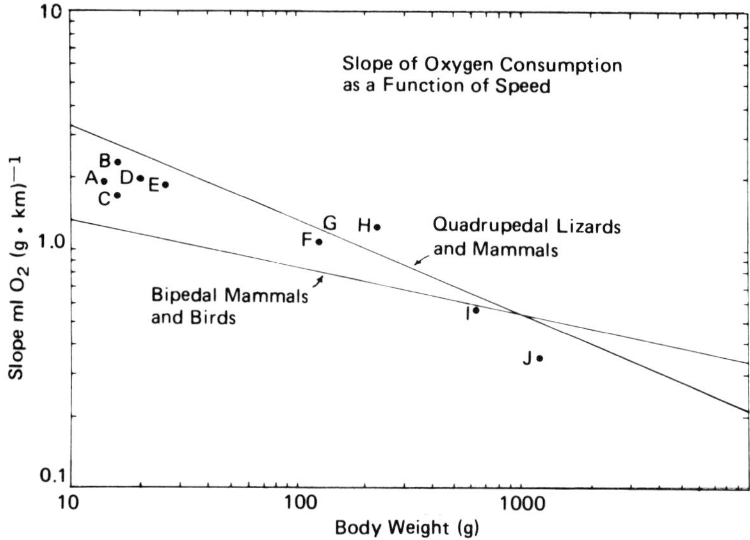

Fig. 21.3. Slope of the energy consumption versus running speed relationship plotted as a function of body weight. Quadrupedal mammal regression line from Taylor *et al.* (1970); bipedal bird and mammal line from Fadak *et al.* (1973) and Taylor and Rowntree (1973a). Lizards: A, B, *Gerrhonotus multicarinatus*; C, J, *Tupinambis nigropunctata*; D, E, H, *Varanus niloticus*; F, *Ctenosaura pectinata*; G, I, *Varanus exanthematicus*. Body temperature in lizards, 30°C.

7. Predator and prey should be scattered through the facies, or, if specimens occur in pockets, the species composition of the pockets should not vary greatly.

If the fossil sample represents a sudden event (e.g., a flood) which kills and buries a certain percentage of all the animals of a given size range in the area, then the preserved predator–prey ratio will reflect the life SC ratio. If the sample is of the victims of normal, day-to-day mortality, mostly predation, then the preserved ratio will reflect the predator–prey SP ratio. The latter could be lower than the SC ratio, owing to the low SP/SC of top carnivores. Predator–prey SC ratios could be inflated in a fossil sample if the predators avoided destroying the carcasses of their own species, as reported for lions (Schaller, 1972).

Running Speed and Exercise Metabolism

Many comparative anatomists and paleontologists have argued that for a given body weight and running speed, the energy expended for locomotion can be modified by morphological adaptation: changing from a wide-track, sprawling gait,

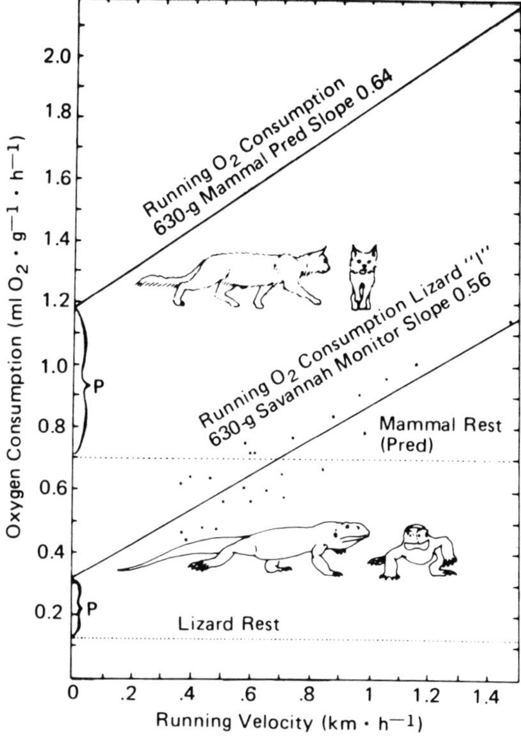

Fig. 21.4. Comparison of energy consumption as a function of running speed in a lizard (body temperature, 30°C) and a mammal of the same weight. Mammal regression line computed from equations of Taylor *et al.* (1970).

typical of most lizards, to a narrow-track, erect gait, typical of cursorial birds and mammals (Bakker, 1971; Romer, 1966) supposedly reduces the cost of locomotion; shifting the muscle bellies and the center of mass of a limb proximally up the limb bone, closer to the pivot point, decreases the moment of inertia and supposedly reduces the energy necessary for the reciprocal acceleration and deceleration of the limb with each stroke (Howell, 1944). Recent experiments show that both of these suggestions are incorrect. Taylor *et al.* (1970, 1971, 1972, 1973a, 1973b) have shown that for a given body weight, energy consumption increases linearly with running speed, and that in quadrupeds the slope of energy consumption as a function of speed decreases with increasing body size as $(W)^{-0.4}$. In bipedal mammals and birds a similar relationship exists, except that the slope decreases as the -0.2 power of body weight (Fig. 21.3; Fadak *et al.*, 1973). Species with heavy-muscle bellies far distal on the limb (man, chimpanzee) do not differ significantly from species with proximal bellies and long, slender extremities (rhea) in the energy expended for locomotion at a given speed and body weight. In quadrupedal lizards the slope of energy expended as a function of running speed is not signifi-

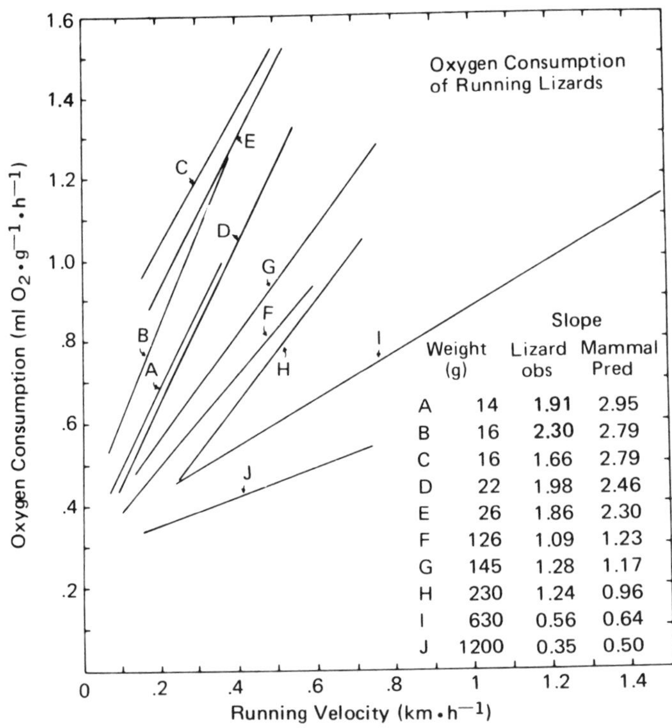

Fig. 21.5. Energy consumption as a function of running speed in lizards A–J (species given in Fig. 21.3). Slopes for mammals of equivalent weight predicted from Taylor *et al.* (1970). Each lizard regression line represents the range for 20–24 runs of 10–30 min each at 30°C body temperature. Observed oxygen debt payback negligible. Correlation coefficient for lizard slopes: A, .872; B, .930; C, .793; D, .909; E, .929; F, .890; G, .880; H, .951; I, .826; J, .841. Calibration procedures as described for mice in Taylor *et al.* (1972).

cantly different from that of mammals of the same weight (Figs. 21.3–5; Bakker, 1972b). The energy expended for locomotion is not less in species with an erect gait (dogs, goats, cats, antelope) than in species with a sprawling or semierect gait (most lizards, ground squirrels). Thus the cost of running can be expressed by two simple formulas: quadrupeds: $M_{run} = 41(W)^{-0.4}(S) + 1.7(M_{std})$; bipeds: $M_{run} = 11(W)^{-0.2}(S) + 1.7(M_{std})$, where M_{run} is in $cal(g \cdot h)^{-1}$ and S is in $km \cdot h^{-1}$. At low speeds, the lizards expend less energy to run than mammals of the same weight, because the $1.7(M_{std})$ term is lower in the lizards, but this difference becomes negligible at high speeds (Fig. 21.4). The two cost-of-locomotion formulas have been derived mainly from experiments on animals supplying all the energy consumed by aerobic, steady-state exercise metabolism, where the observed oxygen-debt payback does not exceed the observed oxygen deficit significantly (Keul *et al.*, 1972). But the contribution of anerobic exercise metabolism does not seem to change the cost of locomotion equations, at least in the two species studied, man and the dog (Margaria *et al.*, 1963; Cerretelli *et al.*, 1964).

Exercise metabolism at top speed can be calculated for fossil tetrapods if top speed can be estimated. For large (> 10-kg) species whose locomotor activity is rather restricted to walking and running over flat, even terrain, top running speed is closely correlated with relative hindlimb length (RHL), here defined as the length (cm) of femur + tibia + longest metatarsal divided by the body weight (kg)

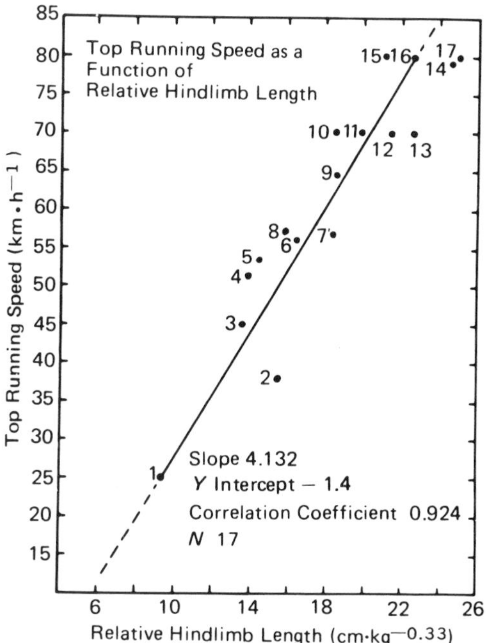

Fig. 21.6. Running speed as a function of relative limb length (length in cm of femur + tibia + tarsus + longest metatarsal ÷ body wt. $^{-0.33}$). Speed data for cursorial mammals from Walker (1968), Schaller (1972), and Howell (1944). Key: 1, hippo; 2, African elephant; 3, black rhino; 4, black bear; 5, lion; 6, Cape buffalo; 7, bison; 8, warthog; 9, wolf; 10, zebra; 11, eland; 12, jackrabbit; 13, topi; 14, Thomson's gazelle; 15, wildebeeste; 16, hartebeeste; 17, Grant's gazelle.

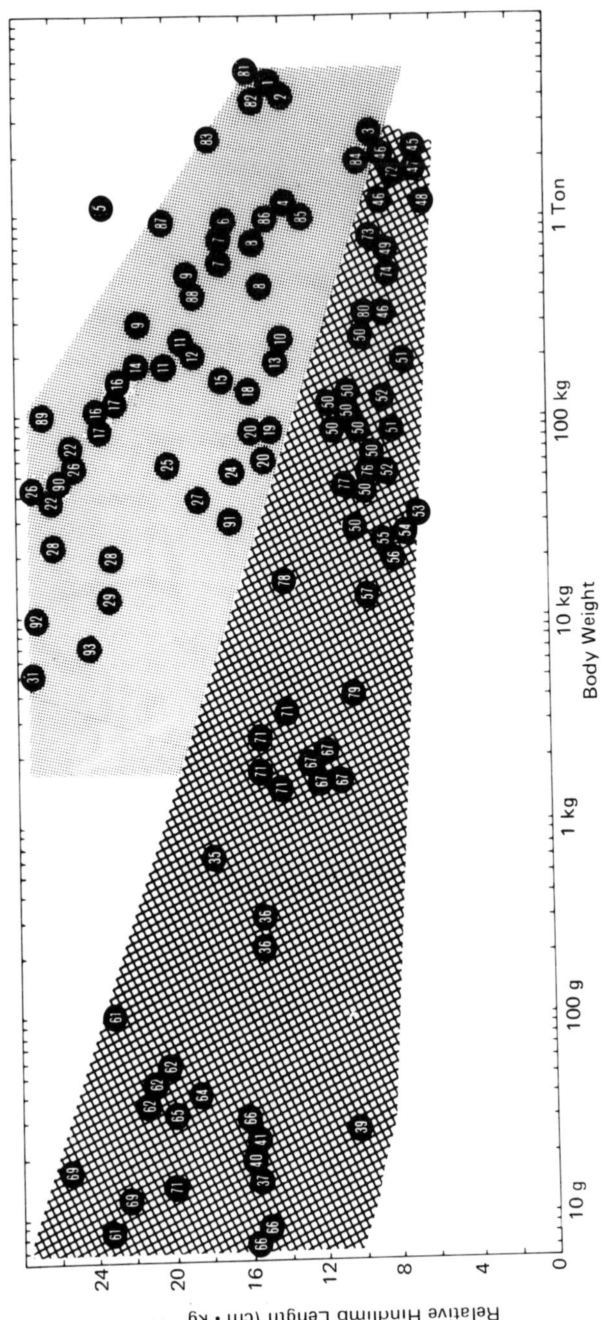
Relative Hindlimb Length (cm • kg⁻.33
0
4
8
12
16
20
24
10 g
100 g
1 kg
10 kg
100 kg
1 Ton
Body Weight

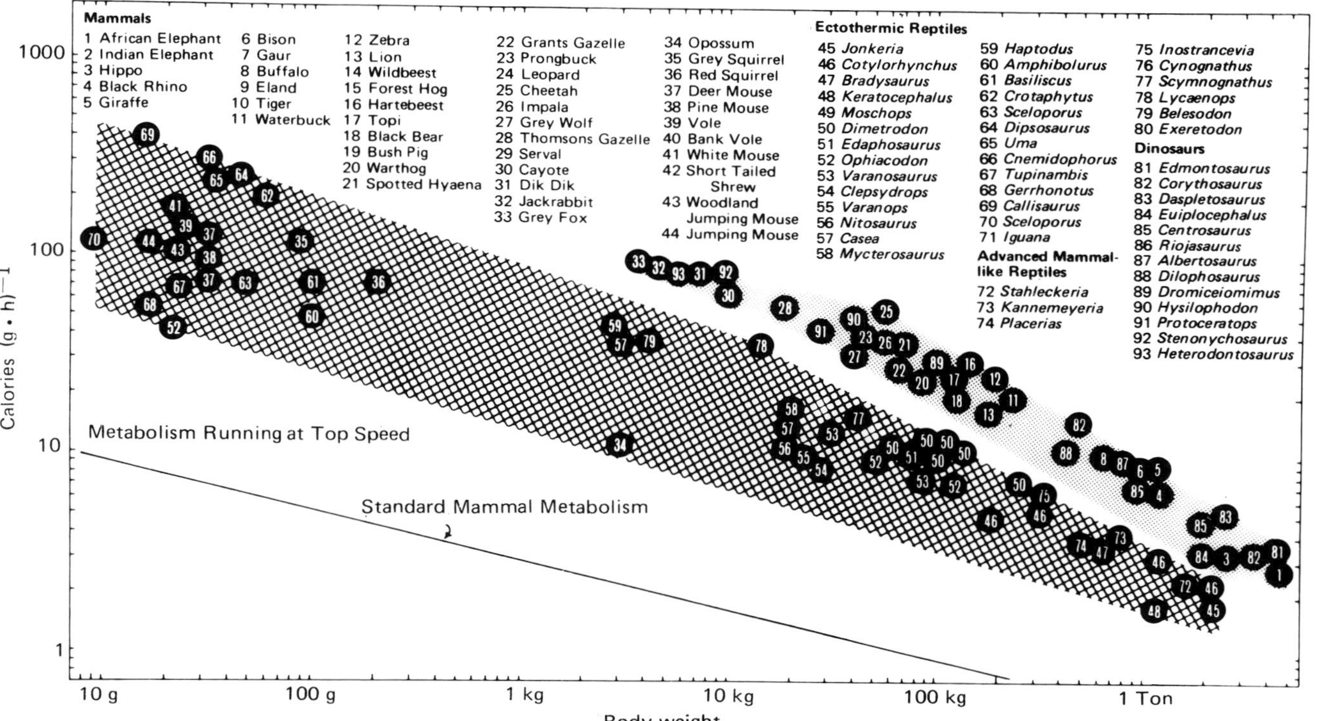

Fig. 21.7. *Upper:* Relative hindlimb length as a function of body weight. Note that, in the small size range, lizards are as long-limbed as in mammals, but in the larger size classes, dinosaurs and cursorial mammals are consistently longer-limbed than non-dinosaurian reptiles. Biped RHL taken as observed RHL $\div 2^{1/3}$. *Lower:* Exercise metabolism at top running speed calculated from the cost of locomotion equations in Table 21.1. Speed data for living tetrapods from Belkin (1961), Oliver (1955), Snyder (1956), Layne and Benton (1954), Schaller (1972), Walker (1968), and Howell (1944). Top running speeds of nonfossorial, nonarboreal fossil tetrapods calculated from running speed as a function of limb length (Fig. 21.6).

to the one-third power (Fig. 21.6). If two animals have identical body proportions but differ in weight, the RHL will remain a constant. The relationship shown in Fig. 21.6 does not hold for species that have greatly lengthened limbs as an adaptation for high browsing (giraffe), or to the cheetah, which increases stride length enormously by vertebral column flexion–extension, or for species specialized for climbing or digging. Finer precision in predicting speed can be obtained by consideration of joint angulation; fast, galloping species have greater flexure at the joints than slow, ambling forms (Bakker, 1971). The highest recorded running speeds for small (<5-kg) tetrapods, both lizards and mammals, are reached by species with RHL in the 18–24 $cm \cdot kg^{-0.33}$ range, the same general RHL seen in the fastest large species (Fig. 21.7), but the speed records for the smaller animals are lower (maximum of 30 $km \cdot h^{-1}$ in *Dipsosaurus*, *Callisaurus*, *Crotaphytus*, squirrels (Belkin, 1961; Layne and Benton, 1954; Snyder, 1956).

Bioenergetic Review of the Fossil Record

Armed with cost-of-locomotion equations and predator–prey analysis, we can follow tetrapod history up the geological column.

Origin of Vertebrates

Date

The earliest well-documented vertebrate fossils are from the Harding Sandstone, late mid-Ordovician, about 475×10^6 yr old.

Environment

The Harding Sandstone seems to represent beach, barrier bar, and other nearshore sediments, deposited in very shallow, agitated, well-aerated water as the sea spread across the western part of the flat, stable craton of North America (Denison, 1956).

Exercise Metabolism

The Harding fish were jawless scavengers, armored with a mosaic of bony plates, and probably lacking a swimbladder. Locomotion was probably limited to benthic wriggles and shovelings analogous to those of the benthic, armored South American catfish (Bunocephalidae; Doryadidae). Dissolved oxygen concentration probably was not a limiting factor in activity; clogging of gills by silt must have been a more serious problem.

Origin of the Lung–Swimbladder

Date

The jawed fishes split early into two great radiations: the sharks, without a swimbladder; and the bony fish (sensu lato), with a swimbladder. Most bony fish belong to two major divisions: the fleshy-fin group (Sarcopterygians); and the ray-fin group (Actinopterygians). Lungs are present in primitive living representatives of the former (lungfish) and the latter (gars; *Polypterus*). In most ray-fins lungs are modified into swimbladders, but the ancestral bony fish may well have had lungs adapted for air breathing. The earliest bony fish (primitive fleshy-fins) are from the Early Devonian; acanthodians, possible bony-fish ancestors, appear in the Late Silurian, ca. 400×10^6 yr B.P.

Environment

Late Silurian acanthodians are marine (Moy-Thomas and Miles, 1971). Many of the earlier Devonian records of both lungfish and crossopterygians are near-shore marine or brackish localities (Thompson, 1969). Thus the vertebrate lung appears to have originated in shallow marine or brackish water.

Exercise Metabolism

The earliest lung-bearing bony fish probably resembled early crossopterygians and early lungfish: moderately elongated, slightly laterally compressed fish adapted for active swimming and equipped with fleshy fins for propping up the body and scuttling along the bottom (Thompson, 1969). The living garpike (*Lepisosteus*) has a similar body form and has a dentition analogous to that of some crossopterygians: close-set, small, sharp, marginal teeth with larger tusks medially. Gars are powerful, stalking predators common in shallow water in warm climates, where seasonal accumulations of vegetation can severely limit the activity of gill-dependent fish. Air breathing in the lung-equipped gar permits the maintenance of exercise levels despite low oxygen tension in the water (Hill *et al.*, 1973). Such an adaptation would have been equally valuable to the Late Silurian and Early Devonian bony fish hunting in the shallow marine or brackish estuaries, lagoons, and mud flats, where stands of primitive vascular plants and stagnant water caused local, seasonal low oxygen tension. Early lungfish and crossopterygians both had large opercula, indicating that normal gill breathing was at least seasonally more important than in the Recent lungfish (Thompson, 1969). Originally, the lung appears to have been a supplement to, not a replacement of, the gills.

Origin of Amphibians

Date

The first known amphibians are from the latest Devonian of East Greenland, ca. 340×10^6 yr B.P. but are already a bit specialized; the crossopterygian/amphibian transition must have occurred somewhat earlier.

Environment

Mid- and Late Devonian crossopterygian fish, close to the ancestry of amphibians, are best known from lake and stream deposits of the Old Red Continent which stretched from New York State across northern Europe to the Uralian geosyncline. The Old Red Continent lay mostly within the Devonian tropical belt. Oxidized deltaic and intermontane sediment plus local lime pellet caliches suggest a warm, seasonally arid climate. At the Devonian–Carboniferous transition, major marine transgressions crossed the Old Red Continent; coal-forming conditions became widespread. Lack of growth rings in the wood of coal forest trees and generally reduced sediments indicate very equable, remarkably continuous growing seasons with high humidity. The Laurasian coal forest belt lay within the Carboniferous tropical zone, and known amphibian localities lie close to the palaeoequator (Panchen, 1973).

Exercise Metabolism

The naked skin of living amphibians is a secondary character, not inherited from Paleozoic forms (Romer, 1972). Devonian crossopterygians and lungfish generally were fully scaled. In the least disturbed carcasses of a variety of Permo-Carboniferous amphibia, remnants are found of what appears to have been a complete body covering of bony scales (Romer, 1972). Living amphibia lack movable ribs attached to a sternum. The ribs of crossopterygians appear to have been short with fibrous articulations with the vertebrae. Long ribs are present in some Recent and Devonian lungfish, but no cartilaginous rods connected the ends of the osseous ribs to a sternum. One of the most dramatic features of the earliest amphibia is the presence of long ossified ribs with mobile, probably synovial joints with the vertebrae, and distal ends pitted for the cartilaginous segments running to the sternum. The rib/vertebral articulations lay in planes set at a large angle to the horizontal; anterior–posterior swing of the arched ribs would expand and contract the pulmonary coelomic spaces. Rib-powered ventilation thus appears at the very beginnings of tetrapod history. Increased tidal volume and higher exercise aerobic metabolism probably were possible with this innovation.

The origin of legs and costal ventilation may be related. The energy expended in swimming is much less than in walking for any given speed and body weight (Schmidt-Nielsen, 1972). To reach even the slow speeds useful for foraging or dispersing across the wet coal forest terrain, the earliest amphibia may have required higher aerobic exercise metabolism than in their crossopterygian ancestors. Once acquired, this increased capacity was useful in other contexts: many Carboniferous amphibian lineages returned to a swimming mode of life, some achieved lengths of 4 m or more; the ability to maintain fairly high levels of exercise completely independent of water oxygen tension, continuously through the year, must have given the amphibians an edge over contemporary fish in many parts of the coal swamp–lake complexes, where reducing conditions frequently prevailed.

Origin of Reptiles and Mammal-like Reptiles

Date

The first reptiles are known from early and middle Late Carboniferous sites, ca. 300×10^6 yr B.P. The first mammal-like reptiles appear almost immediately afterward, and the split between the theropsids (mammals and mammal-like reptiles) and the sauropsids (birds, dinosaurs, and the rest of the Reptilia) began at the very foundations of reptilian phylogeny.

Environment

The stem reptiles (Romeridae) and the earliest pelycosaurian mammal-like reptiles are found together in sediment-filled, hollow, upright tree stumps which acted as pit falls for terrestrial creatures foraging about the floor of the coal swamp forest sand flats (Carroll, 1969). The climate must have been humid and warm, and under the forest canopy the air and ground temperature must have varied little.

Thermoregulation

Under the coal forest canopy, behavioral thermoregulation probably could not have raised body temperature above air and substrate temperature. Thermobiology of these early reptiles must have been like that of living lizards in warm, equable, deep forest tropical habitats (Inger, 1959).

Sphenacodont Communities

Date

The first large tetrapods capable of killing and dismembering prey of body size as large as or larger than themselves were the sphenacodonts, pelycosaurs close to the ancestry of therapsids (advanced mammal-like reptiles). First sphenacodonts appear in the Early Permian, ca. 275×10^6 yr B.P., and this family remained the dominant top carnivores, through the end of the late Early Permian, ca. 240×10^6 yr B.P.

Environment

Small, primitive sphenacodonts (*Haptodus*) come from coal swamp lakes in intermontane basins in France and Germany (Case, 1926). From the Late Carboniferous to the Late Permian and Laurasian climate changed markedly from wet and humid to increasingly arid. By early Late Permian many deltas were adjacent to marine basins receiving thick, bedded evaporites, and the number and extent of freshwater lakes, ponds, and permanent streams shrank. Lungfish that were adapted to seasonal aestivation in burrows replaced nonaestivating species (Olson and Vaughn, 1970). These changes began earlier in some regions (the American Southwest) than in others (West Virginia).

Energy Budgets

Figure 21.8 presents the preserved biomass in the Belle Plains Formation sphenacodont community, *Dimetrodon* top predator. Since skull length varies greatly among Lower Permian tetrapods of the same weight, the census was based on femora and humeri, where the robustness and preservability vary less. Early

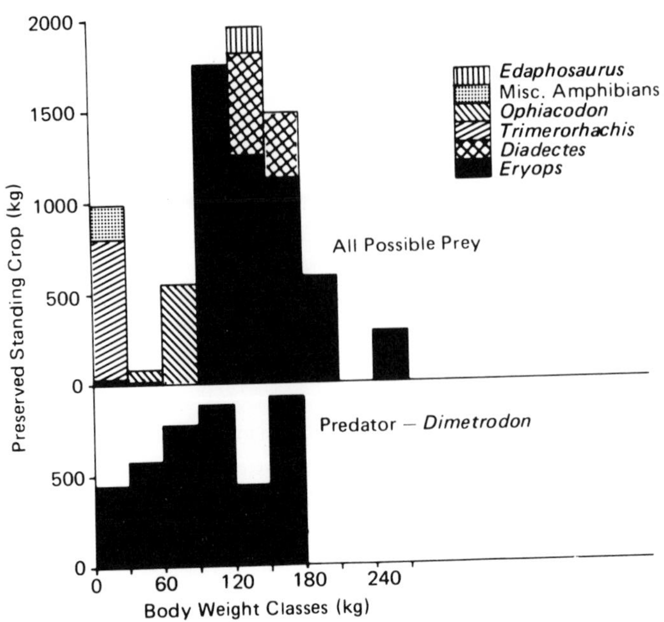

Fig. 21.8. Preserved standing crop of the Early Permian pelycosaurian reptile *Dimetrodon* and all possible prey species from the Belle Plains Formation pond-margin communities. Number of femora and humeri used as census indices. Note very high predator–prey standing crop ratio, about 54 percent.

Permian large herbivores were rare: *Edaphosaurus* is common and well preserved only in bog deposits (Geraldine Bone Bed), where only a few, very small juvenile *Dimetrodon* are found; *Diadectes* is preserved in the pond deposits, which produce large *Dimetrodon*, but is outnumbered overwhelmingly by the latter. As Olson pointed out in a remarkably perceptive paper (Olson, 1966), the prey of *Dimetrodon* in the pond communities was mostly aquatic and semiaquatic amphibians—*Eryops* and *Trimerorhachis*. The preserved predator–prey biomass ratio in the Belle Plains is very high, about 55 percent, and in other *Dimetrodon* communities and in New Mexico, where the closely related *Sphenacodon* was the top predator, ratios are consistently high. Sphenacodonts basically were primitive reptiles, and the predator–prey ratios are similar to those in Recent spider communities or to those of a hypothetical lizard–lizard, predator–prey complex. Sphenacodonts must have had low M_{std} and low yearly heat production.

Thermoregulation

The later species of *Dimetrodon* are common in overbank deposits and must have hunted on open terrain in very warm, arid climates. The adults were too large (100 kg or larger) to hide in burrows and would have been required to thermoregulate at a high body temperature to escape heat death and expand their climate space.

Exercise Metabolism

Large sphenacodonts have limbs much shorter than in cursorial, open-ground mammals and birds of the same weight and about of the same proportions as in the Komodo dragon. Top running speeds and exercise metabolism were on a lizard-like level.

Spread of Herbivory

Date

The percentage of potential prey, available to sphenacodonts, represented by herbivorous tetrapods increases through successively higher formations in the middle Early to early Late Permian of Texas (Olson and Vaughn, 1970). Herbivores make up nearly all the potential prey in the uppermost formations (San Angelo and Flower Pot).

Environment

As mentioned above, the entire Laurasian climate became increasingly arid.

Energy Budgets

Why did herbivory spread during the Early Permian? All the earliest sphenacodont populations depended upon aquatic and semiaquatic fish-eating prey. Thus the primary productivity supporting the sphenacodonts was produced by freshwater plants, mostly algae; terrestrial plants were relatively unimportant. The distribution of herbivory in living lizards provides some insight: A few herbivorous lizard genera are arboreal and favor local mesic habitats—*Iguana*, *Macroscincus*, some giant anoles; but the majority of genera with herbivorous species prefer arid or semiarid climates—*Dipsosaurus*, *Ctenosaura*, *Cyclura*, *Conolophus*, *Dicrodon*, *Angolosaurus*, *Agama*, *Uromastix*. Herbivory also tends to increase with size—in many of these genera the juveniles are more insectivorous than the adults. The secondary productivity of potential arthropod and small vertebrate prey per unit standing crop of vegetation in a desert may not be less than in a better watered situation, but the distance between clumps of vegetation does increase with increasing aridity. Thus, with increasing aridity, a large reptile must travel a greater distance to obtain its total energy budget from animal prey, and the heat loss from exercise metabolism per unit energy assimilated increases. Ecological efficiency (secondary productivity/total assimilation) would fall. Selection may well favor the shift to herbivory in arid climates to keep the exercise metabolism heat loss from inflating the total energy budget. In the Permian, the first common herbivores were captorhinids and caseid pelycosaurs; the latter reached huge size (adult weight up to 1 metric ton) by San Angelo–Flower Pot time.

Origin of Therapsids

Date

The first therapsids (advanced mammal-like reptiles) appear in the earliest Late Permian in Texas, Oklahoma, and the Pri Uralian region of Russia (Olson, 1962), ca. 240×10^6 yr B.P.

Environment

Both the Texas–Oklahoma and the Pri Uralian region lay close to the Permian Equator (Robinson, 1973). The American Southwest was probably more arid than Russia. Late Permian Gondwanaland was recovering from the Permo-Carboniferous glaciation, and virtually no tetrapods are known from early Permian sediments of South Africa. Shortly after therapsids had appeared in Laurasia, South Africa was invaded by a variety of primitive therapsids and some other reptile groups (pareiasaurs). South Africa during this time lay about at latitude 70°S (Robinson, 1973), and must have had a seasonally cool, temperate, or warm temperate climate. Local lime pellet conglomerates suggest some aridity.

Energy Budgets

In living large tetrapods, endotherms can be identified by bone histology: birds and mammals have compacta with either densely spaced, rapidly remodeled haversian systems or without haversian systems but with dense vascularization of several patterns (Currey, 1962; Enlow and Brown, 1957, 1958); reptiles and amphibians have relatively poor vascularization and few, if any, secondary haversian systems. Growth rings are common in the compacta of living lower tetrapods but are absent in endotherms. Unfortunately, the physiological mechanisms responsible for these differences are unclear. Peabody (1958, 1961) suggested that endotherms lack growth rings because growth continues during the winter when ectotherms are dormant, a suggestion developed by Ricqles (1969b). But individual growth rates do fall to zero in many temperate mammals during the winter, and yet growth rings do not form in the compacta; rings do appear in the secondary dentine and cementum of canine and molars (seals, bears, deer; Rausch, 1961). In habitats permitting year-round growth, dentine rings are faint or lacking (Rausch, 1961). Rapid remodeling of Haversian systems may maximize the amount of labile bone available for short-term calcium–phosphate metabolism, but why endotherms would require such a mechanism is not obvious (Bakker, 1972a).

All Early Permian tetrapods have ectotherm-like bone, often with compacta growth rings (Enlow and Brown, 1957; Peabody, 1958). In the early therapsids (deinocephalians, pristerognathids, gorgonopsians, early dicynodonts) endotherm-type histology appears (Ricqles, 1969a). Predator–prey ratios drop from the high sphenacodont level (Fig. 21.9):

1. Tapinocaphalus Zone, South Africa; predators: anteosaur deinocephalians, lycosuchid and pristerognathid therocephalians; prey: titanosuchians, tapino-

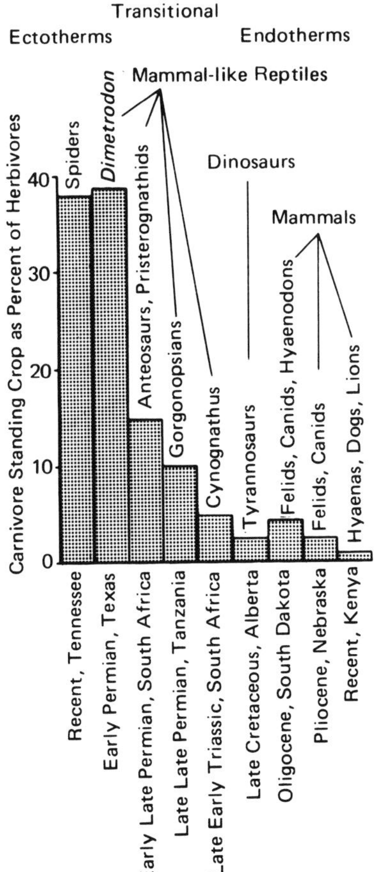

Fig. 21.9. Predator standing crop as a percentage of potential prey standing crop in Recent and fossil communities. Note sudden drop from *Dimetrodon* level to primitive therapsids, and low value for dinosaur communities.

cephalians, and small dicynodonts; predator–prey biomass ratio: 10–20 percent (Bakker, 1972a).

2. North Dvina River (Pri Uralian region); predator: large gorgonopsians; prey: pareiasaurs; predator–prey ratio: 8–10 percent (Amalitzky, 1922; Pravoslavev, 1927a, 1927b).

3. Ruhuhu Valley, Tanzania; predator, large gorgonopsians; prey: large dicynodonts, some pareiasaurs; predator–prey ratio: about 10 percent (R. Parrington, personal communication; Huene, 1942, 1950).

4. Luangwa Valley, Zambia; predator: large gorgonopsians; prey: large dicynodonts; predator–prey ratio: probably around 10 percent in some sites (Drysdale and Kitching, 1963).

Low predator–prey ratios and the change in bone histology strongly suggest that major steps toward mammalian endothermy had occurred; $M_{\rm std}$ and energy budgets were probably higher than in sphenacodonts. The invasion of Gondwanaland in the Late Permian temperate zone by primitive large therapsids may have been facilitated by the appearance of endothermy.

Exercise Metabolism

Morphological changes in early therapsids indicate that the locomotor behavior was different from that of sphenacodonts, but there is no indication that the exercise capacity was increasing. Limb lengths are no greater in primitive therapsids than in sphenacodonts, and top running speeds and exercise metabolism probably were not significantly higher.

Permo-Triassic Extinctions

Many groups become extinct at the Permo-Triassic boundary: gorgonopsians, pareiasaurs, most lineages of the very diverse Permian dicynodonts (ca. 225×10^6 yr B.P.). The earliest Triassic tetrapod faunas are dominated by various species of the peculiar dicynodont *Lystrosaurus*. Small, advanced carnivores are present (cynodonts), but no large species are known, and medium-size carnivores (proterosuchian thecodonts and therocephalians) are represented by only a handful of specimens. *Lystrosaurus* comes in all sizes and in such astronomical numbers that any calculated predator–prey ratio would approach zero. The low diversity of both carnivores and herbivores in the *Lystrosaurus* Zone probably is a result of the Permo-Triassic extinctions; only a few lineages had survived and none of these had yet radiated to replace the lost diversity of the Late Permian. The few carnivores present were almost certainly not food-limited, since the secondary productivity of all the lystrosaurs must have exceeded the energy budgets of the carnivores by several orders of magnitude. Quite probably, this undersaturation of carnivores was also caused by extinction; whatever environmental event killed off the typical Late Permian lines may have suppressed the population size of the carnivores that remained. The nature of the environmental event is not entirely clear. Gondwanaland was moving away from the South Pole so that the overall trend should have been toward warmer climates in the southern land mass. A dramatic worldwide regression of shallow shelf seas occurred at the Permo-Triassic transition, and removal of the thermal buffering effect of shallow sheets of water may have suddenly increased the seasonality.

Advanced Therapsids

Date

The last fauna with top carnivores dominated by theropsids was the *Cynognathus* Zone, known from South Africa and Argentina, which followed the *Lystrosaurus* Zone.

Environment

Equatorward movement of Gondwanaland presumably made the climate less rigorous than in the Permian.

Energy Budgets and Thermoregulation

Preliminary data indicate a low predator–prey ratio, 10 percent or less (Kitching, 1971). The top carnivore was an advanced cynodont, *Cynognathus*; the prey, herbivorous cynodonts and dicynodonts. Some evidence for a layer of dermal musculature, fat, and hair exists: in lizards the facial skin is very thin and closely appressed to the facial bones; the facial nerves and blood vessels exit to the skin via many small foramina; in most mammals, the face is covered by loose dermal musculature controlling the lips, nose, and vibrissae, and most of the nerve and blood vessel openings are concentrated into one or a few concentrated holes in the maxilla. In advanced cynodonts the facial foramina became increasingly mammal-like.

In living herbivorous lizards, the teeth are not used to masticate food but rather merely to cut up leaves and fruit into pieces small enough to swallow. Digestive efficiency (calories ingested − calories egested/calories ingested) in the lizards is as high as in mammals (80–90 percent feeding on arthropods or sweet potato), but the passage of time necessary for assimilation is much greater, 4 days in *Ctenosaura* after a sweet potato meal (Throckmorton, 1973). However, endotherms must cut passage time to supply their very high energy budgets. Most herbivorous mammals employ extensive dental preparation of the food bolus to increase the surface area for digestive enzyme action and separate indigestible fibers from usable vegetable matter. In caseids, the common and diverse pelycosaurian (primitive mammal-like reptile) herbivores of the Early Permian, the tooth mechanics were like those of many living herbivorous lizards; oral preparation was not extensive. However, in advanced, herbivorous cynodonts of the Triassic, postcanine teeth occluded in a shearing and crushing action, jaw musculature was powerful, and oral preparation of the bolus must have been intensive.

Exercise Metabolism

Morphological evolution continued, but limb lengths and limb proportions in large cynodonts indicate no significant increase in maximum capacity for exercise metabolism during running (76 in Fig. 21.7). Reduction of lumbar ribs in advanced cynodonts may indicate the presence of a diaphragm, but whether this feature originally evolved to increase the aerobic exercise metabolism is not clear. The highest recorded aerobic exercise metabolism in primitive living mammals (monotremes, Schmidt-Nielsen *et al.*, 1966; tenrecs, unpublished data at Harvard) is not much greater than that of a varanid lizard of the same weight.

Thecodonts

During the late Early and Middle Triassic, carnivorous cynodonts were replaced gradually by thecodonts, the basal radiation of the archosaurian reptiles.

Several small thecodonts were equipped with very long limbs and must have reached top running speeds greater than that of small cynodonts. Middle Triassic carnivorous thecodonts were roughly on a crocodilian level of organization; living crocodilians have highly subdivided lungs with a peculiar liver-piston ventilation pump (Gans, 1970), which suggests a high aerobic capacity, but no data yet exist about exercise metabolism in crocodilians.

Origin of Mammals

Date

The first known mammal is from the Late Triassic, ca. 200×10^6 yr B.P. The reptile–mammal transition occurred a bit earlier.

Environment

Most Late Triassic vertebrate localities are dominated by oxidized sediment, and some playa-like lakes and dunes have been reported. Early mammals or very advanced cynodonts have been found both in oxidized sediment and in reduced, swamp deposits (*Dromatherium* from a coal swamp within an intermontane basin in North Carolina).

Thermoregulation

The first mammals probably inherited a high, continuous M_{std} from their cynodont ancestors. All the earliest mammals were very small insectivores; by analogy with primitive living mammals, most Mesozoic mammals seem to have been adapted for nocturnal hunting; loss of color vision, highly sensitive olfaction and hearing, long vibrissae, and sensitive palmar surfaces probably represent the original suite of sensory mechanisms (Bakker, 1971). Body temperature probably was regulated at levels below 35°C, as in monotremes and tenrecs. Endothermy and hairy insulation probably permitted Mesozoic mammals to maintain a moderate body temperature and level of activity much more continuously during nocturnal hunting than in their potential ectothermic competitors, lizards and rhynchocephalians.

Exercise Metabolism

As mentioned above, tenrecs and montremes seem to have aerobic exercise levels much below those of most placental mammals, a condition that may be truly primitive.

Dinosaurs

Date

The first known dinosaur is Middle or late Middle Triassic. By the Late Triassic, dinosaurs were the dominant large herbivores and carnivores. From the Late

Triassic, ca. 200 × 10^6 yr B.P., until the end of the Cretaceous, ca. 65 × 10^6 yr B.P., virtually all the terrestrial carnivores and herbivores with an adult weight over 10 kg were dinosaurs.

Environment and the Advantage of Endothermy in Tropical Climates

Fossil floras are cited frequently as evidence of exceptionally warm, equable climates nearly worldwide during the Jurassic and Cretaceous (Spotila *et al.*, 1973). But arguments for extreme Jura-Cretaceous equability can be pushed too far. Two groups of ferns (Marattiaceae and Matoniaceae), now restricted mostly to the tropics, were widespread in the Jura-Cretaceous. However, the survival in the Recent tropics of relics from previously diverse and successful taxa does not necessarily prove that the group has always preferred warm, equable environments; instead, the complex ecology of the tropics may offer the greatest opportunities for finding a refugium to any waning radiation, regardless of the ecology of the group in its more successful days. No evidence exists for continental or alpine glaciation during the Mesozoic, and the predominance of drip points and whole-margin leaves does seem to indicate that severe frost was rare. Nevertheless, a latitudinal temperature gradient did exist throughout the Jura-Cretaceous; in the Juarassic, a northerly floral realm in Siberia was differentiated from the more equatorial fern-cycad belt (Vachrameev, 1972); in the Cretaceous, pollen and megaflora, and marine foraminifera, change along north–south transects (Bandy, 1967; Sloan, 1969). The Jura-Cretaceous climate was not totally equable through time. Oxygen isotope temperatures from marine fossils, pollen, and forams all indicate a drop in temperatures at the Cretaceous–Tertiary boundary (Russell, 1971), which has been cited as the cause of the dinosaur extinctions, but forams and oxygen isotope ratios also indicate a temperature drop of perhaps equal importance within the first half of the Cretaceous, and detailed floral analysis in Asia suggests a series of climatic deteriorations and ameliorations in the mid-Mesozoic (Krassilov, 1973). Very extensive shallow seas in the Cretaceous must have buffered the air temperature of adjacent landmasses, but diverse dinosaur faunas are known deep within Central Asia, thousands of kilometers from the nearest contemporary seaway.

Spotila *et al.* (1973) have quantified the suggestion made by a number of paleontologists that dinosaurs could achieve homeothermy in a warm, equable climate with normal, reptilian M_{std} and gigantic size alone. However, the Jura-Cretaceous climate may well not have been nearly as homogeneous and equable as assumed by the model of Spotila *et al.* (1973); although frost may have been a rare occurrence near epeiric seas, diurnal and seasonal variation probably was severe enough to make endothermy a valuable strategy, even among fairly large animals. Spotila *et al.* (1973) chose to have their dinosaur model regulate around 30°C; however, dinosaur locomotor anatomy indicates high activity, and in both active lizards (varanids) and most large, cursorial ground birds and placental mammals, body temperature is regulated at 38°C or above. Moreover, although many dinosaur species attained adult weights of several tons, many others were mature at 10–100 kg. Small Triassic dinosaurs are known from the South African Triassic, palaeolatitude 55° south; large species are indicated within the Cretaceous Arctic

Circle in the Cretaceous (Russell, 1973). A rigorous, quantitative analytical model of the evolution of world climate has not yet been developed, although Robinson (1973) has made important advances in modeling global rainfall patterns in the Permo-Triassic. Nevertheless, if dinosaurs achieved homeothermy at a body temperature of 37°C throughout their geographical and adult-body-size range, it is almost certain that they must have employed endothermy.

The model of Spotila *et al.* (1973) can be tested in Recent environments. According to the model, in a tropical climate the advantage of endothermy would decrease with increasing body size, and one might predict that the ratio of ectotherms to endotherms in species and biomass would decrease with increasing body size. This does not occur. Even in the tropical rain forests and wet savannahs lying within a few degrees of the equator in Africa, South America, and Southeast Asia, nearly all the terrestrial tetrapods with an adult weight of over 5 kg are mammals and birds; with decreasing body size, the relative abundance of amphibians and reptiles increases. Only in freshwater local habitats do large ectothermic tetrapods enjoy a competitive edge over endotherms; crocodilians and varanid lizards reach larger adult sizes and comprise a greater biomass than otters, aquatic viverrids, etc. Large, fully terrestrial reptiles are dominant only in insular situations—Komodo dragons, giant tortoises on Aldabra and the Galapagos—where potential endothermic competitors and predators have been excluded.

Part of the advantage of endothermy in tropical climates may be what could be called the principle of the occasional or aperiodic thermal disaster. Large lizards probably can maintain fairly constant core temperatures for successive days or even weeks in some tropical locales. But when an unusually cool night or evaporation from the skin after a rain or heavy dew drains the reptile of 10° of heat content, and overcast skies screen out much of the solar radiation, the ectotherm would be helpless to defend itself or its eggs from endotherms operating at a high core temperature. Moreover, the thermal inertia of a very large reptile would require a long reheating period when radiation does become available for raising the core temperature above air temperature. In summary, even if most of the Mesozoic land masses had a frost-free, warm temperate or tropical climate, endothermy, especially when used to maintain a high (37°C or above) body temperature, probably would have been an important advantage to large tetrapods.

Energy Budgets

Large samples from dinosaur communities usually have very low predator–prey ratios:

1. Scores of large, prosauropod dinosaur herbivores are known from Late Triassic sites all over the world, but large carnivores are exceedingly rare.
2. In the Late Jurassic fauna from western North America (Morrison Formation) and East Africa (Tendagaru) predator–prey biomass ratios are under 5 percent.
3. In a mid-Early Cretaceous fauna from Wyoming and Montana (Cloverly Formation), the predator (a "saber-toed" theropod) represents only a small

percentage of the preserved biomass of its potential prey (ornithopods, Ostrom, 1970).

4. In the diverse Late Cretaceous faunas of Canada (Belly River Group), predator–prey biomass ratios are only 3 percent (Bakker, 1972a).

Dinosaur bone histology is endotherm-like; density of Haversian systems and vascularization is sometimes greater than in advanced placental mammals (Currey, 1962). Dinosaurs were almost certainly endothermic.

Exercise Metabolism

Dinosaur locomotor repertoire was severely limited to walking and running over level terrain. Small species resembled ground birds (moas, rheas, etc.); larger

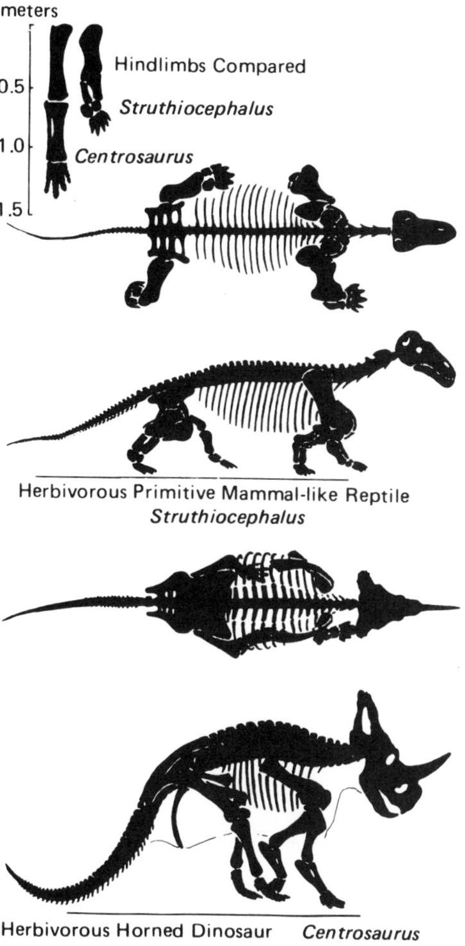

Fig. 21.10. Dinosaur and ecologically equivalent nondinosaurian reptile of the same weight drawn to scale. Note longer limbs and longer distal extremities in the dinosaur.

forms had joint proportions and mechanics analogous to those of rhinos and elephant-type ungulates but without mobile scapulocoracoids (Bakker, 1971). Top running speeds and exercise metabolism must have been much higher than in large therapsids, even cynodonts (Figs. 21.7 and 21.10). In saurischian dinosaurs, vertebral pneumatopores and pneumatocoels demonstrate the presence of an avian-type air sac–lung system, probably with unidirectional flow and high capacity for aerobic exercise metabolism. Birds are basically small, advanced saurischians and probably inherited their high aerobic metabolism from their dinosaur ancestors (Bakker and Galton, 1974).

Tertiary Mammals

Dinosaurs disappeared suddenly at the Cretaceous–Tertiary boundary. By the Eocene, ca. 50×10^6 yr B.P., large cursorial mammals of modern aspect were common: long-limbed carnivores (hyaenodonts), small horses, rhino-like titanotheres, elephant-like uintatheres. Thus by the Eocene, running speeds and exercise metabolism in mammals were far higher than in the advanced therapsids of the Triassic. But Paleocene mammals are peculiar; few of the larger species seem to be highly cursorial, and a great many species of all sizes have what has been interpreted as digging and/or climbing adaptations (Matthew, 1937). The locomotor repertory of Paleocene mammals may reflect their behavioral heritage. Many Mesozoic lines may have been generalized diggers and/or climbers. Or, quite possibly the top running speeds in the Paleocene were limited by low aerobic exercise metabolism. Perhaps the metabolic machinery for a highly cursorial mode of life did not become available until the Eocene.

Conclusions

Efficiency Versus Flow Rates in Bioenergetic Evolution

Paleontologists and anatomists often have argued that the efficiency of performing biological activities (calories of work done or running speed achieved divided by energy expended) can be changed through morphological evolution, rearranging joint–muscle systems. Recent experimental evidence demonstrates that locomotor efficiency (speed divided by energy expenditure rate) at high running speeds is a constant for quadrupeds and bipeds for a given body weight regardless of the taxonomic position, trackway width and posture, limb proportions, and limb moment of inertia. At low speeds lizard-like tetrapods with wide-track, sprawling gaits and limbs with large distal muscle masses are more efficient (i.e., expend less energy for a given running speed) than mammals with a narrow-track gait and long, gracile distal extremities. The efficiency of other types of locomotor activity—acceleration, climbing, digging—most probably also cannot be reduced by morphological adaptations. The morphological evolution observed in the fossil record is the result of selective pressures changing the type of activity performed and the rate at which it is performed. The cursorial adaptations of many dinosaurs, ground birds, and Cenozoic mammals were mechanisms to utilize much higher rates of exercise metabolism to reach greater running speeds than in typical, large, nondinosaurian reptiles, but did not change efficiency. Similarly, evolution of mastication in endotherms has not increased digestive efficiency, but rather has increased the rate of energy assimilation by reducing the gut-passage time.

Competitive Advantage of Endothermy

The high, continuous heat production in birds and mammals is necessary to maintain continuous endothermy, even in warm, equable tropical climates, but greatly reduces ecological efficiency (calories fixed as secondary productivity divided by total calories assimilated). Nevertheless, the fossil record shows that endothermy has always been an overwhelming competitive advantage. For the first hundred million years of tetrapod history, all known vertebrates were ectothermic and could raise body temperature above air temperature by endogenous heat production only during relatively brief periods of high heat production by skeletal muscles during sustained exercise. Continuous heat production appeared in early therapsids, about 240 million years ago, and in a short time therapsids were the dominant terrestrial tetrapods. Therapsids were replaced by dinosaurs in the Triassic, but abundant evidence from predator–prey ratios and bone histology demonstrates that dinosaurs, too, were endotherms. When dinosaurs suddenly disappeared, endothermic mammals took over and have monopolized the niches available to large tetrapods through the last 65 million years. Endothermy is a critical advantage for animals of large body size in all climates; quite probably ambient temperature variations and heat loss from cutaneous evaporation of rainwater and dew prevent large ectotherms from maintaining high core temperatures continuously even in the most equable climates, and whenever thermal disaster strikes the large ectotherms, they are highly vulnerable to endotherm competition and predation.

Competitive Advantage of High Exercise Metabolism

The relatively low aerobic exercise metabolism of most ectothermic tetrapods is at least part of the reason that large ectotherms cannot utilize the high running speeds and exercise metabolic rates common in cursorial mammals and birds. Even in the most advanced therapsids of the Triassic, top running speed was probably little above that of a varanid lizard of comparable size. High running speeds and exercise metabolism appear for the first time in large tetrapods in dinosaurs of the Late Triassic. Endothermy plus high exercise metabolism probably were the key to dinosaur success in the Jurassic and Cretaceous. The first Tertiary mammalian radiation in the Paleocene, just after dinosaurs became extinct, produced few large, fast, highly cursorial species, quite possibly because of the limitations imposed on locomotor behavior by low aerobic exercise metabolism and long recovery times from oxygen debt. By the Eocene, many large, cursorial mammalian lineages were developing, and bioenergetics must have been generally on the same level as in most living cursorial placental mammals.

References

Alexander, C. E., Whitford, W. G.: 1968. Energy requirements of *Uta stansburiana*. Copeia *1968*, 678–683.

Amalitzky, V.: 1922. Diagnoses of new forms of vertebrates and plants from the Upper Permian on the North Dvina. Izv. Rosskiskoy Akad. Nauk *1922*, 329–340.

Asplund, K. K.: 1970. Metabolic scope and body temperatures of whiptail lizards (*Cnemidophorus*). Herpetologica *26*, 403–411.

Auffenberg, W.: 1970. A day with no. 19—report on a study of the Komodo monitor. Animal Kingdom *1970*(6), 18–23.

Avery, R. A.: 1971. Estimates of food consumption by the lizard *Lacerta vivipara* Jacquin. J. Animal Ecol. *40*, 351–365.

Bakker, R. T.: 1971. Dinosaur physiology and the origin of mammals. Evolution *25*, 636–658.

———: 1972a. Anatomical and ecological evidence of endothermy in dinosaurs. Nature *238*, 81–85.

———: 1972b. Locomotor energetics of lizards and mammals compared. Physiologist *15*, 278.

———, Galton, P. G.: 1974. Dinosaur monophyly and a new class of vertebrates. Nature *248*: 168–172.

Bandy, O.: 1967. Cretaceous planktonic foraminifera zonation. Micropaleontology *13*, 1–31.

Bartholomew, G. A., Tucker, V. A.: 1963. Control of changes in body temperature, metabolism, and circulation by the agamid lizard *Amphibolurus barbatus*. Physiol. Zool. *36*, 199–218.

———, Tucker, V. A.: 1964. Size, body temperature, thermal conductance, oxygen consumption, and heart rate in Australian varanid lizards. Physiol. Zool. *37*, 341–354.

———, Tucker, V. A., Lee, A. K.: 1965. Oxygen consumption, thermal conductance, and heart rate in the Australian skink, *Tiliqua scincoides*. Copeia *1965*, 169–173.

Belkin, D. A.: 1961. The running speeds of the lizards *Dipsosaurus dorsalis* and *Callisaurus draconoides*. Copeia *1961*, 223–224.

Bennett, A. F., Licht, P.: 1972. Anerobic metabolism during activity in lizards. J. Comp. Physiol. *81*, 277–288.

Blair, W. F.: 1960. The rusty lizard. Austin, Tex.: Univ. Texas Press.

Brody, S.: 1945. Bioenergetics and growth. New York: Reinhold.

Carey, F. G., Teal, J. M.: 1966. Heat conservation in tuna fish muscle. Proc. Natl. Acad. Sci. U.S. *56*, 1464–1469.

———, Teal, J. M.: 1969. Mako and porbeagle: warm-bodied sharks. Comp. Biochem. Physiol. *28*, 199–204.

Carroll, R. L.: 1969. Problems of the origin of reptiles. Biol. Rev. *44*, 393–432.

Case, E. C.: 1926. Environment of tetrapod life in the Late Paleozoic of regions other than North America. Carnegie Inst. Wash. Publ. *375*, 1–211.

Cerretelli, P., Piiper, J., Mangili, F., Ricci, B.: 1964. Aerobic and anerobic metabolism in exercising dogs. J. Appl. Physiol. *19*, 25–28.

Currey, J. D.: 1962. The histology of the bone of a prosauropod dinosaur. Palaeontog. *5*, 238–246.

Dawson, T. J.: 1972. Primitive mammals and patterns in the evolution of thermoregulation. *In* Essays on temperature regulation (ed. J. Bligh, R. E. Moore), pp. 1–18. Amsterdam: North-Holland.

———: 1973. "Primitive" mammals. *In* Comparative physiology of temperature regulation (ed. G. Whittow), pp. 1–46, vol. III. New York: Academic Press.

Denison, R. H.: 1956. A review of the habitat of the earliest vertebrates. Fieldiana Geol. *11*, 359–457.

Drysdale, A. R., Kitching, J. W.: 1963. A reexamination of the Karroo succession and fossil localities of part of the Upper Luangwa Valley. Geol. Surv. No. Rhod. Mem. *1*, 1–61.

Edelman, I. S., Ismail-Beigi, F.: 1971. Role of ion transport in thyroid calorigenesis. *In* Bioenergetics (ed. R. E. Smith), pp. 67–70, Fed. Am. Soc. Expt. Biol. Dublin: Trinity College.

Enlow, D. H. Brown, S. O.: 1957. A comparative histological study of fossil and Recent bone tissues, Part II. Texas J. Sci. *9*, 186–214.

———, Brown, S. O.: 1958. A comparative histological study of fossil and Recent bone tissues, Part III. Texas J. Sci. *10*, 405–443.

Fadak, M. A., Pinshow, B., Schmidt-Nielsen, K.: 1973. Energy cost of bipedal running. Federation Proc. *422*, 1127.

Freeman, M. R.: 1971. Population characteristics of musk oxen in the Jones Sound region of the Northwest Territories. J. Wildlife Management *35*, 103–108.

French, N. R., McBride, R., Detmer, J.: 1965. Fertility and population density of the black-tailed jack rabbit. J. Wildlife Management *29*, 10–14.

Gans, C.: 1970. Strategy and sequence in the evolution of the external gas exchangers of ectothermal vertebrates. Forma et Functio *3*, 61–104.

Goddard, J.: 1967. Home range, behavior, and recruitment rates of two black rhinoceros populations. E. African Wildlife J. *5*, 133–150.

Golley, F. B.: 1968. Secondary productivity in terrestrial communities. Am. Zool. *8*, 53–59.

Hanks, J., MacIntosh, J.: 1972. Population dynamics of the African elephant (*Loxondonta africana*). J. Zool. *168*. 29–38.

Hemmingsen, A. M.: 1960. Energy metabolism as related to body size and respiratory surfaces, and its evolution. Rept. Steno Mem. Hosp. Nord. Insulin Lab. *9*, 1–110.

Hill, L. G., Schnell, G. D., Echelle, A. A.: 1973. Effects of dissolved oxygen concentrations on locomotory reactions of the spotted gar. Copeia *1973*, 119–124.

Howell, A. B.: 1944. Speed in animals. New York: Hafner.

Huene, F. F. V.: 1942. Die anomodontier des Ruhuhu-Gebietes in der Tübinger sammlung. Palaeontogr. *94*, 219–236.

———: 1950. Die theriodontier des Ostafrikanischen Ruhuhu-Gebietes in der Tübinger sammlung. Neuen Jahr. Geol. Pal. *92*, 47–136.

Hutchison, V. H., Dowling, H. G., Vinegar, A.: 1966. Thermoregulation in a female brooding python, *Python moluris bivittatus*. Science *151*, 694–695.

Inger, R. F.: 1959. Temperature responses and ecological relations of two Borean lizards. Ecology *40*, 127–136.

Jansky, L.: 1965. Adaptability of heat production mechanisms in homeotherms. Acta Univ. Carolinae Biol. *1*, 1–91.

———: 1973. Non-shivering thermogenesis and its thermoregulatory significance. Biol. Rev. *48*, 85–132.

Jordan, P. A., Botkin, D. B.: 1971. Biomass dynamics in a moose population. Ecology *52*, 147–152.

Keul, J., Doll, E., Keppler, D.: 1972. Energy metabolism of human muscle. Baltimore, Md.: Univ. Park Press.

Kitchell, J. F., Windell, J. T.: 1972. Energy budget for the lizard *Anolis carolinensis*. Physiol. Zool. *45*, 178–188.

Kitching, J.: 1971. Fossil localities of the Beaufort Succession. Johannesburg: Univ. Witwatersrand. M.Sc. thesis.

Kleiber, M.: 1961. The fire of life. New York: Wiley.
Krassilov, V. A.: 1973. Climatic changes in Eastern Asia as indicated by fossil floras. I. Early Cretaceous. Palaeog. Palaeocl. Palaeoec. *13*, 261–274.
Layne, J. N., Benton, A. H.: 1954. Some speeds of small mammals. J. Mamm. *35*, 103–104.
McNab, B. K.: 1963. A model of the energy budget of a wild mouse. Ecology *44*, 521–532.
Margaria, R.: 1972. The sources of muscular energy. Sci. Am. *226*, 84–91.
———, Cerretelli, P., Aghemo, P., Sassi, G.: 1963. Energy cost of running. J. Appl. Physiol. *18*, 367–370.
Matthew, W. D.: 1937. Paleocene faunas of the San Juan Basin, New Mexico. Trans. Am. Phil. Soc. *30*, 1–372.
Moberly, W. R.: 1968. The metabolic responses of the common iguana, *Iguana iguana*, to activity under restraint. Comp. Biochem. Physiol. *27*, 1–20.
Moulder, B. C., Reichle, D. E.: 1972. Significance of spider predation in the energy dynamics of forest-floor arthropod communities. Ecol. Monogr. *42*, 473–498.
Moy-Thomas, J. A., Miles, R. S.: 1971. Palaeozoic fishes. Philadelphia: Saunders.
Mrosovsky, N., Pritchard, P. C. H.: 1971. Body temperatures of *Dermochelys coriacea* and other sea turtles. Copeia *1971*, 624–631.
Oliver, J. A.: 1955. The natural history of North American amphibians and reptiles. New York: Van Nostrand Reinhold.
Olson, E. C.: 1962. Late Permian terrestrial vertebrates, U.S.A. and U.S.S.R. Trans. Am. Phil. Soc. *52*, 3–224.
———: 1966. Community evolution and the origin of mammals. Ecology *47*, 291–308.
———, Vaughn, P. P.: 1970. The changes of terrestrial vertebrates and climates during the Permian of North America. Forma et Functio *3*, 113–138.
Organ, J. A.: 1961. Studies of the population dynamics of the salamander genus *Desmognathus* in Virginia. Ecol. Monogr. *31*, 189–220.
Ostrom, J. H.: 1970. Stratigraphy and paleontology of the Cloverly Formation. Bull. Peabody Museum Nat. Hist. *35*, 1–234.
Panchen, A. L.: 1973. Carboniferous tetrapods. *In* Atlas of palaeobiogeography (ed. A. Hallam), pp. 117–125. Amsterdam: Elsevier.
Pasquis, P., Lacaisse, A., Dejours, P.: 1970. Maximal oxygen uptake in four species of small mammals. Respirat. Physiol. *9*, 298–309.
Peabody, F. E.: 1958. A Kansas drought recorded in growth zones of a bullsnake. Copeia *1958*, 91–94.
———: 1961. Annual growth zones in living and fossil vertebrates. J. Morph. *108*, 11–62.
Pravoslavev, P. A.: 1927a. Gorgonopsidae. North Dvina Excav. Prof. V. P. Amalitzky III, Akad. Nauk S.S.S.R (Leningrad), 1–113.
———: 1927b. Gorgonopsids from the North Divina River excavations 1923. Akad. Nauk S.S.S.R (Leningrad), 1–20.
Rausch, R. L.: 1961. Notes on the black bear *Ursus americanus* Pallas in Alaska, with particular reference to detention and growth. Z. Saugetierk. *26*, 77–107.
Ricqles, A.: 1969a. L'histologie osseuse envisagée comme indicateur de la physiologie thermique chez les tétrapodes fossiles. C. R. Acad. Sci. Paris *268*, 782–785.
———: 1969b. Recherches paléohistologiques sur les os longs des tetrapodes. II. Quelques observations sur la structure des os longs des thériodontes. Ann. de Paléon. *55*, 1–52.
Robinson, P. L.: 1973. Palaeoclimatology and Continental Drift. *In* Implications of Continental Drift to the Earth Sciences (ed. D. Tarling and S. Runcorn), pp. 451–476. London, Academic Press.
Romer, A. S.: 1966. Vertebrate paleontology. Chicago: Univ. Chicago Press.
———: 1972. Skin breathing—primary or secondary? Respirat. Physiol. *14*, 183–192.
Russell, D. A.: 1971. The disappearance of the dinosaurs. Can. Geograph. J. *83*, 204–215.

Russell, D. A.: 1973. The environments of Canadian dinosaurs. Can. Geograph. J. *87*, 4–12.

Schaller, G. B.: 1972. The serengeti lion. Chicago: Univ. Chicago Press.

Schmidt-Nielsen, K.: 1972. How animals work. New York: Cambridge Univ. Press.

———, Dawson, T. J., Crawford, E. C.: 1966. Temperature regulation in the Echidna (*Tachyglossus aculeatus*). J. Cellular Physiol. *67*, 63–72.

Sloan, R. E.: 1969. Cretaceous and Paleocene terrestrial communities of Western North America. Proc. N. Am. Paleont. Con. (E), 427–453.

Snyder, R. C.: 1956. Adaptations for bipedal locomotion of lizards. Copeia *1956*, 191–203.

Spotila, J. R., Lommen, P. W., Bakken, G. S., Gates, D. M.: 1973. A mathematical model for body temperatures of large reptiles: implications for dinosaur ecology. Am. Naturalist *107*, 391–404.

Taylor, C. R., Rowntree, V. J.: 1973a. Running on two or four legs: which consumes more energy? Science *179*, 186–187.

———, Rowntree, V. J.: 1973b. Temperature regulation and heat balance in running cheetahs: a strategy for sprinters? Am. J. Physiol. *224*, 848–851.

———, Schmidt-Nielsen, K., Raab, J. L.: 1970. Scaling of energetic cost of running to body size in mammals. Am. J. Physiol. *219*, 1104–1107.

———, Dmi'el, R., Fadak, M., Schmidt-Nielsen, K.: 1971. Energetic cost of running in a large bird, the rhea. Am. J. Physiol. *221*, 597–601.

———, Caldwell, S. C., Rowntree, V. J.: 1972. Running up and down hills, some consequences of size. Science *178*, 1096–1097.

Thompson, K. S.: 1969. The biology of the lobe-finned fishes. Biol. Rev. *44*, 91–154.

Throckmorton, G.: 1973. Digestive efficiency in the herbivorous lizard *Ctenosaura pectinata*. Copeia *1973*, 431–434.

Tinkle, D. W.: 1967. The life and demography of the side-blotched lizard, *Uta stansburiana*. Misc. Publ. Museum Zool. Univ. Michigan *132*, 1–182.

Tucker, V. A.: 1966. Oxygen consumption of a flying bird. Science *154*, 150–151.

Vachrameev, V. A.: 1972. Mesozoic floras of the Southern Hemisphere and their relationship to the floras of the northern continents. Paleont. Zhurnal *3*, 146–161.

Van Hook, R. I: 1971. Energy and nutrient dynamics of spider and orthopteran populations in a grassland ecosystem. Ecol. Monogr. *41*, 1–25.

Varley, G. C.: 1970. The concept of energy flow applied to a woodland community. *In* Animal populations in relation to their food resources (ed. A. Watson), pp. 389–406. Oxford: Blackwell.

Wagener, F.: 1970. Ecosystem concepts in fish and game management. *In* Ecosystem concepts in natural resource management (ed. E. Van Dyne), pp. 283–290. New York: Academic Press.

Walker, E. P.: 1968. Mammals of the world. Baltimore, Md.: Johns Hopkins Univ. Press.

Wunder, B.: 1970. Energetics of running activity in Merriam's chipmunk *Eutamias merriami*. Comp. Biochem. Physiol. *33*, 821–836.

Zweifel, R. G., Lowe, C. H.: 1966. The ecology of a population of *Xantusia vigilis*, the desert night lizard. Novitates *2247*, 1–57.

PART V

Observation of Animal Body Temperatures

Introduction

David M. Gates

Observations of body temperatures of animals from both the laboratory and the field have resulted in a large amount of information. Many questions, both physiological and ecological, are raised by these data. Unique answers are difficult to deduce from information that deals with so many variables. It is important to combine experiment, observation, and theory to obtain a coherent basis for understanding temperature regulation and response in animals. Constant recourse to field data is essential to properly direct theoretical modeling. The papers in this part describe first some field observations concerning small animals (e.g., large insects and small birds) and then laboratory observations of body temperature regulation of reptiles, birds, and rodents.

The theoretical analysis by James Spotila in Chapter 17 of the heat budget of animals as a function of body size showed clearly the high metabolic heat demand in cool environments for very small animals with high body temperatures and the high evaporative water demand in warm environments of small animals with low body temperatures. George A. Bartholomew and Robert J. Epting point out that there are no endothermic vertebrates weighing less than 2–3 g and that all vertebrates of less weight are ectothermic. Because of this cutoff of vertebrate endothermy, the characteristic endothermy of flying insects was of particular interest to Bartholomew and Epting. Here they describe how sphinx moths control rates of heat generation and heat loss before, during, and after flight. It is now timely to do a thorough theoretical analysis of these excellent experimental and observational data to account for all the relationships involved.

The hummingbird, a small animal with a high rate of heat production during active daytime periods, can undergo hypothermic torpor during cool nights. Several species of hummingbirds nest in regions with cool climates and with particularly low nighttime temperatures. William Calder has pioneered studies of the energetics of hummingbird nests and the nestlings near Gothic, Colorado. Edward Southwick began a series of similar studies with the ruby-throated hummingbird nest and nestlings during the summer and spring of 1972 and continued further studies during 1973 at The University of Michigan's Biological Station on Douglas Lake. Measurements of the energy budget of a bird nest are difficult to make. The radiant surface temperatures of nest and surrounding surfaces were easy to determine with a radiometer and are reasonably accurate. The rates of water loss by the birds are calculated using theory, but are also reasonably accurate. Convective heat transfer is difficult to measure and could have considerable error. Although the nest was shaded from the sun, some diffuse sunlight filtered through the canopy and struck the nest. By measuring the heat loss from the nest, Southwick and I attempted to infer the metabolic rates of the fledglings as a function of time of day. These results are reported in the paper presented here. We feel that the nestling stage of birds is likely to be critical in their life cycle from an energetics standpoint. This suggests that much is to be learned concerning the ability of birds to survive extreme environmental conditions by looking closely at bird nest and nestling energetics.

Physiological information concerning body temperatures of animals, their control, and the physiological implications is absolutely essential to any theoretical analysis of the thermodynamics of animals. William Dawson presents a comprehensive review of the preferred body temperatures of reptiles and discusses many of the physiological activities that are affected by these body temperatures.

Important to any consideration of the energetics of homeotherms is the body temperature of an animal as a function of time. A homeotherm may attempt to maintain its body temperature fairly constant as environmental conditions change by varying its metabolic rate and water-loss rate. J. Emil Morhardt worked with our biophysical ecology group during one summer and has long been interested in the energetics of animals. Here he reports work concerning variable set points of body temperature of small birds and rodents.

22

Rates of Post-flight Cooling in Sphinx Moths

George A. Bartholomew and R. J. Epting

Introduction

Effective endothermy depends on the capacity to match rates of heat production and heat loss. Other things being equal, the rate of heat loss of a body is proportional to its surface area, and objects of similar geometry have surface areas that are proportional to the two-third power of their volumes. It is, therefore, not surprising that in both birds and small mammals, rates of weight-specific heat loss are inversely proportional to body weight (Morrison, 1960; Herreid and Kessel, 1967; Lasiewski *et al.*, 1967).

The rate of heat production depends primarily on biochemical rather than physical factors. Most aspects of the cellular physiology of birds and mammals are fundamentally similar. Therefore, the members of these taxa may reasonably be presumed to share an upper limit for rate of heat production. If the upper limit of rate of heat production is fixed while the weight-specific rate of heat loss continues to increase as size decreases, there should be a minimum size below which a bird or mammal cannot produce heat rapidly enough to balance heat loss. At or near this critical size, homeothermy should become difficult or impossible. In any event, natural selection has produced no endothermic vertebrate weighing less than 2–3 g—shrews of the genus *Sorex* and hummingbirds of the genus *Calypte* are familiar examples. There are, of course, many vertebrates weighing less than 2 g, but these are all ectotherms. Moreover, except for shrews, most of the birds and mammals that weigh less than 5 g are heterothermic and maintain high body temperatures only when they are active or when energy is available in excess (see Bartholomew, 1972, for review).

In view of the cutoff at 2- to 3-g for vertebrate endothermy, it is of special interest that numerous insects, many weighing much less than 1 g, generate and maintain high body temperatures during flight. Because the mechanical efficiency of muscle is of the order of 20 percent, most of the energy output of the flight muscles of insects is in the form of heat. Some of this heat is retained in the thorax and allows a spectacular development of endothermy: thoracic temperatures in excess of 35°C are commonplace in large, strong-flying insects, particularly if they have heavy wing loading. Some moths weighing less than 100 mg are conspicuously endothermic in flight (Bartholomew and Heinrich, 1973), as are some bees weighing 200 mg or less (Heinrich, 1972).

The situation summarized above leads one to inquire about the extent to which the endothermy of flying insects depends on high levels of heat production and the extent to which it is dependent on low levels of heat loss. Although it is well established that insect flight muscles produce large quantities of heat—they are among the most metabolically active tissues known—only limited information is available on rates of heat loss in insects. We have, therefore, undertaken to measure rates of cooling of live sphinx moths ranging in weight from less than 200 mg to more than 4 g as their body temperatures declined in still air immediately after flight or following preflight warm-up.

Materials and Methods

The experimental work was carried out at the Tropical Training and Research Center of the Inter-American Institute of Agricultural Sciences of the Organization of American States in Turrialba, Costa Rica. The Center is 9° north of the equator at an altitude of 600 m and receives heavy rainfall throughout the year. The moths were captured on the grounds of the Center at night-lighting stations illuminated from dusk till sunrise by high-intensity mercury-vapor lamps that shone down into the dense forest growing in the steep-walled valley of Rio Reventazon. We visited the lights from 2000 to 2200 h and again just before sunrise. Moths of the desired taxa were selected and each moth put in a separate paper cup covered with mosquito netting held in place with a rubber band. The moths were kept in a lighted room until needed. They usually remained quiet without arousing or fluttering about. Insects captured at night were implanted and measured the following day. Those taken at dawn were implanted on the day of capture. The moths were immobilized either by chilling them to about 10°C or by anesthetizing them with CO_2. After they were immobilized we exposed the cuticle by removing the scales from an area about 1 mm square in the right anterior quarter of the dorsal part of the thorax. A hole was made in the cuticle with a number 27 hypodermic needle, and a 40-gauge copper-constantan thermocouple was inserted to a depth of about 2 mm. The haemolymph coagulated and held the thermocouple leads securely in place.

A few of the implanted moths initiated preflight warm-up as soon as the effects of the immobilizing agent wore off, but most of them remained unaroused until, after an interval of 30 min to several hours, we put them in a darkened container and stimulated them by touching them or by blowing on them gently. During measurement the animals were kept in small cages of mosquito netting in dim light. Temperatures were measured to the nearest 0.1°C with a Model Bat-4 microprobe thermometer (Bailey Instruments Company of Saddlebrook, New Jersey).

Thoracic temperatures (T_b) were monitored during preflight warm-up, flight, and postflight cool-down until they had fallen to within 1°C of ambient temperature (T_a). The moth was then killed, weighed to the nearest milligram on a torsion balance, and placed in an envelope for later identification and photography.

All cooling rates were measured in still air at ambient temperatures between 24 and 26°C. Most of the time it was possible to make the measurements at room temperature, but on a few occasions room temperature exceeded 26°C. Under such circumstances we placed the experimental cages in an improvised water bath maintained at 24–25°C. Measurements were begun near 1000 h and usually were terminated by about 1800 h. The larger moths aroused most readily in the late afternoon or early evening, so they were usually measured last.

After our return to the University of California, the moths were relaxed and spread with their fore- and hindwings overlapping in approximately the position used during flight. They were then photographed together with a millimeter scale. Prints were enlarged to natural size and used for dimensional analysis. Wing area was measured with a planimeter. The thorax was assumed to be a cylinder, and thoracic volumes were computed from lengths and widths measured with calipers to the nearest 0.1 mm.

Data were processed at the UCLA computer center. Identifications were made by Julian P. Donahue of the Entomology Section of the Natural History Museum of Los Angeles County, where all the moths are now on deposit.

Results

When stimulated gently, the sphinx moths usually warmed up to flight temperature, as described by Heinrich and Bartholomew (1971), and took off. As soon as they were airborn, their flight was impeded by the walls of the cage or by the thermocouple leads, and most of the moths alighted at once and started to cool down, with the level of thoracic temperature gradually returning to ambient. Consequently, in most cases preflight warm-up was followed almost immediately by postflight cooling (Fig. 22.1). In some instances, however, moths hovered for several minutes before alighting and starting to cool down. We observed no effect of duration of flight on rate of cooling. Data similar to those shown in Fig. 22.1 were obtained from 41 individual sphingids ranging in weight from 160 to 4010 mg and belonging to the 27 species and 16 genera listed in Table 22.1.

When ambient temperature is constant, the decrease in temperature of a passively cooling object of homogeneous structure is a straight line if plotted on a semilogarithmic grid. This was not true of the sphinx moths we measured. In every case the slope of the initial part of the cooling curve exceeded that of the terminal part (Table 22.1, Figs. 22.2 and 22.3). From this we infer that heat loss during the initial part of postflight cooling was physiologically facilitated, while that during

Table 22.1. Summary of Morphometric and Physiological Data Related to Postflight Cooling of the 41 Individual Sphinx Moths Measured in the Present Study[a]

	Weight (mg)	Thoracic Volume (mm^3)	Initial Temp. (°C)	Conductance (°C min^{-1} °C^{-1})		
				Entire	Terminal	Initial
Agrius cingulatus	965	1207	34.4	0.180	0.177	0.184
Amplypterus sp. a	1367	1192	38.2	0.184	0.154	0.272
	848	1241	34.2	0.131	0.115	0.168
Amplypterus sp. b	970	872	36.7	0.166	0.143	0.226
Callionima inuis	1232	788	39.3	0.249	0.246	0.313
Callionima pan	586	544	40.8	0.193	0.167	0.269
Callionima parce	526	647	39.1	0.205	0.166	0.295
Cautethia spuria	160	239	35.0	0.410	0.378	0.481
	246	219	36.4	0.311	0.267	0.454
	178	254	41.4	0.368	0.302	0.516
	193	250	33.8	0.405	0.382	0.486
Cocytius antaeus	3230	4610	37.4	0.094	0.083	0.196
Epistor ocypete	839	761	43.2	0.193	0.154	0.251
	967	519	41.6	0.246	0.203	0.269
Erinnyis ello	1690	1115	38.3	0.120	0.088	0.355
	1188	1111	35.8	0.168	0.145	0.260
	1591	1377	37.7	0.092	0.078	0.170
Eumorpha achemola	3910	4075	32.6	0.064	0.058	0.090
	4010	4262	43.2	0.090	0.069	0.313
Eumorpha fasciata	1301	805	34.0	0.145	0.124	0.226
Eumorpha triangulum	2350	3010	41.8	0.085	0.067	0.216
Manduca pellenia	2690	2610	36.6	0.159	0.145	0.226
Manduca rustica	3250	3364	38.6	0.104	0.088	0.306
Nyceryx tacita	641	722	39.0	0.166	0.117	0.267
Pachylia ficus	2410	1426	36.2	0.108	0.088	0.223
	1662	1621	41.2	0.134	0.112	0.267
	2440	3133	38.0	0.081	0.064	0.106
	3110	2928	36.0	0.113	0.092	0.193
Perigonia lusca	661	584	42.7	0.177	0.143	0.408
Protambulyx strigilis	1199	1165	35.9	0.191	0.175	0.244
Pseudosphinx tetrio	2090	2577	36.2	0.115	0.097	0.152
Sphinx sp.	1071	1424	36.3	0.193	0.189	0.210
Xylophanes anubus	760	834	36.5	0.161	0.138	0.265
Xylophanes ceratomioides	1518	2213	37.6	0.170	0.161	0.267
Xylophanes chiron	915	1255	39.0	0.134	0.110	0.281
	1415	1189	39.7	0.170	0.152	0.223
Xylophanes pluto	465	706	36.2	0.210	0.184	0.309
	486	461	38.2	0.205	0.161	0.362
	507	617	40.8	0.260	0.228	0.426
Xylophanes thyelia	261	317	37.4	0.283	0.249	0.389
Xylophanes sp.	728	854	37.6	0.207	0.173	0.272

[a] See the text for the method of partitioning conductance.

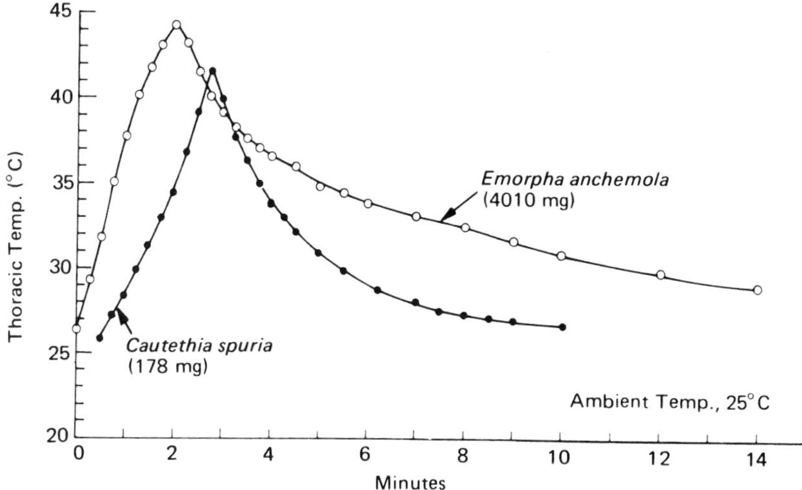

Fig. 22.1. Course of thoracic temperature during preflight warm-up and post-flight cooling in the largest and smallest species measured.

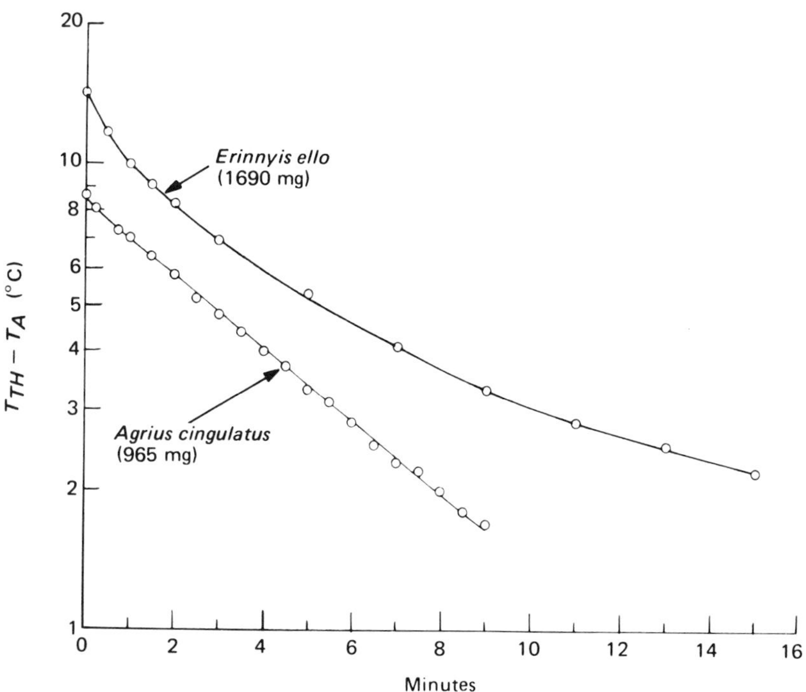

Fig. 22.2. Semilog plot of postflight cooling of two sphingids showing the smallest and largest differences between initial (facilitated) and terminal (passive) cooling rates. See Table 22.1 for conductance values.

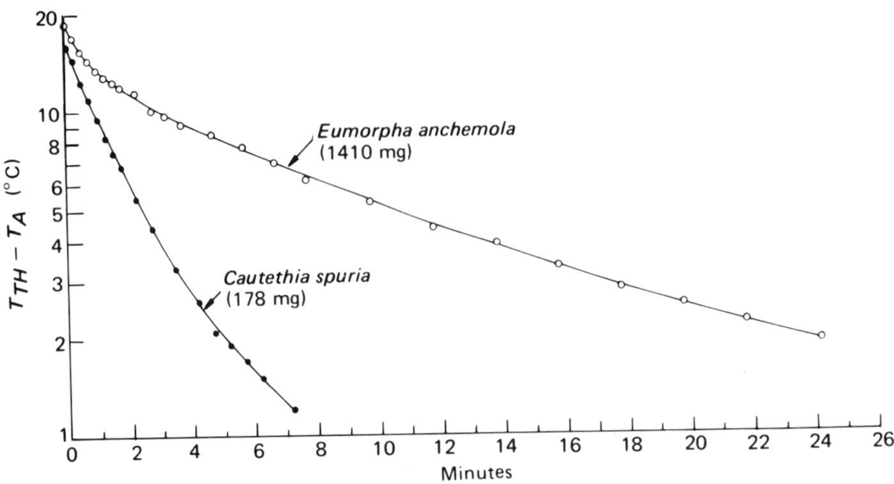

Fig. 22.3. Semilog plot of the postflight cooling of the largest and smallest species measured.

the terminal part was passive. For purposes of analysis we have arbitrarily divided each cooling curve into a facilitated segment (the initial 5°C decrease) and a passive segment (the terminal 5°C decrease). The passive segment is by definition linear on a semilogarithmic grid, whereas the facilitated segment is not. For purposes of comparison, however, we have treated both as linear. For convenience of comparison and plotting, cooling rates have been converted to thermal conductance (see Table 22.1), which is k in the formula

$$\frac{dT}{dt} = k(T_b - T_a)$$

where T = temperature (°C)
t = time (min)
T_b = body (thoracic) temperature
T_a = ambient temperature

Because live sphingids cool faster than dead ones (Heinrich and Bartholomew, 1971), the effect on cooling rate of circulation must be greater than that of oxygen consumption. We have, therefore, ignored the latter in calculating conductance. Rates of facilitated and passive cooling were both inversely dependent on body weight. On a logarithmic grid the relationships of both passive and facilitated conductance to body weight are best described by straight lines (Fig. 22.4), with the fit being better for the former ($r^2 = 0.73$) than for the latter ($r^2 = 0.50$).

The difference between facilitated and passive conductance is strongly correlated with the thoracic temperature at the onset of cooling ($P < 0.001$, $r^2 = 0.27$); the higher the initial temperature, the steeper the slope of the curve of facilitated cooling, and hence the greater the difference between the initial and terminal segments of the cooling curves.

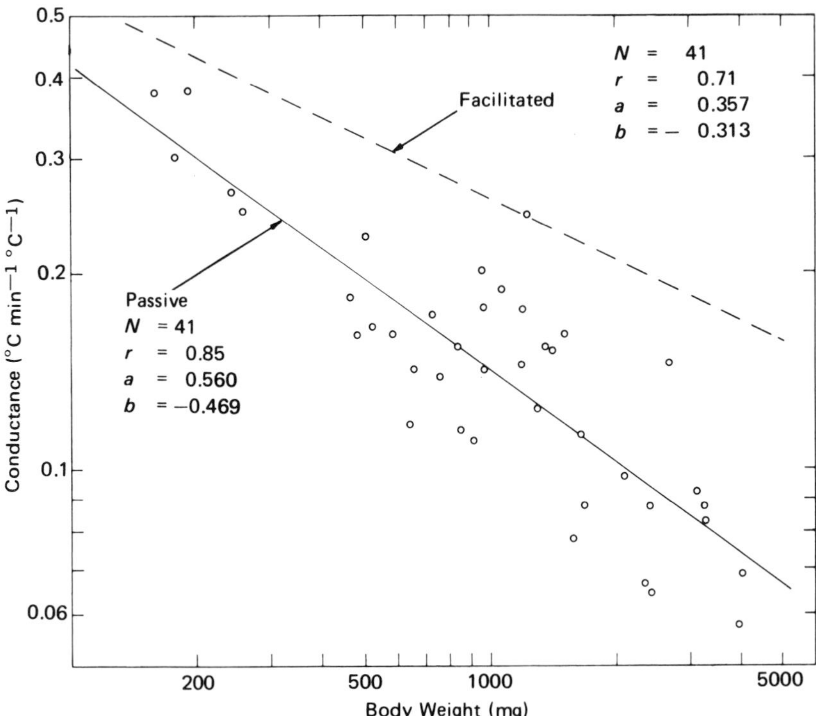

Fig. 22.4. Linear regression of the log-transformed values for thermal conductance and body weight in sphingids. Symbols: *N*, number of individuals; *r*, correlation coefficient; *a*, *y* intercept; *b*, slope. The individual values for facilitated conductance are not shown (see Table 22.1).

The thorax of a sphingid undergoes major changes in temperature before, during, and after flight, but the abdomen remains near ambient temperature. It is, therefore, of interest to relate the mass of the thorax to the rate of cooling. The average slopes of the entire cooling curves, as well as the slopes of their initial and terminal segments, all correlate positively with thoracic volume. However, these correlations are not particularly informative because the slope of the regression of the log-transformed values of thorax volume and total body weight is almost exactly unity (Fig. 22.5). Consequently, our data do not allow the separation of effects of thorax volume from those of total body weight.

Discussion

A remarkable feature of the endothermy of sphingids is their ability to maintain elevated body temperatures although they are generally much smaller than endothermic vertebrates. However, as can be seen from Fig. 22.6, although the slope

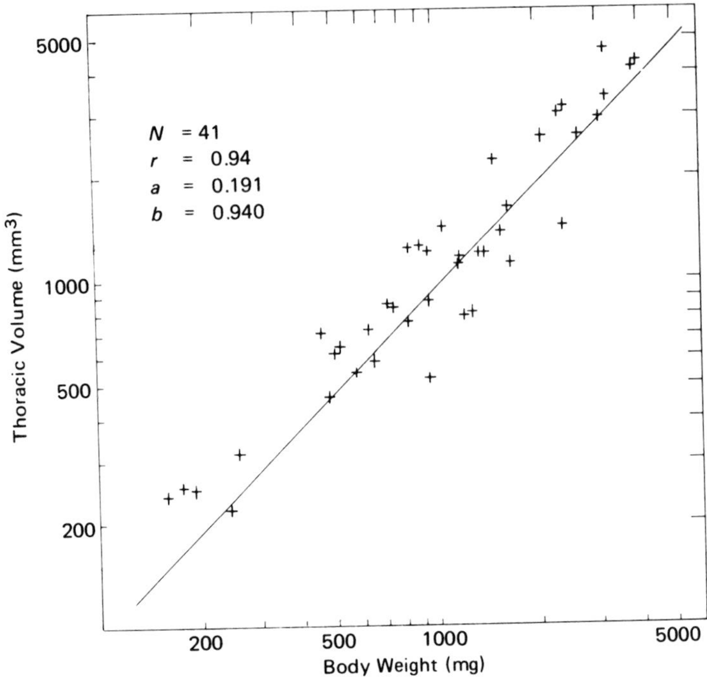

Fig. 22.5. Linear regression of the log-transformed values for thoracic volume and body weight in sphingids (see Table 22.1). Symbols as in Fig. 22.4.

of the regression of weight-specific thermal conductance on body weight in sphingids does not differ significantly from that in birds and mammals, the weight-specific conductance of sphingids is about 60 percent greater than that of either group of vertebrates in the range in which the weights of the three taxa overlap. From this we conclude that the endothermy of sphingids depends primarily on the high levels of heat production that are a by-product of the power requirements for flight and the mechanical inefficiency of the flight machinery, and only secondarily on reduced rates of heat loss.

In contrast to the situation in birds and mammals, body temperature in insects is metabolically elevated significantly above ambient only by means of heat produced by flight muscles. Even in the katydid (*Neoconocephalus robustus*) where high body temperatures are important primarily in relation to singing, the endothermic heat source is the wing musculature (Heath and Josephson, 1970). Endothermy has not so far been demonstrated in any flightless insect.

It is clear that the high value of weight-specific conductance in sphingids precludes the prolonged maintenance of elevated body temperatures except in association with flight, when the metabolic production of large amounts of heat is obligatory. It is noteworthy that even the largest sphingids, which weigh in excess of 6 g, are endothermic only during, or when preparing for, flight.

Despite the primary role of high levels of heat production in the endothermy of sphingids, the importance of thermal conductance should not be discounted.

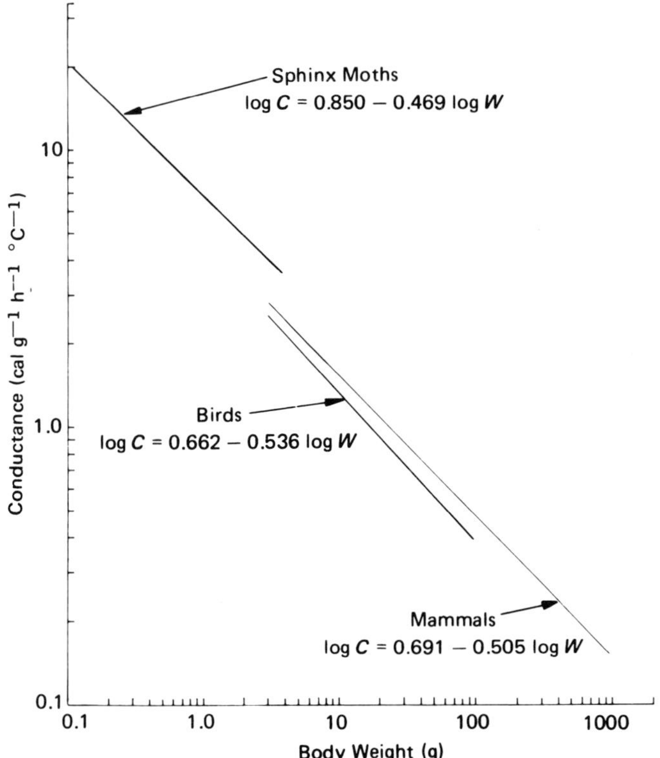

Fig. 22.6. Weight-specific thermal conductance as a function of body weight. The regressions for birds and placental mammals are from Herreid and Kessel (1967). For convenience of comparison the data for sphinx moths have been converted to cal g^{-1} h^{-1} $°C^{-1}$ using the value of 0.83 cal g^{-1} $°C^{-1}$ for the specific heat of their tissues.

Sphingids are covered with a dense layer of scales, which affords readily measurable thermal insulation (Church, 1960; Heinrich and Bartholomew, 1971).

Sphinx moths can control rates of heat loss by variation in circulation (Heinrich, 1970, 1971; Heinrich and Bartholomew, 1971), but they apparently lack mechanisms similar to the pilomotor system of mammals and therefore cannot change the amount of insulation provided by their covering of scales. During post-flight cooling, the conductance of individual sphingids is higher initially than terminally. This presumably is due to the accelerated circulation of haemolymph between thorax and abdomen at the onset of cooling, when the temperature difference between the two regions is maximal. The high rate of heat loss during the early stages of postflight cooling shortens the period during which thoracic temperature remains higher than ambient and thus diminishes the interval during which metabolism is elevated above the resting level. Because of the exponential dependence of rate of energy metabolism on temperature, the high initial rate of heat loss during postflight cooling could result in a significant saving of energy. Such a conclusion is consistent with our observation that the difference between initial and terminal cooling rates is positively correlated with thoracic temperature at the onset of cooling.

Cautethia spuria is among the smallest of the night-flying sphinx moths. *Cocytius antaeus* is the largest of the New World sphingids. Members of these species and the many taxa of intermediate weights also measured in the present study are conspicuously endothermic during flight. It is reasonable to conclude that all night-flying sphingids share this physiological characteristic. This is probably also true of the day-flying forms, but no data are available.

Summary

1. Postflight cooling was studied in 41 sphinx moths belonging to 27 species, ranging in weight from 160 to 4010 mg.

2. In every case the slope of the initial (facilitated) part of the cooling curve was steeper than that of the terminal (passive) part.

3. On a logarithmic grid, both facilitated and passive thermal conductance are linearly and inversely dependent on body weight.

4. The difference between facilitated and passive conductance is positively correlated with thoracic temperature at the onset of cooling.

5. The correlation between thoracic volume and body weight is linear and has a slope of one.

6. During flight all sphingids appear to be endothermic, even though the smallest sphingids for which data are available are only about $\frac{1}{15}$ as heavy as the smallest birds and mammals.

7. The endothermy of sphingids depends primarily on the high level of heat production that is a by-product of the power requirements for flight and only secondarily on reduced thermal conductance.

8. The slope of the regression of weight-specific thermal conductance on body weight in sphingids does not differ significantly from that in birds and mammals, but in sphingids it is about 60 percent greater than that of either group of vertebrates in the range in which their body weights overlap.

Acknowledgments

This study was supported by a grant (GB-32497X) from the National Science Foundation and carried out with the cooperation of the Inter-American Institute of Agricultural Science of the OAS and the Organization for Tropical Studies. We are grateful to Arnold L. Erickson of the IICA and Jorge R. Campabadal of OTS for allowing us to use the facilities of their respective institutions. We are indebted to Victor Becker for sharing some of his knowledge of the insects of Costa Rica with us; to Elizabeth Bartholomew for field assistance; to Timothy M. Casey for help in data processing; and to Julian P. Donahue for identification of specimens.

References

Bartholomew, G. A.: 1972. Aspects of timing and periodicity of heterothermy. *In* Hibernation and hypothermia, perspectives and challenges (ed. F. E. South *et al.*), pp. 663–680. Amsterdam: Elsevier.

———, Heinrich, B.: 1973. A field study of flight temperatures in moths in relation to body weight and wing loading. J. Expt. Biol. *58*, 123–135.

Church, N. S.: 1960. Heat loss and the body temperatures of flying insects. II. Heat conduction within the body and its loss by radiation and convection. J. Expt. Biol. *37*, 186–212.

Heath, J. E., Josephson, R. K.: 1970. Body temperature and singing in the katydid, *Neoconocephalus robustus* (Orthoptera, Tettigoniidae). Biol. Bull. *138*, 272–285.

Heinrich, B.: 1970. Thoracic temperature stabilization by blood circulation in a free-flying moth. Science *168*, 580–582.

———: 1971. Temperature regulation of the sphinx moth, *Manduca sexta*. II. Regulation of heat loss by control of blood circulation. J. Expt. Biol. *54*, 153–166.

———: 1972. Energetics of temperature regulation and foraging in a bumblebee, *Bombus terricola* Kirby. J. Comp. Physiol. *77*, 49–64.

———, Bartholomew, G. A.: 1971. An analysis of pre-flight warm-up in the sphinx moth, *Manduca sexta*. J. Expt. Biol. *55*, 223–239.

Herreid, C. F., Kessel, B.: 1967. Thermal conductance in birds and mammals. Comp. Biochem. Physiol. *21*, 405–414.

Lasiewski, R. C., Weathers, W. W., Bernstein, M. H.: 1967. Physiological reponses of the giant hummingbird, *Patagona gigas*. Comp. Biochem. Physiol. *23*, 797–813.

Morrison, P.: 1960. Some interrelations between weight and hibernation function. Bull. Museum Comp. Zool. *124*, 75–91.

23

Energetics of Occupied Hummingbird Nests

EDWARD E. SOUTHWICK AND DAVID M. GATES

Introduction

Hummingbirds have high weight-specific metabolic requirements combined with the smallest thermal mass among birds (Lasiewski and Dawson, 1967), yet those species which breed in northern climes, such as the ruby-throated hummingbird, *Archilocus colubris*, are exposed to cold night temperatures. The high energy demands required in cold conditions would be intolerable if the birds did not have well-developed means of conserving energy. Adult hummingbirds apparently go into hypothermic torpor, thereby conserving energy (Lasiewski, 1963; Pearson, 1950; Wolf and Hainsworth, 1971). It is unknown if young birds have similarly variable metabolic rates. Although some studies have dealt with egg development and nest temperatures of birds (e.g., Calder, 1973; Howell and Dawson, 1954; Huggins, 1941; Irving and Krog, 1956; Kashkin, 1961), the nest site has not been related to energy-budget considerations, nor has recent literature included adequate discussion of metabolic rates of nestlings (e.g., Calder, 1973, 1971; Dawson and Hudson, 1970). In this study, a primary objective was to describe the changing microclimate throughout a day at an occupied nest of a ruby-throated hummingbird, and to relate this to the energy budget and metabolic rate of the nestlings, incorporating principles developed and used by Gates (1962, 1970), Porter and Gates (1969), Southwick (1971), and Southwick and Mugaas (1971).

Theoretical Background

The continuous maintenance of body temperature by hummingbird nestlings occurs through the flow of energy between them and their immediate environment. The microclimate about the nest is described by all the combinations of absorbed radiation (solar and thermal), ambient air temperature, wind velocity, and water-vapor content of the air which are experienced by the nestlings while in neutral energy balance. The energy balance of the nestlings requires that the total influx of energy must equal the total outflux of energy at constant temperature. The energy budget is written

$$M + Q = \varepsilon\sigma\overline{T}_r^4 + \bar{h}_c(\overline{T}_r - T_a) + E_e + E_s + C + W + S \qquad (1)$$

where M = heat gained through metabolic processes

Q = amount of radiation absorbed at the body surface

$\varepsilon\sigma\overline{T}_r^4$ = amount of radiation given off at the body surface [where ε is the absorptivity of the surface, σ is the Stefan–Boltzmann constant (8.13×10^{-11} cal cm^{-2} min^{-1} °K^{-4}), and $\overline{T}_r$ is the average value of absolute temperature of the radiating surface; in this case, we assume equal contributions from the birds' backs and other portions of the nest surface]

$\bar{h}_c(\overline{T}_r - T_a)$ = heat loss or gain by convection at the body surface, where $\bar{h}_c$ is the average convective coefficient that is a function of air dynamics and body shape and T_a is the absolute temperature of the ambient air

E_e, E_s = heat-loss terms accounting for loss through evaporation of water in respiration and cutaneously, respectively

C = heat loss via conduction

W = work done if the animal is not at rest

S = heat stored if the animal is not in steady state

The young hummingbirds can vary only a few properties physiologically, e.g., metabolism, evaporative water loss, and insulative properties, which affect their radiating surface temperatures. Behaviorally, they might alter their exposure to sun, shade, cold night sky, wind, and air temperature. But as these are young nestlings confined to the nest, they can do little behavioral thermoregulation. Their thermal balance depends on the microhabitat into which they are born.

Development of a Practical Set of Equations

An important question to answer is: Which terms, or which physical variables, are most influential in the thermal balance of the nestlings? If it is assumed that the nestlings are at rest and in steady state over the short time intervals of measurements, there is no change in body temperature and thus no heat stored, and no work done. The conduction term is neglected as minimal heat is lost by this pathway. A cursory examination of the literature reveals that evaporation is almost all pulmonary. Heat loss by respiratory evaporation of water is described by

$$E_e = TV \times BR(\rho_{T_s} - r\rho_{T_a}) \times \frac{578}{A} \qquad (2)$$

where TV = tidal volume
BR = respiratory rate
ρ_{T_s} = saturated air density at the respiratory surface temperature
ρ_{T_a} = saturated air density at ambient temperature
r = relative humidity
578 = number of calories of heat lost in a change of state from 1 g of liquid water to vapor at 35°C
A = total surface area of the nest plus bird backs, yielding evaporative heat-loss units of cal cm^{-2} min^{-1}

To simplify calculations, the occupied nest is taken as a whole, with all energy terms expressed per unit of nest plus bird-back surface area. When comparisons are made to literature values, the energy is expressed in units of surface area of the birds alone. Tidal volumes and breathing rates were estimated from empirical expressions based on the birds' body weights (Calder, 1968; Lasiewski, 1964):

$$TV = 0.018W^{1.0} \tag{3}$$

$$BR = 182W^{-0.33} \tag{4}$$

As evaporative heat loss can be averaged over the total surface area, so also can convective exchange. However, the convection coefficient, $\bar{h}_c$, includes not only the shape and surface characteristics of the nest, but also the fluid properties of the air about the nest. To find this coefficient, it is necessary to utilize empirical formulas common in studies of fluid properties. From fluid dynamics, a relation can be found between the fluid momentum forces and the fluid viscous forces defined by the dimensionless Reynolds number, Re:

$$\text{Re} = \frac{VD\rho}{\eta} \tag{5}$$

where V = air velocity (cm s^{-1})
D = outside diameter of the spherical nest (cm)
ρ = air density at ambient temperature (at 20°C, $\rho = 1.31 \times 10^{-3}$ g cm^{-3})
η = viscosity of air (1.82×10^{-4} poise)

The Reynolds number can be related to another dimensionless fluid property, the Nusselt number, Nu, which is the ratio of convective to conductive heat transfer. For a sphere in any fluid,

$$\text{Nu} = \frac{\bar{h}_c D}{k} \tag{6}$$

where D is the diameter and $k = 3.68 \times 10^{-3}$ cal cm^{-1} min^{-1} C^{-1} for air. This can be rewritten in the form

$$\bar{h}_c = \frac{\text{Nu} \cdot K}{D} \tag{7}$$

The relationship between the Nusselt and Reynolds numbers depends upon the body shape, and for a sphere follows the form given by Kreith (1965):

$$\mathrm{Nu} = 0.37\mathrm{Re}^{0.6} \tag{8}$$

The value calculated for $\bar{h}_c$ in Eq. 7 can then be substituted back into Eq. 2.

The radiation term requires more manipulation as the radiant exchange cannot be simply averaged equally over the whole surface area. Those portions which are influenced by different radiant sources and sinks must be dealt with separately. The nest surface is, therefore, partitioned with respect to radiant exchange with the environment. The partitioning of radiant heat loss for the nest is illustrated in Figs. 23.1 and 23.2.

The total net radiant gain for the nest, ${}_tR_n$, equals the total absorbed radiation, ${}_tR_a$, minus the total radiant loss, ${}_tR$:

$${}_tR_n = {}_tR_a - {}_tR \tag{9}$$

The total net radiation is partitioned to account for the separate effects of the ground, the leaf cover above the nest, and the woods. The total surface area of the nest, A, is the sum of those portions influenced by the ground, leaf canopy, and woods.

$$A = A_g + A_1 + A_w \tag{10}$$

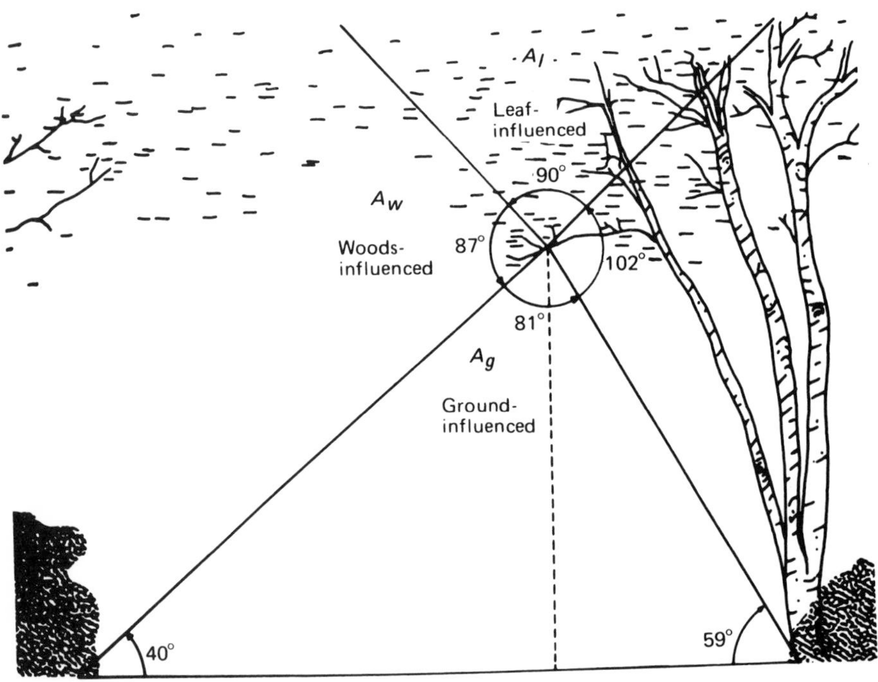

Fig. 23.1. Ruby-throated hummingbird nest site. This site, typical of the species, is 4.5 m above open ground, on a small downward-sloping branch, under a protective leaf canopy. The subtended angles represent partitioning of the site in terms of the radiant exchange with the environment.

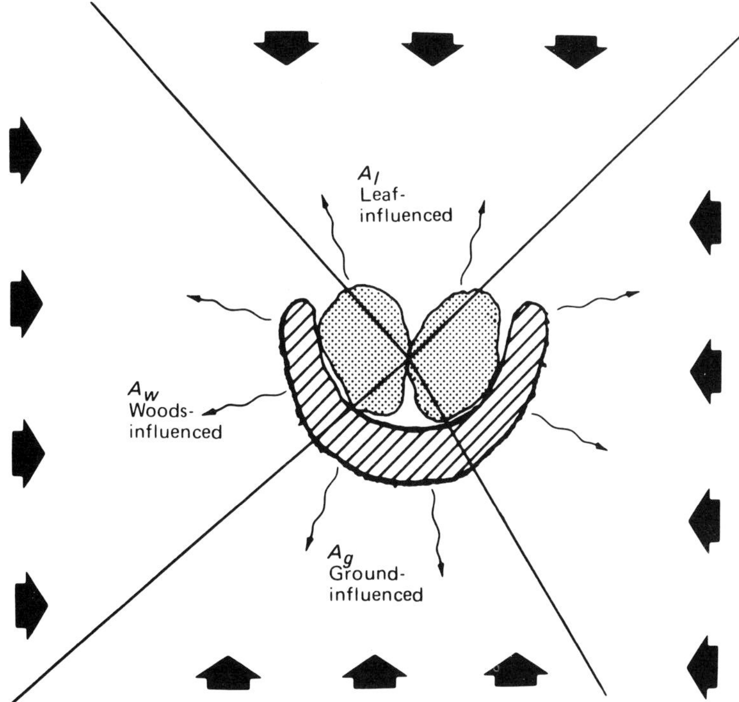

Fig. 23.2. Radiant partitioning of the occupied hummingbird nest. The proportion of the total surface area of the nest influenced by the ground in radiant exchange was about 12 percent; by the leaf canopy, about 15 percent, and by the woods, about 73 percent.

Incorporating this partitioning into Eq. 9 results in

$$ {}_tR_n = A_g(R_g - R_d) + A_1(R_1 - R_b) + A_w(R_w - R_s) \tag{11} $$

The ground-influenced portion, $A_g(R_g - R_d)$, is the radiation from the ground, R_g, toward the nest minus radiation from the nest, R_d, toward the ground times the nest surface area, R_g, influenced by the ground. The other terms are similarly explained, where A_1 is the surface area of the nest exposed to the leaf canopy, R_1 is the incident radiation from the leaf canopy, R_b is the radiation loss from the backs of the birds; A_w is the surface area of the nest exposed to the woods, R_w is the incident radiation from the woods, and R_s is the radiation emitted by the nest surface to the woods. Assuming that the emmisivity to thermal radiation over the surface of the nest and birds is 1.0 (Birkebak, 1966), the total net radiation, ${}_tR_n$, is averaged over the whole surface area, A, to give an average net radiation based on a unit of surface area, $\bar{R}_n$:

$$ \bar{R}_n = \frac{{}_tR_n}{A} \tag{12} $$

$$ \bar{R}_n = \frac{A_g}{A}(R_g - R_d) + \frac{A_1}{A}(R_1 - R_b) + \frac{A_w}{A}(R_w - R_s) \tag{13} $$

The proportion of the ground influence is about 12 percent, i.e., $A_g/A = 0.12$, while the leaf canopy affects 15 percent and the woods accounts for 73 percent of the total radiant exchange surface area. Utilizing the Stefan–Boltzmann radiation law for each portion of the partitioned site and nest, one gets

$$\bar{R}_n = \varepsilon\sigma[0.12(T_g^4 - T_d^4) + 0.15(T_1^4 - T_b^4) + 0.73(T_w^4 - T_s^4)] \tag{14}$$

The energy budget of the nest from Eq. 2 is now written:

$$M = \bar{R}_n + \bar{h}_c(\bar{T}_r - T_a) + E_e \tag{15}$$

$$M = \varepsilon\sigma[0.12(T_g^4 - T_d^4) + 0.15(T_1^4 - T_b^4) + 0.73(T_w^4 - T_s^4)] + 0.37\frac{k}{D}\left(\frac{VD\rho}{\eta}\right)^{0.6}(\bar{T}_r - T_a) + TV \times BR \times (\rho_{T_s} - r\rho_{T_a}) \times \frac{578}{A} \tag{16}$$

where M is in cal cm^{-2} min^{-1}.

Methods

From the theoretical analysis of the energy budget of the nest, we determined the physical variables that needed to be monitored at the site. These included temperature measurements of the air, ground, nest, leaf canopy above the nest, sky (blackbody equivalent), woods, and nestling back surfaces. With these temperatures known, the thermal radiant environment was described. We used a thermocouple thermometer (S B Systems Model 101) and a noncontact infrared thermometer (Barnes PRT-10) for temperature measurements; a pyranometer (Eppley Model 8–48) to monitor the incoming solar radiation; and a hot-wire anemometer (Hastings Air Meter RB-1) to measure the wind velocity at the nest site. Water-vapor content of the air was calculated from wet- and dry-bulb temperatures at the nest site. The measurements of these physical parameters were sufficient to define fully the occupied nest microclimate at any specific time.

The ruby-throated hummingbird nest (Figs. 23.1 and 23.3) was located in northern lower Michigan (latitude 46°33′40″N, longitude 84°40′10″W, altitude 220 m) in the northern fringe of its breeding range (Bent, 1940; Robbins *et al.*, 1966). The nest was about 4.5 m above a dirt road, on a small downward-sloping branch of a paper birch (*Betula papyrifera*), under a complete leaf canopy, and about 60 m from a yellow-bellied sapsucker (*Sphyrapicus varius*) feeding tree. We saw the adult female feed at fresh sapsucker holes and return to the nest. Such a nest site is typical of the species in this area (Bent, 1940; Eyer, 1949; Foster and Tate, 1966; Nelson, 1956). The nest contained two young hummingbirds, a female and a male about 13 and 12 days old, respectively, when the data presented here were taken. As the nestlings maintained their back-surface temperatures well above that

Fig. 23.3. Female ruby-throated hummingbird and two nestlings. The important leaf cover can be seen above the lichen-coated nest.

of the environment (indicating high body temperatures), they were probably capable of homeothermy at this time. This is consistent with observations on other hummingbird species made by Pearson (1953) and by Howell and Dawson (1954). It is also in agreement with the nest attentiveness exhibited by the adult female. She brooded less and roosted more as the nestlings grew older. As she did so, the young had to regulate their own body temperatures or go into hypothermy.

Results and Discussions

Figure 23.4 shows the pertinent radiant temperatures measured during a typical July day. There is always a net heat loss by radiation (the downward-pointing arrows) from the nest to the environment, for the nest is always warmer than (or is at least as warm as) the radiant objects of the environment. The general weather conditions changed from clear in the morning to overcast in the late afternoon, and this is indicated by the increasing sky temperature. The air temperature in the shade was nearly the same as the woods temperature.

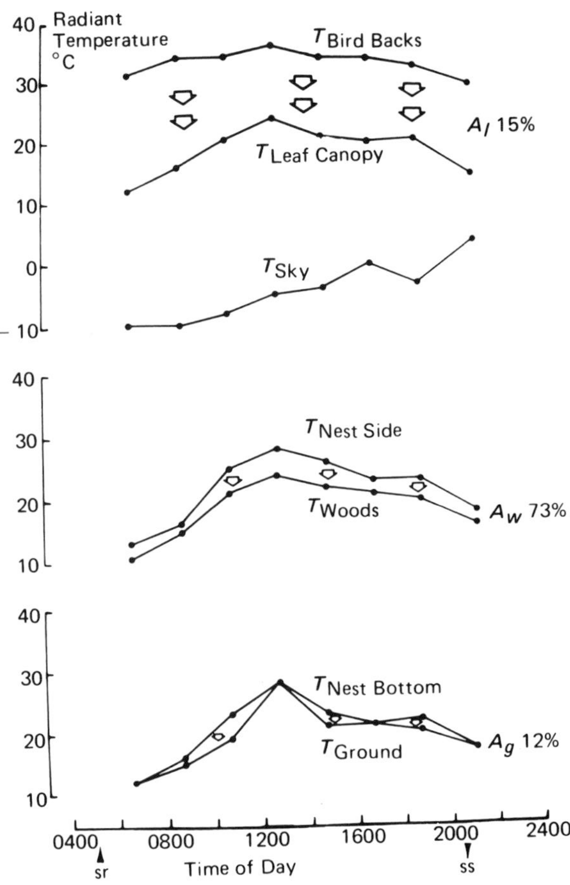

Fig. 23.4. Measured temperatures important in partitional radiant exchange between the nest and environment. The downward-pointing arrows show the radiant exchange (a heat loss) from the nest to the environment. The percentage figures on the right indicate partitioned areas as proportions of the total nest surface area. SR, sunrise; SS, sunset.

The leaf cover in the microhabitat at the nest site is especially significant. The leaves are so arranged that essentially no sky is visible from the nest. This leaf canopy acts as a very important buffer between the high-temperature backs of the nestlings and the cold sky heat sink (which was recorded as low as $-28°C$ on a clear July night). The canopy also shades the birds from direct solar radiation on clear days. (Direct solar radiation was as high as 1.36 cal cm^{-2} min^{-1} in July above the canopy.) Also, the leaves immediately above the nest protect the nestlings in even heavy rain. The rainwater dripped to the side, clearing the nest.

Figures 23.2 and 23.4 show that, although there is a great temperature difference, and thus a large amount of radiant exchange, between the bird backs and the leaf canopy, this affects only about 15 percent of the total surface area. The woods temperature is quite influential in radiant exchange because of the large surface area affected (73 percent). A small change in woods or nest-side temperatures would have a rather large effect on total radiant exchange. Little radiant exchange exists between the nest and the ground because of the small area affected (12 percent) and the small differences in temperatures.

To balance the energy exchange between the nest and the environment, the average net radiant exchange, $\bar{R}_n$, was calculated from the measured temperatures and Eq. 14. Convective heat loss, $\bar{h}_c(\bar{T}_r - T_a)$, was calculated from wind-velocity and air-temperature measurements at the nest site according to Eqs. 6–8. Evaporative heat loss, E_e, was determined from Eq. 2. These combined results are presented graphically in Fig. 23.5. It is apparent here that convective heat loss, which depends on wind speed and air temperature, is the most influential environmental parameter in the energy balance of these hummingbird nestlings at the nest site. With highly variable air movement, the convective loss changed from moment to moment, and the data points taken represent the uppermost levels of loss by this pathway. Evaporative loss is second in importance; and radiant gain or loss is reduced to a minimum by the choice location of the nest. If the nest were not so located and instead were partially exposed to the sky, radiant exchange would be quite different and could easily become the most important parameter, overloading the young with solar influx during a clear day, and resulting in high heat loss to the cold sky heat sink on a clear night.

If the energy-budget equation (Eq. 16) is indeed valid for steady-state conditions, then when all the pathways of net heat loss are summed, the total net energy loss must equal the metabolic heat production of the nestlings. Figure 23.5 shows this metabolic rate as the top curve, expressed in terms of the energy-exchange surface (i.e., total nest surface including surface of backs of nestlings). It is apparent that these nestlings are capable of regulating their metabolism at different rates, much like the highly variable rates shown by adult hummingbirds (Lasiewski, 1963; Pearson, 1950). The highest rate of metabolism occurred in midday when wind was prevalent, with minimum rates at dawn and dusk. The young maintained a relatively constant rate of metabolism of about 0.25 cal cm^{-2} min^{-1} for 4 h between 1400 and 1800. Measurements of solar radiation on a clear day indicate that small but significant amounts of radiation may be transmitted through the leaves to reduce the peak rate of heat loss in midday. The metabolism rate varies in such a way as to maintain a consistently high body temperature, as

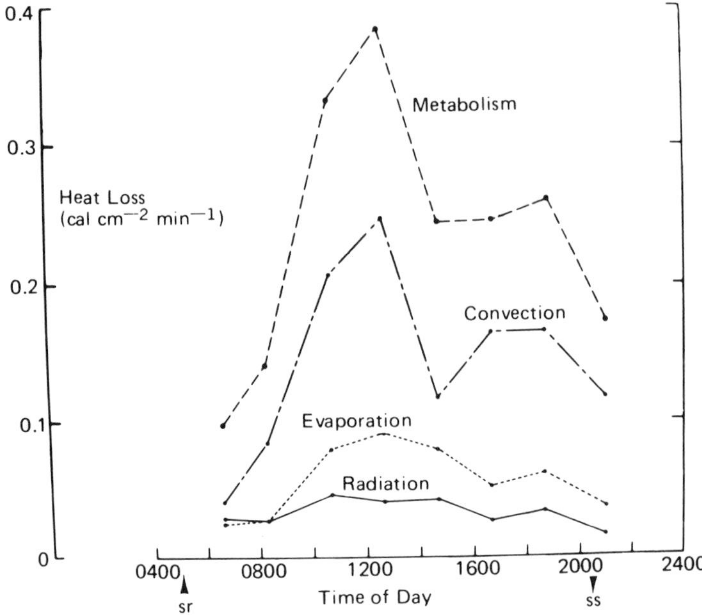

Fig. 23.5. Energy balance of an occupied ruby-throated hummingbird nest. The nest lost heat to the environment by radiation, evaporation, and convection. Summing the energy losses according to Eq. 15 yields metabolism of the nestlings (per unit area of nest) as the uppermost curve. SR, sunrise; SS, sunset.

indicated by the high temperatures on the bird backs. Current work is being done in this area to determine if young hummingbirds go into topor as do the adults, and if they alter their metabolism rates during the night. Calder (1973) has found that, in rare occurrences, the eggs of the broad-tailed hummingbird, *Selasphorus platycercus*, are allowed to cool when the incubating female goes into topor while brooding. Howell and Dawson (1954) did not find topor in nesting Anna hummingbirds, *Calypte anna*.

As a check on the practical validity of Eqs. 15 and 16, an electrical analog was constructed on the vacant hummingbird nest *in situ*. A small resistor was placed inside the nest and covered with a section of a table-tennis ball giving a similar overall shape to the occupied nest. The resistor was supplied with current to heat it up to steady-state conditions with the surface of the table-tennis ball at 34°C, the temperature usually recorded for the backs of the nestlings under similar conditions. The electrical power consumed (from electrical power = voltage × amperage) should be equal to the metabolic rate calculated from Eq. 16. Figure 23.6 shows a schematic of the analog and sample results for a clear night and a clear midday in early August. It must be noted that the nest had been vacated for several days and was not in the maintained condition of an occupied nest. Despite the poor condition, the results are quite close, demonstrating the workability of such an approach. This approach was further verified by a laboratory setup with an artificial nest of styrofoam containing a heated resistor, and comparing values obtained from electrical power consumption and Eq. 16 under various conditions.

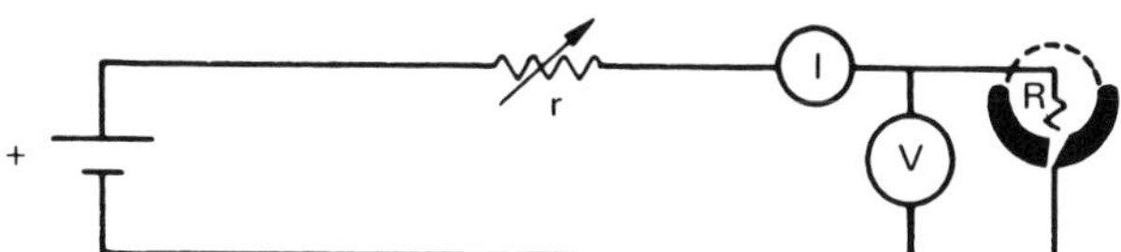

from Electrical Power = Voltage X Current
Energy consumed can be calculated and
converted to the units: cal cm^{-2} min^{-1}

Date	Time	Energy (Electrical)	Energy (Energy-budget Equation)
4 Aug	2200	0.345	0.295
5 Aug	1330	0.252	0.270

Fig. 23.6. Electrical analog of an occupied hummingbird nest. A small resistor was placed in the vacant nest at the natural nest site. It was covered with a portion of a table-tennis ball and heated to a surface temperature of 34°C. The electrical power consumed is similar to the energy calculated from the energy-budget equation (Eq. 16).

Table 23.1 was constructed to show the ranges of metabolism estimated by several methods. The metabolic rates in the table are expressed in terms of the surface areas of the birds. Surface areas were calculated by modeling the animal into a series of cylinders and hemispheres following Birkebak (1966). Literature data in various units were transformed into cal cm^{-2} min^{-1} by assuming that the surface areas were proportional to the two-thirds power of the body weights. Because of the paucity of metabolic data available for *Archilocus colubris*, data from other hummingbird species of similar adult weight are included.

The range of values of resting metabolism from the literature are lower than the values found in the present study. Most of the low values reported were obtained under laboratory conditions, in the dark, and at much higher temperatures. The levels of metabolism obtained in this study correspond better with rates of metabolism in active birds at lower air temperatures during daylight (e.g., *C. costa*, *C. anna*). It is apparent that daytime levels of metabolism in nestlings correspond to active periods, and such levels are probably more prevalent in the natural environment than laboratory-based minimal or basal levels.

The empirical equations based on laboratory data further exemplify this point. The values of basal metabolism predicted for birds of this size (2 g) by the three empirical equations are less than those calculated using the physically based energy-budget equation in this study. The physically based values are more like those measured by oxygen consumption in active birds by Lasiewski (1963) and Pearson (1954). This, too, suggests that the nestlings were in an active state.

The 4-h nearly constant daytime level of metabolism of the nestlings was about twice as high as that estimated for adult *C. anna* in the field, and about $3\frac{1}{2}$ times higher than that of adult *S. sasin* in the laboratory. However, it fell well within the ranges of active adult *C. costa* and the six species tested during daylight hours by Lasiewski (1963). The electrical analog, and sugar consumed by free-flying fledglings in the laboratory, are both quite close to the levels of metabolism calculated in the field from Eq. 20.

Table 23.1. Levels of Metabolism in Nestling Ruby-Throated Hummingbirds

Method of Determination	Conditions	Species	Body Weight (g)	Metablic Rate[a] (cal cm^{-2} min^{-1})
		Present Study		
Energy-budget equation (Eq. 16)	Field, nestlings in nest, daylight, T_a = 15–24°C	*Archilocus colubris*	2.0	0.24–0.95
Energy-budget equation (Eq. 16)	Field, nestlings constant 4-h level, 21–24°C	*A. colubris*	2.0	0.62
Sugar consumption	Lab, captive fledglings, 20–25°C	*A. colubris*	2.5	0.48–0.84
Electrical analog (Fig. 23.6)	Field, in nest, 10–18°C	*A. colubris*	2.0	0.61–0.85
		Literature Records		
Lasiewski (1963), oxygen consumption	Lab, adults, minimal levels, 33–38°C	*A. colubris*	3.2	0.073
Lasiewski (1963), oxygen consumption	Lab, adults, resting at 20°C	*A. alexandri*	3.2	0.20
Lasiewski (1963), oxygen consumption	Lab, adults, resting at 20°C	*Stellula calliope*	3.0	0.19
Lasiewski (1963), oxygen consumption	Lab, adults, resting at 20°	*Calypte costa*	3.2	0.18
Lasiewski (1963), oxygen consumption	Lab, adults, active daytime at 20°C	*C. Costa*	3.2	0.21–0.75
Lasiewski (1963), oxygen consumption	Lab, adults, ranges between 0700 and 1800, 15–20°C	Six S. W. species	3.0–5.4	0.21–0.78[b]
Pearson (1950), oxygen consumption	Lab, adults, average daytime rate	*Selaphorus sasin*	3.2	0.17
Pearson (1954), oxygen consumption	Field estimate from lab data and field behavior, 17°C	*C. anna*	5.3	0.35

Table 23.1.—*continued*

	Predicted Levels of BMR from Empirical Equations (M in kcal day^{-1}, W in kg)			
Body–Proctor equation	1932	$M = 89\ W^{0.64}$	2.0	0.11
King–Farner equation	1961	$M = 74.3\,W^{0.744}$	2.0	0.047
Lasiewski–Dawson equation	1967	$M = 129\,W^{0.72}$	2.0	0.093

[a] All data are expressed in terms of the bird surface area. Literature data were transformed by assuming surface area proportional two-thirds power of body weight.
[b] Metabolic rates do not correlate directly with body weight.

We suggest, as a result of this study, that the energy-budget analysis can be used to determine metabolism of nestling birds under natural conditions.

Acknowledgments

The many discussions within the Biophysical Ecology Group at The University of Michigan, and in particular with Paul Lommen, George Bakken, and James Spotila, were stimulating and refreshing. This study was supported by a grant from the Ford Foundation.

References

Bent, A. C.: 1940. Life histories of North American cuckoos, goatsuckers, hummingbirds, and their allies. U.S. Natl. Museum Bull. *176*(8), 506 pp.

Birkebak, R. C.: 1966. Heat transfer in biological systems. Intern Rev. Gen. Expt. Zool. *2*, 269–344.

Brody, S., Proctor, R. C.: 1932. Growth and development with special reference to domestic animals. XXIII. Relation between basal metabolism and mature body weight in different species of mammals and birds. Missouri Univ. Agric. Expt. Sta. Res. Bull. *166*, 89–101.

Calder, W. A.: 1968. Respiratory and heart rates of birds at rest. Condor *70*, 358–365.

———: 1971. Temperature relationships and nesting of the calliope hummingbird. Condor *73*, 314–321.

———, Booser, J.: 1973. Hypothermia of Broad-tailed hummingbirds during incubation in nature with ecological correlations. Science *180*, 751–754.

Dawson, W. R., Hudson, J. W.: 1970. Comparative physiology of thermoregulation. *In* Comparative physiology of thermoregulation (ed. G. C. Whittow), pp. 224–310. New York: Academic Press.

Eyer, L. E.: 1949. A study of a nest of a ruby-throated hummingbird. Jack-Pine Warbler *27*, 148–158.

Foster, W. L., Tate, J., Jr.: 1966. The activities and coactions of animals at sapsucker trees. Living Bird *5*, 87–113.

Gates, D. M.: 1962. Energy exchange in the biosphere. New York: Harper & Row.

———: 1970. Animal climates (where animals must live). Environ. Res. *3*, 132–144.

Howell, T. R., Dawson, W. R.: 1954. Nest temperatures and attentiveness in the Anna hummingbird. Condor *56*, 93–97.

Huggins, R. A.: 1941. Egg temperatures of wild birds under natural conditions. Ecology *22*, 148–157.

Irving, L., Krog, J.: 1956. Temperature during the development of birds in arctic nests. Physiol. Zool. *29*, 195–205.

Kashkin, V. V.: 1961. Heat exchange of bird eggs during incubation. Biophysics *6*, 57–63.

King, J. R., Farner, D. S.: 1961. Energy metabolism, thermoregulation, and body temperature. *In* Biology and comparative physiology of birds (ed. A. J. Marshal), vol. 2, pp. 215–288. New York: Academic Press.

Kreith, F.: 1965. Principles of heat transfer. Scranton, Pa.: International Textbook.

Lasiewski, R. C.: 1963. Oxygen consumption of torpid, resting, active, and flying hummingbirds. Physiol. Zool. *36*, 122–140.

———: 1964. Body temperatures, heart, and breathing rate, and evaporative water loss in hummingbirds. Physiol. Zool. *37*, 212–223.

———, Dawson, W. R.: 1967. A re-examination of the relation between standard metabolic rate and body weight in birds. Condor *69*, 13–23.

Nelson, T.: 1956. The history of ornithology at the University of Michigan Biological Station, 1909–1955. Minneapolis, Minn.-Burgess.

Pearson, O. P.: 1950. The metabolism of hummingbirds. Condor *52*, 145–152.

———: 1953. Use of caves by hummingbirds and other species at high altitudes in Peru. Condor *55*, 17–20.

———: 1954. The daily energy requirements of a wild Anna hummingbird. Condor *56*, 317–322.

Porter, W. P., Gates, D. M.: 1969. Thermodynamic equilibria of animals with environment. Ecol. Monogr. *39*, 245–270.

Robbins, C. S., Bruum, B., Zim, H. S.: 1966. Birds of North America. New York: Golden Press.

Southwick, E. E.: 1971. Effects of thermal acclimation and daylength on the cold-temperature physiology of the white-crowned sparrow, *Zonotrichia leucophrys gambelii* (Nuttall). Washington State Univ. Ph.D. diss.

———, Mugaas, J. M.: 1971. A hypothetical homeotherm: the honey bee hive. Comp. Biochem. Physiol. *40A*, 935–944.

Wolf, L. L., Hainsworth, F. R.: 1971. Time and energy budgets of territorial hummingbirds. Ecology *52*, 980–988.

24

Factors in the Energy Budget of Mountain Hummingbirds

WILLIAM A. CALDER III

Introduction

The abundant flowers of a compressed summer in the mountains provide an energy resource for exploitation by hummingbirds. However, the cold nights constitute a liability in their total energy budget. Adiabatic cooling, clear dry air, and the heat sink of the cold sky result in chilling conditions for the incubation of eggs by a tiny hummingbird. Howell and Dawson (1954) recorded the ability of the Anna's hummingbird (*Archilochus anna*) to maintain homeothermy while incubating overnight. I was greatly impressed the first time that I saw an even smaller calliope hummingbird (*Stellula calliope*) incubating her eggs in the presunrise cold of Jackson Hole, Wyoming. In a later visit I was equipped to record temperatures from two calliope nests. They also maintained homeothermy all night, despite the colder climate (Calder, 1971), stimulating my interest in heat-exchange principles and problems. It was obvious that ornithologists had given little consideration to physical factors in bird behavior, and that further study of hummingbird nesting would be rewarding. The population of broad-tailed hummingbirds (*Selasphorus platycercus*) at Gothic, Colorado, has been ideal for this. Evidence of the marginal energetic situation for this population may be seen in the occasional recourse to hypothermia during incubation (Calder and Booser, 1973) and the abandonment of live, normal chicks at some late nests when the flower supply declines, simultaneous with influx of competing migrant hummingbirds, in late July and early August (Calder, 1973d). Thus energy conservation in thermoregulation is of major importance.

As is true for any diurnal animal, nightfall marks the onset of fasting. The amount of energy reserves for this fasting is apparently a linear function of body mass ($g^{1.0}$; Tucker, 1971; Hainsworth and Wolf, 1972; Calder, 1973a). The rate at which these reserves are depleted is proportional to $g^{3/4}$ (Kleiber, 1961; Lasiewski and Dawson, 1967; Aschoff and Pohl, 1970a, 1970b; McNab, 1971). Consequently, the endurance of fasting must be

$$\text{time} = \text{amount} \div \text{rate} \propto g^{1.0} \div g^{3/4} = g^{1/4}$$

Thus the smaller the bird, the shorter its fasting endurance. Refueling of small birds may be critical, especially if the climate is cold and metabolic rates are accordingly high. Porter and Gates (1969) developed the concept of the climate space within which an animal must live for steady-state energy balance. With small heat capacity (mass times specific heat capacity) and consequently short transient energy states, hummingbirds should be especially suitable for energetic analysis.

Methods

On a cold night, the greatest demand on these reserves is for thermoregulation, so I have been attempting to measure as many variables of the nocturnal heat exchanges as possible, hoping to account eventually for the entire thermoregulatory budget of the broad-tailed hummingbird. A recent attempt (Calder, 1973b) may have oversimplified the conductive heat loss (see review by Drent, 1973).

In the summer of 1973, I obtained additional data on radiative heat exchange and on hypothermia. Radiation measurements were made with an Eppley Pyrgeometer for hemispherical radiant influx, connected to a recording potentiometer. Surface temperatures were obtained with a Barnes Precision Radiation Thermometer (PRT-5). The PRT-5 sees a 2° field, unlike the 35° field of the PRT10 used previously, thereby yielding more precise data for revising radiation temperatures given in my earlier papers (Calder, 1973b, 1973c). Nest temperatures were simultaneously sensed and recorded from synthetic eggs (Silastic 382), enclosing either thermocouples or thermistors (Calder, 1971).

Results and Discussion

First I will describe the macroenvironment. The Rocky Mountain Biological Laboratory is located at 2900 m elevation in the glacier-carved East River Valley, Gothic Colorado. Nights are generally clear for most of the nesting season, and the valley bottom accumulates cold air draining from the surrounding mountains. Temperatures frequently descend toward or below the freezing point. Only 50 m up the hillsides, minimum air temperatures may be 2.8–8.5° warmer. The heatsink of the night sky in the rarified, high-altitude air is very cold. Surrounding mountains block a significant portion of the celestial hemisphere so that, in an open meadow, the hemispherical long-wave influx on a clear night may be 23.3 mW cm^{-2} (0.335 ly min^{-1}), equivalent to radiation from a $-19.6°$ blackbody. This is similar to the minimum Q_{abs} of the zebra finch, the smallest bird (11.5 g) for which there is a climate-space diagram (Porter and Gates, 1969).

The broad-tailed hummingbirds nest in seral aspen stands (*Populus tremuloides*) on the slopes and in spruce (*Picea engelmannii*) and fir (*Abies concolor*) along the East River and Copper Creek, and also scattered in the aspen stands. The aspens are dense and shed the lower limbs by natural pruning. Branches or stubs suitable for nesting are usually 3–12 m above the ground. The conifers are less densely spaced and usually have boughs suitable for nesting 2 m or less off the ground. In both kinds of trees, the hummingbirds seem to select nest sites where branches overhead provide shelter from rain, sleet, hail, and the heat sink of the cold sky (Calder, 1973c). Observations of 73 nest sites have been made in three summers. The 1972 nests (36) included 3 renestings on identical branches used in 1971, and 1973 nests (27) included 5 nests used in 1971 and/or 1972.

The prerequisite for nesting appears to be the blooming of the low larkspur (*Delphinium nelsoni*). In the aspens the onset of nesting also coincides with expansion of the aspen leaves. At one test site, $\frac{1}{2}$ m below a former nest branch, the aspen trunks and branches increased the nocturnal hemispherical radiant influx 13 percent over the open meadow value. After the leaves had expanded, the hemispherical influx was 26 percent higher than that in the open meadow. By selecting stubs or branches only a few centimeters below overhanging branches or bends in trunks, a still greater portion of the sky is blocked.

Hypothermia

The problem of making ends meet, energetically, is complicated further when the opportunity for feeding during the day is reduced. Constancy of incubation and brooding (percent of daytime spent on the nest or "nest attentiveness") is increased during rainstorms (review: Drent, 1972). The longer the hen must remain on the nest, the less time she has to feed. If inclement weather persists long enough, her energy reserves may not be completely filled by nightfall, a problem that becomes more acute the smaller the bird (see Fig. 24.1).

Considering the scarcity of observations from nature of the energy-sparing mechanism of hypothermic torpor, I feel especially fortunate to have had sensors in the right nests at the right time (Table 24.1). The doubly good fortune of having more frequent rainstorms in the 1973 nesting season has provided a rather tight correlation between missed feeding opportunity and recourse to hypothermia. This confirmed the pattern reported previously (Calder and Booser, 1973).

Radiative Heat Exchange

Nocturnal Surface Temperature of the Bird

The PRT-5 has a sensor head 135 mm in diameter, with a 25-mm window. For it to see only a 1-cm spot on the hen's back, the sensor must be within 28 cm of the bird. It was supported and positioned with a photographic tripod, but the sighting mechanism is 67 mm distant and parallel to the sensor window path. The resulting parallax meant that the only way to aim the sensor was by peaking for

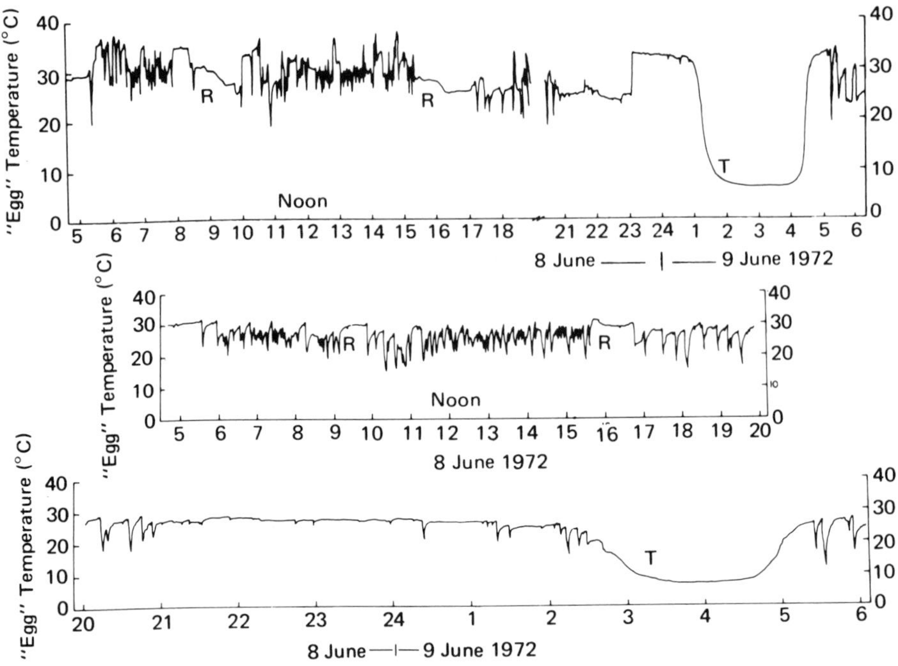

Fig. 24.1. Nest temperatures of the broad-tailed hummingbird obtained from Silastic artificial eggs containing a thermocouple (top) or thermistor (lower). Top: nest 1, Copper Creek: rainstorms (R) 0830 to 0945 and 1520 to 1710 prevented feeding, resulting in torpor entry (T) at 0100, apparent regulation of a low body or nest temperature (ca. 0210 to 0405), and arousal to normothermy at 0500. Middle and bottom traces: nest 5, wet meadow: missed feeding during rain (R) of 0910 to ca. 0950, and 1545 to 1645 was of shorter total duration, correlated with a later entry into hypothermia (T), ca. 0245. (From Calder and Booser, copyright 1973 by the American Association for the Advancement of Science.)

maximum temperature in two axes of tripod-head movement before darkness, so that the hen could settle on the nest and accept the instrumentation. Consequently, the exact part of the dorsal surface seen by the PRT sensor was not determined. These values appear in Table 24.2. The mean value for temperatures taken from 2335 to 0555, after the hen settled to rest and before taking off to feed, was 15.3 ± 1.9°C (Fig. 2). This was warmer than values which I reported earlier from PRT-10 readings (Calder, 1973b, 1973c). The difference is probably due to the wider field of the PRT-10, which might have included adjacent materials. The surfaces of the folded wing and tail would probably be cooler than the feathers of the back (see Veghte and Herreid, 1965), which could explain the cooler dorsal temperatures in nest 10 (Table 24.2) and in the earlier papers cited above. Note that the sensor was apparently squarely aimed to the nest center, as indicated by the 39°C registered when the hen departed to feed.

There seems to be a warming trend in dorsal surface temperature in anticipation of the first feeding trip. If I exclude those temperatures to get a representative value

Table 24.1. Hypothermia of the Nesting Broad-Tailed Hummingbird at Gothic, Colorado, 2900 m

Date	Nest	Stage	Min. Air Temperature (°C)	Min. Nest Temperature (°C)	Fraction of Recorded Nests Showing Hypothermia	Weather Conditions Preceding
9 June 1972[a]	1 spruce	Incubation	2	6.5	$\frac{2}{3}$	Rain 0830–0945, 1520–1710 (1.6 cm, $5\frac{1}{2}$ km distant)
	5 spruce	Incubation	−1	6.5		
5 July 1972[a]	14 spruce	Hatching	3		$\frac{1}{5}$	Interrupted recess pattern, presence of observer?
29 July 1972[b]	28 spruce	Chicks, 9 days of age	5.5	7–7.5	$\frac{1}{4}$	Preceded abandonment of nest and chicks.
30 July 1972[b]	28 spruce	Chicks, 10 days of age	5.5	—	$\frac{1}{4}$	Flower supply fading, 1.3 cm rain, 1625–1700, 1730–1900
29 June 1973	5 spruce	Incubation	1	C	$\frac{2}{3}$	
	9 spruce	Incubation	1	7.5–8		
13 July 1973	1 aspen	Chicks, 6 days of age	4.5	12.5	$\frac{1}{2}$	1.0 cm rain, 1330–1800, heaviest 1425–1510, 0.8 cm rain
	5 spruce	Chicks, 6 days of age	2	11.0		
19 July 1973	1 aspen	Chicks, 12 days of age	1	13.7	$\frac{2}{3}$	
	11 spruce	Incubation	8.5			
20 July 1973	1 aspen	Chicks, 13 days of age	6	11.0	$\frac{1}{5}$	1.1 cm rain (0.4 cm to 0700, 0.2 cm to 0930, 0.5 cm to 2140)

[a] Calder and Booser (1973).
[b] Calder (1973c).
[c] Thermistor placement poor, qualitative indication only.

Table 24.2. Dorsal Surface Temperatures of Nesting Female Hummingbirds Obtained in the Field

Nest	Date, Time	Sky	Air Temp. (°C)	Hemispherical Long-Wave Influx (mW cm^{-2})	Dorsal Surface Temperature[a] (°C)
12	23 July, 2030	Clear	9	33.8	23.5
	23 July, 2335	Clear	7	33.5	17.0
	24 July, 0315	Clear	4	31.9	14.3 (13.5–15)
	24 July, 0530	Clear	2.5	31.7	14.0 (to 16.5 before first departure)
16	27 July, 0520	Clear	4	32.1	15.0
	27 July, 0530–0554	Clear	4	32.1	18.5
	27 July, 0600		(Chicks after ♀ first departure > 30)		
17	29 July, 0520	Clear	4	31.8	15.5 (15.2–16.0)
	29 July, 0535	Clear	4	32.0	16.5 (15.5–17.0)
10	30 July, 0510	Scattered clouds	6.5	32.2	12.0 (11–13)
	30 July, 0530–0555	Scattered clouds	4	32.2	15.3 (13–17.5)
	0557		(Nest cup and chicks upon departure of ♀ = 39°)		
	Mean values ±1 s.d.[b]		4.4 ± 1.4	32.2 ± 0.5	15.3 ± 1.9

[a] Eyeball average of fluctuations (ranges in parentheses).
[b] Excluding 2030 July 23 at nest 12, taken after arrival from last feeding flight, and nest temperatures upon departure of ♀.

for predawn incubation and brooding, the mean value was 14.1°C. Assuming an emissivity of 1.0, the radiation efflux would be 38.5 mW cm^{-2}, slightly higher than my recently published estimates of 36.6 to 38.4 mW cm^{-2} (Calder, 1973b). These values were all obtained at low spruce nests; the logistics of operating the $7000 PRT high in an aspen have not yet been solved.

Hemispherical Long-Wave Influx at the Nest at Night

Considering values for the coldest, predawn hours before the sun hits the nest sites (Table 24.3), it appears that nest-site selection increases the hemispherical radiant influx from that of the open sky plus topography by about 35 percent in either aspen or spruce sites. Thus the difference I reported previously between an aspen and a spruce nest (Calder, 1973b) can be due only to greater air velocity.

The difference between estimated radiation from the hen's dorsal surface of 15.3°C and the hemispherical influx is the net radiative heat loss: 39.2 − 31.4

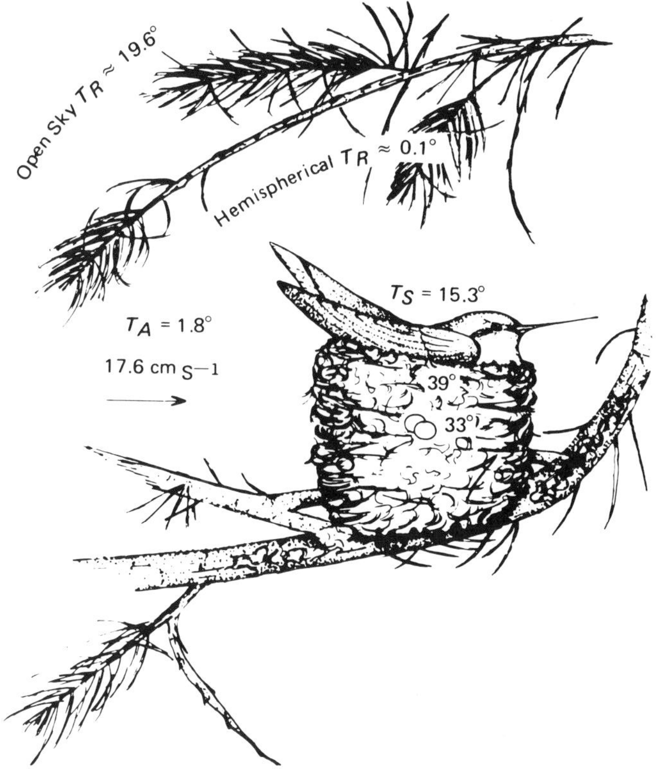

Fig. 24.2. Microenvironmental temperatures during predawn incubation by the broad-tailed hummingbird in Gothic, Colorado, elev. 2900 m. Radiation exchange is between the dorsal surface temperature of 15.3°C and a hemisphere largely shielded by spruce boughs averaging −0.1°C. (From Calder, 1973a.)

= 7.8 mW cm^{-2}, or about 55 percent higher than the mean from my previous estimates. To put this net radiation value in perspective, if one-fourth of the hen's surface is exposed to radiative loss above the nest cup, an estimated 6.2 cm^2 (Calder, 1973b), rate of heat loss by radiation would be 48.3 mW. At an air temperature of 5°C, the standard metabolism reported by Lasiewski for hummingbirds of this size is equivalent to a power output of 320 mW. Thus at the coldest part of the cycle, the radiative heat loss of the broad-tailed hummingbird is about 15 percent of the total heat loss determined for an unsheltered bird in the laboratory exposed to 5°C. If the hummingbird nested on a treetop or in an open meadow, her surface temperature would probably be slightly lower, but ignoring this, the radiation loss could be 6.2 cm(39.2 − 23.3)mW cm^{-2} = 98.7 mW, which is twice that when she nests under branches. If the energy reserves are only barely adequate for the nocturnal fast, as the occasional hypothermia suggests, the hummingbirds probably could not nest successfully in the open. To what extent radiative exchange figured in the natural selection of nesting behavior in which a "roof" is sought, I do not know. However, in addition to protection from predators and precipitation,

Table 24.3 Nocturnal Hemispherical Radiation Influx at Hummingbird Nest Sites[a]

Nest	Date, Time	T_a (°C)	Hemispherical Radiation ($mW\ cm^{-2}$)
2–73 (aspen)	22 June, 0530	2.5	30.19
	23 June, 0540	4	30.61
	24 June, 0530	5	31.03
	27 June, 0625	8	33.33
1–73 (aspen)	6 July, 0650	5.5	32.43
	10 July, 0630	6	31.97
	11 July, 0600	6	32.11
Mean ±1 s.d. of aspen values:		5.3 ± 1.7	31.67 ± 1.11
9–73 (spruce)	2 July, 0630	1.7	30.75
	3 July, 0630	2.0	30.40
	5 July, 0430	2.5	31.24
10–73 (spruce)	22 July, 0500	−0.5	30.13
	30 July, 0530	4	32.25
12–73 (spruce)	23 July, 0135	5.5	31.85
	23 July, 0625	3	30.98
	24 July, 0315	4	31.85
	24 July, 0530	2.5	31.68
16–73 (spruce)	27 July, 0530	4	32.11
17–73 (spruce)	29 July, 0520	4	31.97
Mean ±1 s.d. of spruce values:		3.0 ± 1.6	31.38 ± 0.73

[a] The sky was clear for all these samples.

the "roof" is apparently significant in the energetic economy of hummingbirds, reducing the power output for thermoregulation by 50.4 mW.

Estimates of Heat Output from Hummingbirds Nesting in Nature

From Real Birds

Additional data and better instrumentation make it desirable to revise my previous estimates (Calder, 1973b) for heat output during the coldest hour before sunrise. Drent (1972) argues that the eggs should be considered to be an extension of the body core. The hen increases heat flow from her brooding patch to keep the eggs warm, so the fact that the nest temperatures are in her thermoneutral zone may not mean that a proportion of her metabolic output is at the thermoneutrality rate observed in the laboratory. However, until I have measured the thermal conductances of a series of nests collected over the past summer, I have no better information and will assume as a conservative lower limit that the three-fourths of the body (another estimate, from visual observation) is at thermoneutrality in arriving at the $\dot{H}_g$ value in column 9 of Table 24.4.

The results of this maze of tenuous assumptions are figures of 167.5 mW in a low, streamside, spruce nest and 180.5 mW in a higher nest on an aspen hillside. This suggests that the combined effects of microclimate selection and nesting insulation are to cut the necessary heat output to about one-half that sustained at a similar air temperature in a metabolic chamber (320 mW, estimated from data on rufous and Anna's hummingbirds of similar body size by Lasiewski, 1963). After hatching and as the chicks grow, the female must sit higher in the nest, the walls of which are gradually spread out and lowered, so that she is more exposed to the external climate. This may be counterbalanced by heat liberated by the chicks.

From Electrical Models

Casts of hummingbird carcasses, enclosing a thermistor for temperature sensing and a few turns of "even-ohm" wire for a heater element, were enclosed in broad-tailed hummingbird skins and plumage. These were then placed in real hummingbird nests and located at an actual nesting site used in previous seasons. The electrical input to the heater element was measured with a voltmeter and a milliammeter; the product of the readings was the heat output from the model bird.

At the coldest presunrise hour, model bird 1 (female; dorsal plumage in poor condition, on a Silastic cast) in a nest of spider webbing, lichen, and moss (no. 5-73) maintained a "body" temperature of 37–38°C when the air temperature was 1°C with a power consumption of 311 mW. Model bird 2 [male, plumage in good condition on a cast exposed to the same environment (site 23-72, riparian spruce)] consumed electricity at a rate of 332 mW. These models had steady-state surface temperatures of 21 and 22.5°C, respectively, so a correction was made for "excess" radiation loss to arrive at the above values. I regard these figures as upper limits; they are similar to metabolic-chamber values as converted from Lasiewski (1963) for birds without the benefit of nesting insulation. The differences between electrical simulations and estimates from field environmental and surface temperature data may be due to one or more of the following factors:

1. The models could not employ ptiloerection and could not repair the nest cup to conform exactly to body plus clutch shape, as is done by living females.
2. Some power could have been dissipated by the wires from the volt- and ammeters to the heater element of the "bird."
3. Some heat was probably conducted back from the model along the temperature-probe and power-supply leads.
4. The assumptions about conductance losses from real birds may be too low, as indicated above.

Good agreement, however, may be seen in the absolute increase in heat output of model 2 (last two lines, Table 24.4) when moved out under the open sky (45 mW) and that calculated from dorsal surface temperatures of real birds

Table 24.4 Estimated Heat Losses from Nesting Broad-Tailed Hummingbirds

Min. T_a (°C)	T_s (°C)[a]	V (cm s^{-1})	$\dot{H}_c$ (mW cm^{-1})	$\epsilon\sigma T_s^4$ (mW cm^{-2})	$\dot{H}_{r(\text{in})}$ (mW cm^{-2})[b]	$\dot{H}_c + \dot{H}_{r(\text{net})}$ (mW cm^{-2})	$\dot{H}_c + \dot{H}_{r(\text{net})}$[d] (mW)	$\dot{H}_c + \dot{H}_r + \dot{H}_g$[e] (mW)	$\dot{H}_c + \dot{H}_r + \dot{H}_g + \dot{H}_e$[f] (mW)
					Nest $1\frac{1}{2}$ m above a streambank in a spruce tree				
1.8	15.3	17.6	9.7	39.2	31.4	17.5	108.5	157.4	167.5
					Nest 7 m above a hillside in an aspen grove				
4.6	15.3[c]	53.7	13.0	39.2[c]	31.7	20.5	127.0	175.9	180.5
					Electrical models				
1.0	21.0	(♀ plumage on Silastic cast)		48	—	—	—	330 (311)[g]	—
1.0	22.5	(♂ plumage on Wood's alloy cast)		65	—	—	—	358 (332)[g]	—
1.7	—	(♂ plumage on Wood's alloy cast)		—	24.8	—	—	403	—

[a] From Table 24.2.
[b] From Table 24.3.
[c] Not measured in aspen. Assuming similarity to surface temperature at a spruce nest.
[d] Assumption: one-fourth of body surface exposed to hemispherical radiational and convectional losses above nest walls.
[e] Assumption: three-fourths of body enclosed in nest cup at 33 to 39°C (thermoneutrality).
[f] Extrapolating from evaporation–metabolism ratios reported by Lasiewski (1964).
[g] Values in parentheses have been reduced by an amount equivalent to the difference (from real birds in the $\epsilon\sigma T_s^4$ calculation based upon the differences in T_s (note cols. 2 and 5).

(50.4 mW) discussed above (see hemispherical long-wave influx at the nest at night).

Thus, although my approaches have been crude and assumptions shaky, I think that I have probably bracketed the heat output of the nesting broad-tailed hummingbird as being between the laboratory, uninsulated value, and one-half of that value.

Acknowledgments

This research was supported by the National Geographic Society and the National Science Foundation (GB-38291). I am grateful to William Calder IV for assistance in the field.

References

Aschoff, J, Pohl, H.: 1970a. Rhythmic variations in energy metabolism. Federation Proc. *29*, 1541–1552.

———: 1970b. Der Ruheumsatz von Vögeln als Funktion der Tageszeit und der Körpergrösse. J. Ornithol. *111*, 38–47.

Calder, W. A.: 1971. Temperature relationships and nesting of the calliope hummingbird. Condor *73*, 313–321.

———: 1973a. The consequences of body size for avian energetics. *In* Avian energetics (ed. R. A. Paynter, Jr.). Cambridge, Mass.: Nuttall Ornith. Club. (in press).

———: 1973b. An estimate of the heat balance of a nesting hummingbird in a chilling climate. Comp. Biochem. Physiol. *46A*, 291–300.

———: 1973c. Microhabitat selection during nesting of hummingbirds in the Rocky Mountains. Ecology *54*(1), 127–134.

———: 1973d. The timing of maternal behavior of the broad-tailed hummingbird preceding nest failure. Wilson Bull. *85*, 283–290.

———, Booser, J.: 1973. Hypothermia of broad-tailed hummingbirds during incubation in nature with ecological correlations. Science *180*(May 18), 751–753.

Drent, R.: 1972. Adaptive aspects of the physiology of incubation. Proc. XV Intern. Ornith. Congr., 255–280.

———: 1973. The natural history of incubation. *In* Breeding biology of birds (ed. D. S. Farner), pp. 262–311. Washington, D.C.: Natl. Acad. Sci.

Hainsworth, F. R., Wolf, L. L.: 1972. Crop volume, nectar concentration, and hummingbird energetics. Comp. Biochem. Physiol. *42*, 359.

Howell, T. R., Dawson, W. R.: 1954. Nest temperatures and attentiveness in the Anna Hummingbird. Condor *56*, 93–97.

Kleiber, M.: 1961. The fire of life. New York: Wiley.

Lasiewski, R. E.: 1963. Oxygen consumption of torpid, resting, active and flying hummingbirds. Physiol. Zool. *36*, 122–140.

———: 1964. Body temperatures, heart and breathing rate, and evaporative water loss in hummingbirds. Physiol. Zool. *37*, 212–223.

———, Dawson, W. R.: 1967. A re-examination of the relation between standard metabolic rate and body weight in birds. Condor *69*, 13–23.

McNab, B. K.: 1971. On the ecological significance of Bergmann's rule. Ecology *52*, 845.

Porter, W. P., Gates, D. M.: 1969. Thermodynamic equilibria of animals with environment. Ecol. Monogr. *39*, 227–244.

Tucker, V. A.: 1971. Flight energetics in birds. Am. Zool. *11*, 115–124.

Veghte, J. H., Herreid, C. F.: 1965. Radiometric determination of feather insulation and metabolism of Arctic birds. Physiol. Zool. *38*, 267–275.

25

On the Physiological Significance of the Preferred Body Temperatures of Reptiles

WILLIAM R. DAWSON

Introduction

The thermal relations of reptiles have been extensively analyzed in both field and laboratory during recent years (see Schmidt-Nielsen and Dawson, 1964; Brattstrom, 1965; Mayhew, 1968; Templeton, 1970). Although fundamentally poikilothermic, many of these animals behaviorally achieve some control of their body temperatures during activity. Such behavioral thermoregulation primarily involves basking, postural adjustments, and use of favorable microclimates. This form of regulation is supplemented in certain instances by evaporative cooling (see, e.g., Templeton, 1960; Cott, 1961; Dawson and Templeton, 1963, 1966; Case, 1972; Crawford, 1972), vasomotor responses (see, e.g., Bartholomew and Tucker, 1963, 1964; Bartholomew *et al.*, 1965; Bartholomew and Lasiewski, 1965; Morgareidge and White, 1969a, 1969b; White, 1970; Weathers, 1970, 1971; Weathers and White, 1971; Spray and May, 1972), changes in reflectance (Atsatt, 1939; Norris, 1967; Porter, 1967; Porter and Norris, 1969), and muscular thermogenesis (Hutchison *et al.*, 1966; Vinegar *et al.*, 1970). Combinations of behavioral and physiological responses allow various reptiles during activity under favorable circumstances to maintain body temperature within a preferred range characteristic of the species. This range may be relatively low in animals such as the rhynchocephalian *Sphenodon punctatus* and the gecko *Phyllurus milii* (Brattstrom, 1965; Licht *et al.*, 1966b) to relatively high in various heliothermic lizards (Licht *et al.*, 1966b; DeWitt, 1967; Kemp, 1969).

Now that a wealth of descriptive information is available on the thermal relations of reptiles, it has become feasible to complete several analyses that are fundamental to the full characterization of the thermobiology of this group. The first of these involves quantification of the heat exchanges of various species in nature and under precisely controlled conditions in the laboratory. Studies such as those of Bartlett and Gates (1967), Norris (1967), Porter and Gates (1969), and Beckman *et al.* (1971) comprise a promising beginning in this area. The second type of analysis required concerns the peripheral and central neural mechanisms controlling the various facets of reptilian thermoregulation. Investigations such as those of Hammel *et al.* (1967), Cabanac *et al.* (1967), and Heath *et al.* (1968) constitute valuable efforts in this area. Finally, we need to evaluate fully the physiological consequences of operation at the preferred body temperature for species having well-defined thermal preferenda.[1] Although it appears reasonable to assume that a correlation exists between the general thermal adjustments of such reptiles and their preferred body temperatures and that various physiological processes reach optima at or near these temperatures, analysis of these adjustments and identification of the processes has proceeded very slowly (see Dawson, 1967; Templeton, 1970). However, recent work has provided some new insights concerning these relations. It is with these that this review will deal.

Correlative Information Concerning the Physiological Significance of the Preferred Body Temperature

Resistance to Heat by Intact Animals

Studies of rate processes in intact animals provide several indications of differences in heat resistance among reptiles. Such indications afford some documentation of the existence of interspecific differences in thermal adjustments in members of this class. For example, the lizards *Gerrhonotus multicarinatus* and *Eumeces obsoletus*, with preferred body temperatures of 30.0 and 34.5°C, respectively (Licht, 1964a), undergo heat suppression of resting metabolism when warmed to approximately 40°C (Dawson, 1960; Dawson and Templeton, 1966). In contrast, the thermophilic lizards *Dipsosaurus dorsalis* and *Crotaphytus collaris* with preferred body temperatures of 38–39°C (DeWitt, 1967; the value for *Crotaphytus* is estimated from field observations of Fitch, 1956, and unpublished laboratory observations of Dawson and Templeton) do not suffer such suppression until body temperature reaches levels in excess of 45°C (Dawson and Bartholomew, 1958; Dawson and Templeton, 1963).

Direct estimates of body temperatures required to produce thermal injury in reptiles have principally dealt with two topics: testicular damage and upper lethal body temperatures. The male germ cells of many animals appear less resistant to high temperatures than do somatic cells. Since the gonads of reptiles are situated

[1] "Preferred body temperature" and "thermal preferendum" are used synonymously in this review and refer to the mean (or, in one instance, the modal) body temperature obtained for animals of a particular species while they were in a thermoregulatory state and had access to a full range of biologically relevant thermal conditions, usually in a laboratory thermal gradient. The basis for restricting application of these terms is discussed by DeWitt (1967) and Licht *et al.* (1966b). The term "preferred temperature range" or equivalent is also used sporadically in this review, and it refers to the range in which temperatures used to estimate the above mean are observed to fall. The term "activity temperature" refers to values of body temperature observed for reptiles abroad and active in nature.

within the body cavity at the same temperature as other organs, the sensitivity of the testicular cells may be important in limiting exposure of males to hot conditions. It is of interest in this connection that tolerance of these cells to high temperatures appears to vary in a manner correlated with the thermal relations of the species involved. Severe testicular damage occurred in males of *Xantusia vigilis* with exposure to 36°C for a week (Cowles and Burleson, 1945), whereas similar treatment failed to injure the testes of the more thermophilic *Phrynosoma cornutum* (Mellish, 1936). Maintenance of representatives of three lizards (*Urosaurus ornatus*, *Sceloporus virgatus*, and *S. graciosus*) at body temperatures 1–2°C above the upper limit of their respective ranges of preferred body temperature for 10 h day^{-1} resulted in testicular damage within approximately 3 weeks (Licht, 1965b). No such damage was apparent in other males of these species maintained at their respective thermal preferenda. The effect of the higher temperatures on the testes was accompanied by considerable weight loss, and Licht (1965b) suggested that the damage to the germinal tissue may have involved some general systemic problem rather than a direct consequence of heating. However, a direct effect cannot be ruled out in view of Licht and Basu's (1967) observations of thermal damage in testicular tissue from other species, when this tissue was maintained in organ culture at temperatures exceeding the thermal preferendum by 2–3°C (see below).

Considerable information exists on the upper lethal body temperatures of reptiles. This predominantly concerns the high body temperature at which locomotor disability appears, the critical thermal maximum (CTM). The emphasis accorded this particular function recognizes Cowles and Bogert's point (1944) that the body temperature at which locomotion fails is ecologically the most meaningful measure of heat resistance, since attainment of this temperature in nature would prevent retreat from deleterious conditions. The CTM has been determined for a number of reptiles for which data on thermal preferences are available (see Table 25.1). Considerable interspecific variation is evident, with values for the CTM being correlated with the thermal preferenda or activity temperatures of the animals involved. This comparison could be complicated by the influence of certain factors on the CTM, notably time of day and state of thermal acclimation (Kosh and Hutchison, 1968; Lowe and Vance, 1955; Tremor, 1962; Kour and Hutchison, 1970; Jacobson and Whitford, 1970). Heat resistance of the lizard *Anolis carolinensis* and the CTM of the turtle *Chrysemys picta* are also influenced by photoperiod (Licht, 1968; Hutchison and Kosh, 1965).

Licht *et al.* (1966a) have questioned the validity of using the CTM to define reptilian heat resistance, because it neglects the time factor in heat injury, and because the paralysis serving to indicate the CTM tended to disappear with continued exposure of certain of the Australian lizards they were studying to the temperatures that produced it. This prompted these authors to measure heat resistance in terms of the duration of survival at selected high body temperatures. Resistance defined in this manner appears correlated with the preferred body temperatures of the animals studied (Table 25.1). Variation tending to reduce this correlation in representatives of the agamid genus *Amphibolurus* disappears when these animals are in similar states of thermal acclimation (Bradshaw and Main, 1968; Table 25.1).

Table 25.1 Heat Resistance of Reptiles in Relation to Preferred Body Temperature

Species	Preferred Body Temp. (°C)	Critical Thermal Maximum (°C)	Mean Survival Time (min)			Reference
			40.5°C	43.5°C	46°C	
Crocodilians						
Alligator mississippiensis	32–35[a]	38				Colbert *et al.* (1946)
Turtles						
Pseudemys scripta	24.6–29.1	41.2				Gatten (1973); Brattstrom (1965)[b]
Snakes						
Pituophis melanoleucus	26.7 or 30	40.5				Wilson (1971); Brattstrom (1965)[b]
Lizards						
Amphibolurus barbatua minimus	36.3		S[c]	S	10.4	Licht *et al.* (1966b, 1966a)[b]
A. inermis	36.4		S	S	102.8	
					64[d]	Bradshaw and Main (1968)
A. ornatus	37.1		S	S	25.5	Licht *et al.* (1966b, 1966a)[b]
					62[d]	Bradshaw and Main (1968)
Anolis carolinensis	30–33	41.8				Licht (1968); Brattstrom (1965)[b]
Crotaphytus collaris	37–38[a]	46.5				Brattstrom (1965)
Ctenotus labillardieri	32.2		160	15.6	0	Licht *et al.* (1966b, 1966a)[b]
Dipsosaurus dorsalis	38.5	47.5				DeWitt (1967); Cowles and Bogert (1944)[b]
Egernia stokesii	32.6		S	16	0	Licht *et al.* (1966b, 1966a)[b]
Eumeces obsoletus	34.5	42				Licht (1964a)
Gehyra punctata	34.6		S	97	5	Licht *et al.* (1966b, 1966a)[b]
Gerrhonotus multicarinatus	30.0	39–40.3				Licht (1964a); Brattstrom (1965)
Phyllurus milii	<30		24.5	S		Licht *et al.* (1966b, 1966a)[b]
Sauromalus obesus	37.9[a]	45.6				Brattstrom (1965); Cowles and Bogert (1944)[b]
Sceloporus undulatus	36.3	43.7				Licht (1964a); Cole (1943)[b]
Uma notata	37.5	>45				Licht (1964a); Cowles and Bogert (1944)[b]

[a] Preferred body temperature estimated from determinations of activity temperatures in field.
[b] First reference, source for preferred body temperature; second reference, source for heat resistance.
[c] S indicates survival over 3-h tests. [d] Animals acclimated to 40°C.

Review of reptilian thermal relations indicates that thermophilic reptiles, such as those representing the saurian families Iguanidae, Agamidae, and Teiidae, are among the most heat-resistant (see Cole, 1943; Brattstrom, 1935; Licht *et al.*, 1966a). Less thermophilic animals, such as lizards representing the families Scincidae and Anguidae, are distinctly more sensitive to high body temperatures (Brattstrom, 1965; Licht *et al.*, 1966a), a condition shared generally by snakes, turtles, and crocodilians (Brattstrom, 1965). Geckos comprise a rather heterogeneous assemblage; some have extremely limited tolerances to heat, resembling skinks in this respect, whereas others are more like agamids and iguanids (Licht *et al.*, 1966a). Heat resistance in all these reptiles is strongly correlated with level of preferred body temperature.

Resistance of Reptilian Tissues and Proteins to High Temperatures under in Vitro Conditions

Introduction

Several reports concerning thermal responses of reptilian tissues and proteins in vitro are available (see Dawson, 1967, for partial review). Those including adequate information on the thermal relations of the species involved are particularly useful in evaluating the exact functional significance of reptilian thermal preferences. These particular reports primarily deal with skeletal and cardiac muscle and with certain enzymes. They are supplemented by one study of the thermal resistance of testicular tissue in organ culture.

Muscle

Ushakov and Darevskii's (1960) observations on heat resistance of the lacertid lizards *Eremias strauchii* and *E. pleskii* provided one of the first illustrations of the correlation between thermal resistance of reptilian skeletal muscle and the thermal relations of the animals from which it was obtained. The former animal, which is active at body temperatures of 36–38°C in nature, developed heat rigor at 48.8–49.6°C. Exposure of skeletal muscle excised from it to 47.4°C resulted in loss of excitability after 5 min. In contrast, *E. pleskii* is active at 39–41.5°C in nature and underwent heat rigor at 49.3–49.8°C. Five minutes at 48.5°C was required to eliminate excitability of the skeletal muscle excised from this animal.

The study of lacertids just discussed suggests a correlation between the excitable properties of muscle in vitro and the thermal relations of the intact animals. However, it remains to be determined whether the body temperatures recorded for these animals during activity in nature coincide with the actual thermal preferenda for these species (see Licht *et al.*, 1966b; DeWitt, 1967, for discussions of this problem). A more conclusive correlation between heat resistance of skeletal muscle and preferred body temperature is provided by Licht (1964a) and Licht *et al.* (1969). The earlier study dealt with five North American lizards for which the

Table 25.2 Thermal Dependence of Performance by Saurian Muscle in Vitro

Species	Preferred Body Temp. (°C)	Reference*	Ventricular Muscle[a] Temp. for Max. Contraction (°C)	Ventricular Muscle[a] Lethal Temp. (°C)	Ventricular Muscle[a] Reference	Skeletal Muscle Temp. for Max. Contraction (°C)	Skeletal Muscle Damaging Temp. (°C)	Skeletal Muscle Lethal Temp. (°C)	Skeletal Muscle Reference*
Amphibolurus inermis	36.4	a	16.5–22	> 47	b	30–33	44–48	49.6	b
A. ornatus	36.6	c	15–16.5	> 44	b	30–33	44–48	49.6	b
Crotaphytus collaris	37–38[b]	d	14–19.5	> 44	e				
Diplodactylus spinigerus	35.9	a	14–16.5	> 44	b	34–36	48	49.6	b
Dipsosaurus dorsalis	38.5	f	28–33	> 50	g	31–38	47.5	51	h
Egernia carinata	34.7	a	12–15	> 44	b	28–32	40–44	46.3	b
Eumeces obsoletus	34.6	h	18–26		i	29–33	42	47	h
Gehyra punctata	34.6	a				25–34	41.9–44	46.5	b
Gerrhonotus multicarinatus	30.0	h	3–6		e	28–32	39	45	h
Phyllurus milii	< 30	a	7.5–15	> 44	b	24–29	36–40	44.4	b
Physignathus longirostris	37.1	a				32–36		49.6	b
Sceloporus undulatus	34.6	h				30–34	44	48	h
Uma notata	37.5	h				32–36	47	50.5	h

* (a) Licht *et al.* (1966b); (b) Licht *et al.* (1969); (c) Bradshaw and Main (1968); (d) Brattstrom (1965); (e) Dawson (1967); (f) DeWitt (1967); (g) Dawson and Bartholomew (1958); (h) Licht (1964a); (i) Dawson (1960).

[a] Perfusion fluid utlized for ventricles of *Amphibolorus* spp., *Diplodactylus*, *Egernia*, and *Phyllurus* contained 155 mmoles liter^{-1} of NaCl, whereas that for remaining lizards had 117 mmoles.

[b] Preferred body temperature estimated from determinations of activity temperatures in field.

thermal preferenda had been defined. The high-temperature threshold for irreversible damage and the upper lethal level under the experimental conditions employed varied directly with the preferred body temperatures of the species studied (Table 25.2). The 1969 study concerned the performance of skeletal muscle from seven species of Australian lizards. Licht *et al.* (1969) established a direct correlation between the minimal high temperature producing irreversible tissue damage in vitro and preferred body temperature, and between the upper lethal temperature of this tissue and this thermal preferendum (Table 25.2). These correlations are stronger than any existing between the performance of skeletal muscle in vitro and the geographical ranges of these animals, a situation that Licht *et al.* (1969) interpret as indicating the fundamental role of thermal preference as a component in the physiological adjustments of reptiles to temperature.

The resistance of cardiac muscle of lizards to high temperatures in vitro also appears correlated with level of preferred body temperature. The excised ventricle of the lizard *Dipsosaurus dorsalis* (preferred body temperature, 38.5°C; DeWitt, 1967) remained undamaged through 5-min exposures to temperatures as high as 43°C, and still responded, although damaged, after exposure to 50.2°C for a similar period (Dawson and Bartholomew, 1958). On the other hand, the ventricle of the skink *Eumeces obsoletus* (preferred body temperature, 34.5°C; Licht, 1964a) failed to respond to stimulation after a 5-min exposure to 46°C (Dawson, 1960). Ventricular tissue from five Australian lizards studied by Licht *et al.* (1969) had upper lethal temperatures in vitro that correspond closely with those observed in skeletal muscle from the same species. These lethal levels were in turn correlated with the preferred body temperatures of the various species.

Testicular Tissue

Licht and Basu's (1967) study of the resistance of testicular tissue from the lizards *Uma scoparia* and *Anolis carolinensis* to high temperatures in organ culture provides an indication of how the thermal adjustments of reptilian tissues can parallel the thermal preferences of the intact animal (preferred body temperatures of *Uma* and *Anolis*, 37.5 and 32.5°C, respectively). Measurement of the incorporation of tritiated thymidine by testicular tissue from these animals shows that the temperature at which this tissue begins to degenerate in vitro lies just above the preferred range of body temperature for the particular species.

Thermal Stability of Proteins

Thermal dependence of the function of enzymes and other proteins has received only limited attention in reptiles, but its study promises many insights concerning the functional significance of the thermal relations of these animals. The one study that provides information on both thermostability of an enzyme and on the thermal preferences of the species from which it was obtained deals with myosin adenosinetriphosphatase. The results obtained (Licht, 1964b, 1967c) indicate a strong correlation between these parameters (Table 25.3). At the extremes are samples of myosin ATPase from the lizards *Dipsosaurus dorsalis* and *Phyllurus milii,* which undergo 20 percent denaturation at 45.2 and 37°C, respectively.

Ushakov and associates have provided extensive information on thermostability of proteins in closely related species from different environments. Results for several lizards are included among their observations (see Ushakov, 1967). *Phrynocephalus helioscopus* and *Agama sanguinolenta* were obtained from semidesert regions in Turkmenia. *Phrynocephalus mystaceus* and *A. caucasica* were captured in a sandhill desert of Turkmenia and a montane region of Armenia, respectively. The proteins actomyosin adenosinetriphosphatase, aldolase, acetylcholinesterase, hemoglobin, adenylate kinase, and alkaline phosphatase underwent a 50 percent decrease in activity after 30 min at temperatures 0.8–5.6°C lower in *Agama caucasica* than in *A. sanguinolenta* and 0.8–4.0°C lower in *Phrynocephalus helioscopus* than in *P. mystaceus.* Somewhat lower thermostability is also indicated for *Agama caucasica* and *Phrynocephalus helioscopus* than for their respective congeners in measurements on glycerinated muscle. At the generic level, the proteins of the two *Agama* were found to be more heat-resistant than the corresponding ones from the representatives of *Phrynocephalus.* These results provide further examples of interspecific

Table 25.3 Thermal Characteristics of Two Reptilian Enzymes

Species	Preferred Body Temp.[a] (°C)	Temperature for Enzymatic Characteristics: Maximal Activity[b] (°C)	50% Decline in Rel. Activity (°C)	50% Denatured (°C)
	Myosin ATPase (Licht, 1964b, 1967c)			
Amphibolurus ornatus	36.6	(41)		46
Dipsosaurus dorsalis	38.5	42	47	47.5
Egernia carinata	34.7	(34)		43
Eumeces obsoletus	34.5	(34)		42.5
Gerrhonotus multicarinatus	30.0	33	41	42
Phyllurus milii	< 30	(< 33)		45
	LDH (F. H. Pough, personal communication)			
Ctenosaura pectinata		40		
Dipsosaurus dorsalis	38.5	40		
Gerrhonotus multicarinatus	30.0	25		
Ophisaurus apodus	32[c]	35–40		
Sceloporus occidentalis	30.6–35[d]	35–45		
Uma notata	37.5	35		

[a] For references see Tables 25.1 and 25.2, unless otherwise specified.

[b] Values in parentheses were estimated from data on irreversible denaturation.

[c] This figure is the mean activity temperature found for *Ophisaurus attenuatus* in the field (Fitch, 1956).

[d] Range of values reported by McGinnis (1966), Wilhoft and Anderson (1960), and Mueller (1970).

differences in thermal adjustment among reptiles. However, they are difficult to interpret in the context of the present discussion, since information is not provided on the thermal preferences or acclimation states of the animals from which the proteins were obtained. It is of particular interest that variation was noted in the thermal stability of alkaline phosphatase from *Phrynocephalus* and *Agama* (Ushakov, 1967), since Licht (1964b, 1967c) detected no difference in the thermal dependence of the activity of this enzyme in the lizards he tested, as discussed later in this review.

Physiological Performance of Reptiles at Noninjurious Temperatures

Introduction

Dawson (1967) reviewed several cases in which physiological responses to temperature varied interspecifically among lizards. Some of these in which the thermal preferences of the animals are well defined, and some additional information on reptiles generally, are considered in the ensuing parts of this section.

Performance of Intact Reptiles

Thermal Dependence of Heart Rate in Resting Animals. Studies of the thermal dependence of heart rate in resting individuals have been reported for several species of lizards differing in their thermal preferences (Dawson, 1967; Templeton, 1970). All these animals showed increased sensitivity of this rate function to temperature when cooled sufficiently (i.e., the Q_{10} for heart rate rose markedly). However, this increase developed at higher body temperatures in the thermophilic *Dipsosaurus dorsalis* (Dawson and Bartholomew, 1958) and *Crotaphytus collaris* (Dawson and Templeton, 1963) than in the less-heat-tolerant *Eumeces obsoletus* (Dawson, 1960) and *Gerrhonotus multicarinatus* (Dawson and Templeton, 1966). Adjustment for operation at high temperatures had evidently been attained in *Dipsosaurus* and *Crotaphytus* at the price of increased sensitivity to low temperatures.

Oxygen Consumption by Resting Animals. Dawson (1967) suggested that compensation of the oxygen consumption of resting lizards for temperature might have occurred and that thermophilic species would tend to have lower levels of metabolism at high temperatures than less thermophilic ones. This possibility has recently been reexamined by Bennett and Dawson (1975) through analysis of the relation between weight-independent metabolic rate (i.e., M/W^b, where M is metabolic rate and W is body weight, with b representing the weight regression coefficient for metabolism of lizards, 0.82) and preferred body temperature. The Q_{10} for this relation is approximately 2.5, a value closely matching those observed for the metabolism-temperature relations for individual species, and the variance is small. Bennett and Dawson (1975) conclude that compensation for temperature is not an important determinant of metabolic level in lizards.

Performance of Tissues and Proteins in Vitro

Thermal Dependence of Muscular Contractility. Information on the contractility of reptilian muscle at noninjurious temperatures in vitro affords further examples of correlation between thermal dependence of physiological function and level of preferred body temperature. Licht (1964a) and Licht *et al.* (1969) found generally close agreement between the optimal temperature for tension development in single twitches of skeletal muscle and preferred body temperature in several North American and Australian lizards (Table 25.2). The results of Dawson and Poulson (1958), Dawson and Bartholomew (1958), Dawson (1967), and Licht *et al.* (1969) likewise indicate a direct correlation between the thermal optimum for contractility of excised ventricles in response to widely spaced electrical stimuli in vitro, and preferred body temperature in various lizards. In this case the optimal temperature for contractility fell well below the thermal preferendum in each species (Table 25.2).

Thermal Dependence of Synthetic Activity in Testicular Tissue. In their study of testicular tissue of lizards, Licht and Basu (1967) found that the temperature fostering maximum incorporation of tritiated thymidine and differentiation in vitro agreed closely with the thermal preferendum of the species from which this tissue was obtained. This preferendum is 37.5° in *Uma scoparia* and 32.5° in *Anolis carolinensis*, the two animals on which observations were made.

Thermal Dependence of Protein Function. The major indications of interspecific variability in the thermal dependence of protein function in reptiles concerns hemoglobin and the enzymes myosin ATPase and lactate dehydrogenase (LDH). The oxygen affinity of the bloods of the lizards *Dipsosaurus dorsalis, Uma notata, Uma scoparia, Sceloporus occidentalis*, and *Gerrhonotus multicarinatus* has been found to correlate with degree of thermophily in comparisons at 25 and 35°C (Pough, 1969). *Dipsosaurus*, the most thermophilic animal among these species (listed in order of thermophily), has blood with the highest affinity and *Gerrhonotus*, the least thermophilic, the lowest affinity. Pough (1969) also noted that sensitivity to thermal fluctuation correlates with degree of thermophily, with the more thermophilic *Dipsosaurus* and *Uma* (preferred body temperature of *Dipsosaurus*, 38.5°C, according to DeWitt, 1967, and of *U. notata* and *U. scoparia*, 37.5°C, according to Licht, 1964a, and Licht and Basu, 1967) having the oxygen affinities of their bloods strongly dependent on temperature, in contrast to the situation in *Sceloporus* and *Gerrhonotus* (preferred body temperature of *Sceloporus*, 30.6–35° in Table 25.4, and of *Gerrhonotus*, 30°C, according to Licht, 1964a). The adjustments of blood physiology evident in these lizards result in their having similar P_{50}'s (i.e., P_{O_2} required to produce 50 percent saturation of the blood with oxygen under a particular set of conditions) at temperatures within their respective activity ranges.

Information available on myosin ATPase and LDH suggests that interspecific variation exists in the thermal dependence of the activity by these enzymes. Indeed, the thermal range in which the former attains maximal activity in vitro correlates well with the thermal preferendum in the six species of lizards on which observa-

Table 25.4 Hearing Response of Lizards[a]

Species	Preferred Body Temp.[b] (°C)	Probable Optimal Temperature for Cochlear Potentials (and Optimal range when Known) (°C)
Crotaphytus collaris	37–38[c]	38 (38–42)
Dipsosaurus dorsalis	38.5	38.5 (38.5–40)
Sceloporus occidentalis	30.6–35.0	35
Uma scoparia	37.5[d]	38.5 (38.5–43)
Uta stansburiana	35.4[c]	33

[a] Modified from Werner (1972).
[b] See Tables 25.1 and 25.2 for references regarding preferred body temperatures.
[c] Estimated from observations of activity temperatures of lizards in nature.
[d] Licht and Basu (1967).

tions have been made (Table 25.3). A similar relation may also exist between the temperature at which plasma LDH reaches maximal activity and the respective ranges of body temperature evidently preferred by the lizards *Dipsosaurus dorsalis*, *Uma notata*, *Sceloporus occidentalis*, *Ophisaurus apodus*, *Ctenosaurus pectinata*, and *Gerrhonotus multicarinatus* (F. H. Pough, personal communication; Table 25.3). That the temperatures at which myosin ATPase and LDH attain maximal activity sometimes coincide with the thermal preferendum of the species from which they were obtained is of interest. However, caution should be employed in interpreting such agreement, for the in vitro conditions under which determinations of activity were made undoubtedly differ markedly from those existing in vivo.

The results reviewed here should not be taken to indicate that the thermal dependence of function by all reptilian enzymes varies interspecifically, or that such variation as does occur reflects only the thermal preferendum of the species from which it was obtained. Licht (1964b, 1967c) noted that intestinal alkaline phosphatases from several lizards differing in their thermal preferences all reached maximal activity in vitro near 42°C. In combination with the results for other enzymes discussed previously, this finding tends to support Coulson and Hernandez's (1964) contention that different reptilian enzymes display different thermal sensitivities. In addition to this complication, it also appears certain that the *intraspecific* variability exists in thermal characteristics of enzymes. The LDH in skeletal muscle of northern and southern forms of the garter snake *Thamnophis sirtalis* differs, with that from the northern animals displaying a compensatory shift below 20°C involving an increased affinity for the substrate (Aleksiuk, 1971a). The muscle LDH from southern animals displays a similar shift below 28°C. In this instance, however, it involves a decrease in activation energy at low temperatures. Hepatic malate dehydrogenase shows compensatory increments in enzyme–substrate affinity (i.e., decreased K_m values at low temperatures (Hoskins and

Aleksiuk, 1973). This increased affinity is responsible for the low thermal dependence ($Q_{10} \simeq 1.7$) of the in vitro reaction rate at nonsaturating concentrations of substrate. This enzyme exists in two forms in *Thamnophis sirtalis*, and these are characterized by different affinities for substrate. Changes in the proportions of these two forms might well be involved in thermal acclimation in this species. Several authors point out that differential isozymic activities and substrate affinities may serve to stabilize reaction rates over a broad range of temperature in vivo and thereby lead to a measure of homeostasis in poikilothermic animals (Fry and Hochachka, 1970; Aleksiuk, 1971a, 1971b; Hoskins and Aleksiuk, 1973).

Physiological Optima and Preferred Body Temperatures

Introduction

The discussion thus far has served to indicate the diversity characterizing the responses of reptiles to temperature, and the extent to which these responses tend to be correlated with level of activity temperature or thermal preferendum. It is now appropriate to examine more closely the functional significance of the actual preferendum, by attempting to determine whether specific physiological processes reach optimal levels near or coincident with it. It is gratifying that information bearing on this question has increased substantially since the reviews by Schmidt-Nielsen and Dawson (1964), Dawson (1967), and Templeton (1970).

Thermal Dependence of Sensory, Neural, and Behavioral Function

Following the demonstration by Adrian *et al.* (1938) of the influence of temperature on electrophysiological correlates of hearing in turtles, the thermal relations of various facets of the auditory mechanism have only recently begun to receive the attention they merit. Patterson *et al.* (1968) have analyzed the effects of cochlear temperature on the electrical response of this structure to sounds of 210, 330, or 630 Hz at 60, 70, or 80 dB in the turtle *Pseudemys scripta*. They found a bimodal response with respect to temperature, with electrical output reaching maxima at 19 and 30°C. The significance of the first peak in regard to the thermal relations of this turtle is unclear. The one occurring at 30°C closely agrees with the thermal preferendum observed by Gatten (1973) in recently fed *Pseudemys* tested in experimental thermal gradients. Since these turtles would presumably attain such a temperature in nature by basking, enhancement of auditory sensitivity might well be important in reducing vulnerability to predation.

The thermal dependence of auditory sensitivity has also been investigated in lizards. Campbell (1967, 1969) found that evoked neural responses associated with hearing varied markedly with temperature in eight species representing the saurian families Anguidae, Eublepharidae, Gekkonidae, Iguanidae (four species), and Teiidae. Moreover, the auditory responses reached optima within temperature ranges that correlated well with the respective thermal characteristics of the species in nature. The most direct relation between auditory response and preferred level of body temperature has been established by Werner's (1972) recent study of the effects of temperature on the sensitivity of the inner ear in six iguanid lizards: *Anolis lionotus*, *Crotaphytus collaris*, *Dipsosaurus dorsalis*, *Sceloporus occidentalis*, *Uma scoparia*, and *Uta stansburiana*. Analysis of the sound intensity of a pure tone (50 and 15,000 Hz) required to elicit a standard response of the alternating potentials of the cochlea showed a close correspondence between the minimal temperature for optimal cochlear sensitivity and the preferred temperature (Table 25.4). Werner (1972) determined that the locus of the temperature effects is within the inner ear in these lizards, rather than external to it. He believes that changes in the thermal sensitivity of the hair cells are responsible for these effects.

The thermal dependence of neural function does not appear to have been extensively analyzed in reptiles. Andry *et al.* (1971) present observations on electroencephalographic activity (EEG) and photically evoked responses (ER) from garter snakes (*Thamnophis radix* and *T. sirtalis*). Both the predominant frequencies and the amplitude of the EEG tend to increase with temperature between approximately 5 and 40°C in these animals. The amplitude of the ER attains a maximim near 32.5°C and the latencies for the waves comprising this response reach a minimum between 25 and 30°C. These temperatures are not far from the level apparently preferred by garter snakes. The difference in the temperature relations of the EEG and ER results in the amplitude of the latter greatly exceeding that of the former between 17.5 and 32.5°C. Andry *et al.* (1971) suggest that the relation of the amplitude of ER (as an example of a superimposed signal) to that of the EEG (as indicative of background noise) might be important with respect to learning efficiencies. Learning does appear temperature dependent in reptiles, judging by the observations of Krekorian *et al.* (1968). Their study of maze learning by the lizard *Dipsosaurus dorsalis* indicated that the performance of animals trained at 22°C did not exceed chance levels after 125 trials. Groups trained at 27 and 32°C both learned two types of mazes, with fewer trials and errors characterizing those at the higher temperature. It appears worthwhile to determine whether learning performance improves further in animals trained at temperatures closer to the preferred body temperature of *Dipsosaurus* (38.5°C, DeWitt, 1967). Further work on the temperature dependence of sensory, neural, and behavioral functions of reptiles generally appears highly desirable.

Autotomy in Relation to Temperature

Many lizards readily autotomize their tails when handled, and this capacity probably improves their chances of escaping from predators. It is of interest that the pressure required to produce caudal autotomy in the iguanid lizard *Uta stansburiana* is temperature-dependent, with minimal values occurring across a band extending from 30 to 38°C, the range in which observations on activity temperatures for this species suggest the thermal preferendum would lie (Brattstrom, 1965). It should be worthwhile to determine the thermal dependence of autotomy in other species of lizards that differ from *Uta* in temperature preferences.

Thermal Effects on Digestion and Egestion

Reptiles evidently require temperatures within the upper portion of their tolerance range for digestion and egestion to proceed (Cowles and Bogert, 1944). Putrefaction of the ingested food may ultimately occur at lower temperatures, followed by the animal's being forced to regurgitate it. Cowles and Bogert (1944) noted that such a thermal requirement exists in both herbivorous and carnivorous species. In this connection it is of interest that the snake *Constrictor constrictor* places that segment of its body containing a bolus of food directly below a radiant heat source under laboratory conditions (Regal, 1966). The best indications of an interrelation between digestion and thermal preferendum pertain to turtles. Individuals of *Pseudemys scripta* fed within the preceding 2 days bask significantly more than those deprived of food for more than 48 h (Moll and Legler, 1971). In tests conducted in an experimental thermal gradient, turtles of this species increased their body temperatures an average of 4.5°C after feeding (Gatten, 1973). On the other hand, the terrestrial turtle *Terrapene ornata* maintained essentially the same body temperature within this gradient before and after feeding. However, variation in temperature was substantially reduced following feeding (Gatten, 1973).

The available information on reptilian digestion and egestion suggests that these activities require body temperatures near, if not identical with, the thermal preferendum. It would be of considerable interest to determine the extent to which thermal dependence of secretory activity of the digestive mucosa, peristaltic activity of the gut, and activity of digestive enzymes varies among species differing in their temperature preferences.

Thermal Effects on Reproductive Physiology

In contrast to the damaging effects of temperatures slightly above the preferred range on testicular tissue of lizards both in vivo and in vitro (Licht, 1965b; Licht and Basu, 1967), temperatures within this range appear to play a primary role in controlling annual testicular cycles in these animals. This has been extensively documented in the lizard *Anolis carolinensis* (Licht, 1967a, 1967b, 1969, 1972). Photoperiodism exists in this anole, but it appears to act only in a permissive way, probably serving to modify the thermal response, which functions as the more direct modifier of testicular recrudescence. Day lengths of 12–16 h can be prevented from exerting an effect on this recrudescence during any month of the year if the body temperatures of the anoles are held at 20°C. On the other hand, day lengths of

13 h or longer during early fall stimulate testicular growth in animals maintained at 32°C, near the preferendum for *Anolis carolinensis*. Shorter days at this season lead to inhibition of testicular recrudescence in individuals at 32°C (Licht, 1967a, 1969), and this probably is the response relevant to the natural condition in this species. In anoles with fluctuating body temperatures, photoperiodism depends both on the duration of time spent daily at 32°C (the thermoperiod) and on the extent of nocturnal cooling. Photoperiodically induced responses occur with thermoperiods of 8 h or more and a nocturnal temperature of 20°C. Longer thermoperiods are required for these responses, if nocturnal temperatures are reduced further. Licht (1969) has suggested that some equivalence may exist between heat and light in the photosexual responses of *Anolis*; the critical day length for photostimulation varying with temperature in this animal.

Various testicular activities appear to vary in their sensitivities to temperature in *Anolis carolinensis* (Licht, 1967b, 1972). Low body temperatures inhibit all aspects of testicular development. Moderate temperatures (20°C) well below the thermal preferendum for this anole allow testicular growth but greatly retarded spermatogenic development. Maturation of the testes, spermiation, and androgen production with consequent development of secondary sexual structures and sexual behavior appear largely dependent of some elevation of body temperature to near preferred levels. This requirement appears fairly restrictive, for *Anolis* suffers testicular damage at body temperatures slightly above the preferred range (Licht, 1965b) and undergoes less than optimal response at 28°C, only a few degrees below the thermal preferendum (Licht, 1966).

Licht (1967b) has suggested that attainment of preferred thermal levels may be important for the completion of testicular development and the onset of breeding among male lizards generally. His recent work (Licht, 1973) on *Dipsosaurus dorsalis* and *Xantusia vigilis* provides support for this view. High temperatures appear to provide the primary cue for stimulating spermatogenesis and androgen production in males of both species in spring. This stimulatory effect is terminated in summer, when refractoriness develops. According to Licht's (1973) results, this appears to be governed by an endogenous rhythm. In *Dipsosaurus* (preferred body temperature, 38.5°C, according to DeWitt, 1967), maintenance at 37°C in spring resulted in rapid growth of the testes, which became fully developed histologically after only 20 days. A comparable stage was reached at 28°C by 60 days. However, the disorganized appearance of the germinal epithelium in males held at this temperature suggested that spermatogenesis was somewhat abnormal. Maintenance of *Dipsosaurus* at 20°C produced virtually no change in testis weight up to 60 days, but limited spermatogenesis did occur in this period. In contrast to the results obtained with constant temperatures of 20 and 28°C, exposure of male *Dipsosaurus* to 6 h at 37°C and 18 h at 20 or 28°C daily produced essentially full testicular development and androgen production within 40 days.

Xantusia vigilis (preferred body temperature, 25–30°C, according to Cowles and Burleson, 1945) resembles *Dipsosaurus* in its dependence on warm temperatures for testicular recrudescence in the spring (Licht, 1973). Maintenance of males at 20°C during this season appeared to arrest testes development; the testes remained in the initial condition for 5 weeks. At 30°C spermatogenic and androgenic activity developed rapidly, but the testes did not undergo a normal increase in weight. Testicular development appeared normal in animals subjected to 12 h at 30°C and 12 h at 20°C each day, being completed in 3 weeks. As noted previously, *Xantusia* maintained by Cowles and Burleson (1945) at 36°C for 1 week suffered testicular damage, a point of interest in view of the stimulating effect of 37°C on testis development in the more thermophilic *Dipsosaurus dorsalis.*

Perhaps Licht's (1967b) suggestion concerning the importance of attainment of preferred thermal levels for completion of gonadal recrudescence in male lizards is also applicable to females. Tinkle and Irwin (1965) and Marion (1970) have found that ovarian development is dependent on warm temperatures in *Uta stansburiana* and *Sceloporus undulatus.* Response to such temperatures seems conditioned by an endogenous rhythmicity. Females of both species were refractory to warmth in the fall and early winter, but this condition terminated in midwinter, even if the animals were maintained at unseasonably warm temperatures for several months starting in the fall.

The evidence cited here concerning the importance of existence at or near preferred levels of body temperature for completion of gonadal recrudescence in male lizards provides a further indication of the significance of reptilian thermal preferenda. Fuller understanding of this will require study of additional species, particularly nonsaurians. In these, it will be particularly important to define as precisely as possible the temperature thresholds for the various effects of importance to reproduction, and the relations of these thresholds to the preferenda of the species involved.

Immunological Responses and Temperature

Capacities for antibody production and the titers established appear strongly affected by temperature in reptiles. When maintained at 25°C, the lizard *Dipsosaurus dorsalis* (preferred body temperature, 38.5°C, according to DeWitt, 1967) produced only a weak antibody response in the 14–21 days after receiving *Salmonella typhosa* H antigen. However, animals of this species at 35°C showed high levels of the serum antibody to this antigen 1 week after injection, and the titers remained high for at least 4 weeks (Evans and Cowles, 1959; Evans, 1963a). Immunization of *Dipsosaurus* maintained at 40°C did not improve the antibody response further (Evans, 1963b). Indeed, the titers attained were distinctly lower than at 35°C. A similar set of responses was noted for another thermophilic lizard, *Sauromalus obesus.* Again, antibody production at 25°C was negligible, that at 35°C was well developed, and that at 40° intermediate (and possibly complicated by the fact that the animals did not eat and became emaciated over the test period). The preferred body temperature of *Sauromalus* probably lies near 38°C, like that of *Dipsosaurus.* It would be of interest to determine the immunological responses of these animals at this temperature. It would also appear worthwhile to test them with a thermal regimen in which the animals are allowed to cool between daily exposures to 40°C.

A different set of temperature relations is apparent for the immunological response of the less thermophilic lizard *Egernia cunninghami* (preferred body temperature, 32.5°C, according to Wilson, 1971). Individuals of this species maintained at 30°C commenced formation of antibodies within a week following injection of sheep erythrocytes and established near-maximal titers in approximately 2 weeks. Response to such an injection by *Egernia* maintained at 25°C was detectable only after 2 weeks and the maximal titers established were conspicuously lower than those noted in the animals at 30°C. Little production of antibody occurred in the individuals tested at 20°C (Tait, 1969).

The immunological results obtained for *Dipsosaurus*, *Sauromalus*, and *Egernia* provide additional documentation for the existence of interspecific variation in thermal responses of reptiles. More importantly, they hint that immune responses proceed optimally at temperatures in the vicinity of the thermal preferendum. This possibility should be tested further using a wider array of species, including members of other reptilian groups, as well as lizards.

Physiological Correlates of Activity and Temperature

Aerobic Metabolism

Reptilian metabolism has received far less attention in exercising than in resting animals. Not only are metabolic measurements more difficult to perform on active individuals, but activity patterns of reptiles typically involve short bouts of exertion separated by long periods of quiescence, rather than any steady state. Despite these problems, a growing literature exists on the physiological correlates of activity in these animals. Thus far emphasis has centered on determination of rates of oxygen consumption by active individuals at various temperatures (see Bennett and Dawson, 1975).

The bulk of the information on oxygen consumption by active reptiles pertains to lizards, although limited data are also available on representatives of other groups. In the majority of species studied during *maximal* activity at various temperatures, oxygen consumption usually shows a pronounced thermal dependence below the preferred body temperature (Q_{10} = 2–4) combined with a plateau or decline at temperatures that exceed the preferendum. Monitor lizards (*Varanus* spp.) and the turtles *Pseudemys scripta* and *Terrapene ornata* represent exceptions to this; their oxygen consumption during maximal activity continuing to increase with temperature beyond the preferred level (Bartholomew and Tucker, 1964; Bennett, 1972b; Gatten, 1973).

The attention devoted to activity metabolism of reptiles in recent years has largely been stimulated by interest in the determination of the thermal dependence of metabolic scope, i.e., the difference between metabolic rates during rest and maximal activity (see Fry, 1947). Originally it was not possible to identify any relation in reptiles between preferred levels of body temperature and the temperature at which metabolic scope is maximal (Dawson, 1967; Tucker, 1967; Templeton,

1970), owing at least partially to the failure of some investigators to bring their experimental animals to peak levels of activity for the particular test conditions. (The analyses on which this portion of the discussion is based pertains to metabolic scope represented by the increment between resting and active levels of oxygen consumption. This increment will subsequently be referred to as *aerobic scope*, to avoid confusion with considerations arising out of treatment of anaerobic and total metabolism in the ensuing sections of this review.) The apparent absence of such a relation was surprising, since it would seem highly advantageous for reptiles to achieve the greatest aerobic scope, and, presumably, the greatest potential for exertion at body temperatures attained during activity under favorable conditions. It has therefore been of great interest that Wilson's (1971) recent analysis, based on more complete information than previously available, and some subsequent studies (Bennett, 1972b; Bennett and Dawson, 1972; Dmi'el and Borut, 1972; Greenwald, 1971) have identified several species in which close agreement does appear to exist between the temperatures at which aerobic scope is maximal and the preferred body temperature (Table 25.5). Indeed, aerobic scope appears to reach a maximum at body temperatures near or identical with the thermal preferendum in most animals studied (primarily lizards). Monitor lizards (*Varanus* spp.), which have unusually extensive aerobic capacities (Bennett, 1972b), and the turtles *Pseudemys scripta* and *Terrapene ornata* (Gatten, 1973) represent the major exceptions thus far to this pattern; their aerobic scopes continue to increase with temperature beyond the preferred range of body temperatures.

Anaerobic Metabolism

Active reptiles, like other vertebrates, may supplement the catabolism of foodstuffs with oxygen via the mitochondrial enzymes with anaerobic glycolysis. In lizards, at least, the activities of glycolytic enzymes on a protein-specific basis may match or exceed those in rats. On the other hand, the activities of aerobic enzymes from these reptiles show far lower activities than those from mammals (Bennett, 1972a). The energy yield from glycolysis is substantially less than in the aerobic scheme, and the accumulation of lactate may cause important disruptions of blood equilibria (Bennett, 1971).

The total amount of lactate generated by small lizards during activity to exhaustion usually amounts to 1.0–1.3 mg g^{-1} of body weight and has a low thermal dependence (Q_{10} = 1.1–1.3) according to Bennett and Licht (1972). The thermophilic lizard *Dipsosaurus dorsalis* deviates from this pattern. Maximal lactate levels observed in this animal following peak activity (1.8 mg g^{-1}) are significantly higher than those observed in other species and only appear near the preferred body temperature for this species (Bennett and Dawson, 1972).

Only limited information is available on the thermal relations of anaerobiosis in active reptiles other than lizards. Gatten (1973) has determined that lactate concentrations resulting from maximal activity by the turtle *Terrapene ornata* at 30 and 40°C approximate 1.0 mg g^{-1}. In the more aquatic *Pseudemys scripta*, on the other hand, the values at these temperatures averaged 0.73 and 0.37 mg g^{-1},

Table 25.5 Temperatures for Maximal Aerobic Scope and Maximal Heart-Rate Increment in Reptiles

Species	Preferred Body Temp. (°C)	Temp. for Max. Aerobic Scope (°C)	Temp. for Max. Heart Rate Increment (°C)	Reference
		Rhynchocephalians		
Sphenodon punctatus	18–19, 25[a]	25.5	25	Wilson and Lee (1970)
		Turtles		
Pseudemys scripta	24.6, 29.1[b]	40	40	Gatten (1973)
Terrapene ornata	28.3, 29.8[b]	40	40	
		Snakes		
Pituophis melanoleucus	26.7, 30[a]	30.0	30	Greenwald (1971)
Spalerosophis cliffordi	30	30–35		Dmi'el and Borut (1972)
		Lizards		
Amphibolurus barbatus	35.7	35		Wilson (1971)
Cnemidophorus tigiris	40.4	40[c]		Asplund (1970)
Dipsosaurus dorsalis	38.5	40	42	Bennett and Dawson (1972); Licht (1965a)[b]
Egernia cunninghami	32.5	34.5		Wilson (1971)
Iguana iguana	33.4	32.5	43	Moberly (1968)
Physignathus leseueri	30.1	30.0	32	Wilson (1971)
Sauromalus hispidus	37.1	38.0	40	Bennett (1972b)
Tiliqua rugosa	32.6	33.0	34–35	Wilson (1971); Licht (1965a)[d]
T. scincoides	32.6	40.0[c]	30	Bartholomew *et al.* (1965)
Uta stansburiana	35.4[e]	35.0[c]		Alexander and Whitford (1968)
Varanus gouldii	37.5[e]	40.0	40	Bennett (1972b)
Varanus spp.	35.5[e]	40.0		Bartholomew and Tucker (1964)

[a] The higher temperature more accurately reflects the preferred body temperature of this species, according to Wilson (1971).

[b] The latter temperature pertains to recently fed individuals.

[c] More detailed observations on aerobic scope appear desirable in this species, according to Bennett and Dawson (1975).

[d] Licht's (1965a) observations pertain solely to heart-rate increment.

[e] Activity temperature for animals in the field.

respectively. The extent to which the presence of the shell complicates comparisons of the results for these turtles with those for lizards is unknown.

Total Metabolic Scope

Interpretation of total metabolic scope for activity in reptiles requires treatment of its anaerobic as well as aerobic component. Such treatment necessitates expression of oxygen consumption and lactate production in terms of the associated production of adenosine triphosphate. Consideration of these functions in such terms indicates that activity of a number of reptiles is primarily dependent on anaerobiosis (Moberly, 1968; Bennett and Licht, 1972; Bennett and Dawson, 1972; Gatten, 1973). For example, Bennett and Dawson (1972) and Gatten (1973) calculated from simultaneous determinations of oxygen consumption and lactate production that 58–83 percent and 60–75 percent of the energy expended in 2-min bouts of activity is derived anaerobically in the lizard *Dipsosaurus dorsalis* and the turtles *Pseudemys scripta* and *Terrapene ornata*, respectively. Bennett and Licht (1972) estimated that 80–90 percent of the energy expended within the first 30 s of activity by small lizards comes from anaerobic sources. With the relatively low thermal dependence of lactate production in most of these animals, a capacity for a rapid activity response is possible over a wide range of temperatures. The metabolic implications of this are discussed by Bennett and Licht (1972).

Data permitting a full analysis of the thermal dependence of total metabolic scope (i.e., the sum of the aerobic and anaerobic scopes for activity) have thus far been obtained only for the lizard *Dipsosaurus dorsalis* (Bennett and Dawson, 1972), although Gatten's (1973) measurements of the turtles *Pseudemys* and *Terrapene* at 30 and 40°C represent a significant step in this direction. In the case of *Dipsosaurus*, both aerobic scope and lactate production were maximal in the vicinity of 40°C, about a degree beyond the thermal preferendum for this lizard (DeWitt, 1967). Consequently, the total metabolic scope was also greatest in the vicinity of 40°C. In most lizards lactate production has a considerably lower thermal dependence than in *Dipsosaurus* (Bennett and Licht, 1972). It will be of interest to obtain simultaneous measurements of their aerobic and anaerobic metabolism at various temperatures, to determine if their total metabolic scopes show well-defined maxima in the vicinity of their thermal preferenda. It will also be appropriate to bring our knowledge of the physiological correlates of activity for representatives of other reptilian groups up to a par with that pertaining to lizards. Some of these other reptiles may resemble monitor lizards (*Varanus* spp.) in possessing unusually extensive capacities for supporting aerobic metabolism, unlike the case in lizards generally (Bennett, 1972b, and personal communication).

Recovery from Activity

Lactate formed during activity is only slowly eliminated by reptiles. Some indication of this is provided by the fact that green sea turtles (*Chelonia mydas*) require 5 h to eliminate half of the lactate burden acquired during a 30-min dive

(Berkson, 1966). Moberly (1968) found that only half of the maximum concentration of blood lactate had been removed 1 h after activity in the lizard *Iguana iguana* at most body temperatures. Recovery by this animal proceeded most rapidly at 35°C, near the preferred body temperature for this species. The recovery rate at 40°C was intermediate between those at 30 and 25°C. Bennett (1971, 1972b) has provided the most detailed analysis of anaerobiosis in lizards and has separated the oxygen debt of *Sauromalus hispidus* into lactacid and alactacid components. The former followed the same thermal relations regarding clearance as that evident in *Iguana*, being most rapid at 35°C, a couple of degrees below the preferred body temperature (Bennett, 1972b), and slower at 40 than at 30°C. The rate of disappearance of total lactate following activity by the lizard *Anolis carolinensis* has been measured by Bennett and Licht (1972). The concentration of this substance did not decline significantly in the first hour following termination of exercise at 20°C. The initial rate of lactate elimination was most rapid at 30°C, near the thermal preferendum for *Anolis*, but total recovery proceeded most rapidly at 37°C. The elimination of lactate produced in vigorous exertion by reptiles seems slow in comparison with the rapidity with which it can be formed. This seems to be an impediment to sustained activity. The extent of the aerobic scope appears particularly important in this connection. Although this scope does contribute to capacity for activity, its true significance in lizards and perhaps in other reptiles as well may lie in its determination of the rate at which lactate burdens resulting from activity can be eliminated. The information summarized in Table 25.5 indicates that the most rapid rates of elimination would tend to occur near the thermal preferendum in most species on which adequate information has been obtained.

Increments of Heart Rate Between Rest and Activity in Reptiles

The increased consumption of oxygen associated with and following activity by reptiles requires increased transport of oxygen by the cardiovascular system. This has prompted interest in determinations of circulatory activity and in arterovenous differences in oxygen concentration under various conditions. Most measurements on reptiles have thus far dealt with heart rate, primarily because of technical difficulties in applying standard techniques to measure cardiac output to the circulatory systems of these animals. While the information on heart rate provides a very incomplete picture of the circulatory changes occurring in reptiles with activity, the results obtained are of some interest in the context of this consideration of the functional significance of the thermal preferenda of reptiles. These results are most conveniently summarized in terms of the thermal dependence of the increments between heart rates of various animals at rest and during maximal activity. The body temperature at which this increment is maximal in these reptiles is presented in Table 25.5. The correspondence between this temperature and the thermal preferendum is quite close in a number of species, despite Tucker's (1966, 1967) demonstration that the oxygen transport of *Iguana iguana* is increased by augmentation of stroke volume and the utilization coefficient (volume of oxygen extracted from each volume of blood during passage through the tissues), as well as by increasing heart rate.

Excretory Function in Relation to Temperature

Studies of the response of lizards to experimentally imposed water loads indicate that capacity for producing urine dilute with respect to sodium, and hence the ability to conserve this ion, varies with temperature (Shoemaker *et al.*, 1966, 1967). Water loads equaling 10 percent of body weight were administered to the lizards *Tiliqua rugosa*, *Phyllurus milii*, and *Amphibolurus barbatus*. In the first 6 h at the test temperature, the minimal urinary concentrations of sodium were found at 26 and 37°C in *Phyllurus* and *Amphibolurus*, respectively. In these cases, the temperature specified lies very close to the thermal preferendum of the particular species. The minimal concentration of this ion can be produced by *Tiliqua* over a range of 24 to 32°C, the upper temperature lying near the preferred body temperature of this skink (Licht *et al.*, 1966b).

Temperature Dependence of Thyroid Function

Thyroid hormone appears to play the same role in maintenance and stimulation of metabolic rate in reptiles that it does in mammals. Injection of thyroxine increases metabolism in the following lizards: *Anolis carolinensis* (Maher and Levedahl, 1959), *Lacerta muralis* (Maher, 1961), *Eumeces fasciatus* (Maher, 1965), and *Sceloporus cyanogenys* (Wilhoft, 1966). Thyroid injections accelerate metabolism in brain, heart, kidney, liver, and lung (Maher, 1964). In all the species for which information is available, thyroxine was effective in increasing metabolic rate only when body temperature was raised to near the preferred level (30°C or above), having an insignificant effect at cooler temperatures (20°C or below). Increased environmental temperature increases activity by the thyroid in lizards (Eggert, 1936). The pronounced thermal dependence of thyroid function might well be a factor in the lack of metabolic stimulation observed in a turtle following injection of thyroxine (Drexler and Issekutz, 1935). The observations cited here and those mentioned previously regarding thermal requirements for androgen production in lizards suggest that it would be worthwhile to determine the extent to which the secretion and effects of other hormones are related to temperature in reptiles.

Concluding Remarks

Introduction

The relation between physiological responses of reptiles to temperature and the thermal preferences of these animals has been considered extensively. It is appropriate in these concluding remarks to evaluate some of the features of these preferences and to consider the adequacy of the taxonomic coverage allowed by the available data. These tasks are crucial to determine the functional significance of reptilian thermal adjustments and to estimate how broadly the patterns discerned in these adjustments with the information at hand can be applied within the Class Reptilia.

Some Comments on the Thermal Preferendum

The precision of reptilian thermoregulation is markedly affected by environmental circumstances. Licht *et al.* (1966b) have emphasized that definition of the preferred body temperature for a species requires observations in controlled situations affording a wide range of thermal conditions. The thermal preferendum thus identified appears to be a stable physiological characteristic, judging by Licht's (1968) observations on the lizard *Anolis carolinensis.* However, it can change with prolonged maintenance of animals under constant and apparently stressful temperatures (Wilhoft and Anderson, 1960; see also Licht, 1968, pp. 155–156). While the preferendum does not appear labile when the animals are in a thermoregulatory state, the diurnal lizards studied in thermal gradients by Regal (1967) abandoned it at night in favor of cooler conditions. The thermal preferences with which this review has dealt thus appear primarily linked with the active phase of the daily cycle in reptiles. This indicates the desirability of more extensive incorporation of realistic oscillations of temperature into the experimental designs of future physiological studies of members of this group. It also suggests that the definitive treatment of reptilian thermophysiology will in many instances have to involve assessment of the functional significance of periodic retreat to cooler body temperatures, as well as that of operation at temperatures in the vicinity of the preferendum.

The actual thermal preferences of species treated in this review have generally been expressed in terms of the mean body temperatures for active animals in experimental thermal gradients, although the modal temperature was utilized in one case (DeWitt, 1967). It must be emphasized that precision of thermoregulation may vary widely from species to species. Therefore, a range of preferred temperatures might more appropriately be specified. Uncertainty as to whether the variation in preferred temperature evident within particular species reflected individual differences in the preferendum (see, e.g., Bradshaw and Main, 1968) or imprecision of temperature regulation within individuals dictated the use of mean or modal values in this review. The fact that a range of preferred temperatures is the typical case made it possible to include processes reaching optima within approximately 3°C of the mean or mode among the physiological activities considered previously.

Specification of values for thermal preferenda is further complicated by the fact that differentials of several degrees can exist between the head and body of reptiles, apparently as the result of cephalic venous shunts (see, e.g., Heath, 1964, 1966; DeWitt, 1967; Webb *et al.*, 1972; Johnson, 1972). These differences are particularly evident during heating, cooling, and exposure of the animals to thermally stressful conditions. Head temperatures of pythons appear more precisely regulated than body temperature in these instances (Johnson, 1973). This raises the possibility of difficulty in definition of thermal preferenda from information on body temperatures (usually measured cloacally) alone. However, DeWitt (1967) and Gatten (1973) concluded that head–body differences in temperature are probably small or absent in reptiles that have attained their preferred thermal range under conditions allowing them full use of their behavioral capabilities. Moreover, Johnson (1973) noted only small differentials between the temperatures of the head and body of the python *Liasis childreni* under simulated field conditions. Consequently, the information thus far obtained on various reptiles provides a basis for reasonable approximation of their themal preferences. Nevertheless, it would be desirable routinely to monitor both head and body temperatures in future determinations of reptilian thermal preferenda.

Future Studies of Reptilian Thermophysiology

Information permitting assessment of the extent of interspecific variation in physiological responses of reptiles to temperature has increased substantially since the last review devoted exclusively to this topic (Dawson, 1967). It is now becoming appropriate to launch more extensive investigations of the mechanisms fostering this variation. The role of thermal acclimation, a process with which this review has dealt only tangentially, in modifying the thermal responses of animals *in nature* also merits attention. The paucity of information on these topics is hindering formulation of meaningful generalizations about the thermophysiology of reptiles. However, the gaps in the information now available with respect to the major reptilian groups are more serious deterrents to this formulation. We need to bring our knowledge of physiological responses to temperature and thermal preferences of snakes, turtles, crocodilians, and the surviving rhynchocephalian up to a par with that available for lizards. The bulk of this review has dealt with these latter animals and the extent to which the patterns of thermal adjustment discerned for them are applicable to other groups as well will be of considerable interest.

Physiological Significance of Preferred Body Temperatures in Reptiles

Interpretation of the functional significance of the preferred body temperatures of reptiles has proved more difficult than originally anticipated. Among other problems, studies of rate processes in resting animals have not by themselves been as informative as might have been hoped (Dawson, 1967). However, several lines of information now allow definition of the physiological states associated with these preferred thermal levels. With lizards serving as the primary model, it appears that metabolic capacities for activity, elimination of lactate levels incurred during activity, auditory sensitivity, digestion and egestion, immunological response, secretion and action of certain hormones, aspects of renal function, and certain reproductive processes proceed optimally near the thermal preferendum in many species. However, well-documented exceptions exist, e.g., the thermal dependence of aerobic scope in monitor lizards (Bartholomew and Tucker, 1964; Bennett, 1972b) and in certain turtles (Gatten, 1973). The meaning of these exceptions will probably be forthcoming only with more detailed knowledge of the thermal relations, energetics, and behavior of the species involved. Despite these exceptions, instances are now documented in which maintenance of individuals and survival of the species clearly depend on periodic attainment of temperatures within the preferred range. Such dependence more than justifies the often elaborate behavioral and physiological mechanisms leading to the control of body temperature evident in many active reptiles.

Acknowledgments

Preparation of this review was supported in part by a grant from the National Science Foundation (GB-25022). Results of research by Dawson cited here, which was often carried out in collaboration with valued colleagues, was also supported by grants from the National Science Foundation, the Guggenheim Foundation, and the Horace H. Rackham School of Graduate Studies at The University of Michigan. The colleagues referred to—G. A. Bartholomew, A. F. Bennett, P. Licht, A. R. Main, the late W. R. Moberly, T. L. Poulson, V. H. Shoemaker, and the late J. R. Templeton—have contributed a great deal to the development of my views on reptilian thermophysiology and I am pleased to have this opportunity to acknowledge their contributions.

Dedication

This review is dedicated to Raymond B. Cowles, whose work has supplied direction to much of the research with which it deals. On a more personal note, my contacts with him at the University of California at Los Angeles at the beginning of my professional career had a decisive influence on my selection of a pathway to follow in biological research.

References

Adrian, E. D., Craik, K. J. W., Sturdy, R. S.: 1938. The electrical response of the auditory mechanism in cold-blooded vertebrates. Proc. Roy. Soc. London Ser. B *125*, 435–455.

Aleksiuk, M.: 1971a. An isoenzymic basis for instantaneous cold compensation in reptiles: lactate dehydrogenase kinetics in *Thamnophis sirtalis*. Comp. Biochem. Physiol. *40B*, 671–681.

———: 1971b. Temperature-dependent shifts in the metabolism of a cool temperate reptile, *Thamnophis sirtalis parietalis*. Comp. Biochem. Physiol. *39A*, 495–503.

Alexander, C. E., Whitford, W. G.: 1968. Energy requirements of *Uta stansburiana*. Copeia *1968*, 678–683.

Andry, M. L., Luttges, M. W., Gamow, R. I.: 1971. Temperature effects on spontaneous and evoked neural activity in the garter snake. Expt. Neurol. *31*, 32–44.

Asplund, K. K.: 1970. Metabolic scope and body temperatures of whiptail lizards (*Cnemidophorus*). Herpetologica *26*, 403–411.

Atsatt, S. R.: 1939. Color changes as controlled by temperature and light in the lizards of the desert regions of southern California. Univ. Calif. Los Angeles Biol. Sci. Publ. *1*, 237–276.

Bartholomew, G. A., Lasiewski, R. C.: 1965. Heating and cooling rates, heart rate and simulated diving in the Galápagos marine iguana. Comp. Biochem. Physiol. *16*, 573–582.

———, Tucker, V. A.: 1963. Control of changes in body temperature, metabolism, and circulation by the agamid lizard, *Amphibolorus barbatus*. Physiol. Zool. *36*, 199–218.

———, Tucker, V. A.: 1964. Size, body temperature, thermal conductance, oxygen consumption, and heart rate in Australian varanid lizards. Physiol. Zool. *37*, 341–354.

———, Tucker, V. A., Lee, A. K.: 1965. Oxygen consumption, thermal conductance, and heart rate in the Australian skink *Tiliqua scincoides*. Copeia *1965*, 169–173.

Bartlett, P. N., Gates, D. M.: 1967. The energy budget of a lizard on a tree trunk. Ecology *48*, 315–322.

Beckman, W. A., Mitchell, J. W., Porter, W. P.: 1971. Thermal model for prediction of a desert iguana's daily and seasonal behavior. Am. Soc. Mech. Engrs. Publ. *71-WA/HT-35*.

Bennett, A. F.: 1971. Oxygen transport and energy metabolism in two species of lizards, *Sauromalus hispidus* and *Varanus gouldii*. Ann Arbor, Mich.: Univ. Michigan. Ph.D. diss.

———: 1972a. A comparison of activities of metabolic enzymes in lizards and rats. Comp. Biochem. Physiol. *42B*, 637–647.

———: 1972b. The effect of activity on oxygen consumption, oxygen debt, and heart rate in the lizards *Varanus gouldii* and *Sauromalus hispidus*. J. Comp. Physiol. *79*, 259–280.

———, Dawson, W. R.: 1972. Aerobic and anaerobic metabolism during activity in the lizard *Dipsosaurus dorsalis*. J. Comp. Physiol. *81*, 289–299.

———, Dawson, W. R.: 1975. Metabolism. *In* Biology of the reptilia (ed. C. Gans, W. R. Dawson), vol. 5, Physiology A. New York: Academic Press.

———, Licht, P.: 1972. Anaerobic metabolism during activity in lizards. J. Comp. Physiol. *81*, 277–288.

Berkson, H.: 1966. Physiological adjustments to prolonged diving in the Pacific green turtle (*Chelonia mydas agassizii*). Comp. Biochem. Physiol. *18*, 101–119.

Bradshaw, S. D., Main, A. R.: 1968. Behavioural attitudes and regulation of temperature in *Amphibolurus* lizards. J. Zool. London *154*, 193–221.

Brattstrom, B. H.: 1965. Body temperatures of reptiles. Am. Midland Naturalist *73*, 376–422.

Cabanac, M., Hammel, T., Hardy, J. D.: 1967. *Tiliqua scincoides*: temperature sensitive units in lizard brain. Science *158*, 1050–1051.

Campbell, H. W.: 1967. The effects of temperature on the auditory sensitivity of lizards. Los Angeles, Calif.: Univ. California. Ph.D. diss.

———: 1969. The effects of temperature on the auditory sensitivity of lizards. Physiol. Zool. *42*, 183–210.

Case, T. J.: 1972. Thermoregulation and evaporative cooling in the chuckwalla, *Sauromalus obesus*. Copeia *1972*, 145–150.

Colbert, E. H., Cowles, R. B., Bogert, C. M.: 1946. Temperature tolerances in the American alligator and their bearing on the habits, evolution, and extinction of the dinosaurs. Bull. Am. Museum Nat. Hist. *86*, 327–373.

Cole, L. C.: 1943. Experiments on toleration of high temperature in lizards with reference to adaptive coloration. Ecology *24*, 94–108.

Cott, H. B.: 1961. Scientific results of an inquiry into the ecology and economic status of the Nile crocodile (*Crocodilus niloticus*) in Uganda and Northern Rhodesia. Trans. Zool. Soc. London *29*, 211–356.

Coulson, R. A., Hernandez, T.: 1964. Biochemistry of the alligator: a study of metabolism in slow motion. Baton Rouge, La.: Louisiana State Univ. Press.

Cowles, R. B., Bogert, C. M.: 1944. A preliminary study of the thermal requirements of desert reptiles. Bull. Am. Museum Nat. Hist. *83*, 265–296.

———, Burleson, G. L.: 1945. The sterilizing effect of high temperature on the male germ-plasm of the yucca night lizard, *Xantusia vigilis*. Am. Naturalist *79*, 417–435.

Crawford, E. C., Jr.: 1972. Brain and body temperatures in a panting lizard. Science *177*, 431–433.

Dawson, W. R.: 1960. Physiological responses to temperature in the lizard *Eumeces obsoletus*. Physiol. Zool. *33*, 87–103.

———: 1967. Interspecific variation in physiological responses of lizards to temperature. *In* Lizard ecology: a symposium (ed. W. W. Milstead), pp. 230–257. Columbia, Mo.: Univ. Missouri Press.

———, Bartholomew, G. A.: 1958. Metabolic and cardiac responses to temperature in the lizard *Dipsosaurus dorsalis*. Physiol. Zool. *31*, 100–111.

———, Poulson, T. L.: 1958. Effects of temperature on ventricular contraction in several lizards. Anat. Record *131*, 545.

———, Templeton, J. R.: 1963. Physiological responses to temperature in the lizard *Crotaphytus collaris*. Physiol. Zool. *36*, 219–236.

———, Templeton, J. R.: 1966. Physiological responses to temperature in the alligator lizard, *Gerrhonotus multicarinatus*. Ecology *47*, 759–765.

DeWitt, C. B.: 1967. Precision of thermoregulation and its relation to environmental factors in the desert iguana, *Dipsosaurus dorsalis*. Physiol. Zool. *40*, 49–66.

Dmi'el, R., Borut, A.: 1972. Thermal behavior, heat exchange, and metabolism in the desert snake *Spalerosophis cliffordi*. Physiol. Zool. *45*, 78–94.

Drexler, E., Issekutz, B. V.: 1935. Die Wirkung des Thyroxins auf den Stoffwechsel kaltblütiger Wirbeltiere. Naunyn-Schmiedebergs Arch. Expt. Path. Pharmak. *177*, 435–441.

Eggert, B.: 1936. Zur Morphologie und Physiologie der Eidechsen-Schilddrüse. II. Über die Wirkung von hohen und niedrigen Temperaturen, von Thyroxin und von thyreotropem Hormon auf die Schilddrüse. Z. Wiss. Zool. *147*, 537–594.

Evans, E. E.: 1963a. Antibody response in amphibia and reptilia. Federation Proc. *22*, 1132–1137.

———: 1963b. Comparative immunology. Antibody response in *Dipsosaurus dorsalis* at different temperatures. Proc. Soc. Expt. Biol. Med. *112*, 531–533.
———, Cowles, R. B.: 1959. Effect of temperature on antibody synthesis in the reptile, *Dipsosaurus dorsalis*. Proc. Soc. Expt. Biol. Med. *101*, 482–483.
Fitch, H. S.: 1956. Temperature responses in free-living amphibians and reptiles of northeastern Kansas. Univ. Kansas Publ. Museum Nat. Hist. *8*, 417–476.
Fry, F. E. J.: 1947. Effects of the environment on animal activity. Publ. Ontario Fisheries Res. Lab. 68, 1–62.
———, Hochachka, P. W.: 1970. Fish. *In* Comparative physiology of thermoregulation (ed. G. C. Whittow), vol. I, pp. 79–134. New York: Academic Press.
Gatten, R. E., Jr.: 1973. Aerobic and anaerobic metabolism during activity in the turtles *Pseudemys scripta* and *Terrapene ornata*. Ann Arbor, Mich.: Univ. Michigan. Ph.D. diss.
Greenwald, O. E.: 1971. The effect of body temperature on oxygen consumption and heart rate in the Sonora gopher snake, *Pituophis catenifer affinis* Hallowell. Copeia *1971*, 98–106.
Hammel, H. T., Caldwell, F. T., Jr., Abrams, R. M.: 1967. Regulation of body temperature in the blue-tongued lizard. Science *156*, 1260–1262.
Heath, J. E.: 1964. Head-temperature differences in horned lizards. Physiol. Zool. *37* 273–279.
———: 1966. Venous shunts in the cephalic sinuses of horned lizards. Physiol. Zool. *39*, 30–35.
———, Gasdorf, E., Northcutt, R. G.: 1968. The effect of thermal stimulation of the anterior hypothalamus on blood pressure in the turtle. Comp. Biochem. Physiol. *26*, 509–518.
Hoskins, M. A. H., Aleksiuk, M.: 1973. Effects of temperature on the kinetics of malate dehydrogenase from a cold climate reptile. *Thamnophis sirtalis parietalis*. Comp. Biochem. Physiol. *45B*, 343–353.
Hutchison, V. H., Kosh, R. J.: 1965. The effect of photoperiod on the critical thermal maxima of painted turtles (*Chrysemys picta*). Herpetologica *20*, 233–238.
———, Dowling, H. G., Vinegar, A.: 1966. Thermoregulation in a brooding female Indian python, *Python molurus bivittatus*. Science *151*, 694–696.
Jacobson, E. R., Whitford, W. G.: 1970. The effect of acclimation on physiological responses to temperature in the snakes, *Thamnophis proximus* and *Natrix rhombifera*. Comp. Biochem. Physiol. *35*, 439–449.
Johnson, C. R.: 1972. Head-body temperature differences in *Varanus gouldii* (Sauria: Varanidae). Comp. Biochem. Physiol. *43A*, 1025–1029.
———: 1973. Thermoregulation in pythons. II. Temperature differences and thermal preferenda in Australian pythons. Comp. Biochem. Physiol. *45A*, 1065–1087.
Kemp, F. D.: 1969. Thermoregulatory operant behavior in the *Dipsosaurus dorsalis* as a function of body temperature, substrate temperature, and heating rate. J. Biol. Psychol. *11*, 36–39.
Kosh, R. J., Hutchison, V. H.: 1968. Daily rhythmicity of temperature tolerance in eastern painted turtles, *Chrysemys picta*. Copeia *1968*, 244–246.
Kour, E. L., Hutchison, V. H.: 1970. Critical thermal tolerances and heating and cooling rates of lizards from diverse habitats. Copeia *1970*, 219–229.
Krekorian, C. O., Vance, V. J., Richardson, A. M.: 1968. Temperature-dependent maze learning in the desert iguana, *Dipsosaurus dorsalis*. Animal Behavior *16*, 429–436.

Licht, P.: 1964a. A comparative study of the thermal dependence of contractility in saurian skeletal muscle. Comp. Biochem. Physiol. *13*, 27–34.

———: 1964b. The temperature dependence of myosin-adenosinetriphosphatase and alkaline phosphatase in lizards. Comp. Biochem. Physiol. *12*, 331–340.

———: 1965a. Effects of temperature on heart rates of lizards during rest and activity. Physiol. Zool. *38*, 129–137.

———: 1965b. The relation between preferred body temperatures and testicular heat sensitivity in lizards. Copeia *1965*. 428–436.

———: 1966. Reproduction in lizards: influence of temperature on photoperiodism in testicular recrudescence. Science *154*, 1668–1670.

———: 1967a. Environmental control of annual testicular cycles in the lizard *Anolis carolinensis*. I. Interaction of light and temperature in the initiation of testicular recrudescence. J. Expt. Zool. *165*, 505–516.

———: 1967b. Environmental control of annual testicular cycles in the lizard *Anolis carolinensis*. II. Seasonal variations in effects of photoperiod and temperature on testicular recrudescence. J. Expt. Zool. *166*, 243–253.

———: 1967c. Thermal adaptation in the enzymes of lizards in relation to preferred body temperatures. *In* Molecular mechanisms of temperature adaptation (ed. C. L. Prosser), pp. 131–145. Am. Assoc. Advan. Sci. Publ. *84*.

———: 1968. Response of the thermal preferendum and heat resistance to thermal acclimation under different photoperiods in the lizard *Anolis carolinensis*. Am. Midland Naturalist *79*, 149–158.

———: 1969. Environmental control of annual testicular cycles in the lizard *Anolis carolinensis*. III. Temperature thresholds for photoperiodism. J. Expt. Zool. *172*, 311–322.

———: 1972. Environmental physiology of reptilian breeding cycles: role of temperature. Gen. Comp. Endocrin. Suppl. *3*, 477–488.

———: 1973. Environmental influences on the testis cycles of the lizards *Dipsosaurus dorsalis* and *Xantusia vigilis*. Comp. Biochem. Physiol. *45A*, 7–20.

———, Basu, S. L.: 1967. Influence of temperature on lizard testes. Nature *213*, 672–674.

———, Dawson, W. R., Shoemaker, V. H.: 1966a. Heat resistance of some Australian lizards. Copeia *1966*, 162–169.

———, Dawson, W. R., Shoemaker, V. H., Main, A. R.: 1966b. Observations on the thermal relations of Western Australian lizards. Copeia *1966*, 97–110.

———, Dawson, W. R., Shoemaker, V. H.: 1969. Thermal adjustments in cardiac and skeletal muscles of lizards. Z. Vergleich. Physiol. *65*, 1–14.

Lowe, C. H., Jr., Vance, V. J.: 1955. Acclimation of the critical thermal maximum of the lizard *Urosaurus ornatus*. Science *122*, 73–74.

McGinnis, S. M.: 1966. *Sceloporus occidentalis*: preferred body temperature of the western fence lizard. Science *152*, 1090–1091.

Maher, M. J.: 1961. The effect of environmental temperature on metabolic response to thyroxine in the lizard, *Lacerta muralis*. Am. Zoologist *1*, 461.

———: 1964. Metabolic response of isolated lizard tissues to thyroxine administered in vivo. Endocrinology *74*, 994–995.

———: 1965. The role of the thyroid gland in the oxygen consumption of lizards. Gen. Comp. Endocrin. *5*, 320–325.

———, Levedahl, B. H.: 1959. The effect of the thyroid gland on the oxidative metabolism of the lizard, *Anolis carolinensis*. J. Expt. Zool. *140*, 169–189.

Marion, K. R.: 1970. Temperature as the reproductive cue for the female fence lizard. Copeia *1970*, 562–564.
Mayhew, W. W.: 1968. Biology of desert amphibians and reptiles. *In* Desert biology (ed. G. W. Brown, Jr.), vol. I, pp. 195–356. New York: Academic Press.
Mellish, C. H.: 1936. The effects of anterior pituitary extract and certain environmental conditions on the genital system of the horned toad (*Phrynosoma cornutum* Harlan). Anat. Record *67*, 23–33.
Moberly, W. R.: 1968. The metabolic responses of the common iguana, *Iguana iguana*, to activity under restraint. Comp. Biochem. Physiol. *27*, 1–20.
Moll, E. O., Legler, J. M.: 1971. Life history of a neotropical slider turtle, *Pseudemys scripta* (Schoepff) in Panama. Bull. Los Angeles Co. Museum Nat. Sci. *11*, 1–102.
Morgareidge, K. R., White, F. N.: 1969a. Cutaneous vascular changes during heating and cooling in the Galápagos marine iguana. Nature *223*, 587–591.
———, White, F. N.: 1969b. Nature of heat induced cutaneous vascular response in *Iguana iguana*. Physiologist *12*, 306.
Mueller, C. F.: 1970. Energy utilization in the lizards *Sceloporus graciosus* and *S. occidentalis*. J. Herpetol. *4*, 131–134.
Norris, K. S.: 1967. Color adaptation in desert reptiles and its thermal relationships. *In* Lizard ecology: a symposium (ed. W. W. Milstead), pp. 162–229. Columbia, Mo.: Univ. Missouri Press.
Patterson, W. C., Evering, F. C., McNall, C. L.: 1968. The relationship of temperature to the cochlear response in a poikilotherm. J. Auditory Res. *8*, 439–448.
Porter, W. P.: 1967. Solar radiation through the living body wall of vertebrates, with emphasis on desert reptiles. Ecol. Monogr. *37*, 273–296.
———, Gates, D. M.: 1969. Thermodynamic equilibria of animals with environment. Ecol. Monogr. *39*, 245–270.
———, Norris, K. S.: 1969. Lizard reflectivity change and its effect on light transmission through the body wall. Science *163*, 482–484.
Pough, F. H.: 1969. Environmental adaptations in the blood of lizards. Comp. Biochem. Physiol. *31*, 885–901.
Regal, P. J.: 1966. Thermophilic response following feeding in certain reptiles. Copeia *1966*, 588–590.
———: 1967. Voluntary hypothermia in reptiles. Science *155*, 1551–1553.
Schmidt-Nielsen, K., Dawson, W. R.: 1964. Terrestrial animals in dry heat: desert reptiles. *In* Handbook of physiology, Sec. 4, Adaptation to the environment (ed. D. B. Dill), pp. 467–480. Washington, D.C.: Am. Physiol. Soc.
Shoemaker, V. H., Licht, P., Dawson, W. R.: 1966. Effects of temperature on kidney function in the lizard *Tiliqua rugosa*. Physiol. Zool. *39*, 244–252.
———, Licht, P., Dawson, W. R.: 1967. Thermal dependence of water and electrolyte excretion in two species of lizards. Comp. Biochem. Physiol. *23*, 255–262.
Spray, D. C., May, M. L.: 1972. Heating and cooling rates in four species of turtles. Comp. Biochem. Physiol. *41A*, 507–522.
Tait, N. N.: 1969. The effect of temperature on the immune response in cold-blooded vertebrates. Physiol. Zool. *42*, 29–35.
Templeton, J. R.: 1960. Respiration and water loss at the higher temperatures in the desert iguana, *Dipsosaurus dorsalis*. Physiol. Zool. *33*, 136–145.
———: 1970. Reptiles. *In* Comparative physiology of thermoregulation (ed. G. C. Whittow), vol. I, pp. 167–221. New York: Academic Press.
Tinkle, D. W., Irwin, L. N.: 1965. Lizard reproduction: refractory period and response to warmth in *Uta stansburiana* females. Science *148*, 1613–1614.

Tremor, J. W.: 1962. The critical thermal maximum of the iguanid lizard *Urosaurus ornatus*. Dissertation Abstr. *23*, 1462.

Tucker, V. A.: 1966. Oxygen transport by the circulatory system of the green iguana (*Iguana iguana*) at different body temperatures. J. Expt. Biol. *44*, 77–92.

———: 1967. The role of the cardiovascular system in oxygen transport and thermoregulation in lizards. *In* Lizard ecology: a symposium (ed. W. W. Milstead), pp. 258–269. Columbia, Mo.: Univ. Missouri Press.

Ushakov, B. P.: 1967. Coupled evolutionary changes in protein thermostability. *In* Molecular mechanisms of temperature adaptation (ed. C. L. Prosser), Am. Assoc. Adv. Sci. Publ. *84*, 107–129.

———, Darevskii, I. S.: 1960. Comparison of the heat tolerance of muscle fibers and behavior toward temperature in two sympatric species of semidesert lizards. Am. Inst. Biol. Sci. translation of Dokl. Akad. Nauk S.S.S.R. *128*, 770–773.

Vinegar, A., Hutchison, V. H., Dowling, H. G.: 1970. Metabolism, energetics, and thermoregulation during brooding of snakes of the genus *Python*. Zoologica *55*, 19–48.

Weathers, W. W.: 1970. Physiological thermoregulation in the lizard *Dipsosaurus dorsalis*. Copeia *1970*, 549–557.

———: 1971. Some cardiovascular aspects of temperature regulation in the lizard *Dipsosaurus dorsalis*. Comp. Biochem. Physiol. *40A*, 503–515.

———, White, F. N.: 1971. Physiological thermoregulation in turtles. Am. J. Physiol. *221*, 704–710.

Webb, G. J. W., Johnson, C. R., Firth, B. T.: 1972. Head-body temperature differences in lizards. Physiol. Zool. *45*, 130–142.

Werner, Y. L.: 1972. Temperature effects on inner ear sensitivity in six species of iguanid lizards. J. Herpetol. *6*, 147–177.

White, F. N.: 1970. Central vascular shunts and their control in reptiles. Federation Proc. *29*, 1149–1153.

Wilhoft, D. C.: 1966. The metabolic response to thyroxine of lizards maintained in a thermal gradient. Gen. Comp. Endocrin. *7*, 445–451.

———, Anderson, J. D.: 1960. Effect of acclimation on the preferred body temperature of the lizard, *Sceloporus occidentalis*. Science *131*, 610–611.

Wilson, K. J.: 1971. The relationships of activity, energy metabolism, and body temperature in four species of lizards. Clayton, Vic., Austral.: Monash Univ. Ph.D. diss.

———, Lee, A. K.: 1970. Changes in oxygen consumption and heart-rate with activity and body temperature in the tuatara, *Sphenodon punctatum*. Comp. Biochem. Physiol. *33*, 311–322.

26

Preferred Body Temperatures of Small Birds and Rodents: Behavioral and Physiological Determinations of Variable Set Points

J. Emil Morhardt

Introduction

The levels of body temperature (T_b) of homeotherms, and the amount of metabolically derived energy that homeotherms expend to prevent fluctuations of T_b, are of immense importance to overall energy balance. High T_b's may result in high levels of energy consumption simply because chemical reactions tend to proceed more rapidly at higher temperatures. High T_b's in the presence of low environmental temperatures (T_a) may require additional energy expenditures to compensate for loss of heat to the environment. For small homeotherms the problem of energy exchange with the environment is exacerbated by a variety of physical factors, all of which lead to a closer coupling of the animal to the effective T_a (see S. Morhardt and Gates, 1974; Porter and Gates, 1969; Heller and Gates, 1971).

At lower T_a's, the smaller the gradient between the T_b and the effective T_a, the less time and energy must a homeotherm expend in thermoregulatory activities. From the standpoint of saving energy, it is therefore most expedient for a homeotherm to have a T_b sufficiently close to the effective T_a that insulation allows the T_b to remain at the desired level while the metabolic rate remains at standard levels. This can be accomplished by keeping T_b constant and behaviorally selecting the combination of environmental and insulative factors that result in minimum energy expenditures, or by selecting a T_b which, for a particular environment, results in minimal energy expenditures, or by some combination of the two general methods.

Many homeotherms periodically either abandon temperature regulation (Heller and Hammel, 1972) or regulate their body temperatures at much lower levels than normal (Morhardt, 1970; Hudson, 1973), presumably to save energy and usually at the expense of any other activity. In this paper I shall discuss less profound fluctuations of T_b which occur many times a day in the normal lives of small homeotherms, but which nevertheless may be very important in terms of total energy expenditure. In particular, I shall discuss the concept of highly variable set points for T_b regulation.

The T_b set point (T_{set}) as used here is defined as that T_b threshold below which the metabolic rate increases to offset any further declines of T_b, or below which some behavioral means of thermoregulation is initiated.

First, I shall consider changes of T_b which are effected either behaviorally or physiologically but which, because they have changed in the absence of any externally imposed change in the thermal environment, suggest a change in the T_{set}. As a special case of these results, I shall include some data from my laboratory that show pronounced changes in abdominal temperature (T_{ab}) of rodents when they have behavioral control over the environmental temperature.

Second, I shall discuss some results of experiments in which the skin temperature (T_s) of small birds and rodents was manipulated in such a way that measurements of oxygen consumption ($\dot{V}_{O_2}$) revealed changes of T_{set} over relatively short periods of time.

Variations of Body Temperature in Small Homeotherms in the Absence of External Environmental Temperature Loads

There are many observations of diurnal variations of T_b in small homeotherms (see Folk, 1957; King and Farner, 1961; Hart, 1971; Dawson and Hudson, 1970; for reviews). Continuous observations of the hypothalamic temperatures (T_h) of monkeys (Hamilton, 1963), cats (Adams, 1963), and rats (Abrams and Hammel, 1964, 1965) show fluctuations of up to 2°C in the absence of strenuous activity. On close inspection, increases of T_b appear to occur when animals become alert and ready for activity and not necessarily as a result of the activity. In further support of this view, there is evidence that trained rats, when placed on an unmoving treadmill on which they were accustomed to running, underwent changes in colonic temperature from 38.2 to 39.4°C (Gollnick and Ianuzzo, 1968), and that there were rapid increases in the brain temperatures of free-moving rats upon the appearance of distinct synchronization of the regular hippocampal rhythm of the EEG normally associated with behavioral arousal (Kovalzon, 1973). All these increases in T_b occurred in the absence of any important change in the energy-exchange characteristics of the environment and probably resulted from increased $\dot{V}_{O_2}$. Either the T_{set} or the gain of the physiological temperature controller shifted upward, very likely due to an increase in the level of arousal of the animal.

Body Temperature Maintained by Behavioral Thermoregulation

Rats placed in a cold operant chamber in which infrared radiation was supplied when a lever was pressed increased their T_h initially and then maintained it with a certain amount of variability by altering bar-press rates (Carlisle, 1968).

We have placed shaved albino rats and ground squirrels (*Spermophilus beldingi*) in an operant chamber through which a cold (4°C) wind was blowing (8 mph). In this situation, if heat was not provided, T_{ab} measured by implanted telemetry transmitters fell rapidly. When the animals were trained to push a lever that heated the airstream, the initial response was a rapid increase in T_{ab}. The rate of increase was similar from day to day. The final T_{ab} reached varied from day to day and between animals. At night the T_{ab}'s were initially at higher levels, reflecting the diurnal variation in T_{ab}. The T_{ab}'s were always significantly higher after 30 min in the chamber than at the start. The mean levels of T_{ab} maintained at night were higher than during the day for all three rats (Fig. 26.1).

These data suggest that there is an upward shift of preferred T_b associated with the activity of bar pressing in the chamber, and that it is higher at night, reflecting

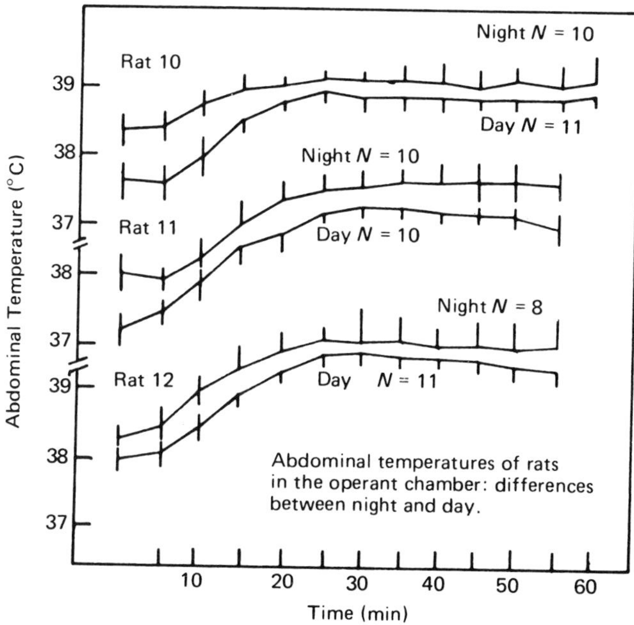

Fig. 26.1. Abdominal temperatures of three albino rats in a wind-tunnel chamber in which the rats controlled the T_a. At time = 0, the rats were removed from their cages and placed in the chamber. The vertical bars are the 95 percent confidence intervals of the mean temperatures at 5-min intervals on N different days or nights. Day runs were made between 0800 and 1130. Night runs were made between 2000 and 2330.

a circadian rhythm. The results are not conclusive because day-to-day variability within the animals is pronounced.

The observation that a particular T_b, for example T_{ab}, remains constant or changes from one level to another is not evidence that it is the regulated temperature (Mitchell *et al.*, 1972). If any T_b changes systematically, however, temperature regulation is proceeding in an altered manner, either as a result of changes of gain in the system or as a result of changes of T_{set}. In view of the corroborating evidence below, the change appears to be one of T_{set}.

An even more striking example of a shift in preferred T_b on being placed in the behavioral chamber was shown by the ground squirrels (Fig. 26.2). When placed in the chamber, they also manipulated the lever and increased the T_{ab} but tended to increase T_{ab} to higher initial levels than did the rats, and then, over the course of an hour or two, appeared to regulate the T_{ab} about a linearly decreasing T_{set}. The slope of the decline with time was similar from day to day within individuals.

In these experiments, if the animals continued to hold the lever down, the T_a

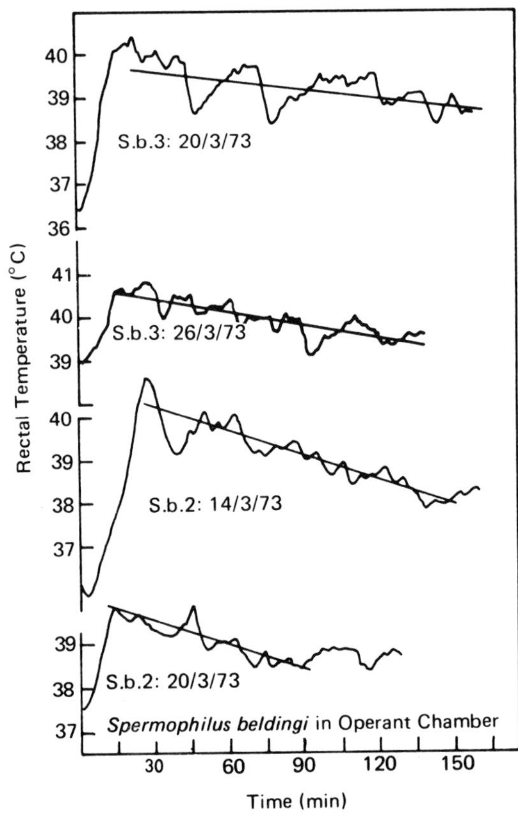

Fig. 26.2. Abdominal temperatures of two ground squirrels from the time they were placed in the wind-tunnel chamber. The variable lines are the abdominal temperatures at 1-min intervals. The straight lines are the least-squares regression lines with time on the X axis and T_{ab} on the Y axis.

rapidly reached 50°C, then more gradually approached 60°C. To thermoregulate, the animals had to continue pressing and releasing the lever. The average T_a maintained in the chamber by the rats was related to the T_{ab}, as shown in Fig. 26.3. Averages of T_a and T_{ab} for 5-min periods were determined by planimetry under three different conditions. In the first, T_{ab} was low, either because the rat had just been placed in the chamber or because the heaters had been turned off for a period of time. In the second, the T_{ab} was higher than usual because the heaters had been kept on even though the rat was off the lever. The third group was a control in which 5-min periods were picked subsequent to a period in which the T_{ab} had been steady for at least 10 min and the rat had been utilizing the lever regularly. The relationship between T_a and T_{ab} under these conditions is hyperbolic and may be interpreted as showing that if T_{ab} is too high, low air and skin temperatures will be endured until T_{ab} falls to its normal range. If T_{ab} is below the preferred level, T_a will be kept above, but not much above, the preferred T_{ab} until the preferred level is reached. Once the preferred T_{ab} is reached, the T_a is maintained slightly below T_{ab} and tends to become more variable.

These data suggest that when rats are placed in a situation in which the amount of metabolic energy required to maintain T_{ab} constant is the same whether the T_{ab}

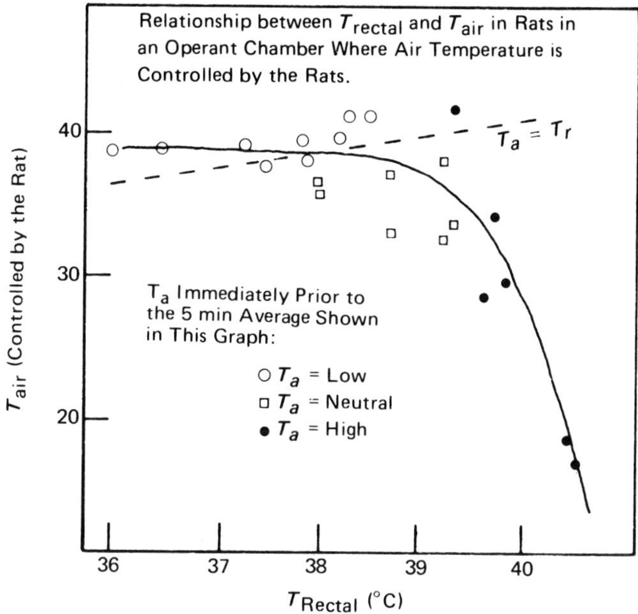

Fig. 26.3. Average T_a and T_{ab} from albino rats in the wind-tunnel chamber. Each point is the 5-min average of T_a and T_{ab} following an experimental manipulation. Open circles represent periods following inactivation of the heaters and consequent forcing of the T_{ab} to values lower than those preferred. Closed circles represent periods following elevation of T_a by the experimenter and consequent elevation of the T_{ab} above preferred levels. Open squares are control periods following a period of regulation of T_a by the animals. The solid hyperbolic line is a third-order least-squares regression. The straight dashed line indicates equivalence of T_a and T_{ab}.

is high or low, they still raise the T_{ab} significantly above the levels found in resting animals. This indicates a shift upward of T_{set} independent of significant exercise. The data in Fig. 26.3 are particularly interesting because they strongly link the behavorial response to an internal temperature. It is unfortunate that we were unable to measure average T_s or subcutaneous temperature accurately since other investigators have linked behavioral thermoregulatory responses to these variables (Lipton *et al.*, 1970; Weiss and Laties, 1961). We did not wish to attach thermocouples, thereby restraining the animals, but we did attempt infrared thermometry of the shaved skin surface. Unfortunately, the temperature of the skin surface varied widely at different sites, particularly in the presence of forced convection, and we had no confidence in our ability to estimate average T_s this way. The infrared thermometry did indicate, however, that the skin surface temperature paralleled changes of T_a, and we felt that average T_a served as an index of T_s.

Physiological Evidence for Variable T_{set}

Physiological responses are often easier to quantify than behavioral ones, particularly when they are of a transient nature. Additionally, when it is necessary to restrain an animal so that a critical temperature can be measured or controlled, it may be better to rely on responses that are not under conscious control and are less likely to be interfered with by restraint.

We have chosen to control the T_s of small rodents and birds by partial immersion in a stirred water bath. We have immersed juncos (*Junco hyemalis*), ground squirrels (*Spermophilus beldingi*), albino rats, and two genera of pocket mice (*Heteromys desmarestianus* and *Liomys salvini*) to their necks in a water bath and have manipulated bath temperature (designated T_s on the assumption that bath temperature is very close to skin temperature) while observing rectal or cloacal temperature (T_r) and $\dot{V}_{O_2}$. $\dot{V}_{O_2}$ was measured by drawing air past the animal's head and measuring the percentage of oxygen remaining. The birds were restrained with a piece of fish net attached to the underside of a board through which only their heads protruded. The rodents were restrained in hardware-cloth cylinders ($\frac{1}{2}$-in. mesh), their heads through a hole in a plastic plate at the top. Once installed in these holders and lowered into the water bath, most of the animals became calm. Two basic experimental approaches were used in these studies. The first, termed a steady-state experiment, involved starting with the T_s high, and then lowering it in 1 or 2°C increments, waiting after each incremental drop for T_r and $\dot{V}_{O_2}$ to come to equilibrium. This protocol is similar to that followed in determining the effects of T_a on $\dot{V}_{O_2}$ except that here T_s is controlled more or less directly rather than through an insulating layer of air trapped in fur or feathers. The second type of experiment involved rapid alterations of T_s timed to result in a variety of combinations of T_s and T_r.

The steady-state experiments cannot distinguish between the relative influences of T_s and T_r on the final $\dot{V}_{O_2}$ since T_s and T_r do not vary independently. The results are interesting, however, because there are clear thresholds for both T_s and T_r

(designated T_{set_s} and T_{set_r}) analogous to the lower critical temperatures of animals during similar experiments done in air. There are occasionally large differences of these set points for the same animal on different days. Figure 26.4 shows such a T_{set} change for a junco. This is the most striking example so far observed, and it serves to illustrate the nature and importance of set point changes to overall energy metabolism. There was a 4°C difference in skin T_{set} (T_{set_s}) between the two days. On the first day, a skin temperature of 36°C resulted in a V_{O_2} more than twice the lowest observed. On the second day, a T_s of 36°C did not cause an increase of $\dot{V}_{O_2}$ above standard levels. Once this T_{set_s} threshold was reached, the metabolic response to further decreases of T_s proceeded similarly on both days. The slopes of the regressions of $\dot{V}_{O_2}$ on T_s (designated β_e to indicate that T_r and T_s are in equilibrium and did not vary independently as would be implied by the use of β alone) are not very different between the two days. This means that the large differences in $\dot{V}_{O_2}$ between the two days at any T_s below 40°C are due to a change in T_{set}. Changes in T_{set} can result in large changes of $\dot{V}_{O_2}$, even though at the time of the T_{set} change there is no change in T_b. The magnitude of a $\dot{V}_{O_2}$ change is identical whether the driving force is a change of T_{set} or an actual change in T_b. The magnitude of the change is inversely proportional to the slope of the regression of $\dot{V}_{O_2}$ on T_b. The T_b's measured in these experiments were T_s and T_r. The slopes of the regressions of $\dot{V}_{O_2}$ on T_s and T_r are β_e and α_e, respectively, and differ between species. Samples of these equilibrium regressions for juncos, pocket mice, and rats are shown in Figs. 26.5 and 26.6 and average values of β_e, α_e, T_{set_s} and T_{set_r} are shown in Table 26.1. The values in Table 26.1 were calculated for each animal by the method shown in Fig. 26.7. Fifth-order regression lines were drawn through 6-min means of T_s

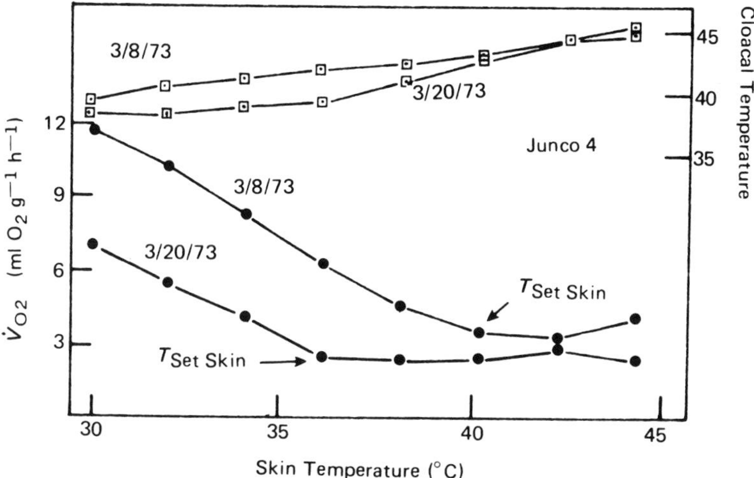

Fig. 26.4. Steady-state type of experiments from a junco on two different days. The bird was immersed to the neck in a water bath. The open squares are the cloacal temperatures. The filled circles are the oxygen consumption. Each filled circle is a 6-min average of $\dot{V}_{O_2}$. The cloacal temperature changed less than 0.1°C during each 6-min period of measurement. The 95 percent confidence intervals of each of the data points shown on the figure are very narrow and are within the confines of the symbol.

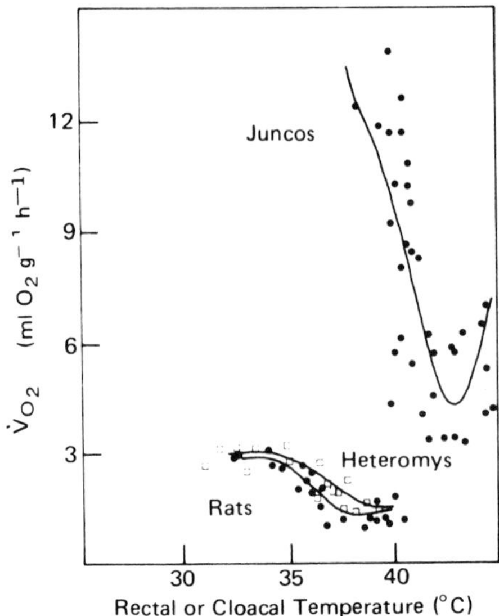

Fig. 26.5. Sample regressions of $\dot{V}_{O_2}$ on rectal or cloacal temperature from 4 juncos, 4 albino rats, and 4 pocket mice. The data for each species were pooled and each regression line is a fifth-order least-squares regression line fitted by computer. The upper closed circles are junco data, the lower closed circles are rat data, and the open squares are pocket mouse data.

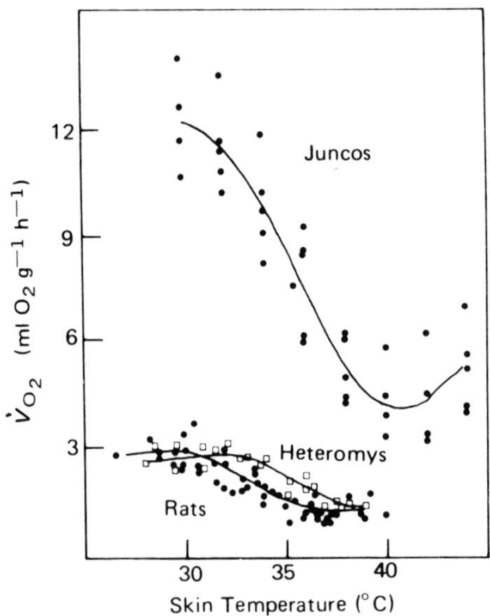

Fig. 26.6. Sample regressions of $\dot{V}_{O_2}$ on skin temperature. The animals and symbols are from the same experiments shown in Fig. 26.5.

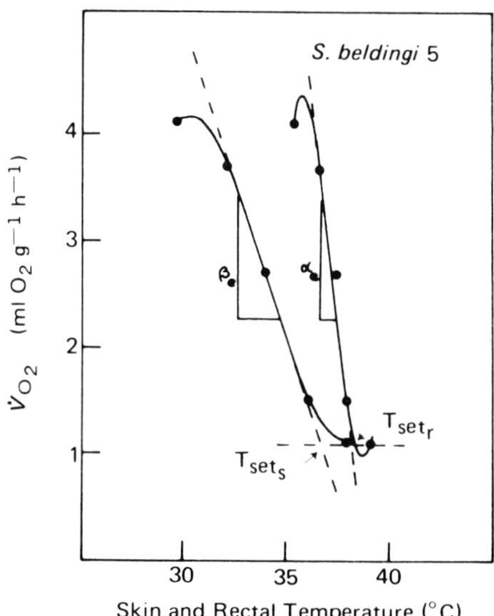

Fig. 26.7. Method for determining the set points and proportionality constants in Table 26.1. A fifth-order least-squares regression line is fitted to 6-min equilibrium values of $\dot{V}_{O_2}$ plotted against equilibrium skin and rectal temperatures. A straight line is drawn through the linear portions of the regression and the slope of the straight line is the proportionality constant α_e or β_e. T_{set} is the temperature at which this straight line intersects the lowest observed 6-min average $\dot{V}_{O_2}$.

and T_r at equilibrium and a straight line was drawn through the linear portions of the regressions. Although these experiments do not allow one to distinguish whether the driving force for $\dot{V}_{O_2}$ is T_s, T_r or some other T_b, they do show that over a range of T_s, there is a tendency to keep T_r above and independent of T_s by increased $\dot{V}_{O_2}$. The degree to which this metabolic effort succeeds is shown in Fig. 26.8, in which T_r is plotted as a function of T_s. The juncos are far more successful at keeping T_r

Table 26.1. Slopes of the Regressions of $\dot{V}_{O_2}$ on T_s, (β_e), and on T_r, (α_e), and the Intersections of These Regressions with Basal $\dot{V}_{O_2}$ (T_{set_s} and T_{set_r}) as Determined by the Method Shown in Fig. 26.8[a]

		Skin		Core	
Species	N	β_e (ml O_2 g^{-1} h^{-1} °C^{-1})	T_{set_s} (°C)	α_e (ml O_2 g^{-1} h^{-1} °C^{-1})	T_{set_r} (°C)
Rats	6	−0.36 ± 0.01	35.6 ± 0.4	−0.65 ± 0.02	37.7 ± 0.1
Heteromys	5	−0.39 ± 0.05	36.7 ± 1.0	−0.49 ± 0.11	37.9 ± 1.0
Juncos	6	−1.31 ± 0.26	39.2 ± 1.5	−5.50 ± 1.50	41.5 ± 0.8

[a] The values following the ± are the 95% confidence intervals, $t_{0.05}s_x$. N is the number included, one experiment on each animal.

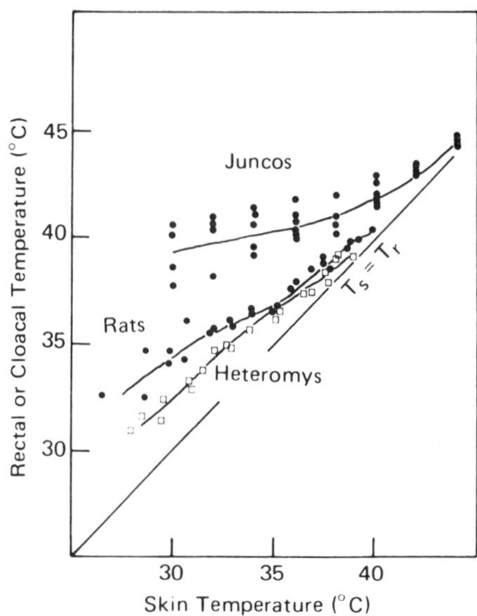

Fig. 26.8. Sample regressions of rectal or clocal temperature on skin temperature. The animals and symbols are from the same experiments shown in Fig. 26.5.

independent of T_s than are the rats and pocket mice. This is not entirely due to greater insulation (although that may play a minor part, even in the water bath) but rather to much greater increases in heat production in response to lowering of T_b, or in other words, much larger values of α_e or β_e. The α_e and β_e values for juncos are 10 and 4 times the respective values for rodents.

It is interesting that the animals could have kept T_r more independent of T_s than they did by having higher values of α_e and β_e up to the point that maximum $\dot{V}_{O_2}$ was reached. There is no obvious reason why they did not. The explanation may be that the physiological temperature controller responds in a manner appropriate to normal values of insulation in air but not of sufficient magnitude to compensate for the rate of heat loss in a water bath. In the air, some species appear to have a sufficiently high β_e to cause T_r to go up rather than down as T_a is lowered (Gonzalez *et al.*, 1971); others have a lower T_r at equilibrium as T_s is lowered (Stitt and Hardy, 1971). It is likely that in water, with its much higher conductivity than air, the gain of the control pathways is insufficient to offset the heat loss.

Oddly enough, the individual animal that showed nearly perfectly stable T_r in these steady-state experiments was a ground squirrel of the species which had highly variable T_r's when it had control of the T_a in the behavioral experiments. Figure 26.9 is a tracing of a steady-state experiment on *S. beldingi* no. 5. The value of β_e for this animal was -0.29, less even than that for the rats. In this case the great stability of T_r was achieved not only by increases of $\dot{V}_{O_2}$, but also by pronounced peripheral vasoconstriction, leaving the lower portion of the body cool while the T_r measured near the liver remained high and constant. At the end of the run when T_s was raised, T_r fell suddenly, indicating a partial resumption of perfusion of peripheral tissues.

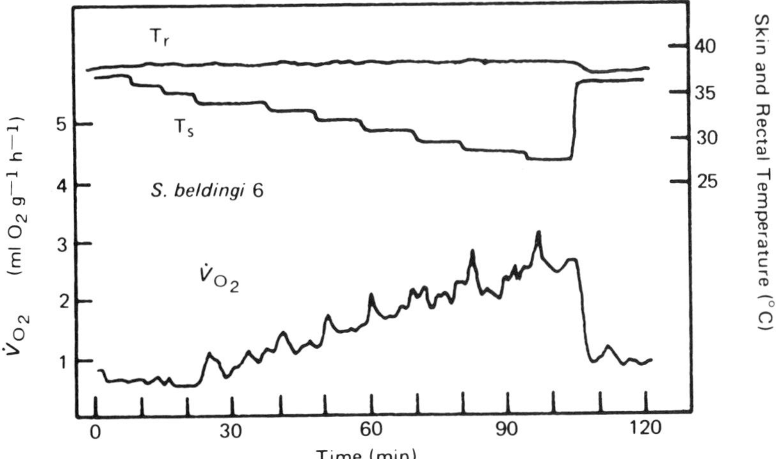

Fig. 26.9. Sample steady-state experiment from a ground squirrel. The T_r remains much more independent of the T_s than in any other animal examined. Note the small spikes of $\dot{V}_{O_2}$ each time the T_s is lowered. These represent responses to rate of change of T_s.

Although these experiments show day-to-day changes in set points, they are not capable of showing changes of T_{set} within a single experiment over a short period of time. Figure 26.10A is a tracing of an experiment on a junco which shows a change of T_{set} during the course of a 90-min experiment. The T_{set} shift shown here is largely one of T_s and constitutes a shift upward of T_{set_s} from 40.2°C at the beginning of the experiment to 43.0°C at the end. In Fig. 26.10A at the start (1), $\dot{V}_{O_2}$ was at the lowest level normally observed in juncos, indicating that T_b must have been above T_{set}. When T_s was lowered abruptly, there was an immediate increase in $\dot{V}_{O_2}$ before there was much decrease in T_r, suggesting that this $\dot{V}_{O_2}$ increase was driven by the skin thermoreceptors. Furthermore, as T_r fell from 42 to 41°C, there was no increase in $\dot{V}_{O_2}$, as if T_r were not important in driving $\dot{V}_{O_2}$. Upon return of T_s to its initial level, $\dot{V}_{O_2}$ rapidly returned toward its initial level but then stabilized at a level somewhat higher than previously (3), despite the fact that T_r and T_s were similar to their initial values. When T_s was again lowered by the same amount, $\dot{V}_{O_2}$ again increased, but the level (4) was higher than before (2) for the same combination of T_s and T_r, and as T_r fell to a slightly lower level than previously, $\dot{V}_{O_2}$ increased proportionally to the decline of T_r as though T_{set_r} had been passed at 41°C. Increase of T_s to its initial level resulted again in a higher $\dot{V}_{O_2}$ (5), suggesting that T_{set_s} shifted upward and that the T_s at (5) was below T_{set_s}. To validate this possibility, T_s was raised and the response was an immediate decline of $\dot{V}_{O_2}$ to basal levels (6) followed by an increase when T_s was again lowered to the initial level. In Fig. 26.10B, the average levels of $\dot{V}_{O_2}$ in (1)–(7) are plotted against the T_s at the time they were taken and the picture of a shifting T_{set_s} emerges. The solid line (1)–(2) is the original regression of $\dot{V}_{O_2}$ on T_s. The dashed line (2)–(3)–(4) indicates the intermediate excursions and the line (4)–(5)–(6)–(7) is the sequence which validates the shift of T_{set_s}. There is very little change of β between

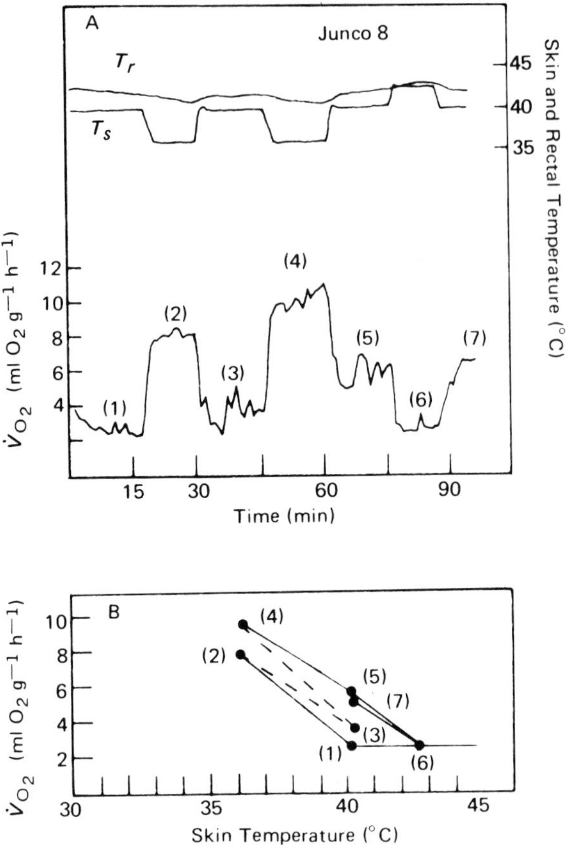

Fig. 26.10. A. Experiment using a junco immersed to the neck in the water bath. The numbers in parentheses are referred to in the text and identify portions of the experiment. Mean values of T_s and $\dot{V}_{O_2}$ corresponding to each of these numbers are plotted in Fig. 26.10B. B. Values of $\dot{V}_{O_2}$ from Fig. 26.10A plotted as functions of T_s. The numbers represent sequential steps in the experiment. The line (1)–(2) is the initial regression establishing β and T_{set}. The dashed line (2)–(3) shows intermediate excursions, and the final line (4)–(7) is the final regression showing a change in T_{set_s} but little change in β.

(1)–(2) and (4)–(6), but there is a change of T_{set_s} from 40.2 to 43°C. T_r appears to play very little part in this sequence since T_r appears to be usually above T_{set_r}. In this junco, T_s appears to be able to drive $\dot{V}_{O_2}$ even when T_r is above T_{set_r}.

In a similar experiment on a pocket mouse (Fig. 26.11A), the influence of T_r is very obvious, and in the pocket mouse T_s does not appear to drive $\dot{V}_{O_2}$ if T_r is greater than T_{set_r}. Initially, $\dot{V}_{O_2}$ was near its standard value. The abrupt drop of T_s did not result in any change in $\dot{V}_{O_2}$ for almost 6 min, until T_r fell below T_{set_r} at 36.2°C at (1); then $\dot{V}_{O_2}$ increased at a rate approximately proportional to the further decrease in T_r, leveling as T_r leveled at (2). An increased T_s caused a rapid decline in $\dot{V}_{O_2}$ to basal levels, even though T_r was still low. When T_r was again above the previous T_{set_r} of 36.2°C, T_s was lowered and again there is a delay while T_r fell, this

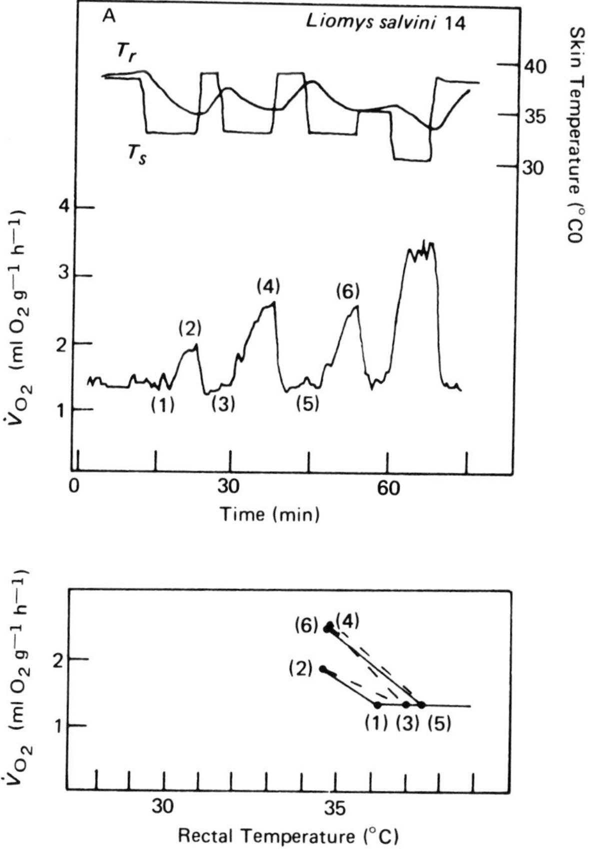

Fig. 26.11. A. Experiment using a pocket mouse immersed to the neck in a water bath. The numbers in parentheses are referred to in the text and identify portions of the experiment. Mean values of T_r and $\dot{V}_{O_2}$ corresponding to each of these numbers are plotted in Fig. 26.11B. B. Values of $\dot{V}_{O_2}$ from Fig. 26.11A plotted as functions of T_r. The line (1)–(2) established the original regression, α, and the original T_{set_r}. The dashed line (2)–(3)–(4) shows intermediate excursions, and the line (5)–(6) is the final regression with α similar to the first, but T_{set_r} at a higher level.

time at T_{set_r} of 37.0°C, before there was an increase in $\dot{V}_{O_2}$. Again the increasing $\dot{V}_{O_2}$ was proportional to the decline in T_r, but this time it leveled at a much higher $\dot{V}_{O_2}$ at (4). A return of T_s to the original level resulted in an immediate return to basal $\dot{V}_{O_2}$ although T_r was still low, and a third lowering of T_s produced a response similar to the first two except that T_{set_r} was 0.5°C higher this time at 37.5°C. When T_s was raised to only 34.7°C, this was enough to cause a return of $\dot{V}_{O_2}$ to basal levels, indicating that T_{set_s} was independent of T_r and was between 32.5 and 34.6°C. A subsequent lowering of T_s to 29.9°C resulted in a $\dot{V}_{O_2}$ higher than previously observed, further confirming that there was a proportional response to changes of T_s below T_{set_s}. In this final manipulation, T_r fell substantially, to 33°C, but the decline had little effect on $\dot{V}_{O_2}$. The analytical expression for this kind of response to T_r and T_s is of the form

$$\dot{V}_{O_2} - \dot{V}_{O_2\text{basal}} = \alpha(T_{\text{set}_r} - T_r)\,\beta(T_{\text{set}_s} - T_s) > 0$$

One of the more interesting implications is that if either T_r or T_s is above its respective T_{set}, the metabolic rate will remain at basal levels regardless of how low the other is driven, a result observed in Fig. 26.11.

Figure 26.11B is an analysis of the *Liomys* data done in a way similar to that for the junco in Fig. 26.10B but using as basal $\dot{V}_{O_2}$ the rate immediately prior to the time T_{set_r} was reached. The shift upward of T_{set_r} in Fig. 26.11B is similar to the shift upward of T_{set_s} in the junco, and as in the junco, there is little or no change in the slope, α, but only in the T_{set_r}.

Conclusions

The fact that the T_{ab} selected behaviorally initially increases in a predictable manner in rats and squirrels and later decreases as a function of time in squirrels is strongly indicative of an upward shift of T_{set} during increased alertness.

The upward shift in T_{set_s} in juncos and T_{set_r} in pocket mice during transient changes of T_s likewise suggests that some input other than temperature is raising these thermoregulatory thresholds. The fact that the proportionality constants α and β do not change much during these experiments supports the hypothesis that the increased $\dot{V}_{O_2}$ results from a change of T_{set} rather than the change of gain observed in studies on human subjects before and after exercise (Mitchell *et al.*, 1972).

The physiological significance of these data is that the great variability of T_b of small birds and mammals can be explained, at least partially, by changes in the set point, or preferred level of body temperature. Low open-loop gain of the thermoregulatory pathways (Heath *et al.*, 1972) is not solely responsible.

A consequence of shifts of set point important to ecologists is that much metabolic energy may be utilized by small animals adjusting the T_b to a variable T_{set} in the absence of any change in the external environment. At the same time, much energy may be saved by having a low preferred T_b when inactive and unalert.

Acknowledgments

The original work presented in this paper was done in conjunction with a number of people: the behavioral experiments using rats, with William H.-K. Chen and Jonathan Kessel; the behavioral experiments using ground squirrels, with Patrick Molt; the water-bath experiments on juncos, with James A. McCrum and Charles Miller; and the water-bath experiments on pocket mice, with Theodore

H. Fleming. The preparation of this manuscript and some of the work reported herein was supported by a Faculty Summer Fellowship from Washington University. H. C. Heller kindly read the manuscript and suggested several changes.

References

Abrams, R., Hammel, H. T.: 1964. Hypothalamic temperature in unanesthetized albino rats during feeding and sleeping. Am. J. Physiol. *206*, 641–646.

———, Hammel, H. T.: 1965. Cyclic variations in hypothalamic temperature in unanesthetized rats. Am. J. Physiol. *208*, 698–702.

Adams, T.: 1963. Hypothalamic temperature in the cat during feeding and sleep. Science *139*, 109–610.

Carlisle, H. J.: 1968. Peripheral thermal stimulation and thermoregulatory behavior. J. Comp. Physiol. Psychol. *66*, 507–510.

Dawson, W. R., Hudson, J. W.: 1970. Birds. *In* Comparative physiology of thermoregulation (ed. G. C. Whittow), vol. I. New York: Academic Press.

Folk, G. E., Jr.: 1957. Twenty-four hour rhythms of mammals in a cold environment. Am. Naturalist *91*, 153–166.

Gollnick, P. D., Ianuzzo, C. D.: 1968. Colonic temperature response of rats during exercise. J. Appl. Physiol. *24*, 747–750.

Gonzalez, R. R., Kluger, M. J., Hardy, J. D.: 1971. Partitional calorimetry of the New Zealand white rabbit at temperatures 5°C-35°C. J. Appl. Physiol. *31*, 728–734.

Hamilton, C. L.: 1963. Interactions of food intake and temperature regulation in the rat. J. Comp. Physiol. Psychol. *56*, 476–588.

Hart, J. S.: 1971. Rodents. *In* Comparative physiology of thermoregulation (ed. G. C. Whittow), vol. II. New York: Academic Press.

Heath, J. E., Williams, B. A., Mills, S. H., Kluger, M. J.: 1972. The responsiveness of the preoptic-anterior hypothalamus to temperature in vertebrates. *In* Hibernation and hypothermia, perspectives and challenges (ed. F. E. South, J. P. Hannon, J. R. Willis, E. T. Pengelley, N. R. Alpert). Amsterdam: Elsevier

Heller, H. C., Gates, D. M.: 1971. Altitudinal zonation of chipmunks (*Eutamias*): energy budgets. Ecology *52*, 424–433.

———, Hammel, H. T.: 1972. CNS control of body temperature during hibernation. Comp. Biochem. Physiol. *41A*, 349–359.

Hudson, J. W.: 1973. Torpor. *In* Comparative physiology of thermoregulation (ed. G. C. Whittow), vol. III. New York: Academic Press.

King, J. R., Farner, D. S.: 1961. Energy metabolism, thermoregulation and body temperature. *In* Biology and comparative physiology of birds (ed. A. J. Marshall), vol. II. New York: Academic Press.

Kovalzon, V. M.: 1973. Brain temperature variations during natural sleep and arousal in white rats. Physiol. Behavior *10*, 667–670.

Lipton, J. M., Avery, D. D., Marotto, D. R.: 1970. Determinants of behavioral thermoregulation against heat: thermal intensity and skin temperature levels. Physiol. Behavior *5*, 1083–1088.

Mitchell, D., Atkins, A. R., Wyndham, C. H.: 1972. Mathematical and physical models of thermoregulation. *In* Essays on temperature regulation (ed. J. Bligh, R. Moore). New York: American Elsevier.

Morhardt, J. E.: 1970. Body temperature of white-footed mice (*Peromyscus* sp.) during daily torpor. Comp Biochem. Physiol. *33*, 423–439.

Morhardt, S. S., Gates, D. M.: 1974. Energy exchange analysis of the Belding ground squirrel and its habitat. Ecol. Monogr. *44*, 17–44.

Porter, W. P., Gates, D. M.: 1969. Thermodynamic equilibria of animals with the environment. Ecol. Monogr. *39*, 245–270.

Stitt, J. T., Hardy, J. D.: 1971. Thermoregulation in the squirrel monkey (*Saimiri sciureus*). J. Appl. Physiol. *31*, 48–54.

Weiss, B., Laties, V. G.: 1961. Behavioral thermoregulation. Science *133*, 1338–1344.

PART VI

Energy-Transfer Studies of Animals

Introduction

David. M. Gates

A great deal of experimental work is needed with various kinds of animals before our understanding of animal energetics is very good. It has been difficult to demonstrate the precise application of the theory of heat transfer to animals because of a lack of adequate experimental data. In addition, much of the experimental information currently in the literature is piecemeal, and generally one piece of data is unrelated to another piece of information even concerning the same animal. Now needed are very thorough studies of a few animals. However, it is critical that the theory of energy balance and heat transfer for an animal be formulated and used to guide the experimentalist. Only then will the data gathered in the laboratory and in the field concerning a particular animal be coherent and useful in terms of a unifying theory which, we may hope, will be able to predict animal behavior and response to environmental factors.

Matthew Kluger has studied the rabbit for several years and made notable contributions toward an understanding of its heat regulation and thermal physiology. The rabbit is a convenient animal for study because of its relatively docile behavior, and is important because of its widespread distribution throughout the world. Kluger is interested in particular in the mechanisms of thermal regulation by the New Zealand white rabbit to varying ambient temperatures and describes this work in this section. It is important for our understanding of energy exchange and temperature control in animals to separate the behavioral thermoregulatory response from the physiological thermoregulatory mechanisms. Kluger's experiments give us insight into this important problem.

The white-tailed deer is another animal being studied as thoroughly as possible. Aaron Moen first became interested in the energetics and heat budget of the deer while he was a graduate student at the University of Minnesota. This was at a time when I gave my first series of lectures on biophysical ecology at Minnesota, published in 1962 as *Energy Exchange in the Biosphere.* Also at the University of Minnesota at that time was Richard Birkebak, a graduate student in mechanical engineering, and a research group at the University Museum with a strong interest in animal microclimates and energetics. The University of Minnesota has always had one of the finest groups of heat-transfer engineers in the world. All these factors made it extremely propitious for Aaron Moen to tackle the difficult study of the energy budget of deer in the field. Since going to Cornell University, he has continued this intensive interest in deer and has established an entire laboratory for this purpose. Moen and Nadine Jacobsen describe their laboratory research in this paper. Moen is spending this winter at the University of Minnesota Forestry and Biological Station on Lake Itasca to study the deer in the field. Here again is an excellent example of the progress that can be made when laboratory experiments and field observations are closely coordinated and theoretical analysis is an integral part of the program.

Richard Birkebak has continued his interest in the heat exchange by animals. His career is an outstanding example of the success that can be achieved by persons thoroughly trained in mechanical and heat-transfer engineering when they apply their skills to a field such as animal ecology. Because of his rigorous training in mathematics and engineering, he is able to do the difficult theoretical analysis involved in the heat-transfer problems of animals and at the same time can design sophisticated laboratory and field experiments. Several years ago Birkebak did the first analysis of heat transfer for an animal using a "distributed-parameter" model in contrast to the "lumped-parameter" models we have had to use more generally. He applied this analysis to birds. Recently Birkebak has done extensive field work on the flight habits and energetics of the birds on Guam.

Animal fur and feathers play a very critical role in the heat exchange between an animal and its environment. What at first appears to be a fairly simple situation involving thermal insulation by fur or feathers soon is seen as an extremely complex problem in heat transfer because of the simultaneous interaction of thermal conduction, radiative transfer, and moisture exchange. When one begins to question the role of fur or feathers in the thermal ecology of an animal, it quickly becomes apparent that very detailed studies are required to understand the mechanisms involved. It is not sufficient to do only experiments to explain the process of heat transfer in fur because of the complexity of the problem. Theoretical analysis of considerable difficulty is required, combined with careful laboratory measurements. Here Birkebak and one of his former students describe the work they have done on heat transfer in fur. This is followed by another important contribution by the consortium of engineers and biologists at the University of Wisconsin. Warren Porter, trained in biology and animal ecology at the University of California in Los Angeles, had sufficient mathematics and physics in his education to allow him to do theoretical analysis and particularly to appreciate its importance to animal ecology. Porter spent two years with me at the Missouri Botanical Garden, during which time we did our first work on the thermodynamics of animals and originated the climate-space diagram. At the University of Wisconsin, where the Biotron, a superb controlled-environment facility, makes it possible to reproduce in the laboratory the habitat conditions that exist for any particular animal in the field, Porter has stimulated the interest of engineers in this subject. Engineers Dale J. Skuldt, William Beckman, and John Mitchell have collaborated with Warren Porter to work on the heat-transfer process in fur, and this important work is described in their paper.

The methods of biophysical ecology are just as applicable to aquatic ecosystems as they are to terrestrial situations. In certain respects the aquatic environment is somewhat simpler than the terrestrial, and in other respects it is more complex. An organism in the aquatic situation is strongly coupled by conduction to its environment; radiative exchange of energy is small or nonexistent; and evaporative exchange does not occur. However, the chemical environment of aquatic ecosystems is often extremely complex. The problems of aquatic ecology are awaiting a strong application of biophysical methodology. An intermediate situation recently captured our attention concerning the ecology of the intertidal zone. Samuel Johnson joined our group last year after completing his Ph.D. degree at the Hopkins Marine Laboratory of Stanford University. Johnson was persuaded that the ecology of the limpet would be a challenging problem in biophysical ecology. After a visit to the coast, where I had a good opportunity to see limpets in their natural habitat, I became convinced of the excitement and importance of a study of their energy budgets and microclimatology. The limpets exist on the rocks of the intertidal zone and exhibit a fairly striking preference for certain habitat conditions. The preferences differ among species. Here on the rocks are sloping surfaces with all the aspects of any hill or mountain, e.g., east-facing, south-facing, horizontal, vertical, etc. The limpets show preference for position above mean sea level. All this habitat selection occurs on quite a small scale and is thereby readily amenable to measurement. In his paper here, Johnson describes some of his studies of the limpet and its habitat, but the full energy-budget analysis is still being worked out. Nevertheless, from the discussion in his paper, one can appreciate the many details that must be dealt with to complete the analysis and interpretation of the energy budget of the limpet.

27

Energy Balance in the Resting and Exercising Rabbit

MATTHEW J. KLUGER

Introduction

During the past several years, my colleagues and I have been investigating the thermoregulatory responses of the New Zealand white rabbit (*Oryctolagus cuniculus*) to various environmental and internal temperature stresses. Areas of thermal sensitivity (both peripheral and internal), as well as the heat-dissipating and -conserving responses initiated by stimulation of these areas, have been studied. The research described here considers those mechanisms that enable the rabbit to regulate its body temperature over a wide range of ambient temperatures while resting, or during and following exercise.

Energy Balance of Rabbits at Rest

In a partitional calorimetric study, the physiological thermoregulatory responses of resting rabbits (acclimated to room temperature) in ambient temperatures of 5–35°C were measured (Gonzalez *et al.*, 1971). In this study each rabbit was restrained so as to eliminate behavioral thermoregulatory responses such as selecting a more comfortable microclimate, huddling (which decreases the exposed surface area), and licking of the forelimbs (which increases evaporative heat loss). Measurements were taken over a 30-min period of steady-state heat production as determined by a rectal temperature that varied no more than O.1°C. In an ambient temperature of 5–30°C, rectal temperature remained relatively constant, with a slight increase at the lower ambient temperatures of 10°C and 5°C. Mean skin temperature, as expected, varied directly with ambient temperature (Fig. 27.1). Since core temperature changed little over these ranges of ambient temperature and yet the metabolic heat production increased in the cold and peripheral blood flow

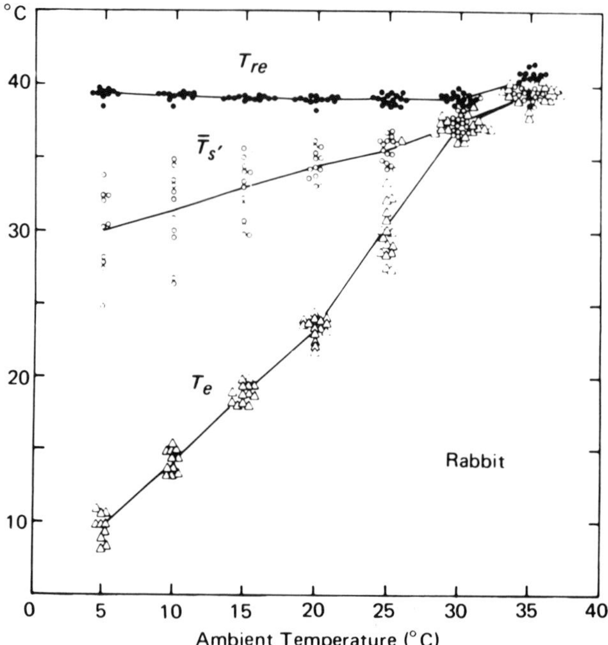

Fig. 27.1. Plot of rectal temperature (T_{re}), mean skin temperature excluding ear temperature ($\overline{T}_s'$), and ear skin temperature (T_e) against ambient temperature in resting New Zealand white rabbits. (From Gonzalez *et al.*, 1971.)

(skin conductance) and respiratory rate increased in the heat, it was concluded that these thermoregulatory responses were initiated by peripheral thermal receptors (Figs. 27.2–27.4).

As observed in Figs. 27.1–27.4, under these experimental conditions, the rabbit is in thermoneutrality in an ambient temperature of about 20–30°C. Its metabolic heat production is minimal at these temperatures, and core temperature is maintained constant by varying its vasomotor tone ("zone of vasomotor regulation"). As ambient temperature approaches or goes above 30°C, the rabbit's skin conductance rises rapidly and remains elevated, indicating that it is peripherally vasodilated. Above an ambient temperature of 30°C, the rabbit pants with a respiratory rate of over 200 breaths per minute. As can be observed in Fig. 27.1 and partially obscured by Fig. 27.4, most of the vasomotor control in the rabbit exists in its vascular and sparsely furred ears. Mean skin temperature, excluding the ears, increases relatively linearly with increasing ambient temperature. Ear skin temperature, however, shows a sharp break in the slope of ear skin temperature versus ambient temperature above 20°C, corresponding to the large increase in ear blood flow. (For an indication of the magnitude of the temperature changes and heat loss from the ears of the rabbit when core temperature is raised, see Fig. 27.7.) Below this zone of thermal neutrality, the rabbit increases its metabolic heat production and respiratory rate and skin conductance becomes minimal and constant (Figs. 27.2–27.4).

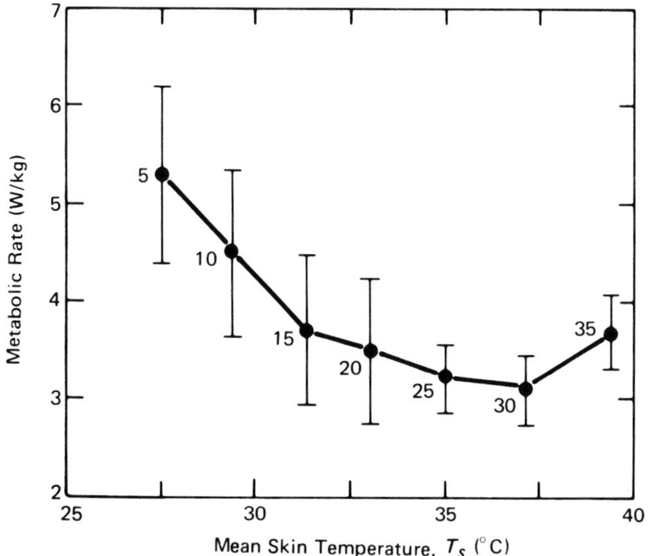

Fig. 27.2. Metabolic heat production in Wkg^{-1} (mean ± S.D.) versus mean skin temperature (T_s) in resting New Zealand white rabbits. Numbers to the left of each data point are equal to the ambient temperature (from 5 to 35°C). (Based on Gonzalez *et al.*, 1971.)

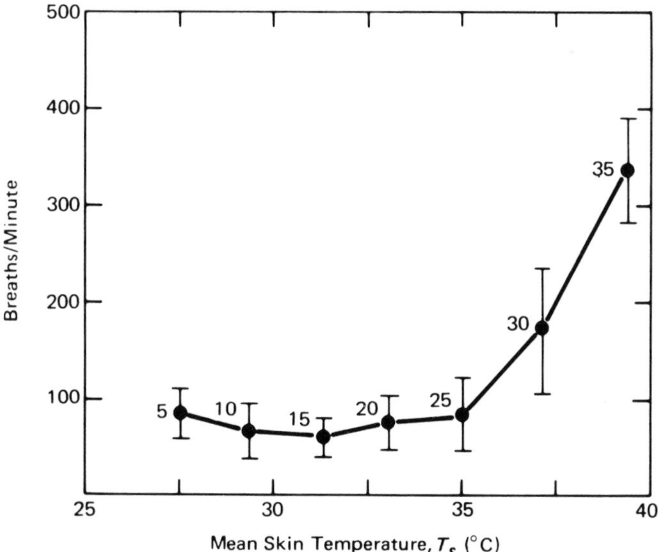

Fig. 27.3. Respiratory rate (mean ± S.D.) versus mean skin temperature in resting New Zealand white rabbits. Numbers to the left of each data point is equal to the ambient temperature (from 5 to 35°C). (Based on Gonzalez *et al.*, 1971.)

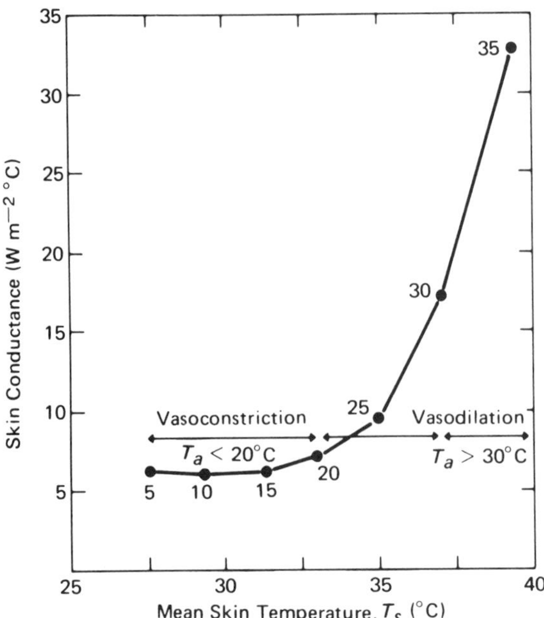

Fig. 27.4. Skin conductance versus mean skin temperature in resting New Zealand white rabbits in ambient temperatures of 5 to 35°C. Below an ambient temperature of 20°C, the rabbit is peripherally vasoconstricted (skin conductance is minimal and constant). Approaching 30°C and above the rabbit is peripherally vasodilated (skin conductance rises rapidly). The zone of vasomotor regulation lies in the intermediate range. (Based on Gonzalez *et al.*, 1971.)

The peripheral receptors for these thermal responses were subsequently found to exist in the ears and skin beneath the heavily furred regions (Kluger *et al.*, 1971, 1972b). Other peripheral areas, such as the nasal region, are probably also thermally sensitive (Hensel and Kenshalo, 1969). Warming or cooling of the ears and skin of the back of these rabbits led to significant changes in metabolic heat production. For example, at a low ambient temperature, warming the ears resulted in a decline in shivering metabolism and subsequently a decrease in core temperature (rectal and brain). While the sensitivity of the ears and skin beneath the heavily furred back was of similar magnitude, it was noted that under natural conditions the ear skin temperature changed over a much wider range than did skin temperature beneath the heavy fur (see Fig. 27.1). Therefore, the ears, in effect, yielded greater information about the thermal environment.

At an ambient temperature of 35°C, both core and skin temperatures were elevated. The percentage of thermal contribution attributed to either core or peripheral receptors is unknown.

During periods of steady-state heat production, the heat produced by an organism equals that lost. This heat loss can be partitioned into wet (evaporative) and dry (radiative, convective, and conductive) heat loss, so that

$$M = h(\overline{T}_s - T_a) + E \qquad (1)$$

where M = rate of metabolic heat production (W/kg or W/m^2)
$\bar{T}_s$ = mean skin temperature
T_a = ambient temperature
E = rate of evaporative heat loss (W/kg or W/m^2)
h = combined heat-transfer coefficient for dry heat loss (W kg^{-1} °C^{-1} or W m^{-2} °C^{-1}

Since the value of h should be a constant under these experimental conditions, it becomes an extremely useful term in partitioning heat transfer. Rearranging Eq. 1 results in

$$h = \frac{M - E}{\bar{T}_s - T_a} \qquad (\text{W kg}^{-1}\,{}^\circ\text{C}^{-1} \text{ or W m}^{-2}\,{}^\circ\text{C}^{-1}) \tag{2}$$

The value of h obtained in these studies was 0.16 W kg^{-1} °C^{-1} or 2.3 W m^{-2} °C^{-1}. Partitioning heat loss into dry heat loss and wet heat loss (both from the respiratory tract and from the skin surface) results in the summary figure (Fig. 27.5) for the steady-state responses in these rabbits.

Energy Balance of Rabbits During Exercise

Core temperature (as well as skin temperature) can be changed in an organism by such artificial means as perfusing thermodes implanted in the hypothalamus (Hammel *et al.*, 1963; Kluger *et al.*, 1973), spinal cord (Guieu and Hardy, 1970; Jessen and Mayer, 1971) or elsewhere with water at different temperatures. Data obtained from these types of studies indicate which areas are thermally sensitive only under these experimental conditions. They do not indicate whether using more natural means of thermal stimulation (described below) would yield similar results. This point will be examined in some detail later.

Core temperature can be changed naturally in the rabbit by exposing it to high (Johnson *et al.*, 1958; Gonzalez *et al.*, 1971) or low (Hart and Heroux, 1955) ambient temperatures or by exercise (Hart and Heroux, 1955; Kluger *et al.*, 1973). Under natural conditions, a high ambient temperature is actively avoided by most rabbits by selecting a suitable microclimate (Schmidt-Nielsen, 1964). When exposure to a high ambient temperature is unavoidable, the rabbit increases its respiratory rate, mean skin temperature, and rectal temperature, as described earlier for the resting New Zealand white rabbit in a 35°C ambient temperature (Gonzalez *et al.*, 1971). Behavioral responses such as licking of the forelimbs have also been reported (Kluger *et al.*, 1973). Exposure to the cold generally presents little difficulty to the rabbit. Because of its heavy fur (low heat-transfer coefficient), even the laboratory rabbit acclimated to room temperature can still maintain a relatively constant core temperature above an ambient temperature of about −35°C while at rest and −20°C during exercise (Hart and Heroux, 1955). To investigate the role of internal thermal sensors leading to thermoregulatory responses in the rabbit, using more or less natural means, exercise was employed to raise core temperature. During

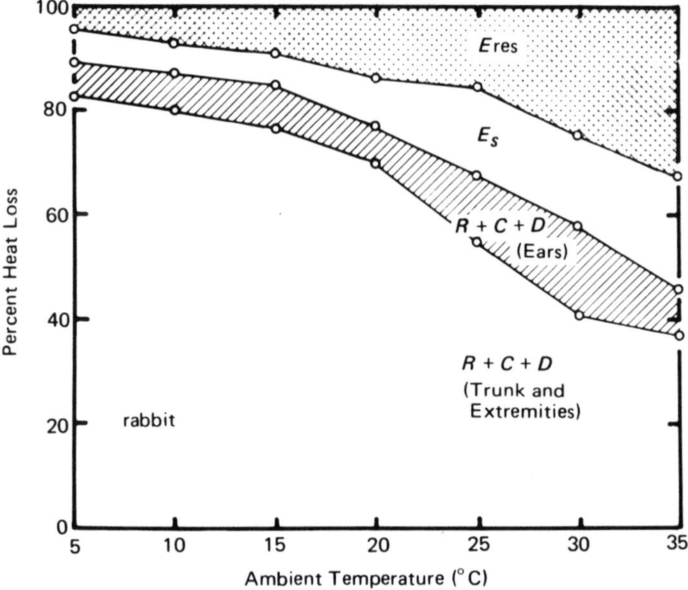

Fig. 27.5. Plot of percent heat loss versus ambient temperature in resting New Zealand white rabbits. Heat loss has been partitioned into wet heat loss from the respiratory tract (E_{res}) and from the skin surface (E_s), and dry heat loss from the ears and from the rabbit excluding the ears (R = radiation, C = convection, D = conduction). (Based on Gonzalez *et al.*, 1971.)

exercise, the rate of rise in core temperature of the New Zealand rabbit depends on, among other things, the environmental temperature and the speed of running or work load (Table 27.1). The data in Table 27.1 are not intended to represent actual increases in core temperature over 1 hour's duration of exercise, for the exercise period was considerably shorter than 1 h. The rates of increase were simply estimated by taking the difference in rectal temperature before and after exercise and calculated on a 1-h basis. These data show that, even at relatively low ambient temperatures and running speed, heat is stored by the rabbit. Mean skin temperature remains fairly low during and following exercise at low (10°C) or moderate

Table 27.1 Rate of Increase in Core Temperature (°C h^{-1}) in New Zealand White Rabbits at Different Ambient Temperatures and Work Loads[a]

	Ambient Temperature		
Running Speed (km h^{-1})	**10°C**	**20°C**	**30°C**
0.7	0.4	2.8	5.2
1.4	4.4	6.6	9.8

[a] See the text for further explanation.

(20°C) ambient temperatures (Kluger *et al.*, 1973). Therefore, the thermal responses initiated by the rabbit during or following exercise at these ambient temperatures are due to stimulation of internal thermoreceptors rather than a combination of internal and peripheral receptors as is found during exposure to high ambient temperatures.

During exercise at a speed of 1.4 km h^{-1}, the rabbits in the three ambient temperatures studied (10°C, 20°C, and 30°C) incurred a substantial oxygen debt, as indicated by high levels of blood lactate (Kluger *et al.*, 1972a). The levels of blood lactate that are associated with fatigue in the rabbit (ca. 100 mg percent) are similar to that observed in man during submaximal or maximal exercise (Karlsson, 1971).

The heat produced by the exercising rabbit must equal that stored, lost to its surroundings, and dissipated as external work. Since the rabbits were exercised mostly on the level (running in place in an exercise wheel), the work term is minimal and is therefore excluded. Therefore,

$$\text{H.P.} = \text{H.L.} + S \tag{3}$$

where H.P. = rate of heat production (W kg^{-1})
H.L. = rate of heat loss (W kg^{-1})
S. = rate of heat storage (W kg^{-1})

Because there is a substantial rate of anaerobic metabolism, measurement of oxygen uptake alone cannot be used to determine the heat production in the exercising rabbit. It becomes necessary to calculate the energy equivalent of the rate of lactic acid accumulation, as well, to determine the rate of heat production (Kluger *et al.*, 1972a). Expanding both sides of the above equation allows one to partition the avenues of heat transfer in the exercising rabbit in more detail than given in Eq. 3:

$$M + L = h'(\overline{T}_s - T_a) + E + S \tag{4}$$

where M = rate of aerobic metabolic heat production (W kg^{-1}) based on oxygen uptake
L = rate of anaerobic metabolic heat production (W kg^{-1}) based on lactic acid production
$h'(T_s - T_a)$ = rate of dry heat loss via radiation and convection
h' = combined heat transfer coefficient for exercising rabbits ($\approx 2h$)
$\overline{T}_s$ = mean skin temperature
T_a = ambient temperature
E = rate of evaporative heat loss (W kg^{-1})
S = rate of heat storage (W kg^{-1})

The results of these calculations for the exercising as well as for the resting rabbit are compared in Fig. 27.6.

As observed from Fig. 27.6, as ambient temperature increases, heat storage increases and dry heat loss decreases. The capacity markedly to increase radiative and convective heat loss in the rabbit is best observed in the exercising rabbit in a low ambient temperature. One reason for this large increase in dry heat loss, as

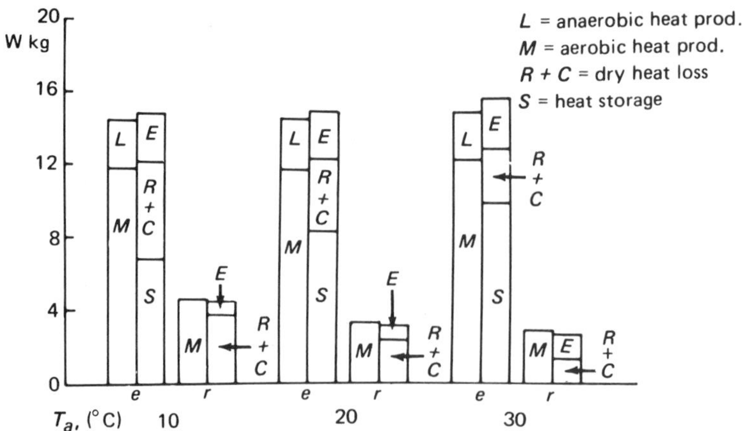

Fig. 27.6. Mean heat production, loss, and storage in three exercising (e) and four resting (r) New Zealand white rabbits at ambient temperatures of 10, 20, and 30°C. (From Kluger *et al.*, 1972a)

core temperature is raised in the cold, is illustrated in Fig. 27.7. The ears of these rabbits are large and sparsely furred, and therefore a considerable amount of heat can be lost from these membranous structures when their blood flow increases. The large capacity for heat loss by the ears is reflected by the heat-transfer coefficient, which, when calculated from cooling curves, is about 9 $W\ m^{-2}\ C^{-1}$ (Kluger *et al.*, 1971), a value some four times greater than that for the whole rabbit (Gonzalez *et al.*, 1971). In an ambient temperature of 30°C, there is already maximal vasodilation in these appendages, so further vasomotor change cannot be elicited by the exercise; however, in an ambient temperature of 10°C, a slight elevation in core temperature leads to a dissipation of approximately 50 percent of the resting metabolic heat production of these rabbits through the ears alone.

Thermal Responses in Rabbits Following Exercise

If the hypothalamus of a resting rabbit (Kluger *et al.*, 1973) or dog (Hammel *et al.*, 1963) or ground squirrel (Williams and Heath, 1971) or any other organism with a thermoresponsive hypothalamus is warmed or cooled, predictable thermal responses occur. As ambient or skin temperature is raised, the "hypothalamic" threshold or set point for the thermal responses is lowered. In other words, in a warm environment it takes less hypothalamic warming to initiate a panting or peripheral vasodilatory response. Recently, this same phenomenon was shown for the extrahypothalamic deep-body thermoreceptors in the rabbit (Kluger *et al.*, 1973). That is, holding hypothalamic temperature constant and raising extrahypothalamic deep-body temperature by exercise led to a decline in the core threshold for both peripheral vasodilation and panting as ambient temperature was raised.

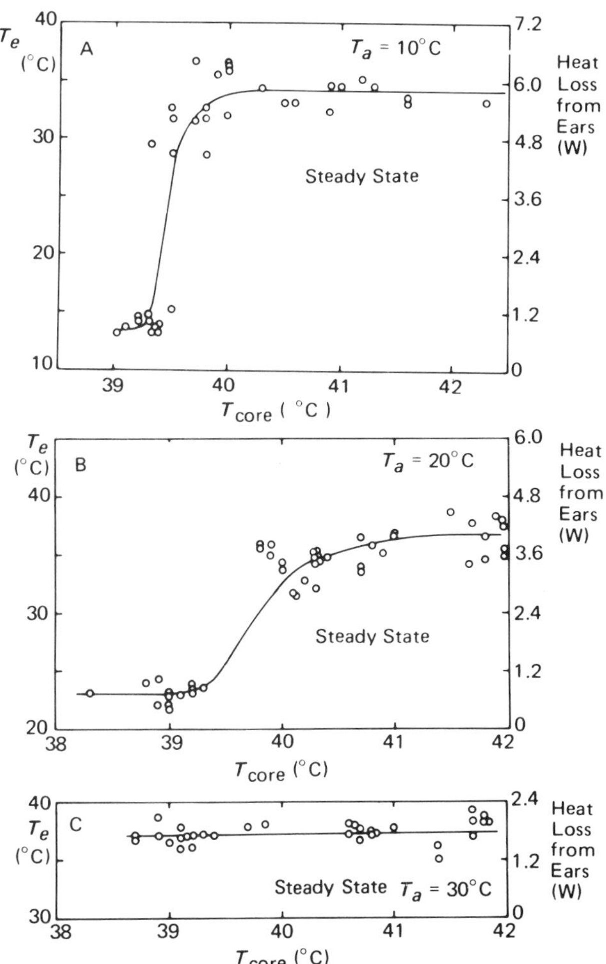

Fig. 27.7. Steady-state values of ear temperature (T_e) and heat loss (in watts) at different levels of core temperature in New Zealand white rabbits at A: 10°C T_a; B: 20°C T_a; C: 30°C T_a. (From Kluger *et al.*, 1973.)

When rabbits are exercised at varying ambient temperatures without the use of artificial heating devices, a completely unexpected phenomenon occurs (Kluger, 1974). In these experiments the rabbits ran for long periods (30 min) at low running speeds (0.7 km h^{-1}) which caused body temperature to rise slowly (see Table 27.1). Following exercise, eight rabbits were peripherally vasodilated at all ambient temperatures (10°C, 20°C, and 30°C) and were panting in the neutral (20°C) and high (30°C) ambient temperature; and as a result, core temperature fell toward normal. In an ambient temperature of 20°C, the rabbits panted following exercise until rectal temperature ($\approx$ hypothalamic temperature) fell from 40.5°C to below 39.9°C $\pm$ 0.16 S.E.M. However, in an ambient temperature of 30°C, the rabbits stopped panting, not at a lower rectal temperature, as might be expected, but after rectal temperature fell from 42.0°C to 40.8°C ± 0.22 S.E.M. (Fig. 27.8). This statistically significant ($p < 0.01$) rise in the core temperature at which this heat-dissipating response ceases brings into question the physiological significance of the results obtained by the many investigators employing artificial thermal stimulation to study thermoregulatory responses rather than using a more natural means such as exercise.

Recently it was shown by Turlejska-Stelmasiak (1973) that, in rabbits partially dehydrated (1–3 days without water), the hypothalamic threshold for panting is elevated, and this same phenomenon might be at work in my studies. For although measurements of body weight before and after experimentation reveal a comparable 1–2 percent total body weight loss in the two ambient temperatures studied in my experiments, undoubtedly more water is lost from panting and from passive diffusion through the skin at 30°C than at 20°C. During natural elevation of core temperature (exercise) this internal thermal information along with this slight dehydration is integrated to result in a cessation of panting in a high environmental temperature. As the lethal core temperature in New Zealand white rabbits is over 42°C (a temperature commonly reached by these rabbits following 30 min of exercise in a 30°C ambient temperature) and panting does not stop until a core temperature of 40.8°C is passed, the rabbit is in no danger of death due to hyperthermia.

The cessation of panting at a higher core temperature in the heat, rather than at a lower core temperature (as predicted from studies using implanted thermodes

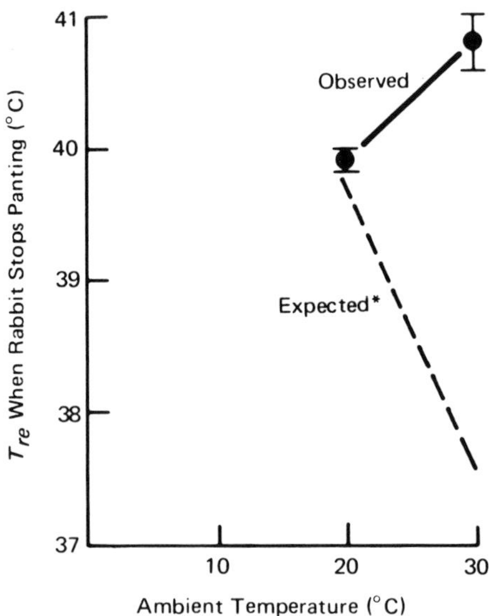

Fig. 27.8. Rectal temperatures when eight New Zealand white rabbits stopped panting (mean ± S.E.M.) following 30 min of exercise at ambient temperatures of 20 and 30°C. The dashed line represents the changes observed in the hypothalamic threshold for this response when rectal temperature or extrahypothalamic deep-body temperature was at a neutral (39.2°C) level. (From Kluger, 1974.)

to simulate exercise), is apparently an adaptation to conserve water and to reduce the energy expenditure associated with thermoregulatory panting. This phenomenon is, in many ways, similar to that reported by Schmidt-Nielsen *et al.* (1957) for the camel, and by Taylor (1970) for many other African ungulates.

References

Gonzalez, R. R., Kluger, M. J., Hardy, J. D.: 1971. Partitional calorimetry of the New Zealand white rabbit at temperatures 5°C–35°C. J. Appl. Physiol. *31*, 728–734.

Guieu, J. D., Hardy, J. D.: 1970. Effects of preoptic and spinal cord temperature in control of thermal polypnea. J. Appl. Physiol. *28*, 540–542.

Hammel, H. T., Jackson, D. C., Stolwijk, J. A. J., Hardy, J. D., Stromme, S. B.: 1963. Temperature regulation by hypothalamic proportional control with an adjustable set point. J. Appl. Physiol. *18*, 1147–1154.

Hart, J. S., Heroux, O.: 1955. Exercise and temperature regulation in lemmings and rabbits. Can. J. Biochem. Physiol. *33*, 428–435.

Hensel, H., Kenshalo, D. R.: 1969. Warm receptors in the nasal region of cats. J. Physiol. London *204*, 99–112.

Jessen, C., Mayer, E. T.: 1971. Spinal cord and hypothalamus core sensors of temperature in the conscious dog. Pflugers Arch. *324*, 189–204.

Johnson, H. D., Cheng, C. S., Ragsdale, A. C.: 1958. Environmental physiology and shelter engineering with special reference to domestic animals. XLVI. Comparison of the effect of environmental temperature on rabbits and cattle. Part 2. Influence of rising environmental temperature on the physiological reactions of rabbits and cattle. Univ. Mo. Agric. Expt. Sta. Res. Bull. *648*.

Karlsson, J.: 1971. Lactate and phosphagen concentrations in working muscle of man. Acta Physiol. Scand. Suppl. *358*, 1–72.

Kluger, M. J.: 1974. Effects of ambient and core temperature on the thermoregulatory set-point in the rabbit. Am. J. Physiol. *226*, 817–820.

———, Gonzalez, R. R., Mitchell, J. W., Hardy, J. W.: 1971. The rabbit ear as a temperature sensor. Life Sci. *10*, 895–899.

———, Nadel, E. R., Hitchcock, M., Stolwijk, J. A. J.: 1972a. Energy balance and lactic acid production in the exercising rabbit. Am. J. Physiol. *223*, 1451–1454.

———, Gonzalez, R. R., Hardy, J. D.: 1972b. Peripheral thermal sensitivity in the rabbit. Am. J. Physiol. *222*, 1031–1034.

———, Gonzalez, R. R., Stolwijk, J. A. J.: 1973. Temperature regulation in the exercising rabbit. Am. J. Physiol. *224*, 130–135.

Schmidt-Nielsen, K.: 1964. Desert animals, physiological problems of heat and water. New York: Oxford Univ. Press.

———, Jarnum, S. A., Houpt, T. R.: 1957. Body temperature of the camel and its relation to water economy. Am. J. Physiol. *188*, 103–112.

Taylor, C. R.: 1970. Dehydration and heat: effects on temperature regulation of East African ungulates. Am. J. Physiol. *219*, 1136–1139.

Turlejska-Stelmasiak, E.: 1973. Thermo-regulatory responses to hypothalamic heating in dehydrated rabbits. Experientia *29*, 51–52.

Williams, B. A., Heath, J. E.: 1971. Thermoregulatory responses of a hibernator to preoptic and environmental temperatures. Am. J. Physiol. *221*, 1134–1138.

28

Thermal Exchange, Physiology, and Behavior of White-Tailed Deer

Aaron N. Moen and Nadine K. Jacobsen

Introduction

"Life is a process that involves the expenditure of energy for the redistribution of matter" (Moen, 1973, p. 3). This process is self-sustaining in living organisms, driven by the metabolic machinery of the organism, and influenced by the energy flux and matter in the operational environment of each organism. The rates at which the various processes occur vary, depending on characteristics of both the organism and environment, and on the interaction between the two.

Time, in a biological sense, is a measure of the intensity of life. Animals exhibit metabolic and behavioral variations during each 24-h period. Some animals are active for just part of the year, hibernating for several months during the winter. While hibernating, metabolic processes are reduced to such low levels that several days or weeks may be physiologically equivalent to but a single day during the active part of the year. Other organisms, such as the white-tailed deer (*Odocoileus virginianus*), exhibit seasonal variations that are not as pronounced as those of hibernators. Environmental transients in the form of weather, other organisms, and physical components of the animal's environment are superimposed on daily and annual rhythms.

Energy Transactions

An organism sustains life by the periodic ingestion of matter that is altered chemically by enzymes secreted into the gastrointestinal tract of monogastric organisms, or by microorganisms (bacteria and protozoa) in the ruminants. The products of digestion (glucose, for example, in monogastric animals, and volatile fatty acids in ruminants) are sources of energy for the metabolism of the organism.

A free-ranging organism has an ecological metabolic rate (Moen, 1973, p. 274) that encompasses all life processes, including activity, growth, reproductive functions such as egg laying, gestation, and lactation, and the support of pathogens and parasites. Many of the energy-requiring processes are sequential over the annual cycle. In willow ptarmigan (*Lagopus lagopus*), molting and egg laying, for example, are not simultaneous processes, resulting in a leveling out of the energy requirements over time (West, 1968). In white-tailed deer, the winter-to-summer molt occurs prior to the last few weeks of gestation when the requirements rise rapidly. Lactation is perhaps the costliest biological process during the entire annual cycle (see Moen, 1973, p. 374), but this occurs in the early summer when forage quantity and quality are also at their highest level.

All organisms must expend energy for the mechanics of life. Sessile animals have the mechanics of blood flow, movement of food through the gastrointestinal tract, and other internal functions that involve the expenditure of energy. Mobile animals have these functions, plus the added requirements for individual activities such as resting, walking, running, flying, swimming, and others (see Tucker, 1969). The physical characteristics of the habitat in which an animal lives affect the energy cost of transport. Hills, for example, increase the cost of walking, because vertical ascent is several times more costly than locomotion on a horizontal plane. There is an upper limit to the distance an animal can travel to food and water; when this distance is exceeded, the animal will not use the area unless physiological compensations can be employed. Some animals live in groups or herds, resulting in an energy expenditure that is partly dependent on the social interactions within the group. The energy expenditure of male elk, for example, is related to their social status in the herd as the subdominant animals spend more time moving about seeking breeding opportunities than do the dominant males. The latter, however, expend more energy actually breeding. These, and similar interactions, are discussed further in Moen (1973).

Energy Interactions

The energy distributed throughout the area where an organism lives interacts with that organism in relation to the chemical, physical, and thermal characteristics of both organism and environment. To illustrate, radiant energy from the sun, both direct and diffuse, reaches the surface of an animal and is reflected, transmitted, or absorbed. The absorbed energy interacts with the animal, warming the animal's surface, which in turn results in an emission of more infrared radiation as the radiant surface temperature increases. Some of the radiation absorbed by the hair coat is dissipated by the effects of wind. Some is conducted by the hair shafts deeper into the hair layer. The net exchange may be either positive or negative, depending on the physical and thermal characteristics of the surfaces (both animal and environment) participating in infrared exchange, and on the interactions among all forms of heat exchange.

Interactions among the various forms of heat exchange in the hair–air interface between animal and environment result in a distinctive atmosphere around an animal called a thermal boundary region. The characteristics of this boundary region are dependent on the thermal characteristics of the animal and the ambient air. Air motion occurs when temperature gradients through this thermal boundary region result in air-density differences and natural convection. External pressure differences may also cause air motion around an animal; this results in forced convection.

Contact between an animal and its substrate results in the exchange of heat by conduction. This occurs as an animal stands on a substrate, although the surface area of hooves is usually small, or as an animal reclines. A deer bedded in the snow melts the snow under its body, and an incubating bird loses heat by conduction to its eggs and to the nest and environs.

The vapor-pressure gradient in the thermal boundary region and on the internal respiratory surfaces of an animal, coupled with the dynamic exchange of water vapor (mass transport), determines the extent of evaporation losses. These can be significant in either winter or summer, depending on the compensatory relationships of the vapor-pressure gradients within mass-transport phenomena. These are inexorably linked to the thermal and mechanical characteristics of the entire thermal boundary region.

Behavioral interactions among animals in a group may result in a reduction of energy expenditure. Winter concentration areas of white-tailed deer, for example, are characterized by the development of hard-packed trails that facilitate movement. This also serves to restrict the availability of forage, however, so the net benefit to the animals is not directly related to the added benefit of packed snow.

Interactions also occur between thermal and chemical energies. The heat dissipated due to thermal exchange in different weather conditions comes from that absorbed and stored in relation to the thermal capacity of the tissue and from the metabolic processes of the animal and, in the case of ruminants, from the intermediary microorganisms. Heat energy released by metabolic processes of the host is called heat of nutrient metabolism, and that by rumen microorganisms, heat of fermentation. Internal friction due to vascular and muscular mechanics also results in heat energy.

Some animals maintain a fairly stable balance between the heat energy retained and the heat energy dissipated. When this occurs, their body temperature remains fairly stable. An animal may permit heat dissipation to exceed heat retention, resulting in a fall in body temperature. The opposite may also be true, resulting in a rise in body temperature. Such heat-exchange transients are adaptations for survival, and are a part of the energy-conservation strategies available to different organisms.

Such strategies are complex interactions of all forms of energy transactions. Consider, for example, a deer bedded in the snow. "Cold" weather may result in an accelerated dissipation of heat energy. This can be compensated for by increased muscular activity, which is an exothermic process. Energy is required for muscular contraction, however, and this must come from nutrient metabolism. Nutrient metabolism depends on a supply of ingested energy or on the mobilization of a fat reserve. Ingestion requires muscular activity as the animal forages. Active foraging places the animal in weather conditions that may result in a greater dissipation of heat since the animal may be more exposed to the wind. Internal thermoregulatory mechanisms may be employed then, such as muscular contraction that results in piloerection. This increases the depth of the hair coat, but it also decreases the density. This decrease permits more penetration of wind, resulting in a decrease in the conductivity of the coat per unit depth. Further, the muscular contraction necessary for piloerection is an energy-requiring process, contributing to an increase in the total energy requirements. The compensatory effects of these reactions—and many more reactions could be mentioned—are necessary considerations in a complete bioenergetics analysis. Further, the animal in vivo employs these strategies to varying extents at different times of the day and year; daily and annual rhythms determine, in part, the propensity for a particular combination of responses.

Factors to Consider in Physical Thermal Models

The two main components of a physical thermal model are the organism and the surrounding energy flux. Characteristics of each can be measured, but to develop a realistic physical model, it is critical to understand the energy flux through the thermal boundary region surrounding the organism. This is the transition region between the two components—the region influenced by both organism and the surrounding energy flux. When the animal has a higher heat content than the surrounding environment (the more usual case), the animal is the heat source and the environment the heat sink. In reality, this is an n-dimensional system, however, and the animal may be a source of one kind of heat energy and a sink for another. Radiant energy is an example; the animal is a source of infrared energy dissipated to a clear blue sky, but a sink for the solar radiation, both direct and diffuse, that strikes its surface.

The hair layer is a prominent component of the thermal boundary region. Its physical characteristics, such as depth, density, and morphology, affect the structure of the thermal boundary region. The thermal relationships between the hair and the ambient atmosphere are dynamic, and there is no physical plane that can be identified as a line of demarcation between the two. The physical surface does not suffice, as both radiant energy and wind penetrate the hair layer. Thus one must recognize a *thermal surface*, but only as a concept until the mechanisms of energy transfer in the hair layer and the thermal boundary region are understood. Further, radiant energy flux and convective energy flux interact with the hair differently. This leads to the conclusion that there is a *radiant thermal surface* and a *convective thermal surface*; these do not necessarily coincide. Further, radiant energy flux may be positive with respect to the animal, as when exposed to bright sunlight, but the convective flux may be negative.

The overall conductance of the hair layer and of the entire thermal boundary region is a dynamic property that is an integration of the radiation conductance, conduction conductance, convection conductance, and evaporation effects. Each of these is dynamic, depending on the interaction between energy flux in the hair layer and the energy flux in the ambient atmosphere.

Given the dynamic exchange characteristics discussed above—and more that have not been mentioned—one is forced to conclude that the measurement of conductance of a hair layer, often expressed as conductivity (k) or heat flux per unit depth, is unrealistic unless the thermal dynamics occurring throughout the hair layer are considered. This is not the case in closed-chamber measurements of k. The resulting measurements permit no consideration of the interaction among radiation, convection, conduction, and evaporation. They are likely to be quite valid for the conductivity of the hair layer in direct contact with the substrate, however, as when a deer beds on the snow, providing the compression effect of the deer weight: bed area is considered. This is necessary because compression of the hair affects its conductivity. Measurements of compression effects are seldom reported for animal coats, although the principle is well recognized in the building trade and in the insulation qualities of sleeping bags.

The morphology of the hair layer in the thermal boundary region is another characteristic of extreme importance in thermal transfer. The hair coat is not smooth, and does not behave thermally like a smooth cylinder. Each of the hairs is potentially a tiny "cooling fin" projecting from the skin surface. A very slight air motion may result in a high rate of convection from the individual hairs because cylinders with very small diameters are very quickly linked to the cooling effect of air flow. The hairs do not act individually, however, because the heat removed from one may be absorbed (advection) in part, by the next hair on the downwind side. Thus there is dynamic two-stage convection at the surface of an animal, with the individual hairs as one stage and the overall hair surface as another. The relative importance of each of these depends on the wind velocity, turbulence level, and the orientation of the hair(s).

Wind-flow characteristics in the hair layer are unknown because of the difficulty in measuring them. Hot-wire anemometers are inadequate, even with elements of just 0.001-in. diameter and high-frequency responses, because the hair density and interhair spaces are considerably smaller than that. Wind flow in natural habitats is often described by a mean-velocity equation (see Sellers, 1965). A roughness parameter can be used to represent the effect of different vegetation types, but vegetative patterns seldom present the necessary fetch for the valid application of the equation. Further, wind flow in the field is almost always turbulent (Moen, in press, "Turbulence and the visualization of wind flow"), and this results in a dynamic air flow in the vegetation and around the bodies of animals that has not, as yet, been described for anything but the simplest situations.

The general distribution of radiant energy flux has been described fairly well by meteorologists. The interaction between radiation flux and the surface of an animal is very complex, however, because of the geometry of the animal. One can conceptualize a "solar radiation profile" (Moen, 1973, p. 89), where part of an animal may be exposed to direct solar radiation, and all of it to indirect solar radiation and to infrared flux, except for the body surface in direct contact with the substrate. The angles of incidence over the surface are many, however, and these angles affect the amount of radiation absorbed. Color patterns are also an important consideration for wavelengths from 0.3 to 0.7 μm.

When the dynamic characteristics of air flow in the thermal boundary region are coupled with the radiation profile considerations, the thermal analysis becomes exceedingly complex. Conduction exchange can be added to the radiation–convection analysis, and while it is, perhaps, the simplest mode of heat transfer, the consideration of substrate characteristics, hair compression, and the distribution of temperatures on the skin–hair surface and hair–substrate surface make this addition complicated enough.

Evaporative losses are also difficult to analyze because one is dealing not merely with vapor-pressure gradients, but with mass transport and temperature-dependent features. Since evaporation losses are so closely related to the temperature field and air flow in the thermal boundary region, this region should be studied in detail before analyzing energy dissipation by evaporation.

Simplified Assumptions and General Conclusions from Thermal Modeling

Thermal models have been described for a number of species, including chipmunk (Heller and Gates, 1971), fossorial rodents (McNab, 1966), tammar wallaby (Dawson *et al.*, 1969), wild mouse (McNab, 1963), Canada geese (Birkebak, *et al.*, 1966), white-tailed deer (Moen, 1966; Stevens, 1972), sharp-tailed grouse (Evans, 1971), and several other animals (Porter and Gates, 1969).

Several different assumptions have been made in these models of thermal energy budgets. Common ones include the following: the animal is a collection of cylinders and cones; posture is static, either bedded or standing; convection coefficients for smooth cylinders are used; evaporation losses are minimal in cold weather; metabolic heat production is related to $W^{0.75}$; and others. Many of these assumptions are valid for certain situations in the life of the animal, and many of them provide useful insights into the pattern of thermal relationships, if not the absolute thermal relationships.

Continued research on white-tailed deer over the past several years has resulted in several generalizations concerning the physical heat exchange of this free-ranging animal. One early conclusion, based on convection and radiation measurements from a deer simulator formulated as a cylindrical animal in a standing posture under clear skies at night (Moen, 1966), was that wind effects increase and radiation effects decrease as the ambient temperature decreases (Moen, 1968c). This conclusion stimulated further analyses of wind effects using a Thermal Environment Simulation Tunnel (TEST) and a Thermal Micro-Simulation Tunnel (TMST) at the BioThermal Laboratory. These two wind tunnels permit control over wind and radiation, and a new series of analyses has been completed. The initial generalization that convection increases in importance as the air temperature decreases

Table 28.1 Percent of the Total Heat Loss Attributed to Natural Convection for White-Tailed Deer and Sharp-Tailed Grouse

	Free Convection As Percent of Total Heat Loss	
T_a	White-Tailed Deer[a]	Sharp-Tailed Grouse[b]
+20	15	43
+10	27	48
0	36	53
−10	43	58
−20	51	63

[a] Calculated from data in Stevens, 1972 (p. 132).
[b] Calculated from data in Evans, 1971 (p. 81).

was again confirmed for both deer and sharp-tailed grouse (*Pedioecetes phasianellus*) (Table 28.1). The effects of forced convection (Table 28.2) are even greater than those shown for natural convection in Table 28.1.

Measurements begun in Minnesota (Moen, 1968c) of radiant surface temperatures on captive deer were continued on live deer at the laboratory, with instrumentation having a more rapid response time, permitting a more thorough scanning of the animal's surface. Additional use was made of simulators that permitted control over wind and radiation variables. These are described in Moen (1973). The measured data from simulators were within the minimum and maximum values measured on the trunks of live deer, indicating that the simulation technique compared realistically with in vivo measurements. The a and b values in the regression equations for calculating average radiant temperatures on seven body regions are shown in Table 28.3; additional data are given in Moen (1973).

A second thermal model of deer (Stevens, 1972) included additional considerations beyond those of Moen (1966). Air flow, for example, was represented by a vertical profile, and the heights of the legs, trunks, necks, and heads of standing deer were determined for profiles over substrates having different roughness characteristics. Radiant temperatures of simulators and live deer discussed above

Table 28.2 Percent of the Total Heat Loss Attributed to Convection from White-Tailed Deer Exposed to an Air Temperature of −20°C and Three Wind Velocities

U (mi h^{-1})	Convection As Percent of Total Heat Loss	
2	30[a]	60[b]
6	50	75
10	60	85

[a] Calculated from data in Moen, 1968b.
[b] Calculated from data in Stevens, 1972 (p. 135).

Table 28.3 Data Necessary to Calculate the Average Radiant Temperature (T_r)[a] of Seven Body Regions of a Standing White-Tailed Deer Exposed to Still Air

Body Region	a	b
Head	10.67	0.66
Neck	8.57	0.76
Trunk	9.01	0.78
Upper front leg	10.75	0.74
Lower front leg	8.66	0.87
Upper hind leg	9.20	0.74
Lower hind leg	8.58	0.83

[a] $T_r = a + bT_a$, where T_a = °C.

were used in the calculation of heat exchange. Hair depths over different body parts were used in the calculation of conductive heat flux through the hair layer. Analyses with this model yielded several new generalizations, which appear to be reasonable in relation to the principles of thermal engineering.

It was shown again that convective losses increase and radiation losses decrease as air temperature (T_a) decreases. To illustrate, at −20°C and no wind, convective heat loss from a standing deer weighing 60 kg was predicted to be about 50 percent of the total nonevaporative heat loss, decreasing to less than 20 percent of the total at +20°C. In a wind of 14 mi h^{-1}, almost 90 percent of the nonevaporative heat loss was due to convection when T_a = −20°C, and 75 percent when T_a = +20°C.

It was also shown that the body trunk of a 60-kg deer is the greatest source of heat loss, with other body parts contributing much less (Table 28.4). These values incorporate the effect of differences in both gross body-surface area and in the vertical wind profiles. When the effect of gross body-surface area is removed by expressing percent heat loss from each body part on a unit-area basis, the trunk is one of the lowest contributors (Table 28.5). The lower parts of the legs are high contributors, even though they are in the low-velocity portion of the vertical wind-velocity profile. This can be attributed to their efficiency as convectors because of their small diameters.

Table 28.4 Percent Heat Loss from Different Body Parts of a 60-kg Deer Exposed to Two Wind Velocities

U (mi h^{-1})	Head	Ears	Neck	Trunk	Upper Front Leg	Lower Front Leg	Upper Hind Leg	Lower Hind Leg
0	6.6	6.4	5.6	44.6	6.6	11.2	7.3	11.6
14	5.7	5.9	5.5	44.5	6.0	12.8	6.6	12.8

Table 28.5 Percent Heat Loss from the Different Body Parts of a 60-kg Deer Expressed on a Unit-Area Basis

U (mi h^{-1})	Head	Ears	Neck	Trunk	Upper Front Leg	Lower Front Leg	Upper Hind Leg	Lower Hind Leg
0	9.1	23.2	6.0	7.2	12.2	21.6	7.2	13.4
14	8.3	21.5	6.3	7.1	11.2	24.5	6.7	14.5

The effect of body size has been analyzed with additional assumptions by Stevens (1972), who reached conclusions similar to those of Moen (1966). Basically, one predicts that smaller deer have less favorable thermal relationships than larger deer. The results of Moen are based primarily on the greater potential for metabolic heat production of a larger deer, and of Stevens, on surface-area and geometry considerations in relation to deer weight. Stevens also points out that the surface area of the legs of a 20-kg deer is 8 percent more of the total surface area than that of a 100-kg deer, contributing to the higher heat losses of smaller deer on a unit-area basis. The smaller deer do have the advantage of being in the lower-velocity portion of the wind-velocity profile, however, compensating in part for their less favorable area/weight ratio.

The interaction between convection from the individual hairs and from the overall hair surface was mentioned earlier. This idea is discussed further in Stevens (1972) and Moen (1973), who express a "theoretical convective diameter" of a deer. Basically, this calculation involves solving for diameter, given the wind velocity and the convective heat loss per unit area. The "theoretical convective diameter" in free convection of a 60-kg deer is 0.8 cm at $T_a = -20$°C and 4 cm at $T_a = +20$°C. It rises sharply to 9.8 cm and 34.7 cm in a wind of 1 mi h^{-1} at -20°C and $+20$°C, respectively, and then decreases in a not quite linear fashion to, at a final velocity of 14 mi h^{-1}, 2.0- and 5.3-cm diameters, which are close to those at 0 mi h^{-1}. Similar calculations for the trunk and the lower part of the hind leg show the same pattern, although the peak is reached at $U = 4$ to 5 mi h^{-1} for the trunk, and less than 1 mi h^{-1} for the lower hind leg. While this parameter is entirely theoretical and of little practical importance numerically, it does illustrate the two-stage characteristic of convection from a hairy surface.

The effect of the interaction between radiation and convection on heat flux from a hair surface is another important consideration. Since the hairs are efficient convectors, radiant energy absorbed by the hairs is subject to loss by convection before it can effectively reduce the temperature gradient in the hair layer. Conversely, a cold night sky, which acts as a radiant heat sink, may tend to cause a reduction in the radiant surface temperature of an animal, but air movement in the thermal boundary region becomes a heat source (advection) rather than a mechanism for convective heat loss. This principle dictates the use of shielded or ventilated radiation-measurement instruments which either eliminate (or at least reduce to a minimum) or stabilize convection effects on the sensor. Stevens (1972) has determined that air movements of less than 4 mi h^{-1} may ameliorate the effects of variations in the radiation flux tested on deer hair. This, coupled with the considerations of the "theoretical convective diameter" discussed earlier, calls attention to the dynamic effect of low wind velocities on the hair surface of deer. The benefits of overhead cover are not as great as the differences in radiant energy flux, such as those described by Moen (1968c) in three different cover types.

The recognition of the importance of the lower wind velocities on thermal exchange on the hair surface can be related to another universal phenomenon in natural habitats, that of wind turbulence. There are reductions in wind velocities in various cover types, with changes also in turbulence scale and intensity. These turbulence characteristics are dependent on the characteristics of the ground surface and its vegetative cover. Kestin (1966) points out that turbulence increases convective losses, resulting in a higher heat loss per unit velocity. This effect, coupled with the effect of lower velocities discussed above and the complex geometry of an animal's surface (which ensures turbulent flow around the animal), lends additional evidence to the importance of wind as a component of the thermal environment of an animal.

The heat loss from deer in given weather regimes can be predicted with the model containing the considerations discussed previously. Conversely, the conditions predicted to elicit responses by the deer that result in a constant heat loss can be determined. A series of responses at various air temperatures, given a wind velocity of 14 mi h^{-1} at 2 m in an open field with uniform radiation flux characteristic of a cloudy sky at night, are shown in Table 28.6 for a 60-kg deer. The sequence may not be in correct biological order, and, further, the order may vary between animals. But such predictions do provide insight into the relative importance of different responses, and provide a starting point for comparing in vivo responses actually observed with those predicted.

The foregoing conclusions on thermal exchange of white-tailed deer illustrate the kinds of outputs one can derive from physical models. Many additional conclusions are possible, and many aspects of thermal exchange can be studied in different ways or in greater detail. For example, Wathen *et al.* (1971) have deter-

Table 28.6 Responses at Various Air Temperatures Predicted to Result in a Heat Loss Equal to 140 kcal per $W^{0.75}$ of a Deer in an Open Field with a Wind Velocity of 14 mi h^{-1} at 2 M

Responses	T_a (°C)
None of the responses listed below	+15
Vasoconstriction	+13
Piloerection 1.5 times normal lie	+ 8
Piloerection 1.8 times normal lie	+ 3
Piloerection (1.8) and vasoconstriction	− 3
Bedded posture, head up, no piloerection	−15
Bedded posture, head down, no piloerection	−21
Bedded posture, head up, piloerection 1.8	−45
Bedded posture, head down, piloerection 1.8	−53

mined convection coefficients for the ears of jackrabbits; Davis (1972) has completed theoretical work on the energy transfer in the fur of various species; Spotila *et al.* (1973) have used a mathematical model to predict body temperatures of large reptiles; Gessaman (1973) has edited a series of papers on thermal modeling; and many others could be mentioned.

One conclusion that appears to be of utmost significance when analyzing our own work and the reports of others is that the interaction between the thermal, mechanical, and metabolic forces is most dynamic and the system is truly *n*-dimensional. The concept of a "critical thermal environment" (Moen, 1968a), which recognizes the interactive nature of energy transactions, applies, since no single parameter can be considered to be at a critical value unless all other parameters are defined.

The usefulness of proceeding with more detailed analyses into the physical aspects of heat exchange is best determined after the actual, in vivo responses of the animal are considered. In the natural environment, an animal can exhibit its full repertory of responses with a minimum of interference. Accordingly, a series of physiological and behavioral analyses using radiotelemetry techniques has been completed at the BioThermal Laboratory (see Jacobsen, 1973). The deer used in the experiments are permitted free range within pens varying in size from $\frac{1}{4}$ to 6 acres. We are confident that the deer have enough freedom from constraints to employ several thermoregulatory responses not possible in chamber-type experiments; our observations indicate that these animals exhibit behavior patterns similar to those of free-ranging deer in the wild, but on a smaller scale.

Physiological and Behavioral Factors to Consider in Thermal Models

Biochemical reactions proceed in living tissues at rates dependent on the thermodynamic properties of the tissues involved, on their temperatures, and on the feedback influents characteristic of living systems. These influents may be rhythmic or sporadic. Some rhythms follow 24-h patterns, and others follow yearly patterns. Sporadic influents may arise because of stimuli received from external sources which trigger the release of stimuli internally. Adrenalin, for example, is released when an animal is frightened. There is interaction between the influents of external and internal origin; levels of sensitivity vary in relation to temporal rhythms.

Such variations in biochemical reactions in white-tailed deer result in an ecological metabolism that is an expression of the effect of all factors that are a part of the life of an animal in its natural habitat throughout the annual cycle (Fig. 28.1). Molting, gestation, lactation, tissue growth, and activity patterns all vary, depending on both biotic and abiotic factors in the operational environment. The concept of ecological metabolism allows for the consideration of the sequence of these functions and the compensatory factors employed by the animal to maintain

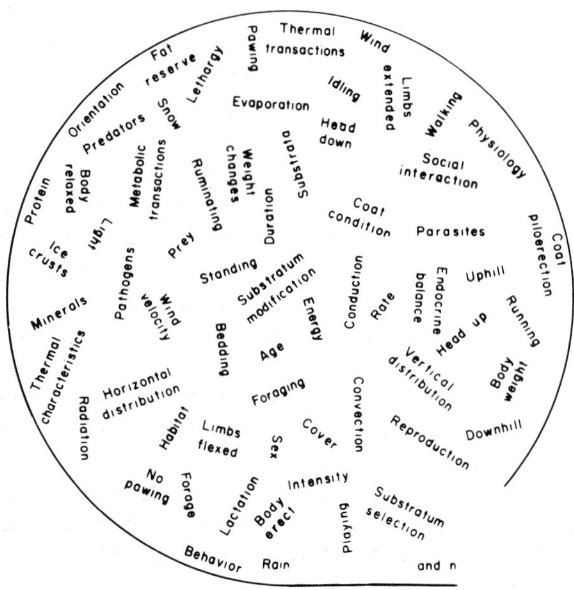

Fig. 28.1. Alternative characteristics and responses of white-tailed deer in relation to their environment.

its productivity over its entire life span, but regulating its productivity within the annual cycle so that it has a supply of nutrients (energy and matter) when they are needed for particular functions.

Metabolic variations are just one set of many alternative responses of a deer to variations in its environment. These responses occur along gradients, and the location of a deer along a gradient of responses varies in time and space. The bedded posture of a deer may result in a maximum bed area, produced by lateral extension of the limbs and body, or a minimum bed area when the limbs are tightly flexed beneath the erect body. The minimum bed surface area of a 50-kg deer effective in conductive transactions with the substratum may be only one-fourth of the maximum surface area observed when the deer is in full lateral recumbency. The changes associated with body conformation also occur; the radiation and convective profiles of the bedded deer are variable.

Substratum selection for the bedding site also affects the rate and direction of thermal transactions. Snow is a variable physical feature. Further, deer may modify the physical depth of snow over their bedding site by pawing it away, exposing the frozen soil beneath it. They also may return to an old bedsite, removing the new snow accumulation from its icy contours before rebedding on it. Such pawing and digging results in "snow wells" that may extend 5–20 in. into the snow. The depression is shaped to accommodate the tightly curled body by repeated turning in the bedsite area and pawing away of the snow until the soil or vegetation is exposed. Bedding in the depression modifies thermal transactions occurring between the substratum and the ventral hair surface of the deer, radiation exchange between the deer and the walls of the snow well, the radiation profile of the deer exposed to the sky, and convective exchange by modifying its location along the velocity profile. A deer bedded on top of a snow surface would be approximately 33 cm higher along the velocity profile, would have a larger radiation profile exposed to the sky, would have a different loss of heat by conduction to the substratum, and may be in a lower ambient temperature regime.

The thermal integrity of the hair coat is an important determinant of alternatives selected. One deer bedded in February in $-9°C$ and 3–9 mi h^{-1} wind velocity had a telemetered heart rate 100 percent higher and a breathing rate 23 percent higher than on another day 2 weeks earlier when weather conditions were similar except for a sleet-like snow driven by the wind into the hair coat. Further, the deer shivered throughout the snow storm.

Characteristics of the deer hair coat which are related to its potential range of thermal responses vary in time and space, however. Physical depth of the hair coat varies regionally from the thick hair of the trunk to about one-sixth that thickness in the lower legs. The hairs of the winter coat reach their maximum physical length and depth over a period of about 4 months. The winter hair coat, which begins to appear in August, is about one-fourth developed in October, one-half in November, and fully developed by late December. Thus the position of the deer along the gradient of coat development affects its range of alternatives and ability to moderate thermal transactions.

Deer in standing posture may exhibit idling behavior at one extreme, or at the other extreme, running activity in response to varying environmental stimuli. Each affects the rate of thermal transactions. When the wind velocity exceeds several mi h^{-1} in winter, deer will bed down or stand idle for longer durations than in lower wind velocities. A running deer increases the movement of air past its hair coat, thus increasing convective thermal transactions. Running activity is maintained by increased metabolic transactions, which in turn affect the distribution of heat in the deer's body. Changes in ventilation of the hair coat or distribution of blood flow further modify the thermally effective surfaces of the deer and its total heat loss.

The duration of bedding is another variable that deer exhibit during the winter. Deer at the BioThermal Laboratory, for example, have remained bedded through the duration of an overnight snowstorm, allowing 4 to 6 inches of snow to accumulate over their tightly curled bodies, further modifying the total insulation interposed between the metabolic tissues of the deer and its thermal environment. Low cardiac and respiratory rates of deer exhibiting this behavior suggest an alternative strategy of energy conservation rather than increased energy expenditure in these conditions.

Similar responses in deer of all weights suggest that the animals have thermal receptors that are not weight-related. Perhaps these receptors are located, in part at least, on the bare skin of the nose or within the nares. Differences in nose sizes are likely of little importance as long as the thermal receptors are developed. The ecological significance of this is of interest since it relates to the position of the animal on a productivity gradient. For example, let us suppose that the physical heat exchange from the thermal surface to the ambient environment is a function, in part, of body size. The smaller animal retains the disadvantages predicted with the physical model. The larger animal does not necessarily wait for colder conditions to employ some of the alternative responses discussed; it employs them as a response to the thermal stimuli received by both small and large alike. Thus the larger animal may be employing responses when they are less necessary than for the smaller animal. This gives the larger animal the added advantage of being equally as responsive physiologically while it is deriving benefits from being physically larger.

The smaller deer has the additional weight-related disadvantages of smaller fat depots and rumen capacity; a small deer is not physiologically equivalent to the large deer in its ability to tolerate weight loss. Many other factors tend to act as constraints on the smaller deer. The above examples and many others are discussed further in Jacobsen (1973), Moen (1973), and Robbins (1973).

Conclusions

The development of more complex physical models of white-tailed deer concomitant with the measurement of physiological and behavioral responses in vivo has resulted in considerably greater insight into the animal-range relationship than was possible before the use of telemetry. The overall analyses become more complex, the research program becomes more unwieldy because of the different instrumentation needed and the greater dependency on natural events, and progress appears to slow because of the need to adapt to the animal's biology. The last characteristic is an extremely important one, however. A wild animal under stress due to handling procedures does not show the same responses that a free-ranging one does. The deer in the experiments at the BioThermal Laboratory are used to gain insight into their responses, not to perform in accord with our convenience.

The extension of this in vivo work to population ecology will focus attention on the living rather than on the dead. In the past, population ecology has often been approached by the analyses of dead animals, collected by trapping, hunting, or some other means. None of these animals can contribute to population dynamics. The living members are constantly moving along a productivity gradient (Moen, 1973), dynamic in space and time. Careful analyses of population productivity, controlled by the physiological and behavioral characteristics of each individual, will result in new insights into the role of a species in the larger ecosystem.

Acknowledgments

Financial support has been provided by the New York State Department of Environmental Conservation through Pittman-Robertson Project W-124-R, the Loyalhannah Foundation, and the Agricultural Experiment Station, Cornell

University. The assistance of W. P. Armstrong, Cheryl Drake, Joanne Lian, and Lynda L. Marrone is acknowledged.

References

Birkebak, R. C., Le Febvre, E. A., Raveling, D. G.: 1966. Estimated heat loss from Canada geese for varying environmental temperatures. Minn. Museum Nat. Hist. Tech. Rept. *11*.

Davis, L. B., Jr.: 1972. Energy transfer in fur. Lexington, Ky.: Univ. Kentucky College Eng. *UKY BU-101*.

Dawson, T. J., Denny, M. J. S., Hulbert, A. J.: 1969. Thermal balance of the Macropodid Marsupial *Macropus eugenii* desmarest. Comp. Biochem. Physiol. *31*, 645–653.

Evans, K. E.: 1971. Energetics of sharp-tailed grouse (*Pedioecetes phasianellus*) during winter in western South Dakota. Ithaca, N.Y.: Cornell Univ. Ph.D. diss.

Gessaman, J. A.: 1973. Ecological energetics of homeotherms. Logan, Utah: Utah State Univ. Press.

Heller, H. C., Gates, D. M.: 1971. Altitudinal zonation of chipmunks (*Eutamias*) energy budgets. Ecology *52*(3), 424–433.

Jacobsen, N. K.: 1973. Physiology, behavior, and thermal transactions of white-tailed deer. Ithaca, N.Y.: Cornell Univ. Ph.D. diss.

Kestin, J.: 1966. The effect of free-stream turbulence on heat transfer rates. *In* Advances in heat transfer (ed. T. F. F. Irving, Jr., J. P. Hartnett), pp. 1–32. New York: Academic Press.

McNab, B. K.: 1963. A model of the energy budget of a wild mouse. Ecology *44*(3), 521–532.

———: 1966. The metabolism of fossorial rodents: a study of convergence. Ecology *47*(5), 712–733.

Moen, A. N.: 1966. Factors affecting the energy exchange of white-tailed deer, western Minnesota. Minneapolis, Minn.: Univ. Minnesota. Ph.D. diss.

———: 1968a. The critical thermal environment: a new look at an old concept. BioScience *18*(11), 1041–1043.

———: 1968b. Energy balance of white-tailed deer in the winter. N. Am. Wildlife and Nat. Resources. Conf. Trans. *33*, 224–236.

———: 1968c. Surface temperatures and radiant heat loss from white-tailed deer. J. Wildlife Management *32*(2), 338–344.

———: 1973. Wildlife ecology: an analytical approach. San Francisco: W. H. Freeman.

———: Turbulence and the visualization of wind flow. In press, Ecology.

Porter, W., Gates, D. M.: 1969. Thermodynamic equilibria of animals with environment. Ecol. Monogr. *39*, 245–270.

Robbins, C. T.: 1973. The biological basis for the determination of carrying capacity. Ithaca, N.Y.: Cornell Univ. Ph.D. diss.

Sellers, W. D.: 1965. Physical climatology. Chicago: Univ. Chicago Press.

Spotila, J. R., Lommen, P. W., Bakken, G. S., Gates, D. M.: 1973. A mathematical model for body temperatures of large reptiles: implications for dinosaur ecology. Am. Naturalist *107*(955), 391–404.

Stevens, D. S.: 1972. Thermal energy exchange and the maintenance of homeothermy in white-tailed deer. Ithaca, N. Y.: Cornell Univ. Ph.D. diss.

Tucker, V. A.: 1969. The energetics of bird flight. Sci. Am. *220*(5), 70–78.

Wathen, P., Mitchell, J. W., Porter, W. P.: 1971. Theoretical and experimental studies of energy exchange from jackrabbit ears and cylindrically shaped appendages. Biophys. J. *11*, 1030–1047.

West, G. C.: 1968. Bioenergetics of captive willow ptarmigan under natural conditions. Ecology *49*, 1035–1045.

29

Convective Energy Transfer in Fur

L. Berkley Davis, Jr., and Richard C. Birkebak

Introduction

Motion of the surrounding air increases the transfer of energy from an animal. To apply thermal modeling techniques, one has to know of the importance of convection within the fur relative to conduction and radiation (see Birkebak, 1966; Birkebak *et al.*, 1966; Gates and Porter, 1969).

To find the amount of energy exchanged between the skin and the environment, equate the energy transfer through the fur with that lost from the surface to the surroundings:

$$q_f = h_f(T_s - T_f) = h(T_f - T_\infty) + \varepsilon_f \sigma (T_f^4 - T_{\infty,r}^4) \tag{1}$$

where q_f = rate of energy transfer per unit area of skin (W m^{-2})

T_s, T_f, T_∞, $T_{\infty,r}$ = temperature (°K) of the skin, the outer surface of the fur, the ambient air, and the external radiation sink

σ = Stefan–Boltzmann constant, 5.67×10^{-8} Wm^{-2} $°K^{-4}$

ε_f = emittance of the fur surface, which has been found to be very close to unity (see Birkebak *et al.*, 1966; Hammel, 1956; Davis, 1972)

The heat-transfer coefficient h (W m^{-2} $°K^{-1}$) applies to convection from the outer surface of the fur, while h_f(W m^{-2} $°K^{-1}$) is a heat-transfer coefficient within the fur and arises from motion of air in the fur layer. The coefficient h_f is the subject of this paper. Radiation and conduction are treated in detail by Davis (1972) and Davis and Birkebak (1974, 1971).

Before discussing the energy transfer within the fur, we shall briefly review the pertinent ideas about convective heat transfer for a common basis of discussion. The third part of this paper reviews the pertinent experimental results available, and the final two parts present the analytical treatment of the results.

Convective Heat Transfer

When a fluid of temperature T_∞ moves over a surface of temperature T_s, the rate of energy transfer per unit area of surface, q, can be calculated from the relation

$$q = h(T_s - T_\infty) \tag{2}$$

Equation 2 is sometimes called Newton's law of cooling. In the present usage, rather than being regarded as a law, it is simply considered the definition of the heat-transfer coefficient. The fundamental problem of convection is to determine the value of h using analytical or experimental methods.

The heat-transfer coefficient is a function of the type of flow field over the surface, fluid properties, shape of the surface, boundary conditions along the surface, and time.

Flow fields are laminar, turbulent, or both. Laminar flow is characterized by smooth, ordered stream lines and is the kind of flow occurring within fur. In turbulent flow, macroscopic "chunks" of fluid move in a rapid and irregular fashion. Unless velocities are small, the motion of air or water in natural systems is turbulent.

Since there is no relative motion of the fluid and the surface just at the boundary, all energy transferred from the surface must be removed by conduction. The use of Fouirer's law gives

$$q = -k \left.\frac{\partial T}{\partial y}\right)_{y=0} \tag{3}$$

where k = thermal conductivity of the fluid
y is measured normal to the surface at the point of interest
$\partial T/\partial y)_{y=0}$ = fluid temperature gradient at the surface

The definition of the heat-transfer coefficient can then be written

$$h = \frac{-k \left.\frac{\partial T}{\partial y}\right)_{y=0}}{T_s - T_\infty} \tag{4}$$

If D is a characteristic dimension of the body, Eq. 4 can be rewritten in dimensionless form

$$\mathrm{Nu} \equiv \frac{hD}{k} = -\left.\frac{\partial T'}{\partial \eta}\right)_{\eta=0} \tag{5}$$

where Nu is the Nusselt number, the dimensionless temperature T' is defined as

$$T' = \frac{T - T_\infty}{T_s - T_\infty} \tag{6}$$

and the dimensionless coordinate η is given by

$$\eta = \frac{y}{D} \tag{7}$$

Forced Convection

A specific example is helpful in interpreting Eq. 5. Consider two cylinders of diameter D_1 and D_2, exposed to a cross flow, with surface temperatures of T_{s1} and T_{s2}. For the Nu's to be equal, the dimensionless temperature gradients at each of the surfaces must be equal. This will occur if the two external flows are similar, as determined by the equality of a dimensionless group called the Reynolds number:

$$\mathrm{Re} = \frac{u_\infty D}{\nu} \tag{8}$$

where u_∞ is the velocity of the free stream and ν is the kinematic viscosity of the fluid. Additionally, a relationship must exist between the properties of the fluids as evidenced by equality of the Prandtl number,

$$\mathrm{Pr} = \frac{\nu}{\alpha} \tag{9}$$

where the thermal diffusivity of the fluid, α, is

$$\alpha = \frac{k}{\rho c_p} \tag{10}$$

with ρ the mass density and c_p the specific heat of the fluid. Then

$$\mathrm{Nu} = fn(\mathrm{Re}, \mathrm{Pr}) \tag{11}$$

and the Nusselt number for flow over any constant temperature circular cylinder in cross flow can be calculated if the functional relationship in Eq. 11 is known. The rate of energy transfer is then

$$q = \mathrm{Nu}\,\frac{k}{D}(T_s - T_\infty) \tag{12}$$

The forms of Eq. 11 for many different geometries and boundary conditions are available in the literature. Additionally, the fact that Nu depends on Re and Pr can be used for experimental design, particularly when analytical solutions are difficult. This technique has not been fully exploited in bioenergetics.

Free Convection

A temperature difference between a body and the surrounding fluid obviously generates a temperature field. If the bulk motion of the fluid is small or if the fluid is stationary, bouyancy forces generated by temperature variations can be effective in creating a velocity field. Just as for forced convection, this fluid motion affects the rate of energy transfer. It has been found that a dimensionless group called the Rayleigh number, Ra, can be used to relate values of Nusselt numbers for similar conditions.

$$\mathrm{Nu} = fn(\mathrm{Ra}, \mathrm{Pr}) \tag{13}$$

where

$$\mathrm{Ra} = \frac{g\beta(T_s - T_\infty)D^3}{\nu\alpha}$$

with g the body force (gravity in the present case) and β the fluid compressibility at constant pressure:

$$\beta = -\frac{1}{\rho}\frac{\partial \rho}{\partial T}\bigg)_{\text{const. press.}} \tag{14}$$

Forms of Eq. 13 are also available in the literature.

In most actual conditions, both forced and free convection occur. However, one or the other mode is dominant if the ratio $\mathrm{Ra}/\mathrm{Re}^2$ is not of order unity.

Energy Equation

In energy transfer through a moving fluid, a balance exists between conduction of energy through the fluid and convection of energy by the moving fluid. This balance is mathematically expressed by the energy equation, here written for two-dimensional flow:

$$\rho c_p\left(u\frac{\partial T}{\partial x} + v\frac{\partial T}{\partial y}\right) = -k\left(\frac{\partial^2 T}{\partial x^2} + \frac{\partial^2 T}{\partial y^2}\right) \tag{15}$$

In this equation, ρ, c_p, and k are fluid properties. The temperature of the fluid at coordinates x and y is T, and u and v, respectively, are the x and y components of velocity.

The left-hand side of Eq. 15 gives the energy convected by the moving fluid through a small volume element surrounding the point x, y; the right-hand side is the conduction energy transfer through the volume element. The latter is driven by temperature gradients, and, in the absence of fluid motion, Eq. 15 reduces to the steady-state conduction equation.

Energy Transfer in Fur

Energy transfer in fur occurs under conditions somewhat different from those in man-made fibrous insulating materials. The most important difference is that the surface of fur is exposed to the ambient environment. For this reason, studies of energy transfer in fur must consider external radiation and flow fields as well as the different factors influencing the total thermal conductivity.

The experiments concerned with total thermal conductivity of fur and the effect of external flows on energy transfer in fur are reviewed in this section. The results and the requirements to be satisfied in a theoretical development are discussed briefly at the end of this section.

Throughout this paper, the terms skin, fur, pelt, and grain of fur denote the following. The skin is the external layer of tissue covering an animal's body. Fur is the mixture of hairs and air covering the body surface of certain mammals. The region occupied by fur is bounded by the skin and by the outer fur surface, through which no hairs pass. The term "pelt" refers to the fur and skin. Hairs grow from the skin in such a fashion that the projections of the axes onto a plane parallel to the skin are all very nearly parallel to one another. The grain of the fur refers to a direction parallel to the average direction of these projections, whose positive sense is from the roots toward the tips of the hairs.

Total Thermal Conductivity

Scholander *et al.* (1950) measured an "insulation" and the fur thickness of the pelts of 18 arctic and 16 tropical mammals. Two samples of pelt from the same animal were placed on each side of a guarded circular plate hung vertically in a room at 0°C. The plate was heated until its surface temperature was 37°C. The insulation of the sample was defined as 37°C, divided by the energy input to the plate. The insulation thus included free convection and radiation effects external to the fur, the skin conductivity and thickness, and the total fur conductivity and thickness.

Temperature profiles in the fur layer were reported for the pelts of two animals. The temperature gradient was found to be steep and nearly constant close to the skin, but was smaller, in absolute value, near the outer surface. This was attributed to fewer hairs in the outer portions of the fur.

Hart (1956) measured insulation values of winter and summer pelts from several arctic animals and found that the value for the summer pelts was from 50 to 90 percent of that for the winter pelts. He ascribed this difference mainly to an increase in the hair length rather than to changes in the size or number of hairs.

Hammel (1955, 1953) has reported extensive measurements of total conductivities near the skin in the fur of different animals. The apparatus, a guarded circular hot plate, was operated in the horizontal position in a chamber filled with any of several gasses: air, CO_2, or Freon-12. The sample was heated from below, and plate temperatures were maintained so that the fur-side skin temperature was near 37°C. Total conductivities were defined as the energy transfer through the fur divided by the temperature gradient within 10 mm of the skin. Quantities corresponding to the same type of insulation values measured by Scholander and some temperature profiles near the skin were also given.

In a personal communication in 1971, Hammel and Thorington cited results which indicate the absence of free convection near the skin. Tests were run with the fur side of the skin facing in several directions—in particular, up or down. Temperature gradients and energy fluxes in the various tests were not significantly different.

Hammel also made measurements to determine the type of energy transfer in fur layers. Either a dead animal (Hammel, 1953) or a pelt (Hammel *et al.*, 1962) was placed on a hot plate and exposed to the clear night sky. The energy input to the hot plate was measured and a radiometer was used to measure the radiation temperature of the sky and the fur. A radiation shield was then placed between the sky and the surface so that the fur was facing a radiation sink at air temperature. With the shield in place, it was found that the energy input to the plate, equal to the total energy loss from the fur, decreased by about 10 percent, while the energy transferred by radiation from the outer surface decreased by 50 percent. The measurements of the difference between the fur surface and the air temperature (Hammel *et al.* 1962) showed that convective energy transfer at the fur surface became more important when the shield was in place. In each case, the temperature on the back side of the skin was maintained constant.

Moen (1968) reported similar measurements on living white-tailed deer and concluded that the difference between the air temperature and the effective temperature at which the fur radiates increases as the air temperature decreases. This was previously noted by Birkebak *et al.* (1964). These experiments indicate that radiation must be considered, but free convection can be neglected, within a fur layer.

Berry and Shanklin (1961) and Berry *et al.* (1963) have attempted to correlate the total thermal conductivity of the hair of live cattle with each of the following: (1) the ratio of fur thickness to the length of hairs, (2) the number of hairs per unit area, (3) the effective density, and (4) the hair diameter. They found the conductivity to depend on all these factors to various degrees, although the manner in which interactions among them occurred was uncertain.

The experimental values of total thermal conductivity calculated from the results of Scholander *et al.*, Hart, and Hammel are given in Table 29.1. Also included in this table are values from work cited in the next section.

Effects of External Flow

Wind passing over an animal's body gives rise to a pressure field that can cause air movement in fur. This will have a large effect on the energy transfer.

To aid in the following discussion, Eq. 1, the energy transfer from an animal, can be written as

$$q_f = (T_s - T_\infty)\Bigg/\left[\frac{1}{h_f} + \frac{1}{h + \sigma(T_f^4 - T_{\infty,r}^4)/(T_f - T_\infty)}\right] \tag{1a}$$

and a total heat-transfer coefficient, taking into account both the effects of the fur and the surroundings, can be defined as

$$h_t = 1\Bigg/\left[\frac{1}{h_f} + \frac{1}{h + \sigma(T_f^4 - T_{\infty,r}^4)/(T_s - T_\infty)}\right] \tag{16}$$

Table 29.1 Total Thermal Conductivity of Animal Fur

Animal	$k \times 10^2$ ($W m^{-1} {}^\circ K^{-1}$)	Animal	$k \times 10^2$ ($W m^{-1} {}^\circ K^{-1}$)
Shrew[a]	2.303	Red fox[d]	4.012
Weasel[a]	2.721	Raccoon[d]	4.838
Lemming[a]	5.024	Rabbit[d]	4.012
Marten[a]	4.187	House cat[d]	3.745
Rabbit[a]	3.559	Muskrat[d]	4.117
White fox[a]	4.396	Oppossum[d]	4.989
Red fox[a]	5.245	Deer[d]	3.919
Beaver[a]	5.245	Skunk[d]	4.014
Dog[a]	5.443	Husky dog[d]	4.222
Wolf[a]	5.443	Snowshoe hare[d]	3.919
Dall sheep[a]	5.652	Lynx[d]	4.222
Caribou[a]	4.187	Caribou[d]	3.745
Grizzly bear[a]	6.059	Bobcat[d]	3.826
Polar bear[a]	6.280	Coyote[d]	3.501
Caribou[b]	4.830	Caribou[e]	4.489
Wolverine[c]	4.396	Deer[e]	3.556
Wolf[c]	3.559	Arctic wolf[e]	4.591
Polar bear[c]	4.396	Raccoon[e]	2.756
		Air[e]	2.617

[a] Scholander *et al.* (1950).
[b] Lentz and Hart (1960).
[c] Hart (1956).
[d] Hammel (1955).
[e] Moote (1955).

Lentz and Hart (1960) measured energy loss from circular samples of fur taken from newborn caribou. The samples were inclined at various angles to a flow whose velocity ranged from zero to 22.9 m s^{-1}. The total heat-transfer coefficient, h_t, including skin, fur, and boundary layer outside the fur, was reported. The largest heat-transfer coefficient occurred for stagnation point flow.

At velocities above 7.6 m s^{-1}, the total heat-transfer coefficients measured for flow with the grain of the fur are not equal to those measured for flow against the grain. As compared to the total heat-transfer coefficients for flow against the grain, the values of h_t for flow with the grain are higher when the sample inclination is 0° and are smaller when the inclination of the sample is 45°. The difference in the measured values of h_t at 45° inclination is evidently caused by flattening of the fur when the flow is with the grain and by penetration of the fur by the free stream when the flow is against the grain. Since the sample holder interfered with the free stream at 0° inclination, it is not possible to determine why the values of h_t are different for that case.

Approximate values of total thermal conductivities can be obtained from these measurements if the heat-transfer coefficient over the fur surface is known. The heat-transfer coefficient for stagnation point flow on a flat circular plate can be calculated (Kays, 1966) from the equation

$$\mathrm{Nu} = 0.93\mathrm{Re}^{1/2}\mathrm{Pr}^{0.4} \tag{17}$$

The Nusselt and Reynolds numbers are based on the radius of the circular plate.

The total energy transfer from the fur surface, q_t, is

$$q_t = (q_R - q_d) + h\,\Delta T \tag{18}$$

where q_R = radiation from the fur surface
q_d = background radiation from the walls of the wind tunnel
h = calculated heat-transfer coefficient
ΔT = temperature difference between the fur surface and the air stream

It is assumed that the effective surface temperature for radiation is equal to the fur-surface temperature and that the wind tunnel is an enclosure at the temperature of the air. Then q_t is given by

$$q_t = \sigma(T_f^4 - T_\infty^4) + h(T_f - T_\infty) \tag{19}$$

where σ = Stefan–Boltzmann constant
T_f = surface temperature
T_∞ = free-stream air temperature
h = heat-transfer coefficient

This equation can be solved for T_f when q_f, T_∞ and h are known. A second equation

$$q_t = k_{\text{eff}} \frac{T_s - T_f}{L} \tag{20}$$

defines the effective thermal conductivity of the fur plus skin. The temperature T_s is that at the backside of the skin and L is the thickness of the fur.

Since q_t, T_w, T_∞, L, and u_∞ are given by Lentz and Hart, the calculations described above can be performed. The values obtained for k_{eff}, the total heat-transfer coefficients reported for three sample inclinations with flow in the direction of the grain, and the heat-transfer coefficient for laminar stagnation point flow on a flat circular plate of the same diameter as the sample are shown in Fig. 29.1. The value of thermal conductivity at the small velocities is given in Table 29.1. The calculations indicate that no substantial penetration of the fur occurs at the lower two air velocities. At higher values of u_∞, k_{eff} increases in proportion to the square root of free-stream velocity u_∞.

The total heat-transfer coefficient shows a distinct change in slope at a particular value of u_∞ above which h_t increases at the same rate as h. This value of u_∞, the penetration velocity, u_p, increases as the inclination of the sample to the free stream decreases.

The thickness of the sample is affected by the external flow. Depending on the degree to which the external air stream flattens the fur, the thermal conductivities calculated above are all too large by a factor equal to the actual fur thickness divided by the fur thickness with u_∞ equal to zero.

The use of the laminar stagnation-point heat-transfer coefficient may be unjustified, for two reasons. First, if external air penetrates the fur to a significant extent, the flow patterns are not those used in deriving the formula for h. Second, the fur surface is not regular, and even if flattened may, especially at higher velocities, cause the flow to become locally turbulent.

Experiments on live sheep performed by Joyce and Blaxter (1964) lend insight

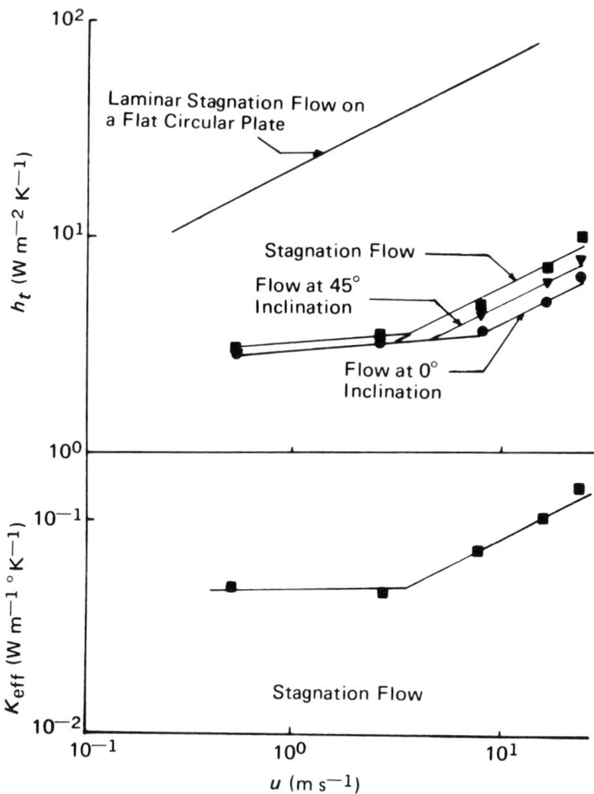

Fig. 29.1. Effect of external flow on the effective thermal conductivities and heat-transfer coefficient of caribou fur.

into the effect of air movement on living animals. In these experiments, sheep were exposed to air streams directed from either side or the rear. Eight thermocouples were affixed to the skin, six on the trunk and two on the legs, and the skin temperatures measured under various external conditions. The metabolic energy generation of the animal was measured in each experiment, and the energy loss from evaporation was estimated from previous experiments. The work was performed in a cold room with known air temperature. Experiments on a given animal were performed for several fleece lengths and, in some cases, with a source of infrared radiation in the cold room. Air velocities were from 0.9 to 4.3 m s^{-1}.

The skin temperature of sheep in a cross wind was found to vary 9°C from the upwind to the downwind sides of the trunk and it was essentially constant on each side of the fetlock. The mean skin temperature generally decreased and the metabolic energy generation increased as the air velocity increased. For a given air temperature, lower values of metabolic energy generation and higher skin temperatures were found at comparable air velocities when the fleece was longer.

The infrared radiation source was found to cause a large decrease in the metabolic energy generation when the velocity of the external air stream and the air temperature were maintained constant.

The total heat-transfer coefficient, including the effect of the fleece and the external air movement, can be obtained by inverting the reported insulation values. It is not possible to calculate the heat-transfer coefficient h since the size of the animals is unknown.

Total heat-transfer coefficients for the cross-flow case are shown versus free-stream velocity in Fig. 29.2. The free and forced convection heat transfer coefficients on a bare circular cylinder of diameter 0.45 m are given for comparison.

The values of h_t were obtained from tests on one of the sheep with fleece of the lengths noted in the figure. The external air temperature, T_∞, was constant.

The variation of h_t with u_∞ is similar to that shown in Fig. 29.1, although u_∞ is too low for the penetration velocity to be well defined.

Moote (1955) measured insulations of various pelts exposed to an external air stream. The air stream was inclined at 45° to the normal of the surface covered by the pelt, and several velocities were used. Temperatures at three locations—the backside of the skin, the outer surface of the fur, and in the external air—and energy loss were measured.

Two insulations, that of the fur and the skin alone and that of the portion of the boundary layer outside the sample are reported and indicate penetration of thick furs for relatively low free-stream velocities; furs that are thinner and have fewer hairs show a rather constant insulation (fur alone) up to wind speeds of about 6.2 m s^{-1}. The results for the insulation of the air alone are in fair agreement with those of Drake (1949) for isothermal inclined plates.

The results for the insulation of the fur and skin alone again indicate that there is some penetration velocity below which the effective thermal conductivity is constant. The magnitude of the penetration velocity is found to be different for different fur samples.

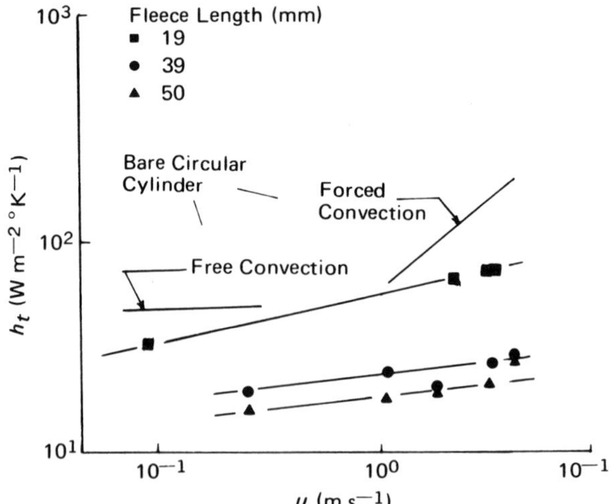

Fig. 29.2. Effect of external flow on total heat-transfer coefficients of the fleece of a sheep.

Thermal conductivities, including both the skin and fur, can be obtained from the data reported and are included in Table 29.1.

Summary: Conduction and Convection in Fur

The mechanisms of energy transfer in fur are examined in this section with the results of the previously cited literature taken into account. Conduction energy transfer takes place in fur through both the hairs and the air. The axes of the hairs are usually at an angle to the direction of energy flow, so that the thermal conductivity of fur is directionally dependent. The contribution of solid conduction is generally not negligible since the arrangement of hairs is uniform and the conductivity of the hairs is higher than that of air. Free convection is evidently not important near the skin. Near the outer surface there may be only a few hairs and free convection will occur.

An external air stream will cause movement of the air in the fur layer. The minimum external velocity necessary for this air movement to cause significant transfer of energy is the penetration velocity. It will depend on the fur structure and orientation of the animal relative to the direction of the free stream.

Energy transfer through the fur by radiation is small but not negligible. Because the outer surface is exposed to a radiation sink at a temperature different from air temperature, radiant energy escaping from deep layers of the fur is not balanced by any back radiation as in the case of fibrous insulation. This influences the temperature profile in the fur. If, for example, the radiation sink temperature is less than the air temperature, less energy must be carried by conduction near the outer surface, and the absolute value of the temperature gradient is smaller.

The structure of fur determines the energy-transfer characteristics. This structure is anisotropic. If the transfer of a quantity G occurs because of spatical variations of a second quantity R, the principal directions are defined as those particular directions in which the amount of G transferred depends only on the gradient of R in the particular direction. The most convenient coordinate system for the treatment of the fur problem does not necessarily coincide with the principal directions of the fur.

The development of equations describing energy transfer in fur must then take into account the anisotropic structure of fur, conduction through both the solid hairs and the air, radiation, and convection. The theory should indicate the relative importance of free convection and the magnitude of the penetration velocity.

Air Motion in Fur

To understand the role of convective heat transfer in fur, we must first have knowledge of the motion of air in the fur layer. In this section we present equations that describe the motion of air in fur and give the results of an experiment to verify the applicability of these equations. The equations are particularly complex and the interested reader is referred to Davis (1972) for greater detail. However, the conclusions drawn about forced convection can be summarized by stating that there exists a particular velocity, called the penetration velocity, u_p, and if the value of the free-stream velocity, u_∞, exceeds u_p, account must be taken of the convective energy transfer within the fur. If u_∞ is less than u_p, the effect of air movement within the fur is of negligible importance in the energy-transfer mechanism. The value of u_p can be predicted from the properties of the fur and air. The conclusion about free convection is that it is not an important mechanism of energy transfer within fur.

These conclusions are demonstrated in the following paragraphs.

Permeability of Fur

A section of fur is indicated schematically in Fig. 29.3. The x direction is aligned with the grain of the fur, the y direction is normal to the skin, and the z direction is normal to both x and y, so x, y, z form a right-handed coordinate system.

Since the axes of the hairs are not parallel to the x, y, z coordinates, fur constitutes an anisotropic medium. In this type of medium there are particular directions, called principal directions, so that a force applied in one of these directions will generate motion in that direction only.[1] The three principal directions of the fur are designated as X_1, X_2, and X_3 in Fig. 29.3. It can be seen that z coincides with

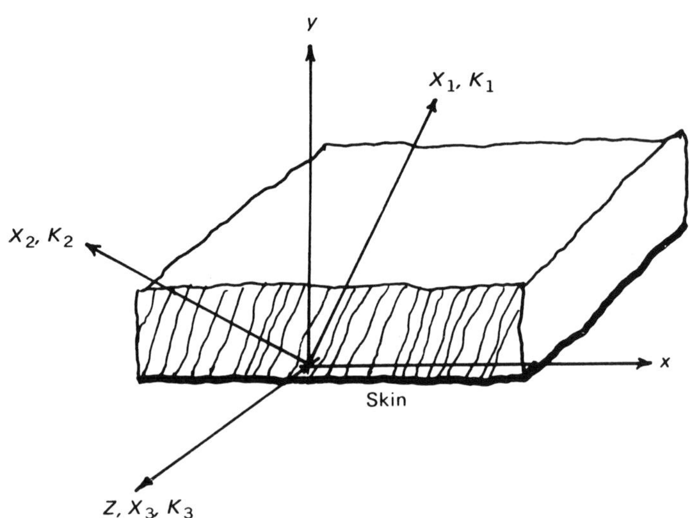

Fig. 29.3. Anisotropic conductivity of fur.

[1] The same applies to temperature gradients and energy transfer.

X_3, but that x and y are not principal directions. Thus a force in the z direction will generate motion in that direction. But a force along either of the x or y coordinates results in fluid motions in both the x and y directions.

The set of equations describing the motion of air in a layer of fur are derived from the Navier–Stokes equations of fluid mechanics applied to very slow flow. The complete derivation of these equations is given in Davis (1972). The equations are similar to those presented by Chan *et al.* (1970) for an isotropic media and are based on Darcy's law.

The results of the derivation are

$$u = \frac{K_x}{\mu}\left(-\frac{\partial p}{\partial x} + \rho g_x + \mu\,\nabla^2 u\right) + \frac{K_{xy}}{\mu}\left(-\frac{\partial p}{\partial y} + \rho g_y + \mu\,\nabla^2 v\right) \tag{21}$$

$$v = \frac{K_{xy}}{\mu}\left(-\frac{\partial p}{\partial x} + \rho g_x + \mu\,\nabla^2 u\right) + \frac{K_v}{\mu}\left(-\frac{\partial p}{\partial y} + \rho g_y + \mu\,\nabla^2 u\right) \tag{22}$$

$$w = \frac{K_z}{\mu}\left(-\frac{\partial p}{\partial z} + \rho g_z + \mu\,\nabla^2 w\right) \tag{23}$$

and

$$\frac{\partial u}{\partial x} + \frac{\partial v}{\partial y} + \frac{\partial w}{\partial z} = 0 \tag{24}$$

where the Laplacian operator, ∇^2, is given by

$$\nabla^2 = \frac{\partial^2}{\partial x^2} + \frac{\partial^2}{\partial y^2} + \frac{\partial^2}{\partial z^2} \tag{25}$$

where u, v, w = components of velocity (m s^{-1}) in the x, y, and z directions
p = fluid pressure (n m^{-2}) at a given point in the fur
ρ = fluid density (kg m^{-3})
μ = fluid viscosity (kg m^{-1} s^{-1})
g_x, g_y, g_z = body-force components in each of the x, y, and z directions (in the present case, only gravity is of interest)

The permeabilities K_x, K_y, K_{xy}, and K_z are discussed below. Fluid properties are assumed constant.

Since compressibility effects are not important in the present application, the appropriate form of the mass continuity equation reduces to Eq. 24. The first term within the parentheses of Eqs. 21, 22, and 23 represent the pressure forces applied to the fluid at a point in the fur, and the second term gives the body forces. The remaining terms arise from viscous forces.

The permeability is best described by briefly considering an isotropic porous material. The velocity of the fluid for any given direction ξ is

$$u_\xi = -\frac{K}{\mu}\frac{\partial p}{\partial \xi} \tag{26}$$

This is Darcy's law when effects of body forces are unimportant. Equation 26 is taken as the definition of the permeability K. For the case of fur, three values of permeability exist, K_1, K_2, and K_3, each applicable to one of the three principal directions X_1, X_2, and X_3. It follows that

$$\begin{aligned} K_x &= K_1 \sin^2 \theta_f + K_2 \cos^2 \theta_f \\ K_y &= K_1 \cos^2 \theta_f + K_2 \sin^2 \theta_f \\ K_{xy} &= (K_1 - K_2) \cos \theta_f \sin \theta_f \\ k_z &= K_3 \end{aligned} \tag{27}$$

The angle θ_f is that between the axis of the hairs and the y direction.

Darcy's law, as given by Eq. 26, suggests that the viscous terms $\nabla^2 u$, $\nabla^2 v$, and $\nabla^2 w$ are not significant in Eqs. 21, 22, and 23. It can be shown that these terms are important only within a distance $K^{1/2}$ from a bounding surface, such as the skin, of the porous material.

A series of experiments, using commercially tanned hides, was performed to measure the permeability of fur. The apparatus was designed so that the v and w velocity components were zero.

A sectional view of the experimental apparatus is shown in Fig. 29.4. Air from a laboratory compressed-air line is introduced at the left, passes through the fur baffle, through the sample, and exhausts to the atmosphere. The pressure drop across the sample is measured with a micromanometer across the pressure taps.

From Eqs. 21–23, the velocity in the x direction is given by

$$u = \frac{K}{\mu}\left(-\frac{\partial p}{\partial x} + \mu \frac{\partial^2 u}{\partial y^2}\right) \tag{28}$$

where

$$K = \frac{K_1 K_2}{K_1 \cos^2 \theta_f + K_2 \sin^2 \theta_f} \tag{29}$$

The mean velocity in the channel for a constant pressure gradient is

$$\bar{u} = -\frac{K}{\mu}\frac{\partial p}{\partial x}\left[1 - \frac{2}{(L/K)\tanh(L/K)} + \frac{2}{(L/k)\sinh(L/K)}\right] \tag{30}$$

Since the pressure gradient $\partial p/\partial x$, $\bar{u}$, and the sample thickness (corrected for skin thickness) L were measured, the permeability K can be calculated from Eq. 30.

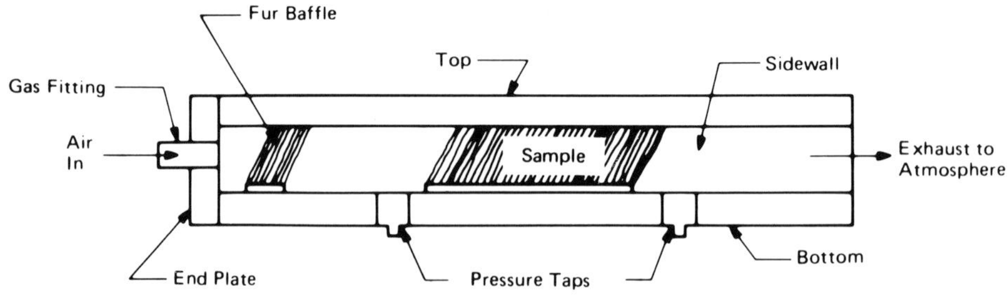

Fig. 29.4. Experimental apparatus for measurement of fur permeability.

Table 29.2 Results of Permeability Measurements

L (mm)	$(\bar{u}A/\Delta p) \times 10^6$ ($m^4\ skg^{-1}$)	$K \times 10^9$ (m^2)
	Deer	
13.9	1.98	2.82
17.1	2.70	3.13
17.8[a]	3.13	3.49
18.6	3.88	4.14
20.8	5.83	5.30
17.8[b]	3.70	4.12
	Rabbit	
4.9	1.33	5.38
7.2	1.99	5.48
7.2	1.87	5.15
7.2	1.95	5.37
12.0	3.47	5.73
	Raccoon	
15.1	4.44	5.83
18.2	7.77	8.39

[a] Not shown in Fig. 29.5.
[b] Flow against grain.

The results are given in Table 29.2. In all cases the effects of $\partial^2 u/\partial y^2$, which generates the second two terms in the brackets, are negligible.

The experimental results are also shown in Fig. 29.5. The product of $\bar{u}$ and the sample cross-sectional area A is plotted versus the pressure drop across the sample. The value of the sample thickness L is shown next to each line. Note, also, that the permeability K is a function of L.

Convection Energy Transfer in Fur

Air movement in the fur will have the greatest effect when it occurs on the trunk of an animal. To assess the importance of convection, the trunk is modeled as a circular cylinder, as shown in Fig. 29.6, the skin temperature T_s is assumed constant, and the grain of the fur is assumed to be coincident with the axis of the cylinder. Effects of curvature are neglected since the thickness of the fur is small compared to the body diameter. The velocity parallel to the axis of the cylinder is assumed to be zero.

The energy equation, Eq. 15, can be written in a dimensionless form by lumping the conduction and radiation heat-transfer terms and writing

$$\rho C_p\left(w\frac{\partial T'}{\partial z} + v\frac{\partial T'}{\partial y}\right) = k_{\text{eff}}\frac{\partial^2 T'}{\partial y^2} \tag{31}$$

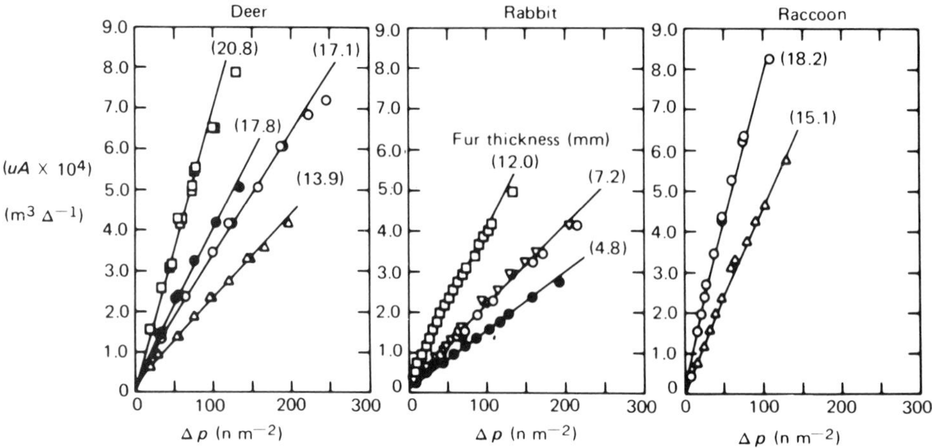

Fig. 29.5. Experimental results of permeability measurements on deer, rabbit, and raccoon fur.

where k_{eff} is the effective thermal conductivity of the volume element. Then, using the continuity equation for an incompressible fluid,

$$\frac{\partial w}{\partial z} + \frac{\partial v}{\partial y} = 0 \tag{32}$$

it follows that the order of magnitude of the velocity component is

$$v = 0\left[\frac{wL}{r_b}\right] \tag{33}$$

where r_b is the body radius and is the characteristic distance in the z direction.

If a reference velocity u_{ref} is chosen so that $u_{\text{ref}} = 0[W]$, then the energy equation becomes (see Davis, 1972)

$$\frac{k_g}{k_{\text{eff}}} \Pr \text{Re}_{1,u_{\text{ref}}} \frac{L}{r_b}\left(\frac{w}{v}\frac{\partial T'}{\partial \xi} + \frac{\partial T'}{\partial \eta}\right) = \frac{\partial^2 T'}{\partial \eta^2} \tag{34}$$

where $\eta = y/L$, $\xi = z/L$, and $\text{Re}_{L,u_{\text{ref}}} = u_{\text{ref}}L/\nu$. The right-hand side of this equation and the term in parentheses on the left-hand side are both of order unity. Thus, if convective effects are important in the transfer of energy in fur, it is necessary that

$$\frac{k_g}{k_{\text{eff}}} \Pr \text{Re}_{L,u_{\text{ref}}} \frac{L}{r_b} = 0[1] \tag{35}$$

For an animal with a body radius r_b of 0.2 m, a fur thickness L of 0.02 m, and for $k_{\text{eff}} = 1.4k_g$, the Reynolds number and reference velocity are equal to

$$\text{Re}_{L,u_{\text{ref}}} = 20 \qquad u_{\text{ref}} = 0.016 \text{ m s}^{-1}$$

The velocity w is of the same order as u_{ref}.

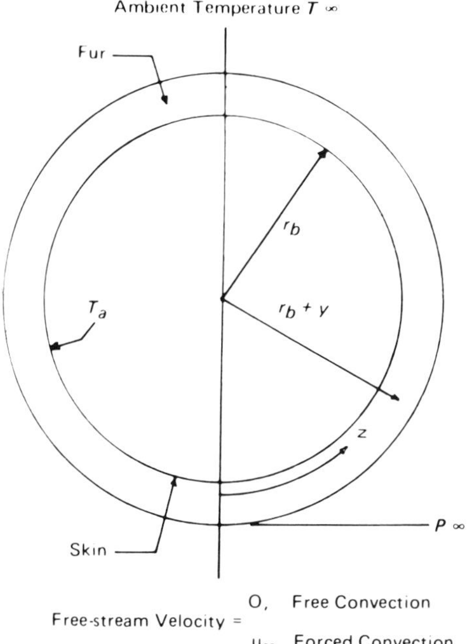

Fig. 29.6. Convection over a horizontal fur-covered layer.

The value of the velocity w which actually occurs in each of the cases of free and forced convection is estimated by applying Eq. 23 to the z direction.

Free Convection

The coordinate system and conditions pertaining to free convection are indicated in Fig. 29.6. The surrounding atmosphere is taken to be isothermal at temperature T_∞ and the pressure is p_∞ at the location indicated. The velocity u_∞ is zero for the case of free convection.

The pressure at any point in the fur is written as

$$p = p_\infty - \left[(r_b + L) - (r_b + y)\cos\left(\frac{z}{r_b}\right)\right]\rho_\infty g + p_d \tag{36}$$

where p_d is a dynamic pressure arising because of the motion of the air and ρ_∞ is the fluid density at a pressure of p_∞ and a temperature of T_∞.

The body-force term acting in the z direction at a given location is

$$\rho g_z = -\rho g \sin\left(\frac{z}{r_b}\right) \tag{37}$$

The compressibility factor at constant pressure, β, is defined as

$$\beta \equiv -\frac{1}{\rho}\frac{\partial \rho}{\partial T}\bigg|_p \cong -\frac{1}{\rho_\infty}\frac{\rho - \rho_\infty}{T - T_\infty} \tag{38}$$

Then, using the pressure, body-force terms, and the approximation for β, the velocity component w, Eq. 23, is given by

$$\frac{\mu w}{K_z} = -\frac{\partial p_d}{\partial z} + \rho_\infty \beta g(T - T_\infty)\sin\left(\frac{z}{r_b}\right) + \mu\,\nabla^2 w \tag{39}$$

and

$$w = -\frac{K_z}{\mu}\frac{\partial p_d}{\partial z} + \left[\frac{g\beta(T_s - T_\infty)L^3}{\nu^2}\frac{K_z}{L^2}\frac{\nu}{L}\right]\frac{T - T_\infty}{T_s - T_\infty}\sin\left(\frac{z}{r_b}\right) + K_z\,\nabla^2 w \tag{40}$$

where $\nu = \mu/\rho_\infty$ is the kinematic viscosity and T_s is the skin temperature. Since viscous effects are very small except near the fur surfaces, and since p_d is proportional to velocity squared, w is of the order of the term in brackets.

Thus the order of magnitude of w is given by

$$w = 0\left[\mathrm{Gr}_L \mathrm{Da}_L \frac{\nu}{L}\right] \tag{41}$$

where $g\beta(T_s - T_\infty)L^3/\nu^2$ has been identified as the Grashof number, Gr, based on L. In terms of the previous discussion given in Davis (1972), $\mathrm{Gr} = \mathrm{RaPr}(k_{\mathrm{eff}}/k_g)$. The velocity w can be estimated using the following data for an animal with a fur thickness of 20.0 mm, a uniform skin temperature of 305°K, a fur permeability of 5×10^{-9} m^2 (see Table 29.2), and for $T_\infty = 255$°K, the Grashof number is $\mathrm{Gr}_L = 9.2 \times 10^4$ and the order of magnitude of w is $w = 0[0.0007 \text{ m s}^{-1}]$. Note that this number is actually independent of fur thickness and that the temperature difference is not small.

The velocity that occurs because of free convection in fur is thus much smaller than that necessary for the convective terms in the energy equation to be important. This leads to the conclusion that heat transfer by free convection is not an important energy-transfer mechanism within the fur of an animal in its normal environment. This is in agreement with the experimental results cited previously. However, free convection does occur over the outer surface of the fur.

Forced Convection

Heat transfer by forced convection is important when the free-stream velocity u_∞ exceeds the penetration velocity, as discussed previously. The magnitude of this penetration velocity is estimated for cross flow over a fur-covered cylinder (Fig. 29.6) (with T_∞ again the temperature of the surrounding atmosphere) by noting that the pressure drop in the z direction is

$$\frac{\partial p}{\partial z} \sim \frac{\rho u_\infty^2/2}{r_b} \tag{42}$$

From Eq. 23, with body force and viscous terms neglected,

$$w = \frac{K}{\mu}\left(-\frac{\partial p}{\partial z}\right) = 0\left[\frac{K}{\mu}\frac{\rho u_\infty^2/2}{r_b}\right] = 0[u_{\mathrm{ref}}] \tag{43}$$

Then it follows

$$\mathrm{Re}_{r_b,u_\infty} = 0\left[\left(2\frac{r_b}{L}\,\mathrm{Re}_{L,u_{\mathrm{ref}}}\frac{r_b^2}{K}\right)^{1/2}\right] \tag{44}$$

and so

$$\mathrm{Re}_{r_b,u_\infty} = 5.66 \times 10^4$$
$$u_\infty = 3.68\ \mathrm{m\ s^{-1}}\ (\text{approx. } 8\ \mathrm{mi\ h^{-1}})$$

with the values of $\mathrm{Re}_{L,u_{\mathrm{ref}}}$, r_b, L, and K used previously.

This velocity is equivalent to a light breeze, and therefore effects of forced convection must be taken into account in fur layers. A method for doing so has been suggested by Cremers and Birkebak (1967) and is described in more detail in the next section.

The effect of the external flow on heat transfer in the fur of an animal is also related to the action of the flow field in deforming the fur. The fur can be flattened, or it can be ruffled. The structure of the fur of some smaller animals can be completely disrupted. The external flow in the natural environment is inherently unsteady, and so the fur is deformed in a time-dependent fashion.

An animal that cannot seek shelter and is exposed to wind under conditions threatening its survival or causing discomfort must orient itself relative to the wind direction to minimize the disturbance of its fur layer. This generally means that the animal faces into the wind. Birds behave similarly; for instance, the albatross habitually sits facing into the wind.

Summary

The energy equation representing heat transfer in fur layers and the terms appearing therein are given by Davis (1972) and Davis and Birkebak (1974, 1971) and have been discussed briefly in the previous sections. All the terms in the energy equation are derived to ensure that the effects of the anisotropy of the fur are completely taken into account. We also concluded that effects of free-convection heat transfer are negligible in the fur and the forced-convection effects are small only at low wind speeds.

Application of the Theory

Methods of applying the equations developed in the previous sections to the problem of energy transfer in fur are described in this section. Two separate cases are considered, one with and the other without forced convection in the fur.

The outer surface of the fur is assumed to coincide with the surface of the underfur in both cases. The effect of the guard hairs outside the underfur surface is thus assumed negligible.

Energy Transfer by Conduction and Radiation

The skin temperatures of animals are constant over large portions of the body surface in the absence of air movement in the fur (Hammel *et al.*, 1962; Lentz and Hart, 1960). It follows then that energy-transfer terms involving temperature gradients parallel to the skin, $\partial T/\partial x$ or $\partial T/\partial z$, are small and can be neglected.

Two methods of applying the theoretical results to animal systems can be considered. The simpler of the two methods gives a value of the effective thermal conductivity of the fur layer, k_{eff}, which is then used in the thermal modeling equations as first applied in Birkebak (1966) and Birkebak *et al.* (1966). It does not, however, specify the temperature field in the fur. The second method involves the solution of the energy equation under the restrictions that $\partial T/\partial x$ and $\partial T/\partial z$ are set equal to zero and that solar radiation is absent (for details see Davis, 1972).

For purposes of calculating energy losses from animal systems, the theoretical value of k_{eff} is of greater utility than knowledge of the temperature profile within the fur layer. The methods of obtaining k_{eff} are presented in papers by Davis (1972) and Davis and Birkebak (1974).

Energy Transfer with Forced Convection in the Fur

The influence of free-stream air movement on the energy transfer in fur layers can be handled in the manner first suggested by Cremers and Birkebak (1967). By definition, the forced-convection heat transfer is given as

$$q_f(u_\infty) = h_f(T_s - T_f) = h_f\,\Delta T_f \tag{45}$$

and the forced-convection Nusselt number is defined as

$$\text{Nu}_f = \frac{h_f L}{k_{\text{eff}}} \tag{46}$$

Combining Eqs. 20, 45, and 46, we obtain

$$q_f(u_\infty) = q_f \text{Nu}_f \tag{47}$$

where q_f is the energy transfer from the fur in the absence of an external flow. The results of Lentz and Hart (1960) and Moote (1955) can be correlated by an expression similar to Eq. 47. The air temperature and external radiation fields are the same for both q_f and $q_f(u_\infty)$.

Based on the discussions given in Davis and Birkebak (1971) and herein, an analogous equation for the effective thermal conductivity is

$$\text{Nu}_f = \frac{k_{\text{eff}}(u_\infty)}{k_{\text{eff}}} = f(\text{Re, Da}) \tag{48a}$$

The Darcy number is (K/L^2) and the Reynolds number, Re, is $(u_\infty l/\nu)$, where l is some reference length related to body size. The left-hand side of this equation is actually a Nusselt Number.

The effective thermal conductivity k_{eff} is found using the method described in the previous section and in Davis (1972) and Davis and Birkebak (1974).

There is a paucity of data which can be used to evaluate the functional form of Eq. 48a. However, realizing the above, we shall use the available results of Moote (1955) and Lentz and Hart (1960) to provide information for an initial formulation of Eq. 48a. The results of Moote (1955) are presented in Table 29.3, recalculated according to the discussion presented in this paper. These results, along with those of Lentz and Hart (1960), were plotted for h_t versus u_∞ on log-log paper. Distinct breaks occur in the curves, as shown in Fig. 29.1. The penetration velocity is determined from these plots where the break occurs. The values of u_p for Moote's (1955) results are presented in Table 29.3. The results of this analysis are shown in

Table 29.3 Forced-Convection Heat Transfer from Fur[a]

	h_t (W m^{-2} s^{-1})	u_∞ (m s^{-1})	$u_\infty^{1/2}$	k_{eff}/L (W m^{-2} s^{-1})	k_{eff} (W m^{-1} s^{-1})
Winter caribou	0.884	0	0	1.012	0.0332
L = 32.8 mm	0.955	2.68	1.64	0.99	0.0323
u_p = 6.9 ms^{-1}	1.050	5.14	2.27	1.11	0.0364
	1.165	7.15	2.67	1.225	0.0402
	1.309	9.39	3.06	1.367	0.0448
	1.619	12.52	3.54	1.647	0.0540
Summer caribou	1.770	0	0	2.274	0.0271
L = 11.9 mm	2.077	2.46	1.57	2.274	0.0271
u_p = 6.5 ms^{-1}	2.171	4.25	2.06	2.274	0.0271
	2.274	6.04	2.46	2.514	0.0299
	2.514	7.60	2.76	2.653	0.0316
	3.412	11.40	3.38	3.412	0.0406
Arctic wolf	1.016	0	0	1.209	0.0331
L = 27.4 mm	1.111	2.01	1.42	1.279	0.0349
u_p = 4.9 ms^{-1}	1.225	3.80	1.95	1.365	0.0374
	1.405	5.81	2.41	1.592	0.0436
	1.619	7.82	2.88	1.837	0.0503
	2.171	12.07	3.47	2.171	0.0595
Muskrat	2.171	0	0	3.184	0.0218
6.86 mm	2.514	2.46	1.57	2.985	0.0205
u_p = 7.6 ms^{-1}	2.653	4.25	2.06	2.985	0.0205
	2.810	6.04	2.46	3.184	0.0218
	3.184	8.05	2.84	3.412	0.0234
	5.307	12.21	3.51	5.430	0.0350
Raccoon	1.327	0	0	1.571	0.0207
13.2 mm	1.447	2.46	1.57	1.541	0.0203
u_p = 2.9 ms^{-1}	1.837	4.25	2.06	1.990	0.0263
	2.171	6.04	2.46	2.388	0.0315
	2.388	7.6	2.76	2.653	0.0350
	2.985	11.62	3.41	3.674	0.0485

[a] After Moote, 1955.

Fig. 29.7. Below a certain penetration velocity, the effective thermal conductivity is constant and for free-stream velocities greater than the penetration velocity, the effective thermal conductivity increases with increasing Re (i.e., u_∞). This indicates that a fundamental change occurs in the mechanism of energy transfer above this penetration velocity. On the basis of these and other results, we suggest that the air in the fur is forced into motion by the action of the external stream.

We are able to correlate all the data as Nu_f versus $(u_\infty/u_p)^{1/2}$. This exponent occurs, to the best of our present knowledge, because of the particular experimental situation (stagnation and 45° inclined flow). We cannot claim that it will always be the correct choice. However, for flow over circular cylinders, $\mathrm{Nu} \sim \mathrm{Re}_d^{1/2}$, and it seems reasonable that our fit is acceptable. For $u_\infty > u_p$, Eq. 48a can be written as

$$\mathrm{Nu}_f = \frac{k_{\mathrm{eff}}(u_\infty)}{k_{\mathrm{eff}}(0)} = -0.43 + 1.43\left(\frac{u_\infty}{u_p}\right)^{1/2} \tag{48b}$$

If we use an average value for $u_p = 5.5\ \mathrm{m\ s^{-1}}$, then Eq. 48b becomes

$$\frac{k_{\mathrm{eff}}(u_\infty)}{k_{\mathrm{eff}}(0)} = -0.43 + 0.612\sqrt{u_\infty} \tag{48c}$$

It is suggested that the function in Eq. 48a be determined from results of body-cast or live-animal experiments. These experiments should be performed for flow parallel to the trunk and for flow across the trunk.

Once this function is known, effective thermal conductivities can be obtained for cases when an external flow causes air movement in the fur.

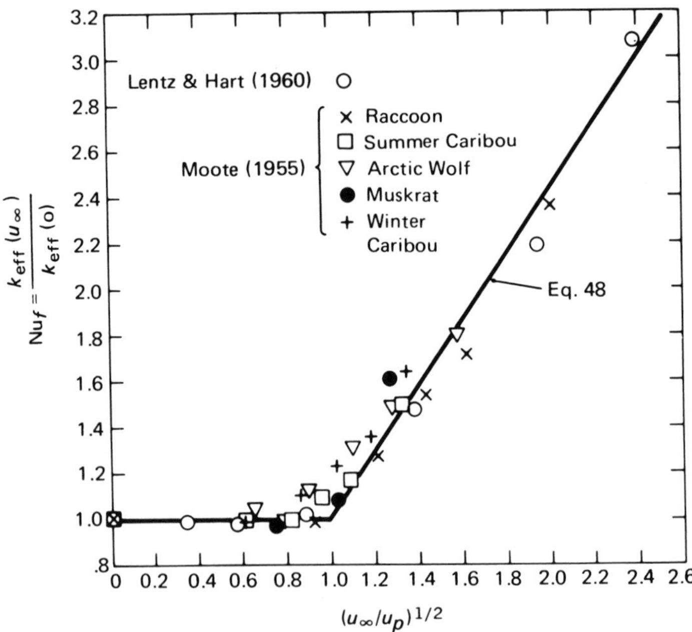

Fig. 29.7. Effect of Nusselt number in fur as a function of external velocity.

Other Considerations

Incident solar radiation can greatly change the energy balance of an animal system. Such effects as those of external radiation fields on the effective thermal conductivity have not been studied experimentally in sufficient detail to be discussed here. However, Davis (1972) treats external radiation fields theoretically.

If solar radiation is negligible or does not penetrate the fur significantly, the effective thermal conductivity, k_{eff} can be evaluated as discussed above and in Davis (1972) and Davis and Birkebak (1974). If, however, solar radiation does penetrate, then account must be taken of the amount and distribution of energy absorbed within the fur. This is quite difficult for convection and will require either special experiments or solution of (a simplified form of) the governing equations.

Conclusions

The conclusions drawn from this study are concerned with the mechanisms of energy transfer and air flow in fur, the validity of the methods described for calculating effective thermal conductivity, and the experimental data that must be obtained if these methods are to be used.

In the absence of forced convection, the most important mechanism of energy transfer through an animal's fur is conduction through the air in the fur. The combined effect of radiation and conduction through the hairs is not negligible and depends on the structure of the fur. Free convection is not an important mechanism of energy transfer in the fur.

Forced convection is most easily taken into account by a method similar to that proposed by Cremers and Birkebak (1967). The air flow in the fur obeys Darcy's law and is driven by the pressure field generated by an external flow field.

Acknowledgment

The financial support of the National Science Foundation, grant GB-15579, is gratefully acknowledged.

References

Berry, I. L., Shanklin, M. D.: 1961. Environmental physiology and shelter engineering. Univ. Missouri Agric. Expt. Sta. Res. Bull. *802*.

———, Shanklin, M. D., Johnson, H. D.: 1963. Thermal insulation of bovine hair coats as related to physical hair characteristics. Am. Soc. Agric. Eng. 56th Ann. Meeting, Paper 63–404.

Birkebak, R. C., Birkebak, R. C., Warner, D. W.: 1964. Total emittance of animal integuments. J. Heat Transfer *86*, 287–288.

———: 1966. Heat transfer in biological systems. *In* Intern. Rev. Gen. and Expt. Zool. (ed. W. S. L. Felts, R. J. Harrison), vol. 2, pp. 268–344. New York: Academic Press.

———, Cremers, C. J., Lefebvre, E. A.: 1966. Thermal modeling applied to animal systems. Trans. Am. Soc. Mech. Engrs., Ser. C, J. Heat Transfer *88*, 125–130.

Cremers, C. J., Birkebak, R. C.: 1967. Forced convection heat transfer from biological surfaces. Am. Soc. Mech. Engrs. Paper *66-WA/HT-55*.

Chan, B. K. C., Ivey, C. M., Barry, J. M.: 1970. Natural convection in enclosed porous media with rectangular boundaries. Trans. Am. Soc. Mech. Engrs., Ser. C, J. Heat Transfer *92*, 21–27.

Davis, L. B.: 1972. Energy transfer in fur. Univ. Kentucky College Eng. *UKY Bu-101*. Ph.D. diss.

Davis, L. B., Birkebak, R. C.: 1971. An analysis of energy transfer through fur, feathers, and other fibrous materials. Univ. Kentucky High Temp. and Thermal Rad. Lab., *HTL-TR-5*.

———, Birkebak, R. C.: 1974. On the transfer of energy in layers of fur. Biophys. J. *14*, 249–268.

Drake, R. M.: 1949. Investigation of the variation of point unit heat transfer coefficients for laminar flow over an inclined flat plate. Trans. Am. Soc. Mech. Engrs., J. Appl. Mech. *71*, 1–8.

Gates, D. M., Porter, W. P.: 1969. Thermodynamic equilibrium of animals with environment. Ecol. Monogr. *39*, 245–270.

Hammel, H. T.: 1953. A study of the role of fur in the physiology of heat regulation in mammals. Cornell Univ. Ph.D. diss.

———: 1955. Thermal properties of fur. Am. J. Physiol. *182*, 369–376.

———: 1956. Infrared emissivities of some arctic fauna. J. Mammalogy *37*, 375–381.

———, Houpt, T. R., Anderson, K. L., Skenneberg, S.: 1962. Thermal and metabolic measurements on a reindeer at rest and in exercise. Tech. Doc. Rep. *AAl-TOR-61-54*, Arctic Aeromed. Lab. Aero-space Med. Div., Air Force Systems Command, Fort Wainwright, Alaska.

Hart, J. S.: 1956. Seasonal changes in insulation of the fur. Can. J. Zool. *34*, 53–57.

Joyce, J. P., Blaxter, K. L.: 1964. The effect of air movement, air temperature and infrared radiation on the energy requirements of sheep. Brit. J. Nutr. *18*, 5–27.

Kays, W. M.: 1966. Convective heat and mass transfer. New York: McGraw-Hill.

Lentz, C. P., Hart, J. S.: 1960. The effect of wind and moisture on heat loss through the fur of newborn caribou. Can. J. Zool. *38*, 679–689.

Moen, A. N.: 1968. Surface temperatures and radiant heat loss from white-tailed deer. J. Wildlife Management *32*, 338–344.

Moote, I.: 1955. The thermal insulation of caribou pelts. Textile Res. J. *25*, 832–837.

Scholander, P. F., Walters, V., Hock, R., Irving, L.: 1950. Body insulation of some arctic and tropical mammals and birds. Biol. Bull. *99*, 225–236.

30

Conduction and Radiation in Artificial Fur

D. J. Skuldt, W. A. Beckman, J. W. Mitchell, and W. P. Porter

Introduction

Energy exchange between an animal and its microclimate has been of interest to zoologists and, more recently, bioengineers. One important aspect is the energy exchange through the fur covering an animal. Zoologists and others have collected data on insulation properties of various animal pelts (Berry and Shanklin, 1961; Hammel, 1955; Scholander *et al.*, 1950; Tregear, 1965). Somewhat related to this has been research investigating the mechanisms of heat transfer in fibrous insulating materials (Davis and Birkebak, 1971; Finck, 1930; Hager and Steele, 1967; Lander, 1954; Rowley *et al.*, 1951; Strong *et al.*, 1960; Verschoor and Greebler, 1952). We attempt here to tie these studies together and establish the relationships between fur properties and the mechanisms of conduction and radiation heat transfer in fur.

The possible avenues of heat transfer in fur are conduction through the hairs, convection and conduction through the air in the fur, radiation between surfaces within the fur and to the environment outside the fur, and evaporation. These different modes are interrelated in a complicated manner. The relative importance of these different mechanisms is determined by the individual hair properties, such as diameter, length, volume fraction, thermal conductivity, and emissivity. The effect of these fur properties on the different modes of heat transfer can be determined by isolating the different avenues. In the present study, radiation and conduction modes have been investigated.

Fur Properties

It is difficult to determine the properties of real fur, owing to the inherent variability in animal pelts. Consequently, we used artificial fur to provide a consistent and repeatable experimental medium in most of our tests. The artificial fur was manufactured by Norwood Mills, Inc., Janesville, Wisconsin. The fur consists of approximately 80 percent underfur and 20 percent guard hairs, in comparison to pelts with an estimated 10 percent guard hairs (Davis and Birkebak, 1971). The hair diameters for the artificial fur were found to be about 26 μm for the underfur, and range from 50 to 80 μm for the guard hairs. Davis and Birkebak (1971) report an average fiber diameter of 20.3 μm for the rabbit, and Tregear (1965) finds the underfur hair and guard diameter to be 15 and 30 μm in diameter, respectively. The underfur of rabbits is somewhat smaller in diameter than that of artificial fur, while the guard-hair diameters were in the same range for the rabbit and artificial fur.

The hair density (number of fibers per unit area) of the artificial fur was determined to be 3600 fibers cm^{-2} with a 95 percent confidence interval of ± 270 fibers cm^{-2}. Values for rabbit fur range from 4100 ± 260 hairs cm^{-2} (Tregear, 1965) to 11,000 hairs cm^{-2} (Davis and Birkebak, 1971). Thus artificial fur is quite similar in its physical aspects to rabbit fur.

The thermal properties of the fibers that make up the fur are accurately known in comparison to those of real fur. The underfur is an acrylic with a thermal conductivity of 0.0298 cal (cm min °C)$^{-1}$. The guard hairs are made of polyacrylonitrile fibers with a conductivity of 0.0372 cal (cm min °C)$^{-1}$ (manufacturer's data).

Theoretical Considerations

Some simple models (Cene and Clark, 1973; Davis and Birkebak, 1971; Hammel, 1955) have been proposed to describe heat transfer in fur. In the absence of convection and evaporation, the total heat flow in fur consists of conduction in the air, conduction in the fiber, and radiation. Thus the total heat flow is given by

$$q_{\text{fur}} = q_{\text{air}} + q_{\text{fibers}} + q_{\text{rad}} \tag{1}$$

Conduction heat flow through fiber and air is described by the basic conduction mechanism

$$q = -\frac{kA\,dT}{dx} \tag{2}$$

where q = heat flow
k = thermal conductivity
A = area
dT/dx = temperature gradient

The conduction heat flux through the air and fibers can be modeled as parallel conductances through the fibers and air. The fur thermal conductivity is then

$$k_{\text{fur}} = k_{\text{fiber}} \frac{A_{\text{fiber}}}{A_{\text{fur}}} + k_{\text{air}} \frac{A_{\text{air}}}{A_{\text{fur}}} \tag{3}$$

The effective thermal conductivity is that which would be measured, and is evaluated as

$$k_{\text{eff}} = -\frac{q_{\text{fur}}}{(dT/dx)A} \tag{4}$$

The effective conductivity can be used to determine a "radiation conductivity." This concept is of value in determining the relative contribution of radiation from experimental data. By definition,

$$k_{\text{rad}} = k_{\text{eff}} - k_{\text{fur}} \tag{5}$$

Models for radiation heat transfer in fibrous insulation have also been proposed (Cena and Clark, 1973; Davis and Birkebak, 1971). Radiation in these materials also depends on the mean temperature, density, fiber diameter, and fiber emissivity. The radiation intensity I in the medium decays exponentially with distance in accordance with Beer's law:

$$\frac{I}{I_0} = e^{-\beta_0 x} \tag{6}$$

where β_0 = extinction coefficient for the fur
I = local intensity
I_0 = incident intensity

Cena and Clark (1973) found the extinction coefficient related to the fur geometry by

$$\beta_0 = nd \tan \phi \tag{7}$$

where n = hair density
d = hair diameter
ϕ = angle the hair makes with the normal to the skin

Cena and Clark have utilized this model to describe radiant energy as decaying exponentially from the skin and being absorbed in planes progressing through the fur. Such a model requires knowledge of the temperature distribution to compute the heat flux; however, the heat flow results from both radiation and conduction and depends on the temperature profile, and a simple integration is thus not possible.

Experimental Methods

The equipment used in the thermal-conductivity tests is on loan from H. T. Hammel, Scripps Institution of Oceanography, La Jolla, California. The facility has been described previously (Hammel, 1955), and is shown schematically in Fig. 30.1. The cylindrical steel chamber is 40 cm in diameter and 38 cm in height, with the capability of either being evacuated or filled with a gas. The chamber is supported on an aluminum frame and may be rotated.

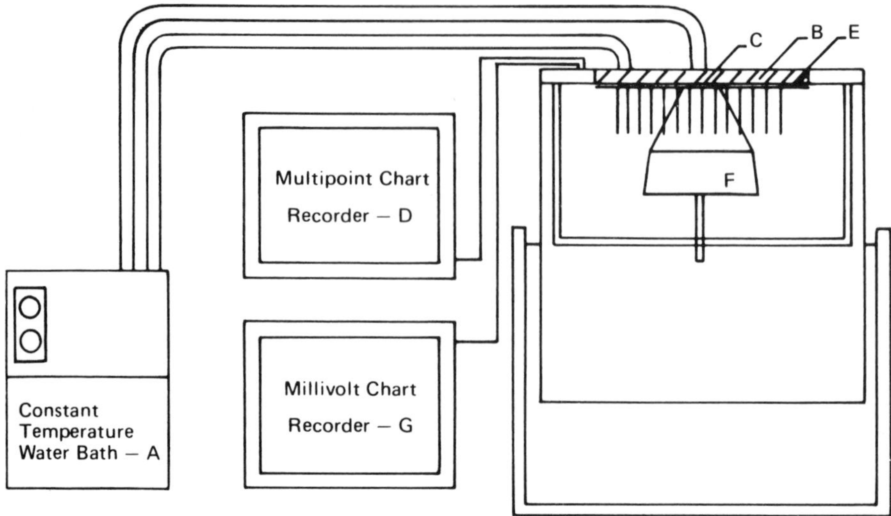

Fig. 30.1. Schematic of inverted test chamber and associated equipment.

Hot water is circulated through a plate at the base of the chamber. A 12.7-cm^2 heat flux meter composed of silver-constantan thermopiles is in direct contact with the circular hot plate. Fur is affixed to the heat-meter surface with double-sided masking tape. A ladder-like array of thermocouples (gradiometer) is placed in the fur to determine the temperature distribution. The copper-constantan thermocouples are 0.19 mm in diameter and spaced as shown in Fig. 30.2. The heat flux

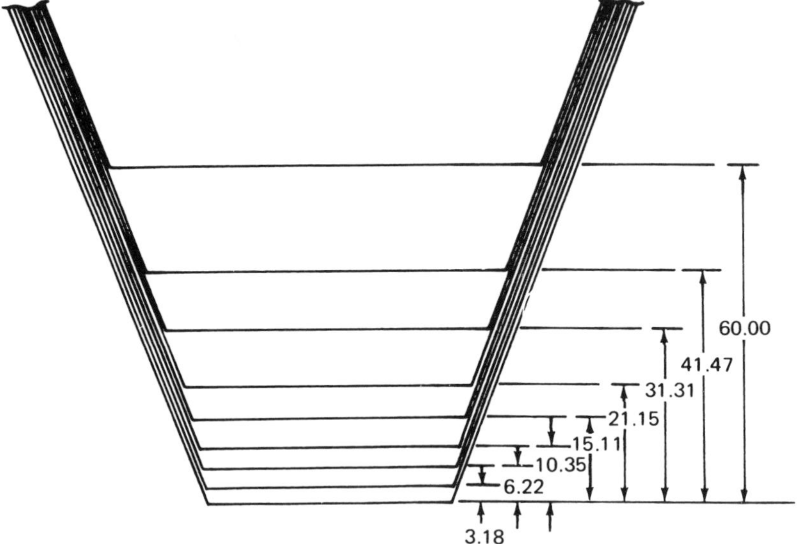

Fig. 30.2. Gradiometer spacings.

meter and thermocouple outputs are recorded continuously on a multipoint-millivolt recorder.

The fiber length and fur density of the samples were varied during testing. An electric hair trimmer was used to shave the fur samples to fiber lengths ranging from 1.27 cm to the original 4.45 cm. The fur density was varied by removing stitches from the samples in a systematic manner to obtain three fiber densities.

The gradiometer was placed in the fur by parting the fur with a wire. The lowest thermocouple was placed in contact with either the skin or the backing. Because the thermocouples were small and because the first could be easily moved and deformed, care was required to get the correct spacing between the first and second thermocouples. The hairs or fibers were arranged to be essentially perpendicular to the skin or backing and essentially parallel to the direction of heat flow. Care was taken to place fibers around the gradiometer and eliminate any gap caused by the part in the fur.

The entire assembly in the chamber was inverted, with the hair hanging downward. This provided a thermally stable environment in the chamber and minimized free convection currents. The water bath was set at the desired temperature, and when steady-state conditions were reached, temperature and heat flux measurements were recorded.

Experimental Results

Temperature profiles were obtained for rabbit fur and for artificial fur samples of different fiber lengths and densities. In Fig. 30.3, temperature profiles for rabbit

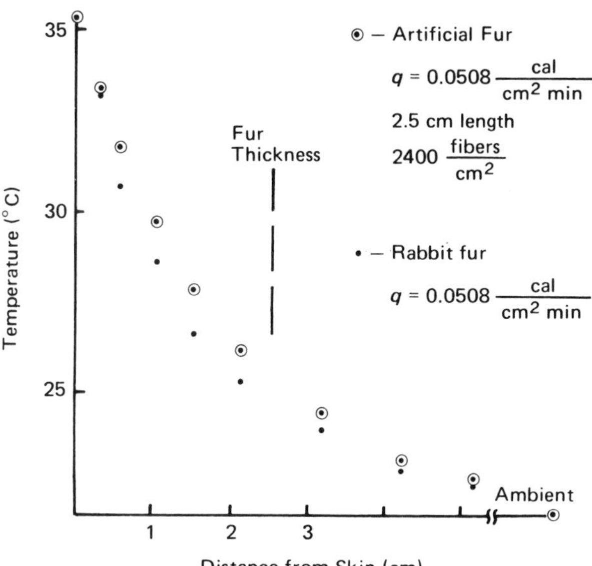

Fig. 30.3. Temperature profiles for artificial and rabbit fur.

and artificial fur are compared. The profiles are quite similar and establish the validity of using artificial fur to simulate real fur.

Forced-convection heat transfer was eliminated by performing the tests in an enclosed chamber with free convection minimized. The magnitude of free convection was determined by placing clear plastic film in direct contact with the outer tips of a few fur samples. The film is virtually transparent to infrared radiation (transmissivity = 0.9), so radiation heat transfer was not appreciably affected. Air conduction and fiber conduction would also not be affected by the film. The temperature profiles for two low-density samples are shown in Fig. 30.4.

As shown in Fig. 30.4, the heat flux decreased slightly with the plastic film on the fur. For the 2400-fibers cm^{-2}, 4.45-cm-length sample, the heat flux decreased 4.5 percent; for the 1200-fibers cm^{-2}, 1.27-cm-length sample, the heat flux decreased 9 percent. We feel that free convection in the outer fur boundary has small influence on the test and that the only significant modes of heat transfer in these tests are conduction and radiation.

The temperature profiles were found to be linear in the region close to the skin. The effective conductivity deep in the fur near the skin was determined from Eq. 4 and found to vary with the fur density and fiber length as shown in Fig. 30.5. It is seen that hair length has a slight effect on the deep-fur conductivity.

Experimentally, the deep-fur conductivity is strongly dependent on fiber density. To compare to these experimental results, predictions of the conduction conductivity were made using Eq. 3, and these are also shown in Fig. 30.5. Two dis-

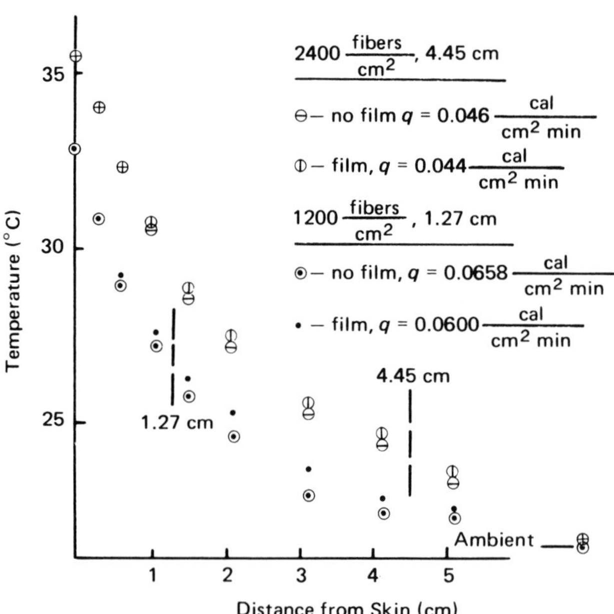

Fig. 30.4. Temperature profiles for artificial fur with and without plastic film.

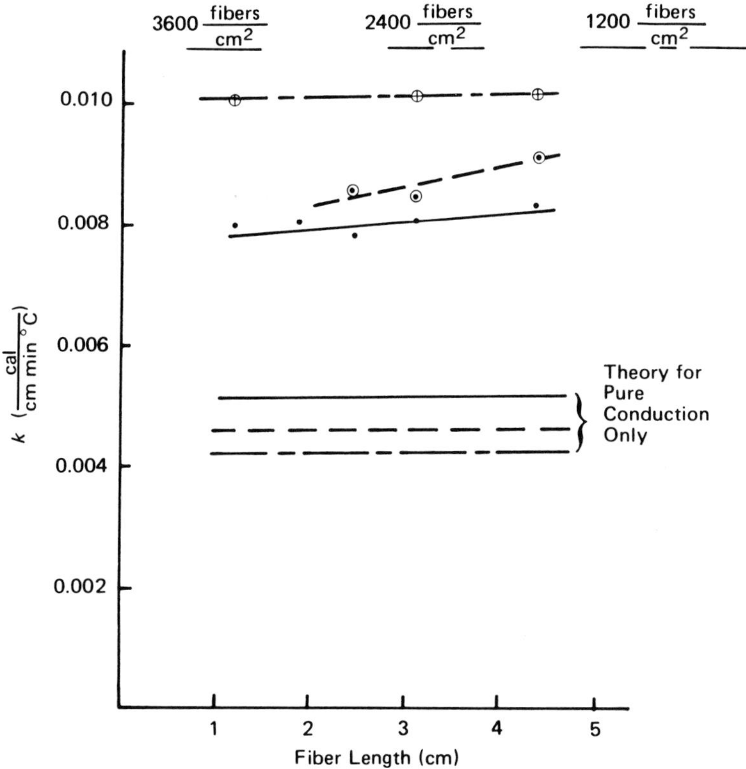

Fig. 30.5. Deep-fur conductivity variation with fur density and fiber length.

crepancies are readily apparent. The actual conductivity is about twice that of the predicted, and indicates that conduction alone is not the only heat transfer mechanism. It is also seen that, for conduction, only Eq. 3 predicts a decrease in conductivity with decreased fiber density due to the lower conductivity of air compared to hair. In contrast to this result, the experiments show increased conductivity with decreased density.

These results agree with the findings of Finck (1930) and Davis and Birkebak (1971) on fibrous insulations. Below a certain density for a given insulation, the conductivity of the material increases as the density decreases, owing to an increase in radiation. However, these results are for optically thick materials, and show the radiation conductivity to be constant with position.

Experimental measurements of the fur conductivity as a function of position are shown in Fig. 30.6. These values are obtained by dividing the heat flux by the local temperature gradient. It is seen that, even though the fur properties remain constant with position, the apparent conductivity increases near the edge of the fur, owing to enhanced radiation transfer. The concept of a radiation conductivity does not appear to be very useful.

To study the influence of radiation further, the emissivity of the surface facing the fur was changed by positioning aluminum foil at the fur fiber tips. The infrared

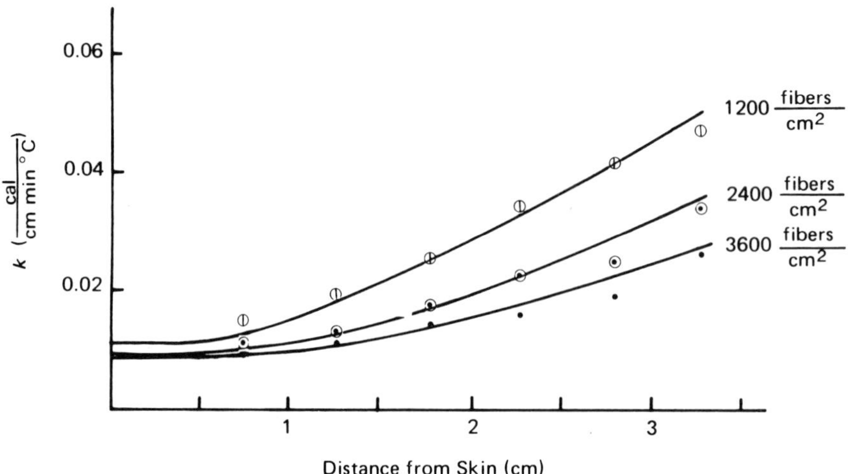

Fig. 30.6. Conductivity as a function of fur length for different densities.

emissivity of the inside of the experimental chamber facing the fur was on the order of 0.9, since the emissivity of both white paint on the chamber walls and thick plexiglass (2.54-cm-thick cover) is approximately 0.9 (Hottel and Sarofim, 1967). When aluminum foil with an emissivity of around 0.1 (reflectivity of 0.9) was placed on the fur, the temperature profiles and heat fluxes were altered significantly as indicated in Fig. 30.7. This experiment, in which radiation is reflected back into

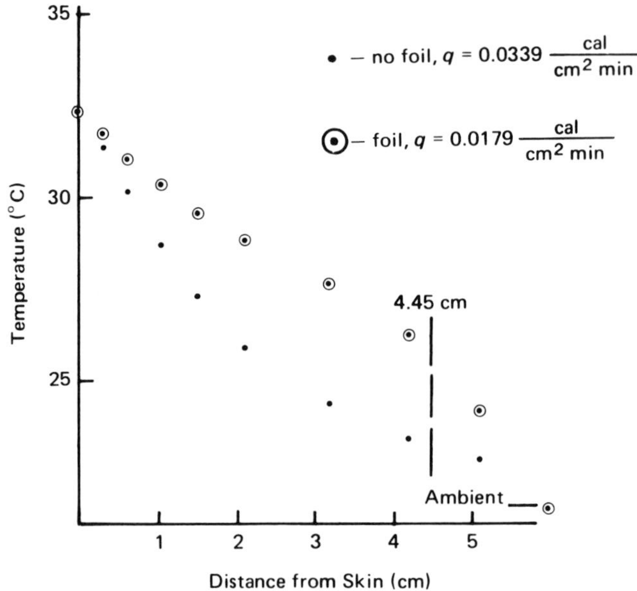

Fig. 30.7. Temperature profiles for artificial fur with and without aluminum foil.

Table 30.1 Extinction Coefficients for Artificial Fur[a]

	Fur Density	
	3600 fibers cm^{-2}	**2400 fibers cm^{-2}**
Backing Orientation	β_0 **(cm^{-1})**	β_0 **(cm^{-1})**
90°	6.17	4.92
60°	4.59	4.24
45°	3.85	3.45

[a] Experimental uncertainty ± 1 cm^{-1}.

the fur, also shows that radiation is an important mode of heat transfer in the outer portion of the fur. It is also seen that the overall fur conductance obtained by dividing the heat flux by the temperature difference between the skin and the air differs by a factor of 2 for the two tests. Thus the radiant environment has a strong effect on heat transfer, and measurements of overall conductance do not represent only fur properties.

Preliminary experiments were performed to evaluate the extinction coefficient β_0 for the fur under test using an infrared spectrograph. The cross-sectional area of the single beam of infrared energy that passed through the samples was approximately 2.5 cm^2. The reference intensity, I_0, was defined as the intensity emerging from a completely trimmed fur sample (fiber length of 0 cm). Two radiation beams of different wave numbers (4200 and 3400) were passed through the sample, and no effect on the extinction coefficient was detected. Fur with different fiber lengths was tested to determine the extinction coefficient. The angles between the beam of energy and the backing of the sample tested were 45°, 60°, and 90°. The effect of fur density and angle is given in Table 30.1. The large error in these coefficients is attributed to the difficulties encountered in trimming the samples to the correct fiber lengths. It is anticipated that these values will aid in formulating models for radiation in fur.

Conclusions

The importance of conduction and radiation heat transfer in fur has been experimentally and analytically investigated. Tests were performed in an enclosed chamber to eliminate forced convection, and free convection was found to account for no more than 10 percent of the heat transfer. Fur conduction and radiation are of about equal magnitude in deep fur, with radiation becoming more important near the edge. Radiation increases as fur density decreases. The use of an overall conductance based on an overall temperature difference between the skin and the environment can lead to significant errors in determining animal heat loss, depending on the radiant environment. These results provide a basis for further modeling of heat-transfer processes in fur.

Acknowledgment

This work was supported by the National Science Foundation, grant GB-31043.

References

Berry, I. L., Shanklin, M. D.: 1961. Environmental physiology and shelter engineering. Univ. Missouri Agric. Expt. Sta. Res. Bull. *802*.

Cena, K., Clark, J. A.: 1973. Thermal radiation from animal coats: coat structure and measurements of radiative temperature. Phys. Med. Biol. *18*, 432–443.

Davis, L. B., Jr., Birkebak, R. C.: 1971. An analysis of energy transfer through fur, feathers, and other fibrous materials. Lexington, Ky.: Univ. Kentucky. Ph.D. diss.

Finck, J. L.: 1930. Mechanism of heat flow in fibrous materials. J. Res. Natl. Bur. Std. *5* (5), 973–984.

Hager, N. E., Jr., Steele, R. C.: 1967. Radiant heat transfer in fibrous thermal insulation. J. Appl. Physics *38*(12), 4663–4668.

Hammel, H. T.: 1955. Thermal properties of fur. Am. J. Physiol. *182*, 369–376.

Hottel, H. C., Sarofim, A. F.: 1967. Radiative transfer. New York: McGraw-Hill.

Lander, R. M.: 1954. Gas is an important factor in the thermal conductivity of most insulating materials, Part II. ASHVE J. Sect. Heating, Piping Air-cond., 121–126 (Dec.).

Rowley, F. B., Jordan, R. C., Lund, C. E., Lander, R. M.: 1951. Gas is an important factor in the thermal conductivity of most insulating materials, Part I. ASHVE J. Sect. Heating, Piping Air-cond., 103–109 (Dec.).

Scholander, P. F., Walters, V., Hock, R., Irving, L.: 1950. Body insulation of some arctic and tropical mammals and birds. Biol. Bull. *99*, 225–236.

Strong, H. M., Bundy, F. P., Bovenkerk, H. P.: 1960. Flat panel vacuum thermal insulation. J. Appl. Physics *31*(1), 39–50.

Tregear, R. T.: 1965. Hair density, wind speed, and heat loss in mammals. J. Appl. Physiol. *20*(4), 796–801.

Verschoor, J. D., Greebler, P.: 1952. Heat transfer by gas conduction and radiation in fibrous insulations. Trans. Am. Soc. Mech. Engrs. *74*, 961–968.

31

Microclimate and Energy Flow in the Marine Rocky Intertidal

Samuel E. Johnson II

Introduction

The rocky intertidal region is a narrow strip between the ocean and land. This region, alternately immersed in water and exposed to air, is often characterized by a great density and diversity of organisms. These are generally organized into distinct horizontal bands or zones (Fig. 31.1). One major cause of this zonation is widely assumed to be the differential responses of intertidal organisms to gradients of physical and chemical conditions resulting from different lengths of exposure and immersion (Stephenson, 1942; Southward, 1958a, 1958b; Newell, 1970). The purpose of this paper is to examine the factors that determine the microclimate conditions of the rocky intertidal and how these may be related to the organism by using an energy-budget approach.

A great number of studies have shown that the tolerance of many marine intertidal organisms to temperature, salinity, and desiccation stress can be correlated with their position in the intertidal, and that organisms living in the higher regions of the intertidal are generally more resistant to environmental stress than those living in lower regions. These studies are reviewed by Gunter (1957), Newell (1970), Kinne (1971), and Vernberg and Vernberg (1972). Although valuable in themselves, their ecological relevance will be greatly enhanced when more is known about the microclimate conditions occurring in the field. It is often important to base physiological studies on naturally occurring conditions. For instance, to examine thermal tolerance limits, the natural heating rates experienced by the organism should be known, as well as the acclimation state of the organism. The latter will vary seasonally and depends on the environmental history of the organism. This type of knowledge may be derived from both long- and short-term monitoring of *in situ* conditions at the intertidal level where the organisms occur, but

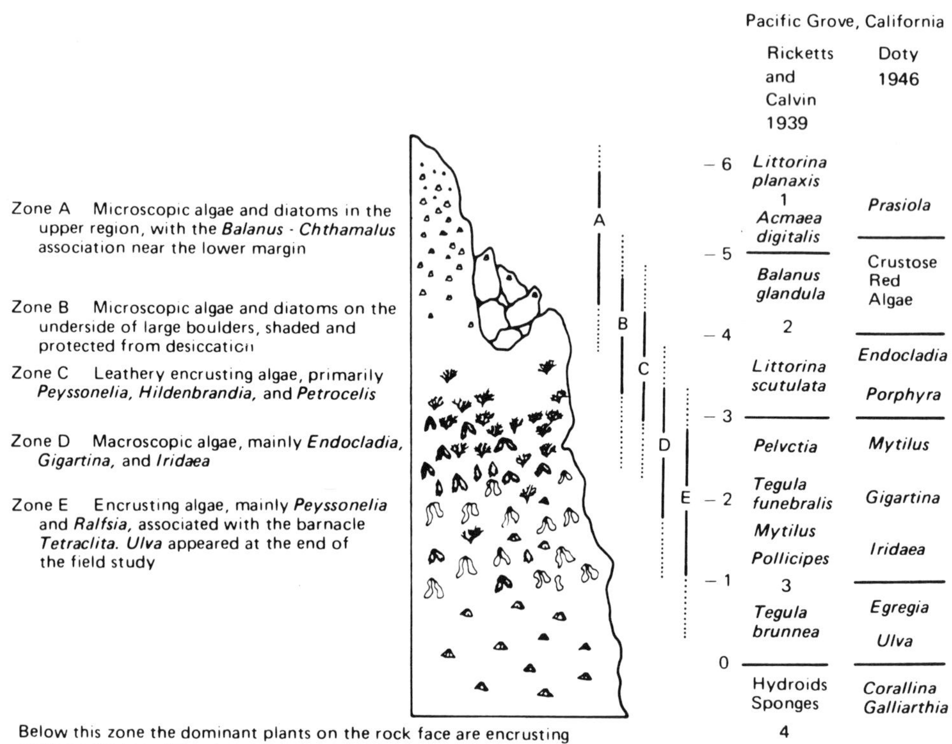

Fig. 31.1. Description and vertical position of rocky intertidal zones at Pescadero Point, Carmel, California, and correlation with the zones of Ricketts and Calvin (1962) and Doty (1946). (Adapted from Johnson, 1968.)

not from spot environmental measurements. Detailed studies of the intertidal microclimate are few.

Glynn (1965) has carried out a time-course study for an 8-h period of temperature changes in and around the red alga *Endocladia muricata*. This study presents an interesting but all too brief glimpse of the microclimate conditions occurring in a specific area of the rocky intertidal region. Southward (1958a) and Gonor (1968) have made substrate-body temperature comparisons for a number of rocky intertidal invertebrates. Work such as this adds to our understanding of the temperature relations of marine organisms, but, like Glynn's work, await full utilization when more complete time-course studies include data on other environmental variables, such as radiation fluxes, wind-flow conditions, and the moisture content of the air. Only the work of Phleger and Bradshaw (1966) presents information on temperatures continuously recorded in a quasi-intertidal situation. Air and water temperatures were continuously recorded in a marine marsh in southern

California for 1 year. The temperatures were not those of the actual substrate but rather of the air and water immediately adjacent to the substrate. This study revealed a surprisingly broad range of annual water temperatures, 5° to 30°C, and a diurnal range of 8°C. It also showed that air temperatures greatly influenced water temperatures at low tide.

The disproportionate number of physiological studies as compared with the number of microclimatological studies can be attributed to two factors. First, most instruments designed for micrometerological studies are designed for fair-weather terrestrial work, or else for continuous monitoring of the subtidal environment (Wadsworth, 1968; Forstner and Rützler, 1970; Monteith, 1972; Platt and Griffiths, 1964). Few instruments are robust enough to survive long exposure to the rigorous environment of the marine intertidal and still give accurate and continuous measurement of the intertidal climatic variables of interest. The ones that do exist, while sturdy, are somewhat inaccurate and difficult to work with (Jones and Demetropoulos, 1968; Muus, 1968; Druehl and Green, 1970; Doty, 1971). With recent advances in solid-state electronics, however, this situation may be changed. For instance, small, sensitive strain gauges may be used to measure wave shock (Forstner and Rützler, 1970), and extremely simple, yet robust circuits exist that can be adapted to measuring wave splash and the time a particular organism or level may be submerged.

Second, until recently, there was no way to describe quantitatively the relationships of the extremely variable and complex environmental factors which determine the microclimate of the rocky intertidal. It is possible, however, to express these factors through a common denominator: energy. This allows the microclimate conditions of the rocky intertidal to be described in an integrated, quantitative fashion.

Microclimate

The actual microclimate conditions occurring in the intertidal region result from the interaction between the marine and terrestrial climate. This interaction has been studied in detail at the Hopkins Marine Station of Stanford University at Pacific Grove, California.

Marine Climate of Monterey Bay

The temperature and salinity conditions of the near-shore surface waters at the Hopkins Marine Station for November 1970 through November 1971 are presented in Fig. 31.2. The data were taken from the 1970–1971 records of daily surface-water temperature and salinity at shore stations from California, Oregon, and Washington, kept by the University of California, Scripps Institute of Oceanography, La Jolla, California. Temperatures range from 9.6 to 17.6°C, and, as Bolin

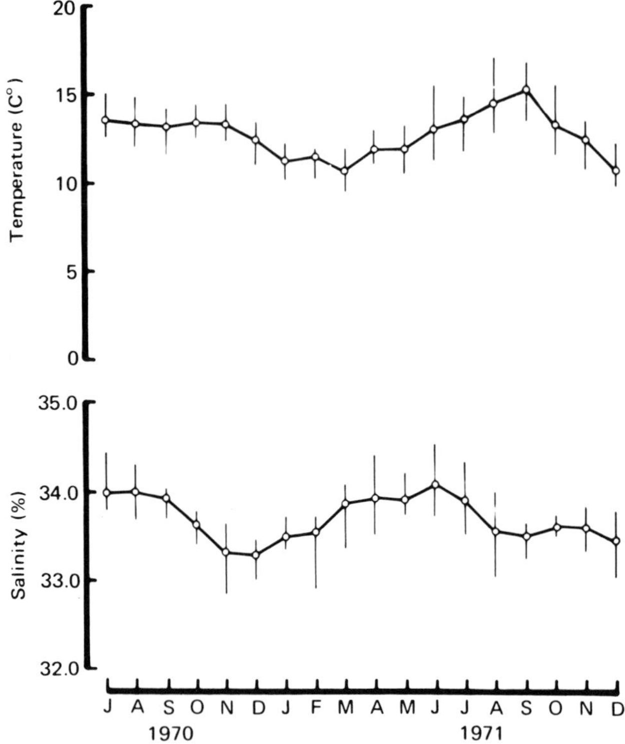

Fig. 31.2. Monthly mean values and ranges of temperature and salinity of the near-shore surface sea water at the Hopkins Marine Station from Nov. 1970 through Nov. 1971, based on daily measurements taken at ca. 0800 h.

and Abbott (1963) found, the highest temperatures occur in the fall and the lowest in the spring. The range for the monthly mean salinity curve is very slight, 33.3 ‰ (parts per thousand) in November 1970 to 33.9 ‰ in June 1971. In spite of this relatively small annual range in temperature and salinity, long-term measurements have shown that three distinct periods of hydrographic conditions can be recognized (Skogsberg, 1936; Bolin and Abbott, 1963; Glynn, 1965):

1. The oceanic period: this occurs in September and October and is characterized by relatively high temperatures, caused, apparently, by the shoreward movement of offshore surface waters, including those of the California Current, which have a higher temperature than the inshore waters.

2. The Davidson Current period: this period, lasting roughly from November through February, is characterized by declining sea water temperatures and a low thermal gradient. The cold water of this period originates from the north-moving Davidson Current, which, because of the Coriolis effect, banks up against the shore, displacing the warmer inshore waters. It is during this period that wave action is greatest, a result of winter storms during the months of December, January, and February.

3. Upwelling period: this period occurs from March to July or August and is the longest of the three periods. It is caused by the inshore upwelling of deep water resulting from a lateral displacement of the surface waters seaward. It is during this period that phytoplankton volumes are highest (Bolin and Abbott, 1963).

Atmospheric Climate of Monterey Bay

Gislen (1943) and Glynn (1965) discuss in general terms the kind of weather intertidal organisms of the Monterey Bay region would experience when exposed during the different seasons of the year. The following description is adapted from that given by Glynn (1965).

1. November, December, January, and February are the months of the year when the lowest temperatures and the greatest rainfall are likely to occur. During this season the single daily submergence occurs mostly during the early hours of the morning, thus tending to decrease the periods of exposure to early morning low temperatures. Afternoon low tides expose the intertidal to moderate temperatures and winds.

2. March, April, May, and June frequently have well-developed westerly winds with moderate temperatures, which become higher toward the summer. The single daily submergence in the early hours of the morning, characteristic of the winter, persists into March and early April. At this time, in the late afternoon especially, the organisms of the high intertidal are subjected to pronounced drying action. In late April, May, and June, a single daily submergence occurs mostly in the late hours of the afternoon and during the evening.

3. July and August are comparatively cool summer months, owing to the fog banks that frequently blanket the region during the morning hours in midsummer. The higher intertidal levels are wetted by the daily high-water tide during the evening and late hours of the afternoon, as in the late spring.

4. September and October are generally warm to hot months with little wind. In early September, the higher intertidal levels are bathed by a single daily submergence in the evening and later afternoon; this then changes to a single daily wetting in the early afternoon by late September and on through October. Early September is the time of the year when prolonged exposure is likely to occur coincident with high temperatures.

Tidal Regime and Climatic Effects

The degree to which the organisms of the rocky intertidal are affected by terrestrial versus marine conditions is determined by the tidal regime of the region (Colman, 1933; Doty, 1946). At Pacific Grove the tides are of the mixed, semidiurnal type, producing each day, in order of decreasing height, a higher high water (HHW), a lower high water (LHW), a higher low water (HLW), and a lower low water (LLW) (Fig. 31.3). The level of mean lower low water (MLLW) (tidal datum: 0.0 ft) is the baseline for all the following height measurements. Mean tide level is +3.0 ft above MLLW (Doty, 1957).

In the past, many ecological studies dealing with the rocky intertidal region have relied on tidal information determined from observations made at selected United States Coast and Geodetic tidal recording stations and corrected for local variation. Although theoretically correct, since the rise and fall of the tide at any given locality is characteristic of the tidal fluctuations over an area of much greater extent, in practice the actual tide level and the period of exposure of any particular level do not necessarily correspond to the predicted values. The discrepancies seen between actual and predicted values have been analyzed by Marmer (1927) and Glynn (1965) and seem to be due primarily to the effects of local sea and atmospheric conditions. Thus to determine accurately the condition of the tide and the degree of exposure experienced by the organisms of the rocky intertidal, it is essential to make the tidal measurements as close as possible to the study site.

The importance of the tidal regime in determining the intertidal environment is, as Lewis (1964) points out, best appreciated from a long-term study of tidal data. One way that has been used to present tidal data is the emersion curve (Fig. 31.4). This gives the total cumulative hours each intertidal level is exposed to terrestrial conditions. Although useful as a general picture of the total exposure at any particular level, this type of presentation does not describe changes in exposure associated with tidal cycles. These include diurnal, spring and neap, and perigean–

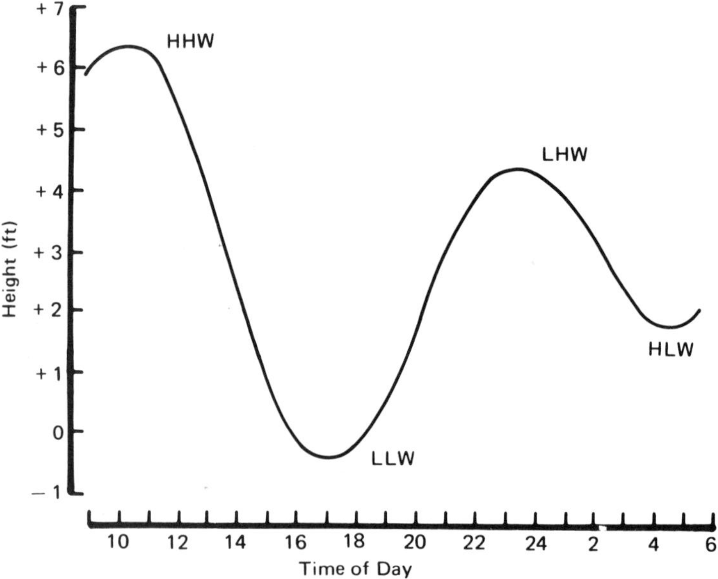

Fig. 31.3. Typical tidal curve at the Hopkins Marine Station (Oct. 5, 1971) showing the position of higher high water (HHW), lower high water (LHW), higher low water (HLW), and lower low water (LLW).

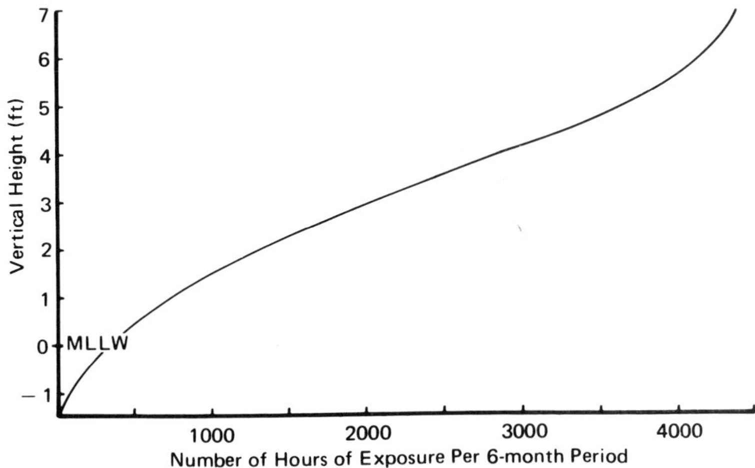

Fig. 31.4. Predicted emersion curve for the 6-month period, Jan. 1–July 1, 1931, for Crissy Warf, San Francisco, California. (From Ricketts and Calvin, 1962.)

apogean tidal cycles, as well as annual tidal cycles associated with the lunar and solar solstices (Marmer, 1927).

Based on tidal measurements made by an automatic recording tide gauge operated by the United States Naval Postgraduate School in Monterey Harbor, 1 mile distant from the Hopkins Marine Station, the mean monthly exposure was determined for three intertidal levels from November 1970 through November 1971 (Fig. 31.5) (Johnson, 1973).

The highest level, +3.9 ft, was exposed for 63.3 percent of the year, and during the months of February, March, April, May, and June was exposed for more than 70 percent of each month. The +2.3-ft level was exposed for 51.2 percent of the year and the lowest level, +1.9 ft, for 26 percent of the year. In 1971 the maximum exposure at the +3.9-ft level occurred in May and lasted for 23.5 h. The longest period of submergence of this level was 20 h in November 1971. The lowest tide occurring during 1971 was –1.9 ft below MLLW and the highest tide was +6.9 ft above MLLW. Both extremes occurred on December 2, 1971. It appears, then, that on the basis of exposure, the highest level is subjected more to terrestrial than marine conditions.

The temperature regime experienced by the upper intertidal region further emphasizes its nonmarine nature. Comparing the weekly maximum and minimum surface-water temperatures and research-area temperatures taken from a thermometer bolted 1 cm above the rock surface +3.8 ft above MLLW, it can be seen from Fig. 31.6 that the sea-water temperatures fall in between the extremes experienced by the research area. When the research-area temperatures are compared with air temperatures measured 10 m above the research area, it is apparent that the temperature regime of the research area is largely controlled by nonmarine conditions (Fig. 31.7).

Figure 31.8 gives the daily maximum rock-surface temperatures for a 2-week

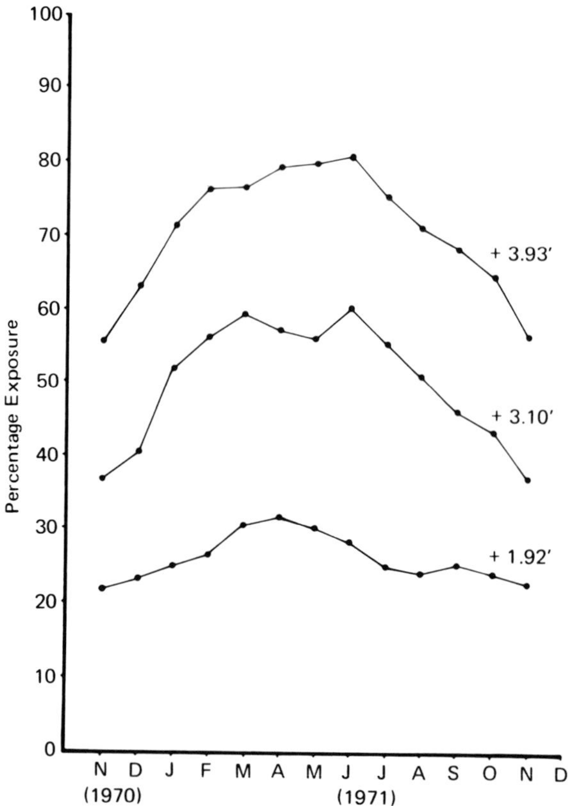

Fig. 31.5. Percentage of the time given areas are exposed to air, per month, for three intertidal levels, +3.9, +3.1, and +1.9 ft above MLLW from Nov. 1970 through Nov. 1971 at the Hopkins Marine Station.

period at the Hopkins Marine Station recorded 1 cm above the rock surface +3.8, +3.1, and +2.1 ft above MLLW. The variation of temperature with intertidal height seen in this figure parallels the exposure gradient seen in Fig. 31.5 and supports the idea that atmospheric climatic conditions are of great importance in determining the temperature regime of the rocky intertidal region.

To understand fully the effect of the atmospheric climate on the microclimate conditions of the intertidal, it is important to know as precisely as possible the length of each exposure period and when it occurs (Doty, 1946; Lewis, 1964). Conditions of greatest stress are most likely to occur when long periods of exposure to atmospheric conditions coincide with periods of high temperature, high winds, and low atmospheric moisture levels. The importance of knowing the atmospheric climatic conditions occurring at the time of exposure is emphasized by a study on the distribution and abundance of the rocky intertidal gammarid amphipod *Oligochinus lighti* (Johnson, 1973). This amphipod lives on the algae *Gigartina papillata* and *Endocladia muricata* in the high rocky intertidal region of Monterey Bay, California. The dramatic seasonal fluctuation in population density exhibited

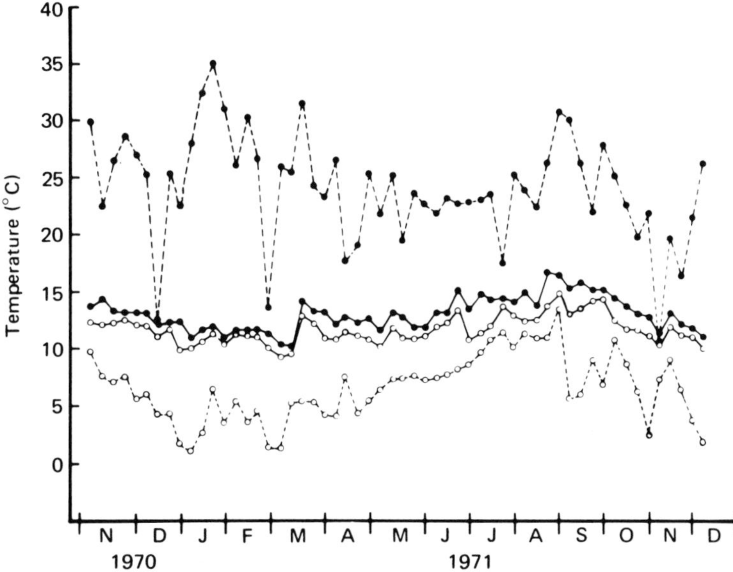

Fig. 31.6. Weekly maximum and minimum surface sea water temperatures and research-area temperatures taken from maximum-minimum thermometers bolted 1 cm above the rock surface +3.8 ft above MLLW, from Nov. 2, 1970, through Dec. 11, 1971. Dashed line, 1 cm above the rock surface; solid line, surface sea water. Filled circle, maximum, open circle, minimum.

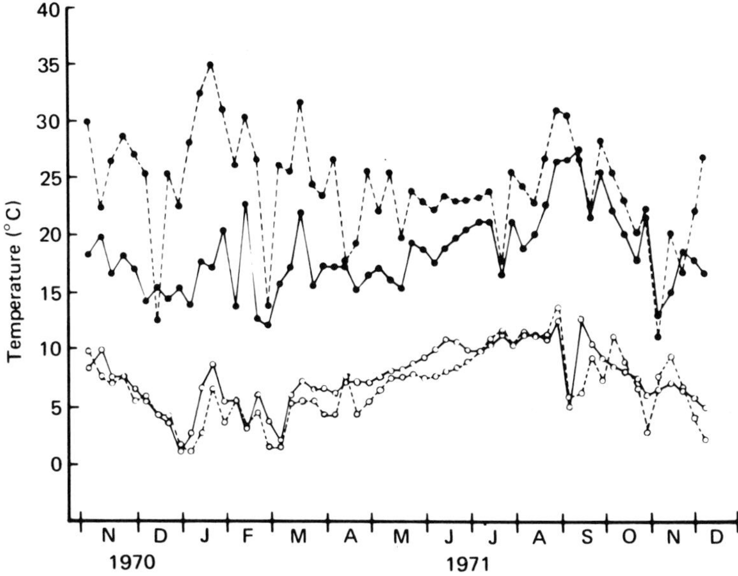

Fig. 31.7. Weekly maximum and minimum research-area temperatures taken +3.8 ft above MLLW, and air temperatures taken 10 m above the research area from Nov. 2, 1970, through Dec. 2, 1971. Dashed line, 1 cm above the rock surface; solid line, 10 m above research area; filled circle, maximum; open circle, minimum.

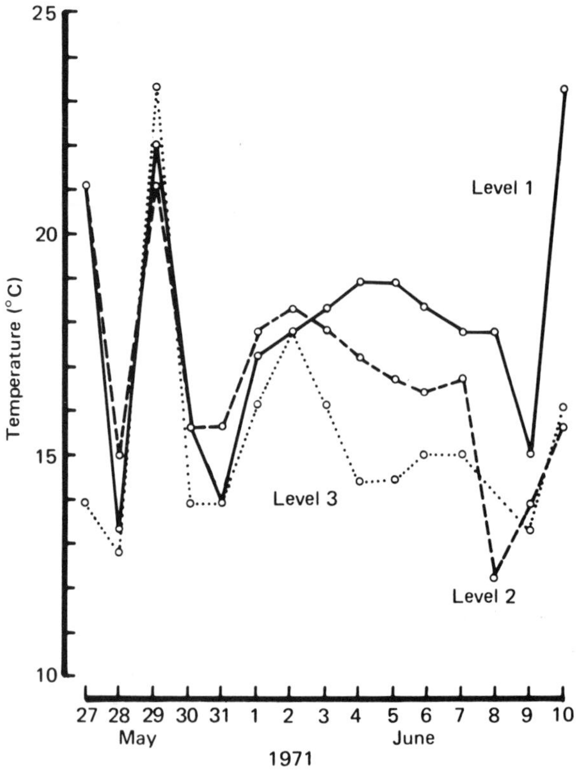

Fig. 31.8. Variation in temperature in the research area with vertical position in the intertidal. Maximum research-area temperatures taken from May 27, 1971, through June 10, 1971, at three intertidal levels: solid line, level 1 (+3.8 ft above MLLW); dashed line, level 2 (+3.1 ft above MLLW); dotted line, level 3 (+2.9 ft above MLLW).

by this species is closely correlated with changes in a number of environmental variables (Fig. 31.9).

In the winter and spring, when the population is decreasing in density, both exposure and wind speed are increasing, the number of exposure periods 16 h or longer that occur during the day are increasing, and the mean maximum temperatures (measured by maximum–minimum thermometers bolted 1 cm above the rock surface) are high.

During the summer months, when the population is increasing, the long periods of exposure during the day occur mostly in the morning (not shown but determined from tidal records), while the high winds, with their drying effects, occur in the afternoons when the population is submerged (Table 31.1). The concurrence of the daytime low tides with the morning fogs characteristic of this region from May through September (Figs. 31.10 and 31.11) results in lower research-area temperatures and a decrease in evaporation due to the supersaturation of the intertidal atmosphere with fog droplets (Figs. 31.12 and 31.13).

From July to October the period of exposure moves into the midmorning and early afternoon hours. Fog appears more sporadically and the organisms are

Table 31.1 Mean Wind Speed for 3-h Periods for Each Month from 1949 Through 1969[a].

Month	Time of Day								
	0000–0159	0300–0559	0600–0859	0900–1159	1200–1459	1500–1759	1800–2059	2100–2359	All hours
Jan.	4.7	4.5	4.4	4.3	6.3	5.5	4.5	4.7	4.3
Feb.	4.4	4.3	4.1	4.5	7.4	7.1	4.7	4.4	5.3
Mar.	5.1	4.5	4.3	6.1	9.6	9.1	5.6	5.1	6.4
Apr.	4.6	4.1	3.9	6.9	10.4	9.7	6.1	4.9	6.7
May	4.5	3.9	4.1	7.8	10.9	10.3	6.9	5.1	7.1
June	4.4	3.6	3.9	7.6	10.6	10.1	6.9	5.1	7.0
July	3.7	3.1	3.2	6.8	10.2	9.6	6.4	4.4	6.4
Aug.	3.2	2.8	2.9	6.3	10.1	9.3	5.9	4.0	5.9
Sept.	3.0	2.3	2.3	5.8	8.9	7.9	4.3	3.5	5.1
Oct.	3.0	2.6	2.5	4.6	7.7	6.6	3.4	3.2	4.5
Nov.	3.6	3.3	3.3	3.7	6.1	4.9	3.7	3.5	4.1
Dec.	3.3	4.0	4.1	3.9	5.8	4.6	4.0	3.7	4.4

[a] All readings taken at the United States Naval Air Station, Monterey, California. Units, mi h^{-1}.

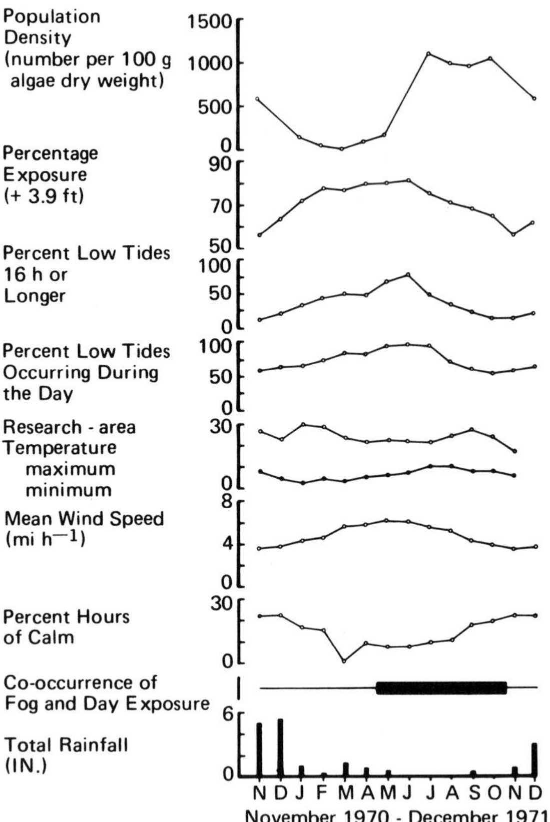

Fig. 31.9. Comparison of the total density of *Oligochinus lighti* from all levels of the research area with nine environmental parameters from Nov. 1970, through Dec. 1971.

subjected to high temperatures and to the declining, but still fairly high winds of the afternoon hours. As observed by Glynn (1965), this combination of conditions produces severe drying of the algae, and is probably the cause of the initial decrease in the population which occurs during this period. In the winter and early spring, the period of exposure moves further into the afternoon and evening. In addition to the problems of exposure, the area is now subjected to hyposaline conditions caused by the winter and spring rains (Fig. 31.14) and to intense wave action associated with the winter storms.

The amount of solar radiation incident on the Hopkins Marine Station from November 1970 through November 1971 is shown in Fig. 31.15. Longer day length and less attenuation of the incoming incident radiation as it passes through the earth's atmosphere, both due to the tilt of the earth's axis, result in the high summer levels observed. That this greater overall energy does not affect the intertidal region (except perhaps for the slight warming effect it has on the surface sea water) during these months is due primarily to the presence of fog in the area when the intertidal region is exposed (Fig. 31.9).

Before considering factors that modify the intertidal microclimate but are not

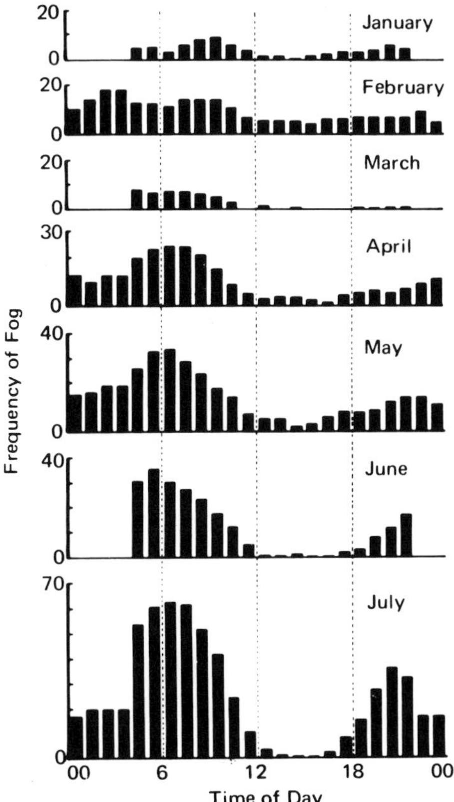

Fig. 31.10. Frequency of occurrence of fog each hour, given as a mean for each month from January through July, taken at the United States Naval Air Facility, Monterey, California, from 1945 through 1953.

associated with the tidal regime, it is important to stress that normal macroclimatological data taken by most maritime weather stations are generally too far removed to describe accurately local conditions at the study site, and the interval between measurements is usually so great that the data are of little value for most intertidal work. This point has been discussed by Lewis (1964) and Hedgepeth and Gonor (1969). The macroclimatological data presented in Figs. 31.10–31.13 represent previously unpublished long-term hourly measurements taken at the United States Naval Postgraduate School and at the United States Naval Air Facility, both in Monterey, California, 1 mi from the Hopkins Marine Station. That these excellent records can be used to describe conditions at the Hopkins Marine Station has been established by Glynn (1965), Darling (1969), and Johnson (1973).

Factors Modifying the Marine Intertidal Microclimate

The effect of the terrestrial (or atmospheric) climate on the organisms of the rocky intertidal during periods of exposure is modified by a number of factors. One

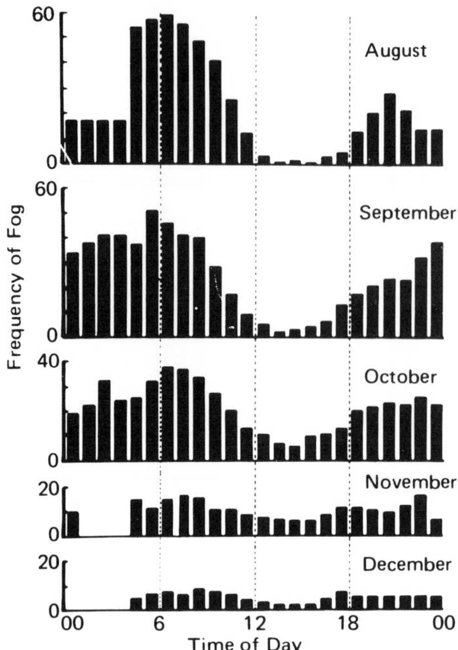

Fig. 31.11. Frequency of occurrence of fog each hour, given as a mean for each month from August through December, taken at the United States Naval Air Facility, Monterey, California, from 1945 through 1953.

of the most obvious is wave splash. By decreasing both the thermal and desiccation stress, this factor tends to extend the upward limit of many intertidal organisms (Newell, 1970). The degree of wave splash experienced by the intertidal is largely determined by the topography of the substrate and the sea conditions. As Lewis (1964) points out, the sharply sloping rock faces within unbroken profiles usually experience the most violent water movement and the greatest wave splash, while for gentle, flatter slopes with irregular profiles, the energy of the waves is rapidly dissipated and little upward surge or splash takes place.

The orientation and nature of the substrate will also affect the rate at which it dries off. Gently sloping, rough and irregular surfaces will retain water much longer than smooth, steep surfaces. Porous rock such as sandstone will retain more water than a more compact rock such as granite or shale. On the other hand, granite and shale, with greater thermal conductivity and thermal capacity than sandstone, will heat up more slowly and provide more stable substrate conditions for the attached organisms (Segal and Dehnel, 1962).

One very important effect of the substrate orientation concerns the amount of insolation its surface receives. Surfaces with greater slopes in general receive much less direct sunlight than those with a more horizontal orientation relative to the sun's rays (Gates, 1965; Van Deventer and Dold, 1966; Buffo *et al.*, 1972). Figure 31.16 shows a temperature profile across a large intertidal boulder at the Hopkins

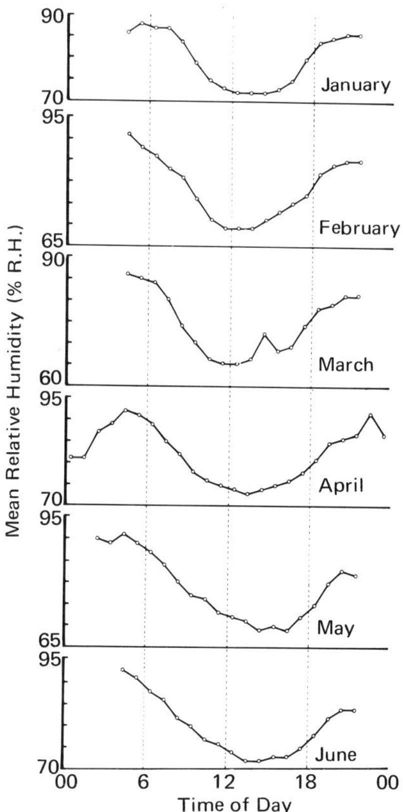

Fig. 31.12. Percent relative humidity each hour, given as a mean for each month from January through June, taken at the United States Naval Air Facility, Monterey, California, from 1945 through 1953.

Marine Station. The areas most nearly perpendicular to the sun's rays show the most elevated temperatures. Two points of great importance to the study of intertidal microclimates are illustrated by this figure. First, tremendous gradients of temperature are present. In the space of a few centimeters, surface substrate temperatures may change by as much as 15°C. Second, substrate and organism temperatures have little relation to the air temperature. This is due to the strong coupling of the substrate (or organism) surface to the radiant energy of the environment. During the day, the surface, when exposed to solar radiation, absorbs radiation and is generally warmer than the adjacent air. At night, if the sky is clear, the surface may be cooler than the air, owing to the reradiation of long-wave radiation to the heat sink of deep space (Gates, 1965). Since little attention has been paid to radiation measurements in intertidal work, many of the air-temperature measurements made in this region are of little use in describing the temperature regime experienced by the organisms that live there. If radiation fluxes cannot be measured, it would be much better to use rock surface–body temperature

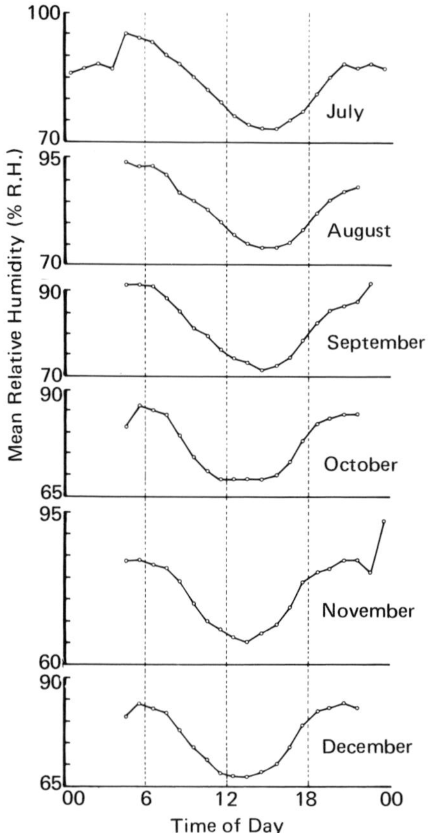

Fig. 31.13. Percent relative humidity each hour, given as a mean for each month from July through December, taken at the United States Naval Air Facility, Monterey, California, from 1945 through 1953.

measurements, for, as Gonor (1968) has shown, these often follow each other quite closely.

Biological factors may also modify the physical environment experienced by organisms of the rocky intertidal. The presence of algae, for instance, may increase protection from insolation and thus reduce desiccation and thermal stress for those organisms living on or under it (Colman, 1940; Wieser, 1952; Morton, 1954; Kensler, 1967; Johnson, 1973). Figure 31.17 presents a temperature profile taken in three adjacent clumps of red algae from the high intertidal region of Monterey Bay, California. As can be seen, only slight differences in size, geometry, or position within the algal clump make a considerable difference in the temperature an organism would experience.

The position an organism occupies in the intertidal is often related to its behavioral response to light, temperature, or desiccation (Newell, 1970). During periods of exposure, mobile animals such as limpets and various snails often retire

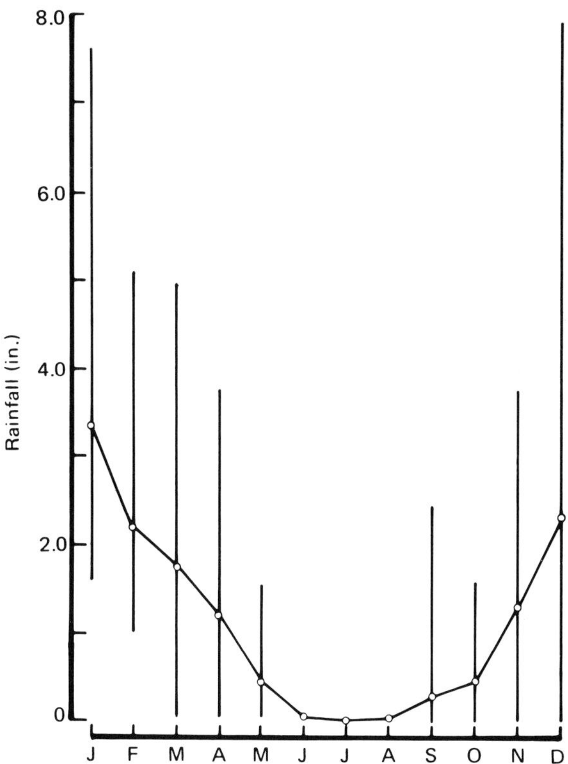

Fig. 31.14. Mean monthly rainfall recorded at the United States Naval Air Facility, Monterey, California, from 1945 through 1969. Vertical lines indicate the range.

to crevices and other shaded surfaces, where conditions of stress are less extreme. Other organisms, such as the Australian sand-dwelling gastropod *Cerethrium clypeomorus moniliferum*, reduce environmental temperatures and increase local humidity gradients by clustering (Moulton, 1962). Similar behavior by the littorine snail *Littorina scutulata* and the turban snail *Tegula funebralis* may serve the same purpose.

Physiological and morphological mechanisms also play a role in modifying the climate conditions experienced by an organism of the intertidal. Lewis (1963) has found that the barnacle *Tetraclita squamosa*, the limpet *Fissurella barbadensis*, and the gastropod *Nerita tesselata* have body temperatures lower than the surrounding environmental temperatures. He suggests that these animals may be using evaporative cooling as a means of thermoregulation. The fiddler crab *Uca* is also believed to utilize evaporative cooling (Edney, 1961, 1962). It has been suggested that morphological variations in shell shape, especially in the limpet genera *Patella* and *Acmaea*, play a role in determining the thermal and desiccation stress an organism may experience and tolerate (Test, 1945; Davies, 1969; Vermeij, 1972). Higher shells have smaller surface/volume ratios, resulting in a reduction in the amount of water lost by evaporation per unit wet weight of the animal (Shotwell,

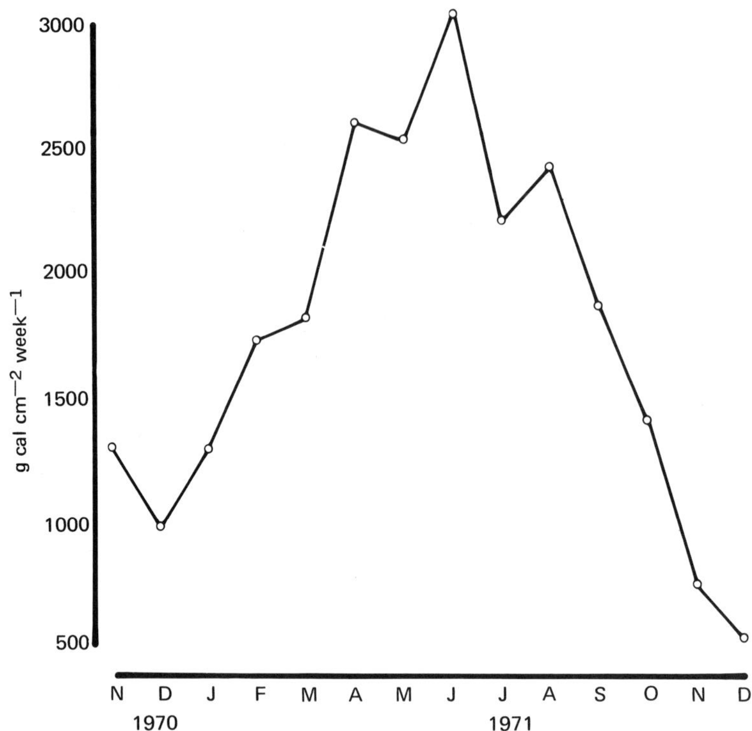

Fig. 31.15. Total incident radiation in g-cal cm^{-2} $week^{-1}$ recorded at the Hopkins Marine Station from Nov. 1970 through Dec. 1971.

1950). The extravisceral water capacity of the limpet *Acmaea limatula* appears to be an important factor in modifying thermal and desiccation stress, with larger animals holding more water than smaller ones (Segal, 1956).

The shape and ornamentation of the shell are also important. An examination of heat transfer in *Acmaea digitalis*, which has a moderately ribbed, spineless shell, and *Acmaea scabra*, which has a heavily ribbed spinose shell, showed that the convection coefficients for similar-sized animals under similar environmental conditions are quite different. This is shown in Fig. 31.18, which presents Nusselt–Reynolds number plots for both species. From this figure it can also be seen that differences in shell height will affect the convection coefficient, with larger animals having larger coefficients, thus losing more heat per unit surface area than smaller animals.

Energy Flow

It is clear from the preceding discussion that the environmental conditions affecting an organism of the rocky intertidal depend upon the complex relationship between the maritime macroclimate and the tidal cycle, modified by the nature and condition of the substrate, the position and orientation of the organism in the

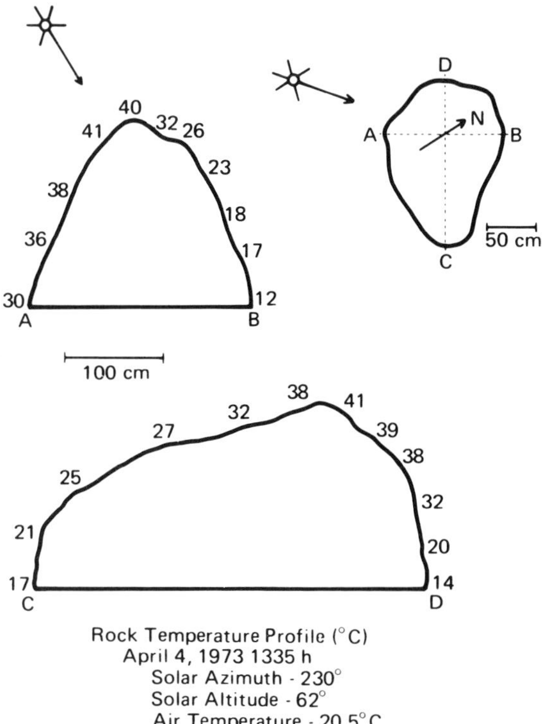

Fig. 31.16. Rock-surface-temperature profile over a large intertidal granite boulder at the Hopkins Marine Station, April 4, 1973, at 1335 h.

intertidal, and its behavioral, physiological, morphological, and physical characteristics. Any attempt to analyze the thermal or desiccation stress experienced by an organism must take all these factors into account. As stated earlier, until recently there was no way to quantify and integrate the extremely variable and complex environmental factors which determine the microclimate surrounding the organisms of the intertidal. Thus comparisons of intertidal habitats such as a steep, quickly draining slope, shaded for most of the day, with a gently sloping, slowly draining, algae-covered surface, were based on density and diversity of species, or on qualitative generalizations about their microclimates. It is possible, however, to compare such habitats by determination of their effect on the energy budgets of the organisms that live there (Lowry, 1967).

The energy budget, as described by Gates (1962) and Lowry (1967), analyzes energy flow between an organism and its environment. The basic tenet of this approach is that an organism must be in equilibrium with its environment. It can neither take in more thermal energy than it can lose, nor lose more than it can take in. If it is not in equilibrium, the organism will eventually heat up or cool down beyond its physiological limits, and die.

The physical processes that govern the energy exchange between an organism and its environment are primarily those of evaporation, conduction, convection, and radiation. These are essentially concerned with heat and mass transfer, the

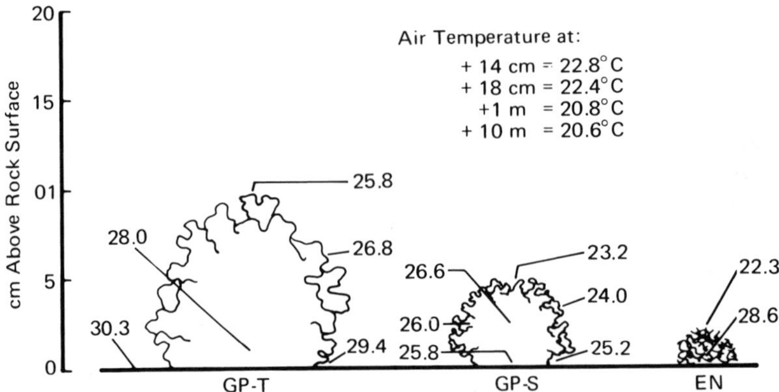

Fig. 31.17. Record of temperatures taken in, on, and around the short and tall forms of *Gigartina papillata* (GP-S and GP-T), and *Endocladia* (EN) at 1430 h, Sept. 15, 1971. All temperatures in Celsius.

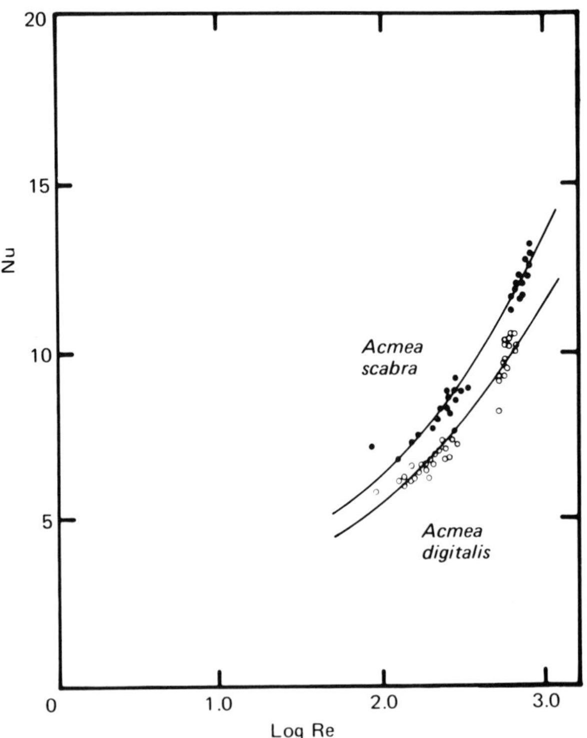

Fig. 31.18. Nusselt–Reynolds number plots for gold-plated silver casts of the limpets *Acmaea scabra* and *Acmaea digitalis*.

theory of which has been well worked out (Eckert and Drake, 1959; Krieth, 1965). Application of this theory to the study of thermal energy budgets and the quantitative analysis of an organism's climate space has been started by Gates (1962), Bartlett and Gates (1967), Birkebak (1966), and Porter and Gates (1969). With the exception of Edney's work (1953) with the supralittoral isopod *Ligia*, and recent work on the thermal ecology of *Uca* (Smith and Miller, 1973), few organisms from marine or aquatic situations has been studied with this approach.

Edney found that for the isopod *Ligia*, evaporation and convection were of equal importance as avenues of heat loss and that heat exchange via conduction was of little consequence. The absorption of incident solar radiation was important in determining the isopod's body temperature, for body temperatures were much higher than air temperatures (28°C versus 21°C). Given an understanding of *Ligia's* energy budget, Edney was able to predict accurately the body temperature and consequent desiccation stress that an isopod would experience under any given set of environmental conditions. This work, although important as a beginning, was done under rather simple circumstances with a largely terrestrial organism. Smith and Miller (1973) have gone into more detail in their analysis of the thermal ecology of two species of *Uca*. Using a heat-balance model with a number of microclimate and organism parameters, they were able to evaluate and predict the responses of the two species of crabs to various combinations of conditions, accounting in an integrated, quantitative manner for many of the modifying factors present in the intertidal.

Energy Budget

In its simplest form, the energy budget for an organism can be written

$$\text{energy in} = \text{energy out}$$

An expansion of this equation describes more fully the avenues of energy exchange available to an organism:

$$Q_{\text{abs}} + M = \varepsilon\alpha T_s^4 \pm \bar{h}_c(T_s - T_a) \pm G(T_b - T_{g_z}) + LM_e$$

where Q_{abs} = amount of radiation absorbed by the organism

M = metabolic energy produced by the organism

$\varepsilon\alpha T_s^4$ = energy reradiated by the organism's surface at a temperature T °K; ε = emissivity of the radiating surface; α = Stefan–Boltzmann constant

$\bar{h}_c(T_s - T_a)$ = energy lost (or gained) by convection; T_s = organism's surface temperature; T_a = air temperature; $\bar{h}_c$ = convective heat transfer coefficient

$G(T_b - T_{g_z})$ = energy lost (or gained) by conduction to the substrate; T_b = body temperature; T_g = substrate temperature at some depth z; G is the conduction coefficient

LM_e = energy lost due to evaporation of water from the organism; L = latent heat of vaporization; M_e = mass of water lost

All terms in this equation are expressed in calories per square centimeter per minute per unit surface area.

Consideration of each term in the energy-budget equation in more detail shows how various modifying factors affect the energy flow between an organism and its environment.

Radiation

The amount of radiation absorbed, reflected, and reradiated by an organism will depend on the source of the incident radiation and its spectral makeup; the orientation of the substrate; the amount of surface area exposed to direct, diffuse, and reflected radiation; and the absorptivity and emissivity of the organism's surface. The shape and orientation of the organism is also important. At noon, when the attenuation of incoming solar radiation is least, the effective area for absorption of short-wave radiation by a conical animal such as a limpet approximates a disk whose diameter is the diameter of the animal's base; the effective area for reradiation of long-wave radiation is the entire conical surface area. As the animal's shell becomes flatter, the ratio between the effective area for the absorption of short-wave radiation and the effective area for reradiation of long-wave radiation approaches 1. The concept of effective area versus actual and total surface area is discussed more fully by Bartlett and Gates (1967).

Metabolism

The amount of heat generated by metabolic processes depends on the activity of the organism and the nature of the food source. The limpet *Patella vulgata* has been found to respire at a maximum rate of 100 μl 0_2 g^{-1} h^{-1} (Davies, 1966). This animal's food consists largely of carbohydrates, which have an approximate caloric value of 4.199 cal g^{-1}, equivalent to 0.82 liter of 0_2 g^{-1} (Prosser and Brown, 1950). A 5-g animal would then produce 25.6×10^{-6} cal h^{-1}. If the surface area of this animal is estimated as 6.5 cm^2 (including the area of the shell and the foot), and it is assumed that all the metabolic energy produced by the animal is degraded to heat, this is 6.58×10^{-8} cal min^{-1} cm^{-2}, which for all but the most unusual circumstances will be a negligible part of the organism's energy budget.

Conduction

The conduction component of the energy budget can more properly be expressed as

$$\frac{k}{z}(T_b - T_{g_z})$$

where k is the coefficient of thermal conductivity and z is the depth where T_g is measured. Conductive heat exchange depends on the thermal, physical, and optical properties of the substrate (thermal conductivity, thermal capacity, density, porosity, absorptivity, and emissivity), on the surface area and orientation of the exposed portion of the substrate, on its mass, and on the temperature gradient within the substrate in the area of contact with the organism. More porous substrates will retain more water and thus stay cooler longer, but when dry, will be effective insulators and cause surface temperatures to be higher than they might be in less porous but more conductive substrates. Substrates with high thermal capacities and high thermal conductivities will provide a more even surface temperature regime in the face of externally varying conditions than substrates of lower thermal capacity and conductivity. The thermal properties, temperature, and surface area of that part of the organism in contact with the substrate are also important in determining the amount of energy exchanged by conduction.

Convection

Convective heat transfer depends on the temperature differential between the organism's surface and the air flowing over it, and on the convective heat-transfer coefficient. The latter is a complicated function of the air flow, the thermal properties of the air, and the geometry of the system (including both the organism and the substrate). It is generally incorporated into the dimensionless Nusselt number (Nu) which is related to the dimensionless Reynolds number (Re) by the equation (Kerslake, 1972)[1]

$$\mathrm{Nu} = a\mathrm{Re}^b \qquad \text{or} \quad \frac{\bar{h}_c D}{k} = a\left(\frac{VD}{\nu}\right)^b$$

where $\bar{h}_c$ = convective heat-transfer coefficient, averaged for the entire surface of the organism (cal cm^{-2} s^{-1} °C^{-1})
D = characteristic linear dimension of the organism (cm)
k = thermal conductivity of the air (cal cm^{-1} s^{-1} °C^{-1})
ν = kinematic viscosity of the air (cm^2 s^{-1})
V = velocity of the air flowing past the organism (cm s^{-1})

The coefficients a and b are constants of proportionality determined experimentally for each organism (Bartlett and Gates, 1967). Both k and ν are temperature-dependent and can be related to the ambient air temperature (C°) by the following equations:

$$k = (5.8 + 0.017T_a) \times 10^{-5}$$
$$v = 0.1346 + 0.0091T_a$$

[1] At a given temperature, the Reynolds number (Re) is proportional to VD and the Nusselt number (Nu) is proportional to $\bar{h}_c D$. V is the wind velocity, D is a characteristic linear dimension of the organism, and h_c is the convective heat-transfer coefficient. A more detailed discussion of convection and the Nusselt–Reynolds number relationship is presented later in this paper and by Kerslake (1972).

In the past, convection coefficients have usually been determined in a wind tunnel under laminar flow conditions with the object of interest held suspended in the center of the tunnel, well away from the boundary-layer conditions near the walls.[2] Under natural field conditions, however, turbulent flow is usually a significant component of the flow that occurs, with laminar flow occurring only in the boundary layer near the surface. As convection coefficients determined under turbulent flow conditions may be 50 percent greater than those determined under laminar conditions, it is extremely important to know the nature of the flow immediately adjacent to the organism. A rough substrate, for instance, will promote greater turbulence resulting in greater heat transfer for organisms in the turbulent region, but will also increase the thickness of the boundary layer within which the flow is essentially laminar. Organisms very near the substrate surface may now be within the boundary layer and experience a decrease in heat loss by convection. The situation is further complicated by changes in existing temperature gradients and thermally induced turbulence patterns. For a general discussion of this problem, see Krieth (1965).

To avoid many of the simplifying assumptions and the resulting inaccuracies associated with the application of laboratory convection coefficients to field situations, a method of accurately measuring convection coefficients in the field has been developed (Walthen *et al.*, 1971; Bakken and Gates, Chapter 16, this volume). This utilizes a gold-plated silver cast of the organism which is electrically heated and used as a constant-current anemometer (Yaglou, 1938). I determined Nusselt–Reynolds number plots for the limpets *Acmaea digitalis* and *Acmaea scabra* presented in Fig. 31.18, using this technique in a laboratory situation where plaster casts of the substrate simulated field turbulence and boundary-layer conditions.

Evaporation

Evaporative water loss, like convective heat loss, depends on the velocity and turbulence of the air flow, the geometry of the system, and on the temperature differential between the organism's evaporating surface and the air. It also depends on the moisture content of the air. The similarity of evaporative water loss to convective heat loss is expressed by Lewis's rule:

$$\bar{h}_d \simeq \frac{\bar{h}_c}{\rho c}$$

[2] When a fluid passes over a solid object, a thin zone develops within which the fluid velocity increases from the solid surface outward. This transition zone is known as the boundary layer. The thickness of this layer, which is defined as the distance from the surface at which the velocity reaches 99 percent of the mainstream velocity, depends on the velocity of the fluid, the geometry of the object over which it flows, and the surface roughness of the object. This layer acts as a buffer zone or resistance between the object and its environment. The thicker it is, the greater the protection provided (Ajiri and Gates, 1971).

where $\bar{h}_d$ = mass-transfer coefficient in (cm $s^-{}_1$)
$\bar{h}_c$ = convective heat-transfer coefficient (cal cm^{-2} s^{-1} $°C^{-1}$)
ρc = thermal capacity of the air at the ambient air temperature (cal deg^{-1} cm^{-3})

The mass of water evaporated from a unit surface area can then be expressed as

$$M_e = \bar{h}_d(C_s - C_a)$$

where M_e = mass of water evaporated
C_s = water-vapor concentration at the evaporating surface (g cm^{-3})
C_a = water-vapor concentration of the air (g cm^{-3})

The energy lost through evaporation is the product of the latent heat of vaporization at the ambient air temperature and the mass of water evaporated (M_e). When the air at the evaporating surface is saturated,

$$C_s = {}_sd_s(T_s) \qquad \text{and} \qquad C_a = \phi_{as}d_a(T_a)$$

where ${}_sd_s$ = saturation density at the temperature of the evaporating surface, T_s (g cm^{-3})
${}_sd_a$ = saturation density of the air at the ambient air temperature, T_a
ϕ_a = relative humidity of the air

The total energy lost by evaporative water loss then becomes

$$Q_{\text{evap}} = Lh_d[{}_sd_s(T_s) - \phi_{as}d_a(T_a)]$$

where L is the latent heat of vaporization at T_s.

Comparing this formulation with that given by Gates (1968) shows that h_d is equivalent to the reciprocal of $r_s - r_a$, where r_s is the resistance to flow from the organism and r_a is the boundary-layer resistance. This latter term is related to the geometry of the system, the wind velocity, and the turbulence of the flow. Kerslake (1972) presents an excellent and more detailed discussion of evaporative water loss from saturated surfaces, including a discussion of the effects of salinity on evaporation.

The manner in which wave splash affects the energy-budget equation is less apparent than for many of the other factors. It will cut down on evaporative water loss by increasing the humidity of the air as well as replacing the water reserves of the organism. It will cool the organism's surface as well as the surface of the substrate, thus altering radiative and conductive heat exchange. Finally, a wet surface will have a different albedo, causing a change in the amount of solar radiation that will be absorbed.

Conclusions

Almost all the physical and biological factors that determine the microclimate conditions affecting an organism of the rocky intertidal region may be accounted for by the energy-budget approach. Once an organism's energy budget and physiological tolerances are known, it is possible to predict the climate space that it must occupy to survive. This, as described by Porter and Gates (1969), is a four-dimensional space formed by radiation, air temperature, wind, and humidity acting simultaneously. Although an organism may occupy a smaller climate space than predicted because of biotic considerations such as food availability, competition, or predator–prey interactions, it cannot occupy a larger climate space than its physical properties (as described by the energy budget) permit. That energy-budget and climate-space analyses can predict the response of organisms to environmental changes has been shown by Bartlett and Gates (1967) and Porter and Gates (1969). Comparison of the predicted climate space with the climate space occupied by the organism in the field should shed light on the factors that regulate the distribution and abundance of natural populations. This approach, for instance, might be fruitfully applied to a study of the competitive interaction between the barnacles *Chthamalus stellatus* and *Balanus balanoides* (Connell, 1961), or the niche differences in the intertidal limpets *Acr aea scabra* and *Acmaea digitalis* (Haven, 1971, 1973). In any case, the sound analytical principles upon which biophysical ecology is based, when applied to investigations of the rocky intertidal region, will result in a more quantitative picture of the microclimate conditions occurring there.

Acknowledgments

This work was carried out at the Hopkins Marine Station of Stanford University, Pacific Grove, California, and at The University of Michigan Biological Station, Douglas Lake, Pellston, Michigan. Support came from National Science Foundation Doctoral Training Grants GZ-788, GZ-1143, and GZ-1542, and from Atomic Energy Commission contract AT-(11-1)-2164. C. L. Taylor and W. Thompson of the United States Naval Postgraduate School in Monterey, California, provided access to the invaluable tide and meterological records used in this work. Thanks are given to G. Bakken for his help with the gold-plated silver casts, and to Jacqueline Wyland for her valuable assistance and suggestions. Special thanks are given to David M. Gates and Welton L. Lee for the guidance and encouragement they provided throughout the course of this work.

References

Ajiri, S. I., Gates, D. M.: 1971. Boundary layer on leaves. Missouri Botan. Garden Bull. *59*, 26–27.

Bartlett, P. N., Gates, D. M.: 1967. The energy budget of a lizard on a tree trunk. Ecology *48*, 315–322.

Birkebak, R. C.: 1966. Heat transfer in biological systems. Intern. Rev. Gen. and Expt. Zool. *2*, 269–344.

Bolin, R. L., Abbott, D. P.: 1963. Studies on the marine climate and phyto-plankton of the central coast of California, 1954–1960. Rept. Calif. Coop. Oceanic Fishery Invest. *9*, 23–45.

Buffo, J., Fritschen, L., Murphy, J.: 1972. Direct solar radiation on various slopes from 0° to 60° north latitude. U.S. Dept. Agric. Forest Serv. Res. Pop. *PNW-142.*

Colman, J.: 1933. The nature of the intertidal zonation of plant and animals. J. Marine Biol. Assoc. U.K. *18*, 435–476.

———: 1940. On the faunas inhabiting intertidal seaweeds. J. Marine Biol. Assoc. U.K. *24*, 129–183.

Connell, J. H.: 1961. The influence of interspecific competition and other factors on the distribution of the barnacle *Chthamalus stellatus.* Ecology *42*, 710–723.

Darling, D.: 1969. Local geographic variation in the macro-climate along the central coast of California. Unpublished student report, on file at the U.S. Naval Postgraduate School, Monterey, Calif.

Davies, P. S.: 1966. Physiological ecology of *Patella.* I. The effect of body size and temperature on metabolic rate. J. Marine Biol. Assoc. U.K. *46*, 647–658.

———: 1969. Effect of environment on metabolic activity and morphology of Mediterranean and British species of *Patella*. Publ. Staz. Zool. Napoli. *37*, 641–656.

Doty, M. S.: 1946. Critical tide factors that are correlated with the vertical distribution of marine algae and other organisms along the Pacific coast. Ecology *27*, 315–328.

———: 1957. Rocky intertidal surfaces. *In* Treatise on marine ecology and paleoecology (ed. J. W. Hedgepeth), vol. I, pp. 535–585. New York: Geol. Soc. America.

———: 1971. Antecedent event influences on benthic marine algae standing crops in Hawaii. J. Expt. Marine Biol. Ecol. *6*(3), 161–166.

Druehl, L. D., Green, P.: 1970. A submersion-emersion sensor for intertidal biological studies. J. Fisheries Res. Board Can. *27*, 401.

Edney, E. B.: 1953. Temperature of woodlice in the sun. J. Expt. Biol. *30*, 331–349.

———: 1961. The water and heat relationships of fiddler crabs (*Uca* spp.). Trans. Roy. Soc. S. Africa *36*(2), 71–91.

———: 1962. Some aspects of the temperature relations of fiddler crabs (*Uca* spp.). Biometeorology *I*, 79–85.

Eckert, E. R. G., Drake, R. M.: 1959. Heat and mass transfer. New York: McGraw-Hill.

Forstner, H., Rützler, K.: 1970. Measurements micro-climate in littoral marine habitats. Oceanog. Marine Biol. Ann. Rev. *8*, 225–249.

Gates, D. M.: 1962. Energy exchange in the biosphere. New York: Harper & Row.

———: 1965. Heat, radiant and sensible. Chap. 1: Radiant energy, its receipt and disposal. Meteor. Monogr. *6*(28), 1–26.

———: 1968. Transpiration and leaf temperature. Ann. Rev. Plant Physiol. *19*, 211–238.

Gislen, T.: 1943. Physiographical and ecological investigations concerning the littoral of the northern Pacific. Lunds Univ. Arsskr. Avd. 2, *39*, 1–63.

Gonor, J. J.: 1968. Temperature relations of central Oregon marine intertidal invertebrates: a prepublication technical report to the Office of Naval Research. Corvallis, Ore.: Oregon State Univ. Dept. Oceanog. Data Rept. *34*, Ref. 68–38.

Glynn, P. W.: 1965. Community composition, structure, and interrelationships in the marine intertidal *Endocladia muricata–Balanus glandula* association in Monterey Bay, California. Beaufortia *12*, 1–198.

Gunter, G.: 1957. Temperature. *In* Treatise on marine ecology and paleoceology (ed. J. Hedgepeth), vol. 1, pp. 159–184. New York: Geol. Soc. America.

Haven, S. B.: 1971. Niche differences in the intertidal limpets *Acmaea scabra* and *Acmaea digitalis* (Gastropoda) in central California. Veliger *13*, 231–248.

———: 1973. Competition for food between the intertidal gastropods *Acmaea scabra* and *Acmaea digitalis*. Ecology *54*, 143–151.

Hedgepeth, J. W., Gonor, J. J.: 1969. Aspects of the potential effect of thermal alteration in marine and estuarine benthos. *In* Biological aspects of thermal pollution (ed. P. A. Krenkl, F. L. Parker), Proc. Natl. Symp. Thermal Pollution. Nashville, Tenn.: Vanderbilt Univ. Press.

Johnson, S.: 1968. Occurrence and behavior of *Hyale grandicornis*: a gammarid amphipod commensal in the genus *Acmaea*. Veliger *11*: 56–60 (Suppl.).

———: 1973. The ecology of *Oligochinus lighti*, a gammarid amphipod from the rocky intertidal region of Monterey Bay, California. Stanford, Calif.: Stanford Univ. Ph.D. diss.

Jones, W. E., Demetropoulos, A.: 1968. Exposure to wave action: measurements of an important ecological parameter on rocky shores of Anglesey. J. Expt. Marine Biol. Ecol. *2*, 46–63.

Kensler, C. B.: 1967. Desiccation resistance of intertidal crevice species as a factor of their zonation. J. Animal Ecol. *36*, 391–406.

Kerslake, D. L.: 1972. The stress of hot environments. New York: Cambridge Univ. Press.

Kinne, O.: 1971. Temperature-invertebrates; salinity-invertebrates. *In* Marine ecology (ed. O. Kinne), vol. I, Part 1, pp. 407–514; Part 2, pp. 821–996. New York: Wiley-Interscience.

Krieth, F.: 1965. Principles of heat transfer. Scranton, Pa.: International Textbook.

Lewis, J. B.: 1963. Environmental and tissue temperatures of some tropical intertidal marine animals. Biol. Bull. *124*, 277–284.

Lewis, J. R.: 1964. The ecology of rocky shores. London: English Universities Press.

Lowry, W. P.: 1967. Weather and life, an introduction to biometeorology. New York: Academic Press.

Marmer, H. A.: 1927. Tidal datum planes. Dept. Com., U.S. Coast and Geodetic Survey Spec. Publ. *135*. Washington: U.S. Govt. Printing Office.

Monteith, J. (ed.): 1972. Survey of instruments for micrometeorology. I.B.P. Handbook 22. Oxford: Blackwell.

Morton, J. E.: 1954. The crevice faunas of the upper intertidal zone at Wembury. J. Marine Biol. Assoc. U.K. *33*, 187–224.

Moulton, J. M.: 1962. Intertidal clustering on an Australian gastropod. Biol. Bull. *123*, 170–178.

Muus, B. J.: 1968. A field method for measuring "exposure" by means of plaster balls. A preliminary account. Sarsia *34*, 61–68.

Newell, R. C.: 1970. Biology of intertidal animals. New York: American Elsevier.

Phleger, F. B., Bradshaw, J. S.: 1966. Sedimentary environments in a marine marsh. Science *154*(3756), 1551–1553.

Platt, R. B., Griffiths, J. F.: 1964. Environmental measurement and interpretation. New York: Van Nostrand Reinhold.

Porter, W. P., Gates, D. M.: 1969. Thermodynamic equilibria of animals with environment. Ecol. Monogr. *39*, 245–270.

Prosser, C. L., Brown, F. A.: 1950. Comparative animal physiology. Philadelphia: Saunders.

Ricketts, E. F., Calvin, J.: 1962. Between pacific tides. Stanford, Calif.: Stanford Univ. Press.

Segal, E.: 1956. Adaptive differences in water-holding capacity in an intertidal gastropod. Ecology *37*, 174–178.

———, Dehnel, P.: 1962. Osmotic behavior in an intertidal limpet *Acmaea limatula*. Biol. Bull. *122*, 417–430.

Shotwell, J. A.: 1950. Distribution of volume and relative linear measurement changes in *Acmaea*, the limpet. Ecology *31*, 51–52.

Skogsberg, T.: 1936. Hydrography of Monterey Bay, California: thermal conditions, 1929–1933. Trans. Am. Phil. Soc. *29*, 1–152.

Smith, W. K., Miller, P. C.: 1973. The thermal ecology of two south Florida fiddler crabs: *Uca rapax* Smith and *U. pugilator* Bosc. Physiol. Zool. *46*, 186–207.

Southward, A. J.: 1958a. Note on the temperature tolerance of some intertidal animals in relation to environmental temperature and geographic distribution. J. Marine Biol. Assoc. U.K. *37*, 49–66.

———: 1958b. The zonation of plants and animals on rocky shores. Biol. Rev. *33*, 137–177.

Stephenson, T. A.: 1942. The causes of the vertical and horizontal distribution of organisms between tidemarks in South Africa. Proc. Linnean Soc. London *154*, 219.

Test, A. R.: 1945. Ecology of California *Acmaea*. Ecology *26*, 395–405.

Van Deventer, E. N., Dold, T. B.: 1966. Some initial studies on diffuse sky and ground reflected solar radiation on vertical surfaces. *In* Biometeorology II (ed. S. W. Tromp, W. H. Weihe), Proc. Third Intern. Biometeor. Congr., pp. 735–742. Oxford: Pergamon Press.

Vermeij, G. J.: 1972. Intraspecific shore-level size gradients in intertidal molluscs. Ecology *53*, 694–700.

Vernberg, W. B., Vernberg, F. J.: 1972. Environmental physiology of marine animals. New York: Springer.

Wadsworth, R. M. (ed.): 1968. The measurement of environmental factors in terrestrial ecology. Brit. Ecol. Soc. Symp. *8*. Oxford: Blackwell.

Walthen, P., Mitchell, J. W., Porter, W. P.: 1971. Theoretical and experimental studies of energy exchange from jackrabbit ears and cylindrically shaped appendages. Biophys. J. *11*, 1030–1047.

Wieser, W.: 1952. Investigations on the microfauna inhabiting seaweeds on rocky coasts. J. Marine Biol. Assoc. U.K. *31*, 145–173.

Yaglou, C. P.: 1938. The heated thermometer anemoneter. J. Ind. Hyg. Toxicol. *20*, 497–510.

32

Conclusions: The Challenge of the Future for Biophysical Ecology

DAVID M. GATES

Plants

Following the presentation of papers, the symposium participants met in small-group workshops to discuss the challenge of the future for research on biophysical ecology. Emerging from these discussions were many ideas and recommendations concerning future work in this field. The participants expressed much concern about the urgent need to understand ecosystems as a whole so that mankind can manage the earth in as productive and stable a manner as possible. Pivotal to this comprehension of ecosystems is the necessity to understand throughly the processes by which green plants function within their habitats. On the one hand, there was recognition of a critical need to understand the detailed mechanisms involving physiological and ecological processes within various components of a plant, and how these processes are integrated into the functional operation of the whole plant. On the other hand, there was a sense of urgency to assimilate whatever information is currently available into large-scale ecosystem modeling so that techniques involving ecosystems management, in particular food production, can be advanced as quickly as possible.

It seems to me that, in terms of urgent problems confronting mankind, we always have more time than we think we do and less time than we want. It is in this sense that I believe we must make every attempt to advance whole ecosystem modeling as quickly as possible by bringing into it all aspects of our current understanding of biophysical ecology and, in addition, whatever empiricism is required. At the same time we must be willing to do the meticulous, laborious laboratory experiments and field observations that are absolutely necessary for better basic understanding of plant processes to have a sound theoretical foundation for plant ecology. Large-scale ecosystem modeling will sooner or later require a solid foundation of information involving detailed physiological and ecological plant processes. One must do both, the large-scale modeling work and the fine-structure analysis of metabolic events, transport flow, and exchange processes. In either the large-scale or the small-scale domain of research, the very best minds, able to utilize the full complement of modern science, are required.

The workshop groups concluded that a chasm exists between experimentalists and modelers which tends to inhibit communication between them. Modeling enables the experimentalist to integrate information, work efficiently, and interpret data. Modeling is particularly important to ecology because of the complex interactions that are the essence of the discipline. Modelers who are not clearly connected with experimentation are quite likely to make unrealistic assumptions, owing to the complexity of biological systems and the need for first-hand experience. Integrated approaches to ecology are recommended involving both experimentation and mathematical modeling.

The development of models that may interface with each other was emphasized as a problem area receiving insufficient attention; the pivotal role of leaf diffusion resistance in models of water relations, photosynthesis, and energy balance provides an example of one of the mechanisms of interfacing. One may conclude from this that even moderately empirical models of photosynthesis should include this feature if they are to be used in ecosystem modeling to facilitate interfacing with other models. Similarly, experimentalists who are primarily interested in photosynthesis should not neglect leaf diffusion resistances because such information facilitates broader interpretation extending to water relations and the energy balance.

Two types of models were defined to facilitate discussion: predictive models, for use in broad ecosystem models, and analytical models, for use in evaluating the interaction of components within the leaf system. Predictive models are needed at this time for various biome studies. Empirical models may have to be used in the initial stages until more theoretical models are developed to improve interfacing and the generality of the predictions. It was recognized that empirical models may continue to be necessary in some broad-scale studies due to constraints set by the extent of input information, balance of information quality within the model, and computer limitations. The need for a more adequate theoretical basis in analytical models of photosynthesis was emphasized.

The components of photosynthetic models were discussed in relation to environmental dependencies to identify areas receiving insufficient attention. Photosynthetic model components important within a leaf were: transport of carbon dioxide, metabolism, and photosystems. Environmental factors of importance are: light, temperature, carbon dioxide and oxygen concentrations, humidity, and wind. From a discussion of these interactions, we concluded the following:

1. Photorespiration is poorly understood, but this probably reflects the complex relation of the process with CO_2 fixation. The area is receiving substantial attention by scientists.

2. Dark respiration has attracted attention from biochemists, but its relations with phenology and development are poorly understood.

3. Experimentation and models of stomatal response to environment are needed

that differentiate between C_3, C_4, and CAM species, because of the pivotal role of this parameter in numerous models.

Interactions between photosynthetic models and models of other plant processes were examined to identify neglected interactions. The photosynthetic model components were the same as before; the plant processes included water relations, energy balance, nutrient relations, salinity, phenology, development, partitioning of carbohydrate, hormones, and other biochemical and biophysical events. The importance of developing models capable of interfacing is apparent from the extensive interactions between models of other plant processes. Relations between models of phenology, development, and photosynthesis appear to be important areas for future research. The bidirectional character of the intermodel relations was emphasized.

Figure 32.1 indicates some aspects of primary production that must be studied further by biophysical ecologists. Considerable judgment is necessary to understand the amount of information required to produce a useful predictive quantitative model. The work of H. Mooney and others on the carbon balance of perennial plants from different ecosystems, for example, provides data that may ultimately be used in predicting fluxes of materials not only between different plant organs but also within ecosystems. Information and models of this type, although often empirical, are of considerable ecological significance and usually are developed prior to mechanistic models. They draw attention to the life cycle and adaptive strategies of plants within and between different ecosystems; they are able to provide predictions of productivity that may be of value, for example, to ecologists studying feeding and grazing or succession; and above all, they combine field observational evidence with quantitative predictions.

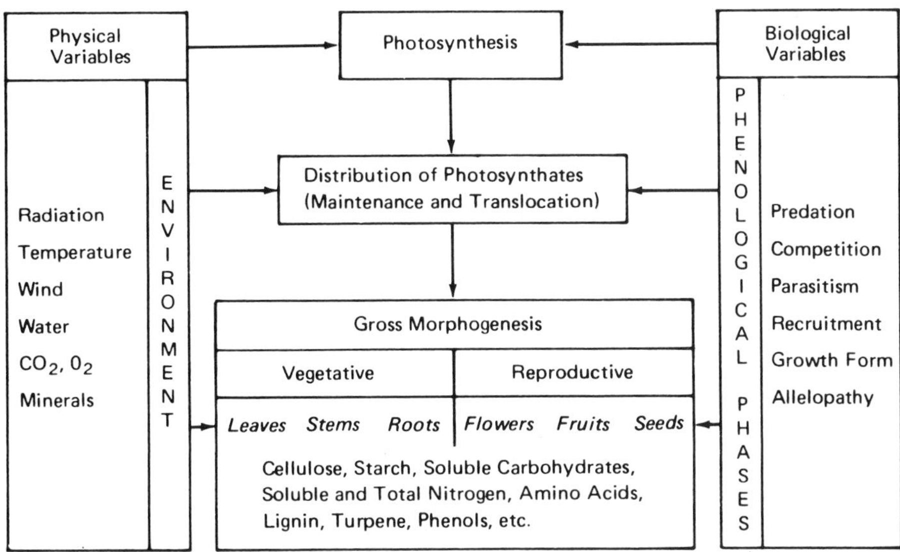

Fig. 32.1. Salient components of primary production.

Emerging from the workshop discussion was recognition that much more consideration should be given to the influence of water and nutrient fluxes within the soil–plant–air continuum on the photosynthetic and respiration processes. Analytical models exist concerning nutrient and water fluxes and, where possible, these should be linked to the leaf photosynthetic models to achieve a more coherent understanding of the interrelationships among energy, water, nutrient fluxes, and primary productivity. Some of the necessary information required to formulate such models already exists, but some critical experiments need to be done before such models can be completed. For example, we need to know the hydraulic conductivity coefficients for roots and how they change with age, water potential gradients, and other environmental variables. We also need to know the effects of changing root hydraulic conductivity on water transport to leaves and the consequent effects on leaf hydration.

Analysis of leaf photosynthesis should be extended to evaluate the effects of nutrients on the growth of the whole plant as well as on net photosynthesis. The elements to be analyzed include nitrogen, phosphorus, and potassium. Quantitative models have already been developed for predicting the movement of ions in the rhizosphere as a consequence of mass flow and diffusion. These movements have been related to the growth and development of crop plants. There may be sufficient information available concerning the fate and movement of certain elements within plants to allow the construction of a model linking nutrient fluxes with growth and photosynthesis.

It is important to apply these models to the different stages of plant development. At present, plant-growth models are concerned with the growth of crop plants, which are short-lived. There is considerable urgency to extend these models to include plants exhibiting the perennial growth habit, as such plants are often dominant components of the major ecosystems of the world. Plants species vary greatly in their photosynthetic capacities. Variability exists in the way photosynthates are partitioned during the plant life-cycle processes. Plant species differ from one another in allocation of resources during these processes.

The results of studies on growth-form strategies involving photosynthesis are encouraging. In addition to photosynthesis studies, more information is needed showing the distribution of the photosynthates through the entire life cycle of plants from widely different ecosystems. This should be done on a growth-form basis since this approach would provide a manageable framework for evaluating the many plant species that make up the ecosystems of the world. Experimental work with native plant species is needed along the lines outlined above to define adaptive variation in plant behavior and its evolutionary or ecological significance.

A quantitative study of the various photosynthates pools is required for the development of predictive plant-growth models. Soluble and total carbohydrates, hemicellulose and cellulose, proteins, lipids, and energy-rich by-products (e.g., turpenes and phenols) should be included in these measurements. These several organic fractions should be determined in various plant tissues and organ systems. This information is much needed by animal and microbial ecologists. Plant and animal ecologists should work together to ensure the interfacing of the various subprocess models required to understand each trophic level.

A much greater effort than ever before must be made to understand the full complement of processes that affect the growth and productivity of entire plants. The level of understanding must be such that accurate quantitative prediction of important growth, development, and productivity events is possible for any set of environmental conditions. When this is achieved, it will then be possible to understand in considerable detail the significant relationships of plant adaptation to habitat conditions, plant competition, succession, and other ecological phenomena.

Animals

An animal interacts with its environment through the flow of energy, the exchange of solids and fluids, and by sensual messages. It is possible to study many of these interactions in a very precise manner according to the basic laws of physics and chemistry. Although most animal–environment interactions are time-dependent, it is convenient for the sake of simplicity to begin such analysis using time-independent, or steady-state, models. Also, because of the complexity of animals in size, shape, and form, it is convenient to consider animals as represented by simple, regular geometrical bodies such as cylinders, cones, or spheres. These assumptions are justified for the first approximation to the formulation of the problems, but as one desires an improved analytical model of animal–environment interactions, one is confronted with time-dependent systems and with complex distributed geometries.

The problem of temperature regulation and heat exchange in animals is one of the first to be investigated using theory from biophysical ecology. The metabolic rate of an animal has relatively little meaning as an isolated single physiological parameter out of context with the many other factors that contribute to the energy level of an animal and thereby to its body temperature. Evaporative water loss also should not be viewed as a single isolated physiological parameter but must be understood in the full complex of factors that contribute to an animal's energy level. These physiological processes can be understood only when the complete array of all events relating to energy flow between an animal and its environment is considered. This can be done only by using analytical or mathematical techniques to formulate a self-consistent model of all physical and physiological events relating directly to the heat exchange and energy level of animal in a specified environment.

An analytical or theoretical model of animal–environment interaction is only as good as the facts upon which it is founded. Just as quantities of experimental and observational data are of questionable value without a good analytical theory into which they can be arranged, so also is theory nearly useless if it does not have adequate recourse to data from the real world. Biophysical ecology has made significant contributions toward an understanding of the physical limitations that contribute to the behavior and distribution of animals, while at the same time recognizing that plant–animal and animal–animal interactions are also exceedingly important. The physical limitations imposed by the environment on an animal create the broad framework within which it must exist while the organism interactions impose additional limitations.

This workshop concluded that the methodology of biophysical ecology will continue to improve and give further insights into the relationships between the biological and the physical components of the ecosystem. Such improvements will result if:

1. Attention is given to the two extremes, ranging from the most general application of biophysical theory to many species, including the identification of constraints, to the identification of specific questions centered on specific organisms.
2. A range of systems is analyzed, varying from terrestrial to aquatic.
3. The technique of modeling is fitted to the animal–environment relationships, and extreme care is taken so that the models and instruments used are retained as tools rather than driving forces.
4. The biophysical approach aims for the inclusion of the analyses of circadian rhythms, annual rhythms, ontogeny, population dynamics, and evolution.
5. Special emphasis is placed on the development of transient models.
6. Models are designed so as to permit the testing of the effect of short-term biophysical events, such as drought and storms, on animals.

We may hope that the models of biophysical ecology will apply to analyses of long-term biological events due to global climatic change. Considerable effort needs to be directed toward the analysis of animals of all ages, especially the young. Once again the importance of extending the analytical models beyond the domain of single animals or of pairs of animal–animal interactions, such as predator–prey relationships, to whole ecosystems was emphasized. It is important to build a solid foundation for large-scale modeling of ecosystems, and this requires meticulous care with the development of detailed analytical models based on good biophysical theory and much experimental and observational evidence at the animal–environment level.

The subject of biophysical ecology is difficult and complex. The education of people in this discipline is hard because it requires aptitude for and training in both biology and physics. Effective teaching methods must be advanced, including the use of audio-visual aids supplemented by adequate written material. Little textual material is available in biophysical ecology. It is not sufficient to teach students the subject matter; they must have considerable aptitude to be able to become self-sustaining for original contributions in biophysical ecology. An ability to do analytical science, combined with a deep interest in physiology and ecology, is necessary. Assuming that these attributes are available among the population of people interested in biophysical ecology, special courses are needed in biophysical ecology which would be taught in 6- or 8-week summer institutes. But much more than this is needed. In some way, one must place in various major universities one or two key people thoroughly trained in biophysical ecology, who in turn can impart the rigorous training necessary to students who have an interest in the subject. Very good theoreticians are needed who can work closely with experimentalists. In part, the tendency has been to have experimentalists only as professors of physiology or ecology. Now we need theorists who collaborate with experimentalists or who will do some experimental work while remaining active in theoretical analysis. Until first-rate theorists are working in the field of biophysical ecology, it will advance slowly. Opportunities for strong theoreticians in biophysical ecology are enormous, and for the right people, the challenge of the subject matter will be unbounded.

In addition to the need for persons trained in both physics and biology, there is a problem with instrumentation for both laboratory and field research. The whole problem of what to measure and how to measure it is complex. The next problem is equally critical, and that is, what to do with the measurements once they are made. To know what to do with the measurements requires good theoretical work in advance concerning the laboratory experiment or field observation. Presumably there is a question asked which requires an experiment or field observation for the answer. Usually this will require some analysis in advance so the experiment is adequately and sufficiently planned. Too often in the field of physiological ecology, experiments or field observations are done without a detailed plan and in particular without sufficient analytical work in advance. Experimental work in biophysical ecology is extremely hard, and to be fruitful, it must be based on a good theoretical foundation.

Once a laboratory experiment or field observation is designed, adequate instrumentation is required. This is a serious problem for many investigators. The type of equipment needed for various kinds of experiments in biophysical ecology must be described. Lists of commercially available equipment must be maintained and levels of performance indicated. Availability of large facilities for research, such as the Biotron at the University of Wisconsin or the phytotrons at Duke University and North Carolina State University, must be made known to young investigators. The problem of adequate equipment for research in biophysical ecology is a difficult one. Excellent environmental control is needed for good research in biophysical ecology. This requires first-rate growth rooms or control chambers, and these are expensive. However, beyond the problems of expensive facilities is the critical problem of knowing precisely what measurements to make and how many data to take. Too often investigators become slaves of their equipment for lack of sufficient advance planning and adequate theoretical analysis. It is easy to take too many data and very hard to take just the optimum amount of data of the correct diversity. More often than not, some very critical measurement is inadvertently overlooked for lack of sufficient analysis in advance.

Epilogue

All organisms live in a physical environment that is continually modified by the interactions between the two. Energy flows through all ecosystems as it passes from a higher to a lower energy state, and in the process it causes matter to cycle in the organic–inorganic system within which highly arranged disorder creates new life. Birth, growth, death, and decay are perpetuated by the current of solar energy that floods the planet Earth. Mankind is inexorably dependent upon these intricate events, which form the fabric of the earth's ecosystems. Yet to sustain a technological society, man has found it necessary to exploit resources and to manipulate and modify the natural arrangement of living things. Because of the enormous amount of stored resources and the vast surface of productive landscape available, man has until now been able to use it for his own enrichment, but in the process he has reduced its organic diversity and dispersed its inorganic resources.

The ultimate challenge for man is whether or not he can maintain a productive global ecosystem and simultaneously sustain a highly consumptive technological society. The mechanisms involved are so vast, the interactions so complex, that the ultimate consequences of exploitation, management, and manipulation are disguised in the apparent comforts of modern society. Many ecologists are absolutely convinced that man's behavior dooms him to ecological catastrophe. At the same time they know that management of the earth's ecosystems, if based on good ecological understanding, could sustain a productive, healthy civilization for a very long time. Much ecological knowledge is currently available to improve our management methods greatly; at the same time, very much more knowledge is necessary to assure the solution of many difficult problems facing us. Sustaining the global ecosystem which contains within it a highly consumptive industrialized society will require the very best possible ecological understanding. Fundamental to this is our ability to maintain primary productivity throughout the world.

The methods of biophysical ecology can provide the knowledge and insight necessary for understanding the complexities of ecosystems. There is a redundancy in the term "biophysical ecology" since the word "ecology" itself means the study of the interactions of organisms with their environment, a study that involves the biological and the physical world simultaneously. Yet the redundancy gives emphasis to the basic biophysical character of these interactions, an emphasis that has not been achieved by ecologists in a manner commensurate with the full capability of modern science. I am confident that the time is at hand for great advances to be made in modern ecological science based on the methods of biophysical ecology. What we have done to date is just a beginning. What could be achieved in the future by young people of brilliant minds, thoroughly trained in the fundamentals of physical and biological sciences, is fantastic indeed. The opportunity is at hand, the theoretical methods are known, and modern instruments and computers are available for rapidly advancing our understanding of the most complex and fundamental of ecological problems. What is utterly frustrating to me is to see a generation or more of scientists bypassing many of the fundamental ecological issues in a hasty attempt to solve quickly the large-scale ecosystem problems. Not that one should delay working with the large-scale system; but the need for some very good, very modern science, involving the mechanisms by which organisms interact with their environment, is urgent. Modern ecology must, by definition, deal with the complete spectrum of events from the macroelements of the physical world to the microelements of the biological world. In doing so, ecologists must be complete scientists fully capable of integrating the total body of relevant scientific knowledge available to modern society.

Index